BARRON'S E-Z ALGEBRA 2

Meg Clemens
Mathematics Department Chair
and Instructor
Hugh C. Williams High School
Canton, New York

Glenn Clemens
Mathematics Instructor
Norwood–Norfolk Central School
Norwood, New York

All inquiries should be addressed to:
Barron's Educational Series, Inc.
250 Wireless Boulevard
Hauppauge, New York 11788
www.barronseduc.com

Library of Congress Control Number: 2012003689

ISBN: 978-1-4380-0039-8

Library of Congress Cataloging-in-Publication Data
Clemens, Meg and Glenn.
Barron's E-Z algebra 2 / Meg Clemens and Glenn Clemens.
p. cm.
Summary: "This book provides a comprehensive review of algebra 2 for advanced high school and junior college students. It includes practice exercises to reinforce concepts and terms reviewed in the book"— Provided by the publisher.
Includes index.
ISBN: 978-1-4380-0039-8 (pbk.)
1. Algebra—Problems, exercises, etc. I. Title. II. Title: E-Z algebra 2. III. Title: Barron's E-Z algebra two. IV. Title: Barron's easy algebra 2. V. Title: Easy algebra two.
QA157.C627 2012
512.9—dc23 2012003689

PRINTED IN THE UNITED STATES OF AMERICA
15 14 13 12 11 10

CONTENTS

INTRODUCTION

This book can be used in many ways:

- As a companion to your Algebra 2 class and textbook;
- As a review book to prepare for an Algebra 2 exam;
- As a refresher before taking a math course that follows Algebra 2;
- As a review of Algebra 2 for anyone who has not taken a math course recently.

The topics in this book were chosen to match the new Common Core Curriculum for a traditional Algebra 2 course. Many states are changing their Algebra 2 courses to match this new curriculum. The Common Core Curriculum includes an emphasis on modeling, applications, and connections among topics. These ideas have been integrated into every chapter in this book.

The Common Core Curriculum does not specify in what order the topics should be covered. We have tried to write the chapters so that they do not have to be read in order. You should be able to read the chapters in the order in which your course covers them.

Although technology, such as graphing calculators and computers, has become an extremely important part of math education, its use in Algebra 2 classes is not yet universal. We have tried to explain as many of the math concepts in this book as possible without relying on technology. Some of the practice exercises do require the use of technology. We expect you have access to it and know, or are learning, how to use it. Even if your present Algebra 2 course does not make use of technology, familiarity with it will be invaluable in future math courses. A separate section on using a graphing calculator to do or check your work is included at the end of the book.

You need to know Algebra 1 to study Algebra 2. Many Algebra 2 topics build on material learned in Algebra 1. In this book, we have briefly reviewed some important Algebra 1 topics to refresh your knowledge. After learning the material in this book, you should have a strong foundation for a precalculus or a college algebra course.

STUDYING MATH

Many students have told us, "You can't study for math!" They think math is something either you are good at or you aren't, and if you aren't, there is not much you can do about it. This is simply not true. You can study for math. If you want to do well at it, you must study. Of course, studying math takes effort, but so does any other worthwhile endeavor. Math is definitely worthwhile!

The following are some tips for studying math. Follow these tips while reading this book and also while learning math in class at school.

Learn the vocabulary Math uses a large number of vocabulary terms that have very precise meanings. These meanings are not always the same in math as in everyday English. You need to know the exact meanings of the words being used to be successful in math.

Flash cards are helpful for many students. For each term, in addition to a definition, list some examples that illustrate the term and also some examples that may appear close but do not satisfy the definition.

Learn the symbols and notation For math, symbols and notation are really just an extension of the vocabulary. You need to know what the symbols mean and how they are used correctly. The same is true for notation. Remember that in math, a change in position can change the meaning. For example, $\sin^2 x$ and $\sin 2x$ do not mean the same thing; neither do a_{n-1} and $a_n - 1$.

Know the formulas Math uses a great number of important and useful formulas. You need to know the ones that apply to what you are currently learning. Whether you need to memorize the formulas may depend on the class you are in and the importance of the formula. Some classes (and some tests) provide official formula sheets. However, you must know what the formulas mean and when and how to use them.

Know the procedures There are procedures in math for solving different kinds of problems. Know the procedures that are important for your class. Know when it is appropriate to use them and how to use them.

Practice If you want to be a good writer, you need to practice writing. If you want to be a good basketball player, you need to get onto the court and practice basketball. If you want to be good at math, you need to take the time to practice. In addition to many worked examples, this book has practice problems throughout each chapter and at the end of the chapters. Do not assume because you read the examples and they made sense that you "know" the material. Practice the problems in the book, and do the homework assigned in your math class. When you are consistently getting problems right that you have done on your own, then you know the material.

Summarize As you study, write down the important ideas in your own words. Write out examples and vocabulary terms. Create a reference sheet of everything you think will be important.

Pace yourself You can learn only so much at once. Then your brain needs time to process what you have learned and tie it together with other things you have learned in the past. Try to study math when you are fresh and mentally alert. When you feel yourself starting to slow down or getting frustrated, take a break and do something else. Come back to the math later when you are fresh again.

Study with a friend Studying with another person or a small group can be a very effective way to learn math and can be fun at the same time. You can learn a lot from seeing how other people solve a problem; you can even learn from their mistakes. Nothing reinforces what you have learned better than explaining it to someone else. Just be sure to choose your study partners wisely. Remember that your goal is to help each other learn math, not prevent each other from learning.

SOLVING MATH PROBLEMS

Many students feel overwhelmed when they approach a new problem, especially if it is even a little unfamiliar or unlike other problems they have done. Do not give up; try the problem. The only way to get good at problem solving is to practice solving problems. If you have learned the terminology, symbols, notation, and procedures, and if you use them appropriately, you will be surprised at what you can accomplish. The following are some strategies to help solve math problems.

Read carefully Read the entire question, especially if it is one of the dreaded word problems. While you are reading, identify what is known in the problem and exactly what you are trying to figure out. If a lot of information is provided, write it down.

Think Yes, this is the hard part. It is also one of the things math is all about. After identifying what you know and what you need to find, consider all the skills, techniques, and procedures you know. Decide how you are going to try to solve the problem.

Show your work Show your work neatly. Your work should clearly show someone, including possibly yourself, exactly how you found your answer. Organize your work logically step by step, and try to resist the temptation to skip steps.

Check your answer If your answer can be checked in an equation, do so. If the answer is at the back of the book, check there (after you have done the problem). If the question was a word problem, take a moment to consider if your answer seems reasonable. If the answer is correct or if you at least have no strong reason to believe it is wrong, great. If not, do not just quit. Go back over the problem and see if you can find a mistake. This is one of the reasons showing your work is so important.

Get help if needed Everyone who does math gets stumped from time to time. Do not give up. Go back and review the relevant section of the book or part of your notes. If that does not work, ask a friend from class. If neither of you can figure out the problem, it may be time to go ask the teacher.

TEST-TAKING TIPS

All math students have to take tests to show that they understand the topics. Here are some tips to help you do your best.

Prepare Do you think you can't prepare in advance for a math test? Go back and reread the section "Studying Math." If possible, find a practice test to do in advance. Remember that test preparation involves more than just studying. Get a good night's sleep the night before, and have a good breakfast.

Read the directions After reading the directions, follow them.

Read the questions carefully Make sure you understand a question before you start working on it. Then make sure you have answered the question when you are done. On multiple-choice questions, read all the choices; do not just pick the first one that seems reasonable. Pay careful attention to the differences between choices that seem similar but not quite the same.

Show your work Most teachers will give partial credit on open-response questions, but you have to give them something to work with. Never leave a question totally blank. Try something.

Manage your time Many tests have a time limit. Know what it is, and keep track of how much time you have remaining. Use your time wisely. Sometimes that means knowing when to move on. If you have given a problem a fair try and it has you stumped, do not use up too much time wrestling with it. Circle it, and move on to other problems you can do. If you have time left when you are done, go back to the circled problems.

Double-check your answers If you have time at the end of the exam, check your answers either by using a calculator method that you did not use before or by redoing the work on a blank part of the test without looking at the previous work. Then compare your check with your original work.

Learning math can be hard work, but it is also very rewarding. Math is important in an incredibly wide variety of career areas you may be interested in. Also, math helps build your problem-solving skills and your self-confidence. Learning and doing math can be gratifying, even fun. It is very satisfying to solve a problem successfully that first appeared difficult. Challenge yourself and learn.

1

Linear Functions

WHAT YOU WILL LEARN

Linear equations are a fundamental part of algebra. In your previous math courses, you probably learned many skills for dealing with linear equations. In this chapter you will learn, or review, how to:

- Solve linear equations and inequalities;
- Write and graph equations of lines and inequalities;
- Solve systems of equations both algebraically and graphically;
- Solve systems of inequalities graphically;
- Solve problems with direct and inverse variation.

SECTIONS IN THIS CHAPTER

- Linear Models
- Solving Linear Equations in One Variable
- Solving Linear Inequalities in One Variable
- Linear Equations in Two Variables
- Linear Inequalities in Two Variables
- Systems of Two Linear Equations
- Systems of Two Linear Inequalities
- Direct and Inverse Variation

Linear Models

One of the important uses of mathematics is to model real-life situations. A mathematical model can be a table, a graph, or an equation used to describe a particular phenomenon and make predictions about it. For example, suppose a small aircraft is flying at an altitude of 10,000 feet and begins to descend at a steady rate of 600 feet per minute. The altitude of the aircraft can be modeled as follows:

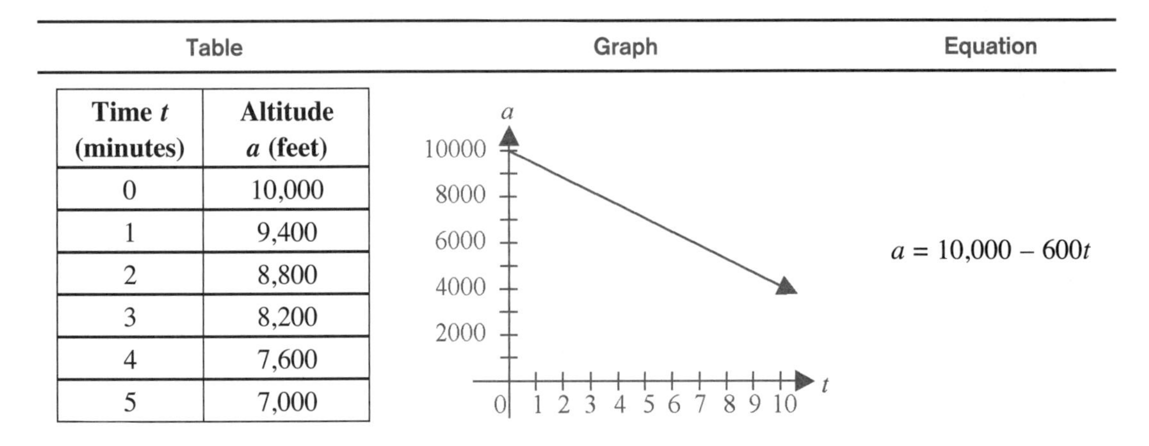

Time t (minutes)	Altitude a (feet)
0	10,000
1	9,400
2	8,800
3	8,200
4	7,600
5	7,000

FIGURE 1.1

The model can be used to answer questions such as "What is the altitude of the aircraft after 4 minutes?" and "How long will the aircraft take to reach an altitude of 2,000 feet?"

The model above is an example of a *linear model.* The defining feature of a linear model is that it has a constant rate of change. This can be seen in the table by noting that the difference in altitude for any two consecutive minutes is –600 feet (negative because the altitude is decreasing). The graph of a linear model will be a line (hence the name). The equation will involve variables to the first power only. Linear models have the advantage of being the simplest of models to use and understand while at the same time being capable of describing a wide variety of real-life problems.

Solving Linear Equations in One Variable

A linear equation in one variable is an equation that can be simplified to $ax + b = c$, where x is a variable and a, b, and c are any numbers. A solution is a value of x that makes the equation true. As long as $a \neq 0$, this equation has exactly one solution. If $a = 0$, there are two possibilities. If $b \neq c$, no value of x will solve the equation. If $b = c$, then all real numbers x satisfy the equation; such equations are called *identities.*

Two main tools can be used to solve linear equations:

1. *The addition property of equality:* Equal quantities may be added to (or subtracted from) both sides of an equation.
2. *The multiplication property of equality:* Both sides of an equation may be multiplied (or divided) by the same (nonzero) quantity.

Note that both of these properties incorporate a general rule of equations you learned in Algebra 1: do the same thing on both sides of the equation. The addition property is used to move all the variable terms to one side of the equation and the constants to the other side. The multiplication property is often used last to finish getting the variable by itself.

EXAMPLE 1.1

Solve $8x - 12 = 3x + 7$ for x.

SOLUTION

$8x - 12 = 3x + 7$

$\underline{-3x \qquad -3x}$ Addition property to move the variables to the left.

$5x - 12 = 7$

$\underline{+12 \quad +12}$ Addition property to move the constants to the right.

$\frac{5x}{5} = \frac{19}{5}$ Multiplication (division) property to get the variable alone.

$x = 3.8$

More complicated equations may require the distributive property to eliminate parentheses. Be careful to multiply signed numbers accurately.

EXAMPLE 1.2

Solve $4x - 2(x + 6) = 3(2x + 7) + 5$ for x.

SOLUTION

$4x - 2(x + 6) = 3(2x + 7) + 5$

$4x - 2x - 12 = 6x + 21 + 5$ Distributive property to eliminate parentheses.

$2x - 12 = 6x + 26$ Combine like terms to simplify the equation.

$\underline{-2x - 26 \quad -2x - 26}$ Addition property to get variables on one side and constants on the other.

$\frac{-38}{4} = \frac{4x}{4}$ Multiplication property to get the variable alone.

$-9.5 = x$ or $x = -9.5$

SEE CALC TIP 1A

COMMON ERROR

Some students will distribute 2 but not –2 and then write $4x - 2x + 12 = 6x + 21 + 5$ as the second step for this example. However, $x + 6$ multiplied by –2 is $-2x - 12$.

Note that the goal is to get all the variables on one side and the constants on the other. However, which side each goes on does not really matter. Some people always move the variable to the left side. Others like to move the variable to whichever side will keep its coefficient positive.

Some equations that may not appear to be linear at first become so while solving them.

EXAMPLE 1.3

Solve $x(x + 3) + 5 = x^2 + 7x - 10$ for x.

SOLUTION

$x(x + 3) + 5 = x^2 + 7x - 10$ — Because of the x^2 term, this equation does not appear to be linear.

$x^2 + 3x + 5 = x^2 + 7x - 10$ — The x^2 terms on both sides will cancel out, leaving a linear equation.

$$\underline{-x^2 - 3x + 10 = -x^2 - 3x + 10}$$

$$\frac{15}{4} = \frac{4x}{4}$$

$$x = 3.75$$

EXAMPLE 1.4

Solve $\dfrac{x + 7}{2x} = \dfrac{4}{5}$ for x.

SOLUTION

$\dfrac{x + 7}{2x} = \dfrac{4}{5}$ — Linear equations do not normally have the variable in a denominator.

$5(x + 7) = 4(2x)$ — Cross-multiplication (an application of the multiplication property) changes this to a linear equation.

$$5x + 35 = 8x$$

$$\underline{-5x \qquad -5x}$$

$$\frac{35}{3} = \frac{3x}{3}$$

$$x = \frac{35}{3} \approx 11.67$$

EXAMPLE 1.5 Solve $4 - (3 - x) = x + 5$ for x.

SOLUTION

SEE CALC TIP 1B

$$\begin{aligned} 4 - (3 - x) &= x + 5 \\ 4 - 3 + x &= x + 5 \\ x + 1 &= x + 5 \\ 1 &= 5 \end{aligned}$$

No value of x makes this equation true. It is an example of an equation with *no solution*.

Solve $2(x + 1) - (x - 2) = x + 4$ for x.

SOLUTION

SEE CALC TIP 1C

$$\begin{aligned} 2(x + 1) - (x - 2) &= x + 4 \\ 2x + 2 - x + 2 &= x + 4 \\ x + 4 &= x + 4 \\ 4 &= 4 \end{aligned}$$

This is true for any value of x. The solution set is all real numbers. This equation is an example of an *identity*.

LITERAL EQUATIONS

Literal equations, or formulas, can be rearranged to solve for a variable.

EXAMPLE 1.7 Solve $P = 2w + 2l$ for w.

SOLUTION

$$\begin{aligned} P &= 2w + 2l \\ P - 2l &= 2w \\ \frac{P - 2l}{2} &= w \end{aligned}$$

COMMON ERROR

$$\frac{P - \cancel{2}l}{\cancel{2}} = w$$

$$P - l = w$$

The 2s cannot be canceled in this fraction. See Chapter 9.

In addition to using the addition and multiplication properties of equality, you may also need to factor out a greatest common factor to solve some literal equations.

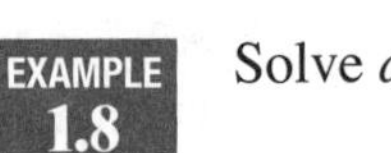

Solve $ax + c = bx + 10$ for x.

SOLUTION

$$ax + c = bx + 10$$

$$ax - bx = 10 - c$$ Move all terms with x to one side. Move all terms without x to the other side.

$$x(a - b) = 10 - c$$ Factor x out of each term on the left side.

$$\frac{x(a - b)}{a - b} = \frac{10 - c}{a - b}$$ Divide each side by $a - b$ to get x by itself.

$$x = \frac{10 - c}{a - b}$$ Note that the solution assumes $a \neq b$. If $a = b$, the equation has no solution. (Unless $c = 10$, in which case the original equation is an identity.)

SECTION EXERCISES

Solve each equation for x.

1–1. $7(x - 3) - 5(2x - 4) = 5$

1–2. $\frac{x}{x + 6} = \frac{x - 2}{x}$

1–3. $3(x + 2) - 11 = 5(x - 4)$

1–4. $\frac{a}{x} = \frac{3}{x - a}$

1–5. Solve $S = 2\pi rh + \pi r^2$ for h.

1–6. $ax + b = c(x + 1)$

Solving Linear Inequalities in One Variable

SIMPLE LINEAR INEQUALITIES

A linear inequality is a relationship involving $<$, $\leq$, $>$, or $\geq$ instead of $=$. Linear inequalities can be solved the same way linear equations are, but with one important modification to the multiplication property. In an inequality, if both sides are multiplied (or divided) by the same *negative* quantity, the direction of the inequality changes. This is illustrated in the next example.

EXAMPLE 1.9 Solve $-3x + 7 < 25$ for x.

SOLUTION

$$\begin{aligned} -3x + 7 &< 25 \\ -7 \quad & \;\; -7 \\ \hline \frac{-3x}{-3} &< \frac{18}{-3} \\ x &> -6 \end{aligned}$$

Note that because we divided both sides by -3, the direction of the inequality changed from $<$ to $>$.

This quirk of inequalities can be avoided by always moving the variable to the side where its coefficient will be positive. However, this can introduce a second difference between equations and inequalities. If the two sides of an inequality are switched right to left, the direction of the inequality must be reversed. For example, to rewrite $3 > x$ with the variable on the left, you must switch the direction of the inequality to get $x < 3$.

EXAMPLE 1.10 Solve $3(2x + 4) \geq 10x - (x - 6)$ for x.

SOLUTION

$$\begin{aligned} 3(2x + 4) &\geq 10x - (x - 6) \\ 6x + 12 &\geq 10x - x + 6 \\ 6x + 12 &\geq 9x + 6 \\ -6x - 6 \; & \;\; -6x - 6 \\ \hline \frac{6}{3} &\geq \frac{3x}{3} \\ 2 \geq x \text{ or } x &\leq 2 \end{aligned}$$

When dividing by a positive quantity, the direction of inequality does not change. When reversing the order of the two sides of an inequality, the direction changes.

INTERVAL NOTATION

Inequalities in one variable represent an interval on the number line. Interval notation is a compact way to write inequalities. Here is what you need to know:

1. Intervals are always written left to right, smaller number to larger number.
2. $[a, b]$ means $\{x \mid a \leq x \leq b\}$. This is called a *closed interval*; the endpoints are included.
3. (a, b) means $\{x \mid a < x < b\}$. This is called an *open interval*; the endpoints are *not* included. You need to use context to avoid confusing it with an ordered pair.
4. $[a, b)$ and $(a, b]$ are called *half-open intervals*. $[a, b)$ means $\{x \mid a \leq x < b\}$, while $(a, b]$ means $\{x \mid a < x \leq b\}$.
5. *Unbounded intervals* are written using ∞ and/or $-\infty$; ∞ and $-\infty$ are not included in the interval. For example, for $x \geq 5$, we write $[5, \infty)$.
6. In interval notation, $\cup$ (*union*) replaces $\vee$ (*or*) and $\cap$ (*intersection*) replaces $\wedge$ (*and*) as notation for writing two or more intervals.

The following table illustrates interval notation.

TABLE 1.1
INTERVAL NOTATION

Inequality Notation	Number Line Graph	Interval Notation
$-2 \leq x \leq 5$	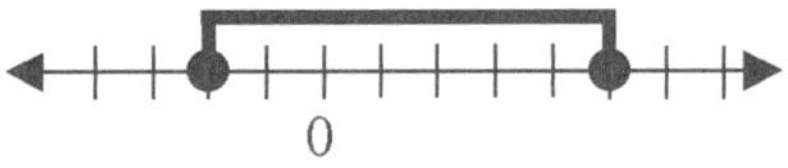 0	$[-2, 5]$
$-2 < x < 5$	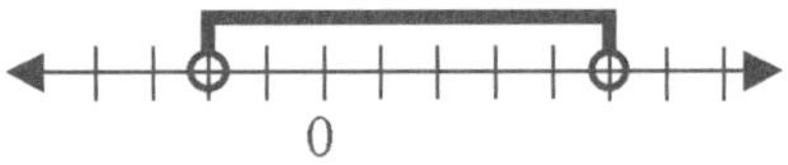 0	$(-2, 5)$
$-2 < x \leq 5$	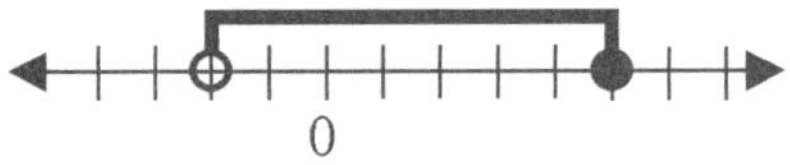 0	$(-2, 5]$
$-2 \leq x < 5$	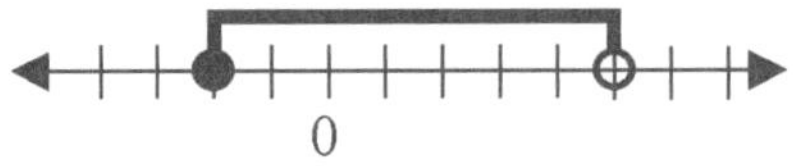 0	$[-2, 5)$
$x \leq 5$	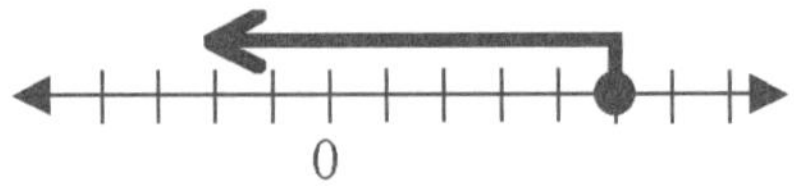 0	$(-\infty, 5]$
$x > -2$	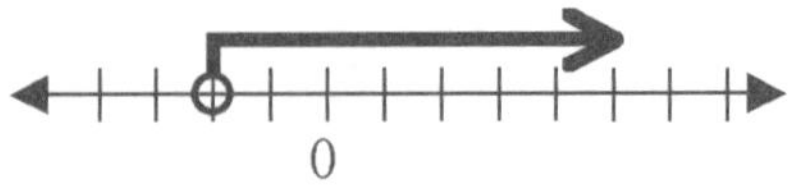 0	$(-2, \infty)$
$x < -2 \vee x \geq 5$	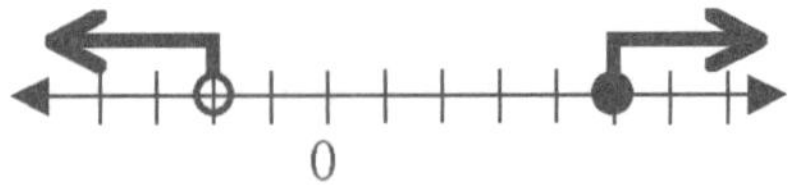 0	$(-\infty, -2) \cup [5, \infty)$

COMPOUND LINEAR INEQUALITIES

A *compound inequality* in one variable consists of two inequalities connected by either "and" ($\wedge$) or "or" ($\vee$). In general, each part may be solved separately and then the final answer written appropriately.

Solve $5x + 7 > 27$ and $2x - 8 < 4$ for x.

SOLUTION

$$\begin{aligned} 5x + 7 > 27 \quad &\text{and} \quad 2x - 8 < 4 \\ 5x > 20 \quad & \quad\quad 2x < 12 \\ x > 4 \quad &\text{and} \quad\quad x < 6 \\ 4 < x &< 6 \end{aligned}$$

EXAMPLE 1.12

Solve $3(x + 1) \le -6$ or $5x - 14 \ge 21$ for x.

SOLUTION

$$\begin{aligned} 3(x + 1) \le -6 \quad &\text{or} \quad 5x - 14 \ge 21 \\ 3x + 3 \le -6 \quad & \quad\quad 5x \ge 35 \\ 3x \le -9 \quad & \quad\quad x \ge 7 \\ x \le -3 & \\ x \le -3 \text{ or } x &\ge 7 \end{aligned}$$

Inequalities joined with "and" are often written together as in the next example. They are solved in the usual way. However, the same operation must be done to all *three* parts of the inequality to isolate the variable in the center.

Solve $-5 \le 4x - 3 \le 5$ for x.

SOLUTION

$$\begin{aligned} -5 \le 4x &- 3 \le 5 \\ \underline{+3 \qquad\quad} &\underline{+3 \;\; +3} \\ \frac{-2}{4} \le \frac{4x}{4} &\le \frac{8}{4} \\ -0.5 \le x &\le 2 \end{aligned}$$

Sometimes compound inequalities, especially ones involving "and" ($\wedge$), may have no real solutions.

Solve $2x - 1 < 3 \wedge 6x - 5 > 19$ for x.

SOLUTION

$$\begin{aligned} 2x - 1 < 3 &\wedge 6x - 5 > 19 \\ 2x < 4 &\quad\; 6x > 24 \\ x < 2 &\wedge \quad x > 4 \end{aligned}$$

No numbers are both less than 2 and greater than 4. So the solution set for this problem is the *empty set*, which may be written as $\{\ \}$ or $\varnothing$ (but *not* as $\{\varnothing\}$).

SECTION EXERCISES

Express each of the following using interval notation.

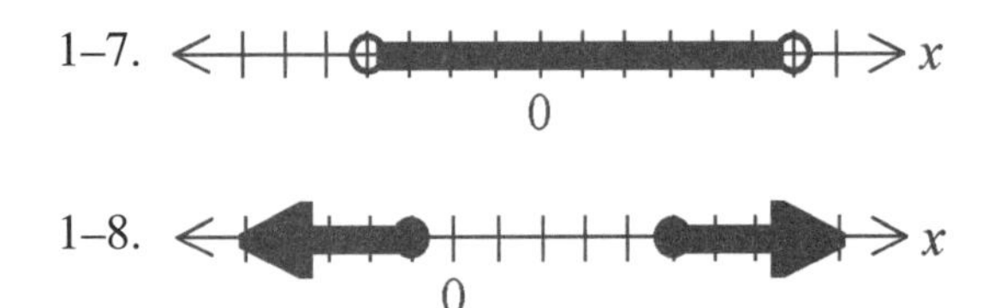

For 9–12, solve for x *and write the solution in interval notation.*

1–9. $2(x - 5) \leq 6x + 2$

1–10. $-6x + 7 \leq 9 - 2(x - 3)$

1–11. $3x - 7 > 8 \vee 5 - 2x \geq 7$

1–12. $-15 \leq 5 - 2x < 25$

Linear Equations in Two Variables

A linear equation in two variables can be simplified to the form $Ax + By = C$ where A, B, and C are real numbers and A and B are not both 0. Solutions to such equations are ordered pairs (x, y). Although linear equations in one variable usually have a single solution, infinite ordered pairs solve a linear equation in two variables. When the solutions are graphed in the x-y plane, they form a straight line.

Graph the line $2x + 3y = 12$.

SOLUTION

There are several ways to graph a line. One that is often convenient is the intercept method. We need only two points to determine the graph of the line. Finding the two intercepts is easy.

y-intercept: set $x = 0$	$2(0) + 3y = 12 \rightarrow y = 4$
x-intercept: set $y = 0$	$2x + 3(0) = 12 \rightarrow x = 6$

Now graph both intercepts, and draw a straight line through them.

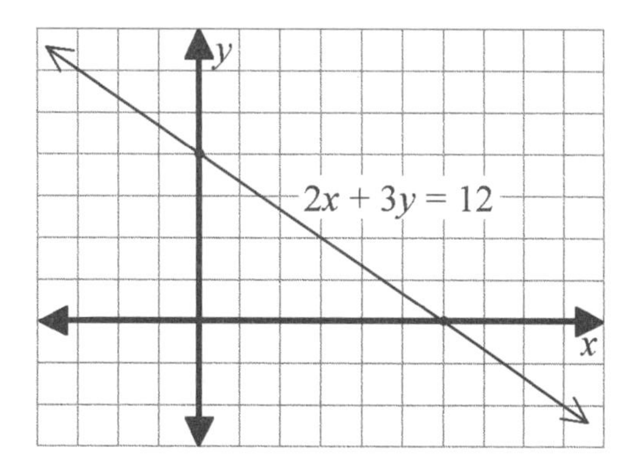

SLOPE

You are probably already familiar with slope. However, it is extremely important, so we will review it here. You probably first learned slope as "rise over run," which is represented graphically and in symbols in Figure 1.2 below.

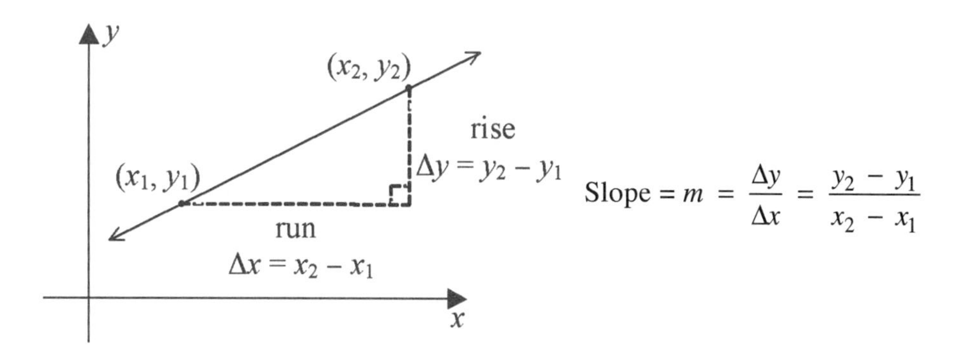

FIGURE 1.2

There are two ways to think about slope. You are probably already familiar with the idea that, graphically, slope measures the steepness and direction of a line. Different categories of slopes are depicted in the following table.

TABLE 1.2
SLOPE OF A LINE

1. If the slope is positive, the line is increasing (rising) from left to right. Larger slopes mean steeper lines (increasing faster).
2. If the slope is negative, the line is decreasing (going down) from left to right. Slopes with larger absolute values have steeper lines (decreasing faster).
3. If the slope is 0, the line is horizontal.
4. The slope of a vertical line is undefined.

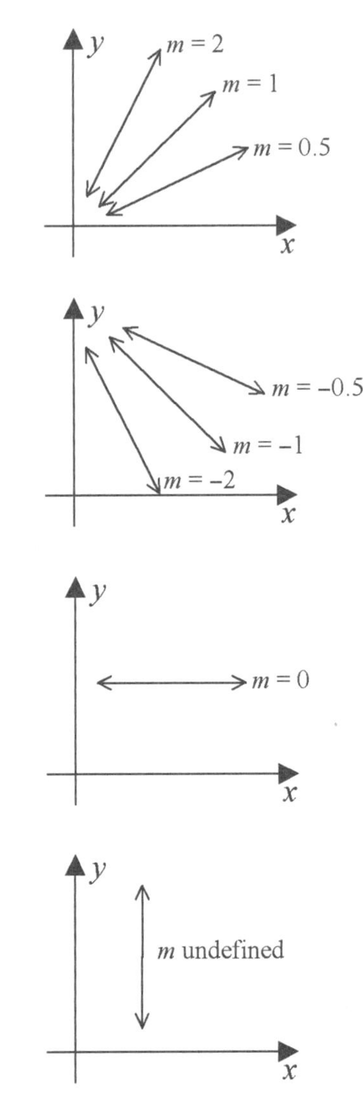

The second interpretation of slope isn't always emphasized in Algebra 1 but is very important. Slope is the *rate of change of* y *with respect to* x. If a line has a slope of m, then each time x increases by one unit, y will change by m units. For example, if $m = 2$, then each time x increases by 1, y will increase by 2. If $m = -\frac{2}{3}$, then each time x increases by 1, y will decrease (because of the negative sign) by $\frac{2}{3}$.

EXAMPLE 1.16 Suppose you are filling a swimming pool with water from a garden hose. You note that at 1 o'clock, the pool has 18 inches of water in it. At 5 o'clock, there are 32 inches of water in the pool. What is the rate of change of the depth of the water?

SOLUTION

A sketch is optional but may help you see what is going on.

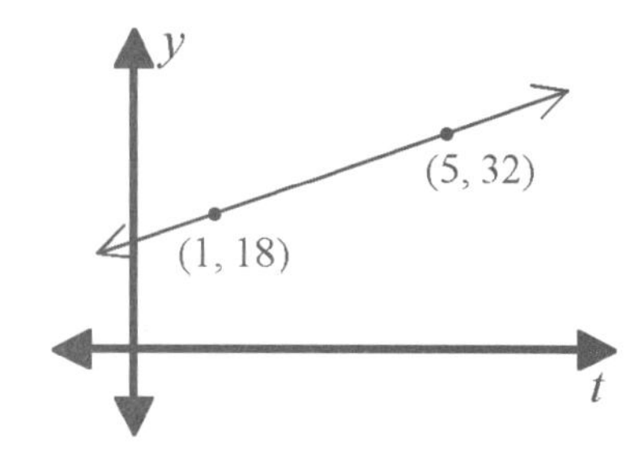

Asking for the rate of change of a line is the same as asking for its slope.

$$\text{Rate of change} = m = \frac{32 - 18}{5 - 1} = \frac{14 \text{ inches}}{4 \text{ hours}} = 3.5 \text{ inches per hour}$$

Note that for rate of change, we usually include units, in this case, inches per hour. The water level in the pool is increasing 3.5 inches each hour.

EQUATIONS OF LINES

The equation $Ax + By = C$ is often called the *general form* of the equation of a line. (Some people prefer $Ax + By + C = 0$. Find out which one your teacher uses.) It is called the general form because any line, whether horizontal, vertical or diagonal, can be written in this form. Unfortunately, the A, B, and C by themselves don't really stand for anything. For lines where $B \neq 0$, in other words, for lines that are not vertical, two other forms of equations of a line are very useful both for graphing lines and for modeling with lines.

SLOPE-INTERCEPT EQUATION OF A LINE

You should know the slope-intercept equation from Algebra 1. Any nonvertical line can be written in *slope-intercept form.*

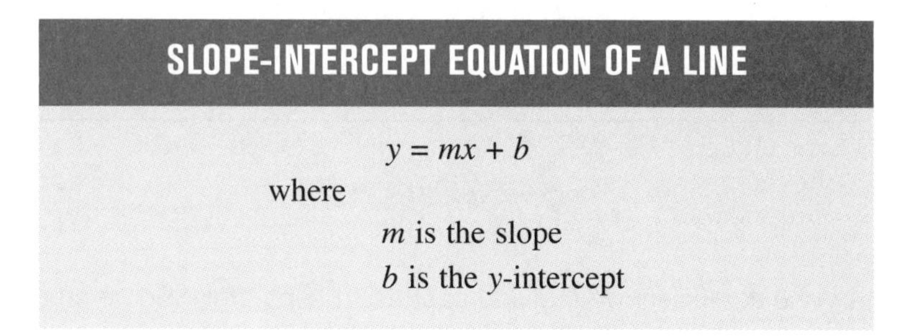

SLOPE-INTERCEPT EQUATION OF A LINE

$$y = mx + b$$

where

m is the slope

b is the y-intercept

Write the equation of the line $3x + 5y = 20$ in slope-intercept form and then graph the line.

SOLUTION

Writing the equation in slope-intercept form means you must solve for y.

$3x + 5y = 20$	
$5y = -3x + 20$	Subtract to move x to the right side.
	Write the term with x to the left of the constant (not at the end).
$\frac{5y}{5} = \frac{-3x}{5} + \frac{20}{5}$	Divide each term by the coefficient of y. Write separately to avoid mistakes.
$y = -\frac{3}{5}x + 4$	Simplify.

From this, we see that the slope is $-\frac{3}{5}$ and the y-intercept is 4. This is easy to graph.

Plot a point at 4 on the y-axis. From there, follow the slope "down 3, right 5" to the next point. Repeat as necessary. It is sometimes convenient to reverse direction and go "up 3, left 5." Connect the points with a straight line.

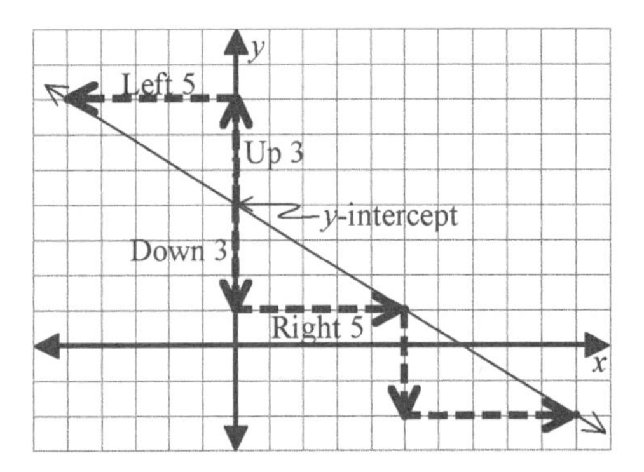

COMMON ERROR

Be sure to plot 4 on the y-axis, not on the x-axis.

COMMON ERROR

When dividing by the coefficient of y, divide each term separately. Some students write $\frac{5y}{5} = \frac{-3x+20}{5}$ but then simplify incorrectly to $y = \frac{-3x}{5} + 20$ or $y = -3x + 4$.

EXAMPLE 1.18 The speed of sound at sea level is a function of temperature. It can be modeled by the equation $S = 0.60T + 331.45$, where S is the speed in meters per second and T is the temperature in degrees Celsius.

a. What is the rate of change of the speed of sound with respect to temperature?

b. What does 331.45 represent in this problem?

SOLUTION

a. Remember that rate of change of a line is the slope. In this problem, the rate of change is 0.60 meters per second per degree Celsius. This means that for each increase of 1°C in temperature, the speed of sound increases by 0.60 meters per second.

b. The number 331.45 is the S-intercept. It is the value of S when $T = 0$. So when the temperature is 0°C, the speed of sound at sea level is 331.45 meters per second.

POINT-SLOPE EQUATION OF A LINE

Given both the slope and the y-intercept of a line, it is easy to write the equation in slope-intercept form. For example, if the slope is 0.75 and the y-intercept is 20, the equation is $y = 0.75x + 20$. If we know the slope and a point other than the y-intercept, the slope-intercept form can still be used but it takes a little more work. For this kind of problem, use the more convenient form of the equation of a line called the point-slope form. The equation comes directly from the slope formula. If (x_1, y_1) is a known point on the line, m is the slope of the line, and all other points on the line are represented by (x, y), then the formula is derived as shown in Figure 1.3 below.

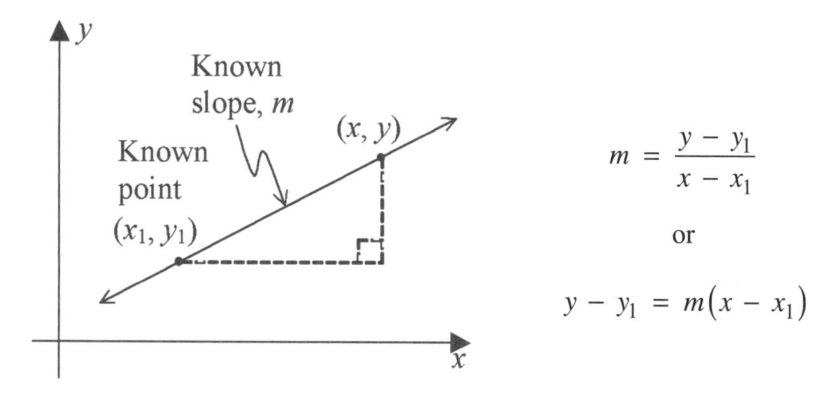

$$m = \frac{y - y_1}{x - x_1}$$

or

$$y - y_1 = m(x - x_1)$$

FIGURE 1.3

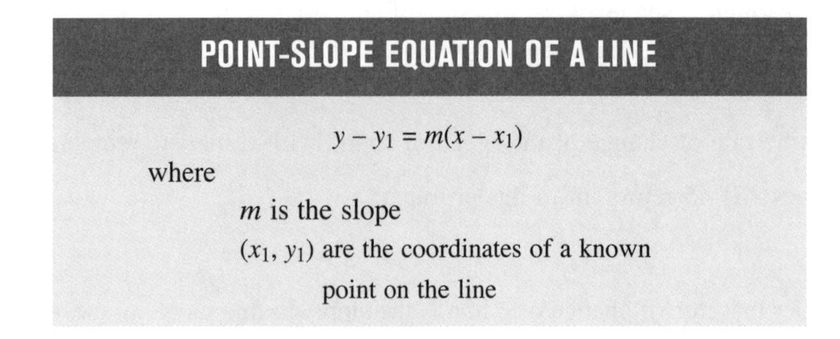

POINT-SLOPE EQUATION OF A LINE

$$y - y_1 = m(x - x_1)$$

where

m is the slope

(x_1, y_1) are the coordinates of a known point on the line

Some people prefer to use this equation with y by itself: $y = m(x - x_1) + y_1$. Use whichever form you find easier to remember.

EXAMPLE 1.19 Write the equation of the line with slope $= -\frac{1}{2}$ that passes through the point (–4, 7).

SOLUTION

Use the point-slope form, $y - y_1 = m(x - x_1)$, with $m = -\frac{1}{2}$, $x_1 = -4$, and $y_1 = 7$.

$$y - y_1 = m(x - x_1)$$

$$y - 7 = -\frac{1}{2}(x - (-4))$$

$$y - 7 = -\frac{1}{2}(x + 4)$$

Many teachers consider $y - 7 = -\frac{1}{2}(x + 4)$ to be an acceptable final answer. If you need to, you can distribute the slope and add 7 to both sides of the equation to rewrite it in $y = mx + b$ form.

$$y - 7 = -\frac{1}{2}(x + 4)$$

$$y - 7 = -\frac{1}{2}x - 2$$

$$y = -\frac{1}{2}x + 5$$

Write the equation of the line passing through the points (–3, 6) and (9, –10).

SOLUTION

First find the slope of the line.

$$m = \frac{\Delta y}{\Delta x} = \frac{-10-6}{9-(-3)} = \frac{-16}{12} = -\frac{4}{3}$$

Now choose either point and use the point-slope formula.

Using the point (–3, 6) gives $y - 6 = -\frac{4}{3}(x + 3)$.

Using the point (9, –10) gives $y + 10 = -\frac{4}{3}(x - 9)$.

We seem to have two different answers to the same question. Both are correct and, if multiplied out and simplified, both are equivalent to $y = -\frac{4}{3}x + 2$.

SEE CALC TIP 1D

COMMON ERROR

$\frac{9-(-3)}{-10-6} = \frac{12}{-16} = -\frac{3}{4}$ is wrong. Be sure to put the *y*-values on top (in the numerator) and the *x*-values on the bottom (in the denominator) of the slope fraction.

COMMON ERROR

$\frac{-10-6}{-3-9} = \frac{-16}{-12} = \frac{4}{3}$ is wrong. Watch the order of subtraction. One set of ordered pairs must be subtracted from the other set of ordered pairs.

EXAMPLE 1.21

Graph the line $y + 4 = \frac{10}{7}(x - 1)$.

SOLUTION

You can rewrite the line in $y = mx + b$ form, but the *y*-intercept turns out to be an inconvenient $-\frac{38}{7}$. Instead, it is easier to note that the line has slope $\frac{10}{7}$ and passes through the point (1, –4). (Note that the signs of the coordinates of the point are the opposite of the signs in the equation. Get used to that. You will see it again in other chapters.) Plot (1, –4). Then count "up 10, right 7" from there to find a second point.

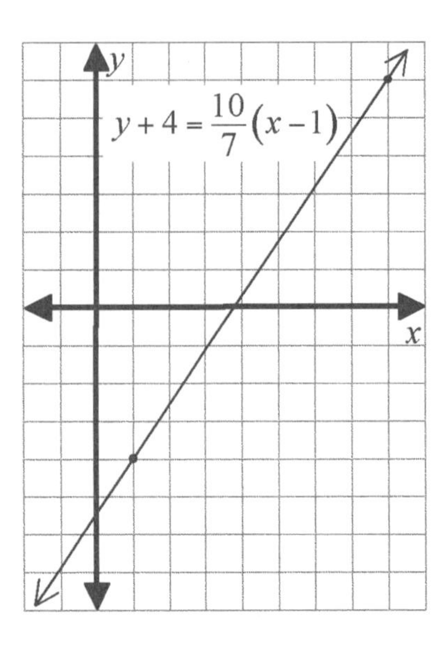

COMMON ERROR

Starting at (–4, 1) is wrong. Don't be fooled by the order in which the numbers appear in the equation. The value for the *x*-coordinate is next to *x* in the equation. The value for the *y*-coordinate is next to *y* in the equation.

EXAMPLE 1.22 Two minutes after an oven is turned on, its temperature is 115°F and is increasing at a rate of 24°F each minute. Write a linear equation to model the temperature in the oven as a function of time in minutes since the oven was turned on.

SOLUTION

Let t represent time in minutes since the oven was turned on and T represent the temperature in the oven in degrees Fahrenheit. At time $t = 2$ minutes, the temperature was 115°F. So we know one point, (2, 115). The rate of change of temperature, 24°F per minute, is the slope. We can use the point-slope formula.

$$T - 115 = 24(t - 2)$$

This equation can be left as is or simplified:

$$T = 24t + 67$$

SECTION EXERCISES

For 13–16, graph and label each line on graph paper.

1–13. $y = 5$

1–14. $x = -3$

1–15. $y = -2x$

1–16. $y = 4 - x$

For 17–18,

a. *express the equation in* $y = mx + b$ *form;*
b. *identify the slope and* y*-intercept;*
c. *sketch the graph.*

1–17. $4x + 3y = 6$

1–18. $3x - 6y = 12$

1–19. Write the equation of the line having the given slope and y-intercept.
a. slope = -2, y-intercept = 6
b. slope = 0, y-intercept = 4

1–20. Find the equation of the line having slope 3 and passing through the point $(4, -3)$.

1–21. Find the equation of the line that passes through the points $(3, 2)$ and $(6, -4)$.

1–22. Find the equation of the line passing through the points $(3, -2)$ and $(3, 4)$.

1–23. When a pleasure boat enters a small lock, there are 120 inches of water (depth) in the lock. Two minutes after the lock begins draining, there are 102 inches of water in the lock.
a. What is the average rate of change of water depth in the lock over the first two minutes?
b. Write a linear model for the depth of water in the lock as a function of time since the lock began to empty.
c. If the lock's lowest water level is 48 inches, how long does it take the lock to reach this water level?

1–24. The following table shows the population of New York State (in thousands) according to U.S. Census data for the last century.

Year	1900	1920	1940	1960	1980	2000
Population (in thousands)	7,268	10,385	13,479	16,782	17,558	18,250

a. What was the average rate of change of New York's population during the 20th century?

b. Use the average rate of change from 1980 to 2000 to write a linear equation for New York's population as a function of time for 1980 to 2000.

c. Using your equation from part *b*, in what year should the population reach 20 million?

Linear Inequalities in Two Variables

Graphing an inequality ($<, \leq, >, \geq$) is different from graphing an equation (=). The solution to an inequality is half a plane, or all the points on one side of the line, indicated graphically by shading. The points on the line are included if the symbol is $\geq$ or $\leq$. This is represented by drawing a solid line. The points on the line are excluded if the symbol is $<$ or $>$. This is represented by drawing a dashed line.

To sketch an inequality, follow these steps:

1. Graph the line (solve for y and then use the slope-intercept method or a table). Sketch a dashed line ($<$ or $>$) or a solid line ($\leq$ or $\geq$) depending on the inequality symbol. Label the equation on the line.
2. Shade the appropriate half plane.
   a. Shade up ($y >$ or $y \geq$) or shade down ($y <$ or $y \leq$) from a horizontal or diagonal line.
   b. Shade right ($x >$ or $x \geq$) or shade left ($x <$ or $x \leq$) from a vertical line.

Some people like to check their shading by testing a particular point. Pick a point that is *not* on the line (the origin is an easy choice as long as it is not on the line) and check it in the original inequality. If the inequality is true for that point, then the half plane that includes that point should be shaded. If the selected point does not check in the inequality, then the half plane not containing that point should be shaded.

EXAMPLE 1.23 Sketch the graph of $3x - 2y < -4$.

SOLUTION

First solve for y.

$$3x - 2y < -4$$
$$-2y < -3x - 4$$
$$y > \frac{3}{2}x + 2$$

Notice how the inequality sign changed direction because you divided by –2.

Now graph the line. Draw a dashed line because of the strict inequality. Then shade the appropriate half plane, in this case above the line because of the > symbol.

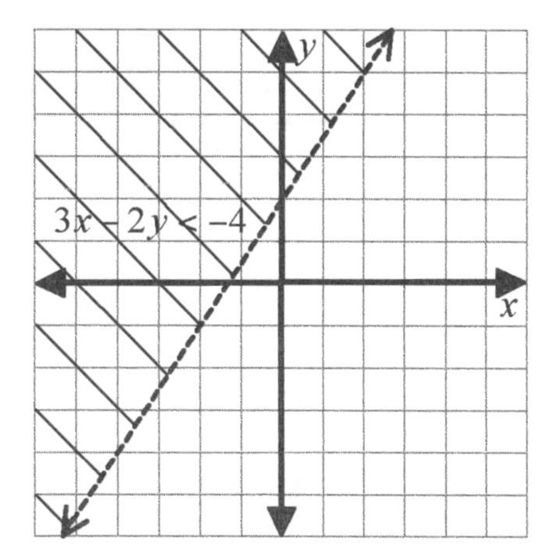

If you want, you can confirm the shading is correct by checking the coordinates of the origin in the inequality: $3(0) - 2(0) < -4$ is false, so in this problem the half plane that does not include the origin should be shaded.

EXAMPLE 1.24 Sketch the graph of $x \geq 3$.

SOLUTION

The line $x = 3$ is vertical. Graph it as a solid line because of the inclusive inequality. For a vertical line, greater than (>) means shade to the right.

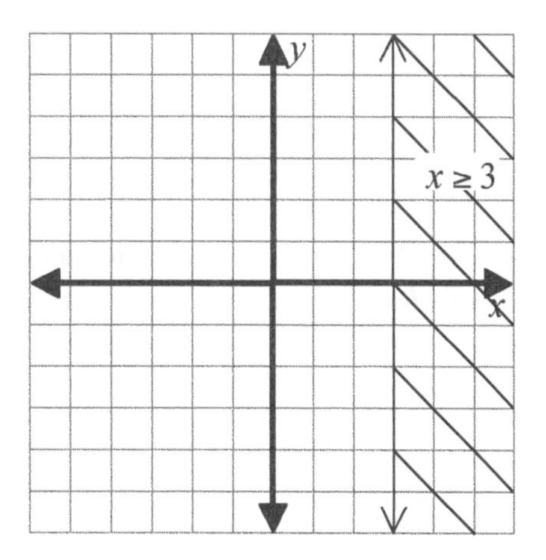

SECTION EXERCISES

For 25–27, graph and label the following inequalities.

1–25. $y \geq x + 1$

1–26. $x > 5$

1–27. $3x - 2y > 2$

Systems of Two Linear Equations

A system of equations involves more than one equation with more than one variable. In this section, we will look at systems of two linear equations in two variables.
A solution to a system of two equations in two variables is an ordered pair that satisfies *both* equations. Typically, there will be only one such solution although other possibilities will be discussed later. The solution can be found either graphically or algebraically.

ALGEBRAIC SOLUTIONS

There are two basic algebraic methods for solving a system of two linear equations, the *elimination method* and the *substitution method*. The method you use is largely a matter of personal preference. However, sometimes the form of the given equations makes one method easier than the other.

ELIMINATION METHOD

For two equations in the form $\begin{cases} Ax + By = C \\ Dx + Ey = F \end{cases}$, the elimination method may be easiest.
The idea of the method is to eliminate one of the variables by adding the two equations. The resulting equation can be solved for the remaining variable. This value can then be substituted into either equation to find the value of the second variable.

Solve the system: $\begin{cases} 5x + 2y = 9 \\ x - 2y = 3 \end{cases}$

SOLUTION

In this case, when the two equations are added, the y-terms will "drop out" of the equation because they have opposite coefficients.

$$\begin{array}{rl} 5x + 2y = & 9 \\ + \; x - 2y = & 3 \\ \hline 6x \qquad = & 12 \\ x = & 2 \end{array} \qquad \begin{array}{r} 5(2) + 2y = 9 \\ 10 + 2y = 9 \\ 2y = -1 \\ y = -0.5 \end{array}$$

The solution is (2, –0.5).

This was an easy example; the coefficients of the y-terms in the two given equations were opposites. When this does not happen, you can multiply one or both of the equations by appropriate constants so that the coefficients of one variable are opposites. Which variable you eliminate does not matter. Sometimes an easy choice will be obvious. If not, you can multiply the first equation by the coefficient of x in the second equation and multiply the second equation by the *opposite* of the coefficient of x in the first equation. This will eliminate x.

EXAMPLE 1.26

Solve the system: $\begin{cases} 2x - 3y = 9 \\ x - 5y = -6 \end{cases}$

SOLUTION

If the second equation is multiplied by –2, the coefficients of x will be opposites.

$$\begin{array}{r} 2x - 3y = 9 \\ -2(x - 5y = -6) \end{array} \rightarrow \begin{array}{r} 2x - 3y = 9 \\ -2x + 10y = 12 \\ \hline 7y = 21 \\ y = 3 \end{array} \qquad \begin{array}{r} x - 5(3) = -6 \\ x - 15 = -6 \\ x = 9 \end{array}$$

The solution is (9, 3).

EXAMPLE 1.27

Solve the system: $\begin{cases} 5y = 11 - 2x \\ 3x = 4y - 18 \end{cases}$

SOLUTION

First, note that the equations must be rearranged into the general form:

$$\begin{aligned} 2x + 5y &= 11 \\ 3x - 4y &= -18 \end{aligned}$$

Now, multiply the first equation by 3 (the coefficient of x in the second equation) and the second equation by –2 (the opposite of the coefficient of x in the first equation) to eliminate x.

$$\begin{aligned} 3(2x + 5y &= 11) \\ -2(3x - 4y &= -18) \end{aligned} \rightarrow \begin{aligned} 6x + 15y &= 33 \\ -6x + 8y &= 36 \\ \hline 23y &= 69 \\ y &= 3 \end{aligned} \qquad \begin{aligned} 2x + 5(3) &= 11 \\ 2x + 15 &= 11 \\ 2x &= -4 \\ x &= -2 \end{aligned}$$

The solution is (–2, 3).

SUBSTITUTION METHOD

This method is often easier if one of the equations already has one variable expressed in terms of the other. That first variable can then be eliminated from the other equation by substitution.

EXAMPLE 1.28 Solve the system: $\begin{cases} 5x - 2y = 16 \\ y = x - 5 \end{cases}$.

SOLUTION

The second equation gives y in terms of x. In words, the second equation says "y is the same as $x - 5$." The expression $(x - 5)$ can be substituted for y in the first equation, leaving an equation with only one variable.

$5x - 2y = 16 \rightarrow 5x - 2(x - 5) = 16$ Note the parentheses around $x - 5$.

$y = (x - 5)$

$$\begin{aligned} 5x - 2x + 10 &= 16 \\ 3x &= 6 \\ x = 2 \rightarrow y &= 2 - 5 = -3. \end{aligned}$$

The solution is (2, –3).

Any system of two linear equations may be solved by either method.

EXAMPLE 1.29 Solve: $\begin{cases} 2x = 3y - 14 \\ x + 3y = 2 \end{cases}$

SOLUTION 1

Use elimination. Rearrange the first equation into general form. Then eliminate y by addition.

$$\begin{aligned} 2x - 3y &= -14 \\ x + 3y &= 2 \\ \hline 3x \quad &= -12 \\ x &= -4 \end{aligned} \quad \rightarrow \quad \begin{aligned} -4 + 3y &= 2 \\ 3y &= 6 \\ y &= 2 \end{aligned}$$

Solution: $(-4, 2)$

SOLUTION 2

Use substitution.

$$\begin{aligned} 2x = 3y - 14 &\rightarrow 2x = 3y - 14 \\ x + 3y = 2 &\rightarrow x = -3y + 2 \end{aligned} \quad \rightarrow \quad \begin{aligned} 2(-3y + 2) &= 3y - 14 \\ -6y + 4 &= 3y - 14 \\ -9y &= -18 \\ y &= 2 \quad \rightarrow x = -3(2) + 2 = -4 \end{aligned}$$

Solution: $(-4, 2)$

GRAPHICAL SOLUTIONS

To solve a system of two equations in two variables graphically:

1. Graph both equations on the same set of axes.
2. Find the point(s) where they intersect.

There are three possibilities when graphing two lines. The lines may intersect in a single point, they may be parallel, or they may turn out to be the same line. These possibilities are shown in the following Table 1.3.

TABLE 1.3
SOLVING SYSTEMS OF EQUATIONS GRAPHICALLY

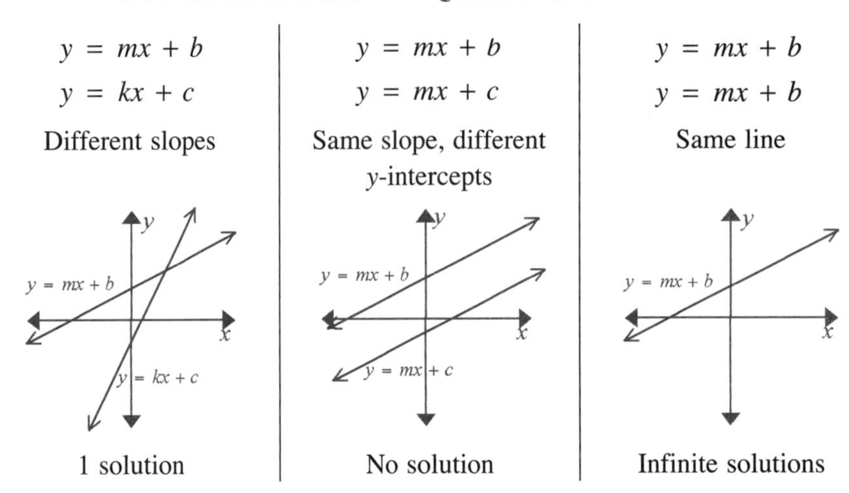

EXAMPLE 1.30 Solve the system of equations graphically: $\begin{cases} y = 2x - 3 \\ 3x + 2y = 8 \end{cases}$

SOLUTION

First solve the second equation for y.

$$\begin{aligned} 3x + 2y &= 8 \\ 2y &= -3x + 8 \\ \frac{2y}{2} &= \frac{-3x}{2} + \frac{8}{2} \\ y &= \frac{-3}{2}x + 4 \end{aligned}$$

Graph each line, label the equations, and write the point of intersection as the answer.

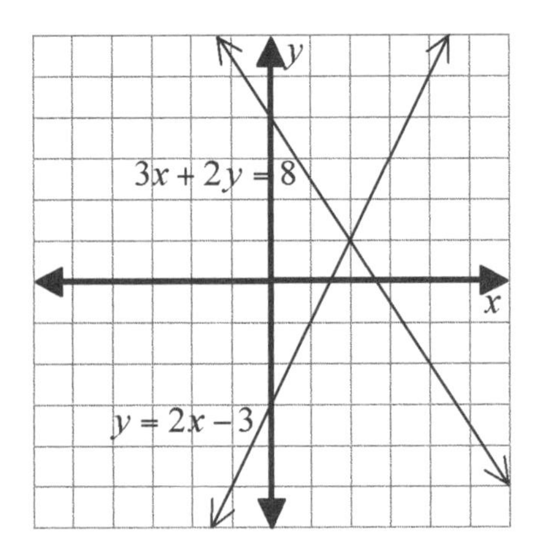

The solution is (2,1).

EXAMPLE 1.31 Solve the system of equations graphically: $\begin{cases} y = \dfrac{1}{2}x + 2 \\ x - 2y = 2 \end{cases}$

SOLUTION

Solving the second equation for y gives $y = \dfrac{1}{2}x - 1$.

At this point, it should be clear that the two lines have the same slope but different y-intercepts. They will be parallel. A graph is not really necessary but is shown below.

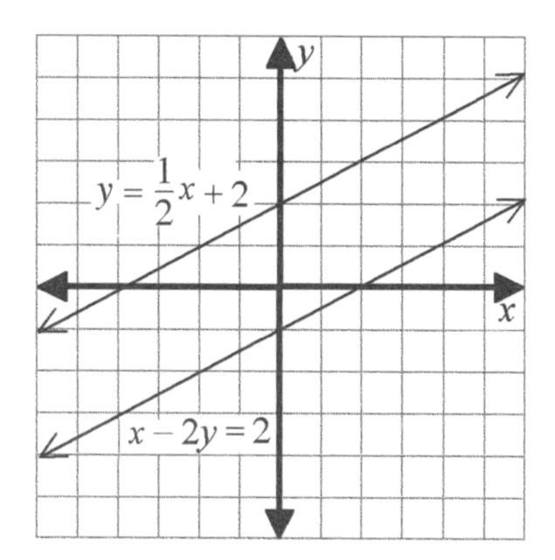

This system of equations has no solution. Such systems are called *inconsistent*.

EXAMPLE 1.32 Solve the system of equations graphically: $\begin{cases} y = -\dfrac{2}{3}x + 1 \\ 2x + 3y = 3 \end{cases}$

SOLUTION

Solving the second equation for y gives $y = -\dfrac{2}{3}x + 1$, which is exactly the same as the first equation. The two equations represent the same line.

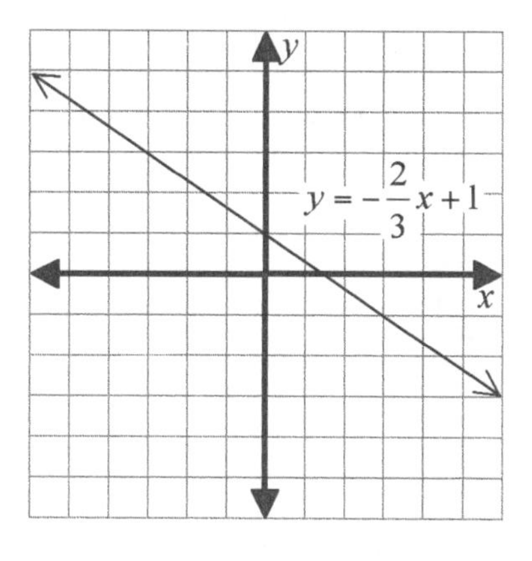

All points on the line $y = -\dfrac{2}{3}x + 1$ are solutions to the system.

SECTION EXERCISES

For 28–30, solve the systems algebraically.

1–28. $\begin{cases} 6x + 2y = 13 \\ y = \frac{1}{2}x - 4 \end{cases}$

1–29. $\begin{cases} x + y = 76 \\ -5x + 6y = 60 \end{cases}$

1–30. $\begin{cases} x = 3y + 1 \\ 2y - x = 7 \end{cases}$

For 31–33, solve the systems graphically.

1–31. $\begin{cases} y = 2x - 1 \\ 2x + 3y = 21 \end{cases}$

1–32. $\begin{cases} 2x + y = 7 \\ y = -2x + 3 \end{cases}$

1–33. $\begin{cases} -6 = 4x - 2y \\ y = 2x + 3 \end{cases}$

Systems of Two Linear Inequalities

When graphing a system of inequalities, the solution set is all the points that satisfy both inequalities. You may want to review how to graph inequalities, which was described earlier in this chapter. Using two different colors to sketch each inequality can help you see the double-colored region that satisfies both inequalities.

EXAMPLE 1.33 Sketch the graphs of $\begin{cases} y < 2x + 1 \\ y \geq -\frac{1}{3}x \end{cases}$ and label the solution set, S.

SOLUTION

For $y < 2x + 1$, graph $y = 2x + 1$ as a dashed line and shade below the line.

For $y \geq -\frac{1}{3}x$, graph $y = -\frac{1}{3}x$ as a solid line and shade above the line.

Label the solution set, S, in the double-shaded region.

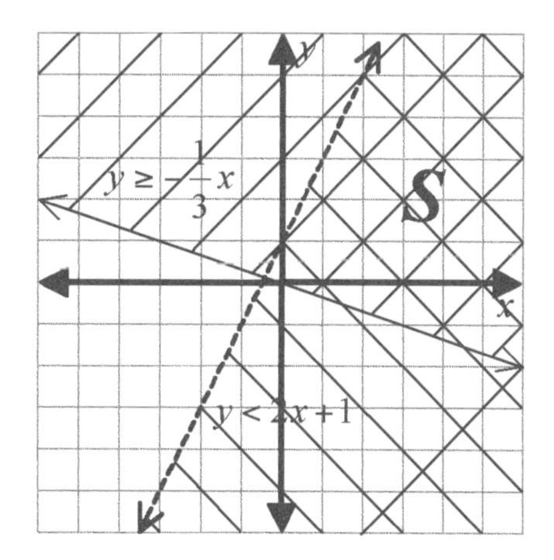

SECTION EXERCISES

Graph the system and label the solution set, S.

1–34. $\begin{cases} 3x + 2y < 8 \\ 2x - y \leq 4 \end{cases}$

Direct and Inverse Variation

Suppose gas costs exactly $4.00 a gallon. The amount you pay to fill your tank is directly related to the amount of gas you buy. The more gas you buy, the more you pay. If you purchase x gallons of gas, the total cost, y, will be $y = 4x$. The graph of this function is shown below.

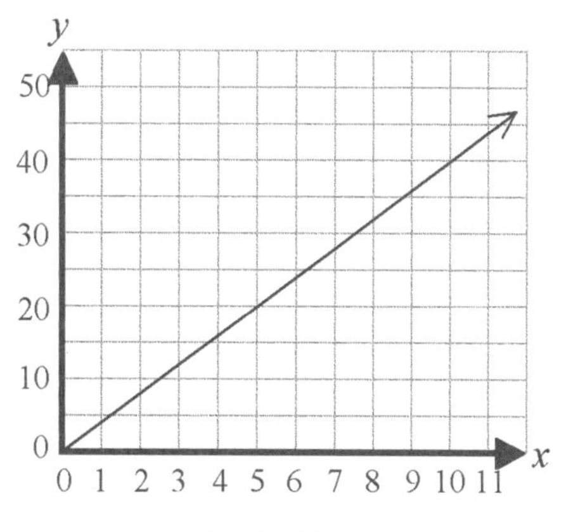

FIGURE 1.4

This is an example of *direct variation*. If y varies directly with x:

1. y and x are always in the same ratio; $\frac{y}{x} = k$, for some value k. This can also be written as $y = kx$.
2. The graph will be a straight line through the origin.
3. When x is doubled, y is doubled. When x is halved, y is halved.

EXAMPLE 1.34

If y varies directly with x and $y = 12$ when $x = 8$:

a. Find y when $x = 20$.

b. Find x when $y = 6$.

SOLUTION

If y varies directly with x, they are always in the same ratio. This means we can set up a proportion: $\frac{y_1}{x_1} = \frac{y_2}{x_2}$.

a.
$$\begin{aligned} \frac{12}{8} &= \frac{y}{20} \\ 8y &= 240 \\ y &= 30 \end{aligned}$$

b.
$$\begin{aligned} \frac{12}{8} &= \frac{6}{x} \\ 12x &= 48 \\ x &= 4 \end{aligned}$$

Now suppose you are going on a 120-mile trip in your car. The time to make the trip is inversely related to your average speed for the trip. The faster you go, the less time it will take. If you travel at x miles per hour, the total time y for the trip will be $y = \frac{120}{x}$. This is shown in the graph below.

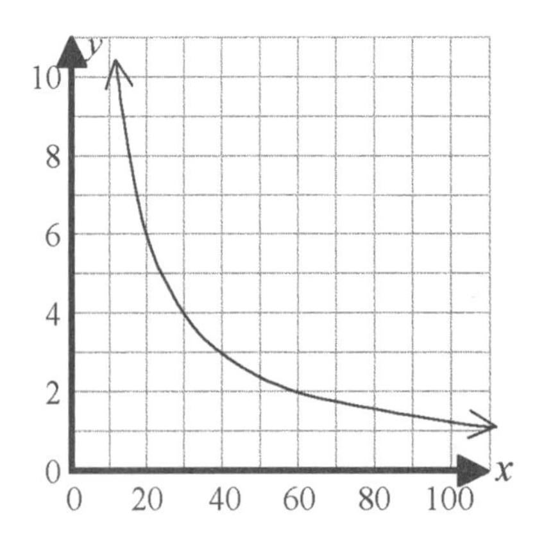

FIGURE 1.5

This is an example of *inverse variation*. If y varies inversely with x:

1. y and x always have the same product, $xy = k$, for some value k. This can also be written as $y = \frac{k}{x}$.
2. The graph is a hyperbola. (Mathematically, the hyperbola has a second part, a reflection through the origin of the part shown above. In many real-life applications, such as speed and time, these negative values do not make sense in the problem.)
3. When x is doubled, y is halved. When x is halved, y is doubled.

EXAMPLE 1.35

If y varies inversely with x and $y = 12$ when $x = 8$:

a. Find y when $x = 20$.

b. Find x when $y = 6$.

SOLUTION

When y varies inversely with x, the product of x and y is a constant: $x_1y_1 = x_2y_2$.

a.
$$\begin{aligned} 20y &= 8(12) \\ 20y &= 96 \\ y &= 4.8 \end{aligned}$$

b.
$$\begin{aligned} 6x &= 8(12) \\ 6x &= 96 \\ x &= 16 \end{aligned}$$

EXAMPLE 1.36

Identify each of the following as direct variation, inverse variation, or neither.

a.

x	y
1	3
2	6
3	9
4	12

b.

x	y
1	24
2	18
3	12
4	6

c.

x	y
1	24
2	12
3	8
4	6

d. $y = 2x + 3$

e. $y = \frac{12}{x}$

f. $y = 0.5x$

SOLUTION

a. It is easy to check that the ratio $\frac{y}{x}$ is a constant: $\frac{3}{1} = \frac{6}{2} = \frac{9}{3} = \frac{12}{4}$. This is direct variation.

b. Neither the ratio $\frac{y}{x}$ nor the product xy is a constant: $\frac{24}{1} \neq \frac{18}{2}$ and $(1)(24) \neq (2)(18)$. This table shows neither direct nor inverse variation.

c. These ordered pairs all have the same product: $(1)(24) = (2)(12) = (3)(8) = (4)(6)$. This is inverse variation.

d. This is a line, so it is not inverse variation. However, it does not pass through the origin, so it is not direct variation either. It is neither.

e. This equation is in the form $y = \frac{k}{x}$, which is inverse variation.

f. This equation is in the form $y = kx$, which is direct variation.

EXAMPLE 1.37 A spring hangs vertically in a science classroom. The amount by which the spring stretches, x, varies directly with the amount of mass, m, hung on the end. If 200 grams of mass make the spring stretch 5 cm, how far will the spring stretch for 250 grams?

SOLUTION

For direct variation, set up a proportion.

$$\frac{200}{5} = \frac{250}{x}$$
$$200x = 1{,}250$$
$$x = 6.25$$

The spring will stretch 6.25 cm.

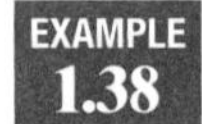

EXAMPLE 1.38 The amount of time, t, to travel a fixed distance varies inversely with speed, v. If a certain trip takes 3.75 hours at 40 mph, how long will the same trip take at 50 mph?

SOLUTION

For inverse variation, the product vt is a constant.

$$50t = 40(3.75)$$
$$50t = 150$$
$$t = 3$$

The trip will take 3 hours.

SECTION EXERCISES

1–35. If y varies inversely with x and y is 12 when x is 4, then

a. what is y when x is 8?
b. what is y when x is 3?
c. what is x when y is 2?

1–36. If y varies directly with x and y is 12 when x is 4, then

a. what is y when x is 8?
b. what is y when x is 3?
c. what is x when y is 6?

1–37. For two kids to balance on a seesaw, their distances from the fulcrum must be inversely proportional to their weights. Madeline weighs 80 pounds and is 4 feet back from the fulcrum. How far back must 50-pound Adelaide sit to balance Madeline?

1–38. In chemistry, you learn about the ideal gas law that tells how gases behave in terms of temperature, pressure, and volume. The following are true:

- If temperature is held constant, volume and pressure vary inversely.
- If pressure is held constant, then volume varies directly with temperature.

a. If the temperature is held constant and the pressure is doubled, what happens to the volume?
b. If the pressure is held constant and the temperature is doubled, what happens to the volume?

1–39. Suppose a 10-liter volume of a certain gas has a temperature of 300 Kelvin (300 K) and a pressure of 200 kilopascals (kPa).

a. If the temperature is held constant and the pressure is increased to 500 kPa, what will be the new volume?
b. If the pressure is held constant and the temperature is increased to 400 K, what will be the volume, to the nearest tenth?

CHAPTER EXERCISES

C1–1. Solve for x: $2(3x + 2) - 5(x - 1) = 13$

C1–2. Solve for x: $a(x + 2) = bx - a$

For 3–5, solve for x *and write the solution in interval notation.*

C1–3. $8 - 2x > 4(x - 4) + 9$

C1–4. $5 - 2x > 11 \vee 5x + 3 \geq 13$

C1–5. $-7 \leq 2x + 3 \leq 7$

C1–6. For $5x - 2y = 8$,

a. express the equation in $y = mx + b$ form;
b. identify the slope and y-intercept;
c. sketch the graph.

C1–7. Write the equation of the line having slope = $\frac{1}{2}$ and y-intercept at the origin.

C1–8. Find the equation of the line that passes through the points (–4, 1) and (3, 4).

C1–9. Skeletal bones can be used to estimate an individual's height when living. In men, a femur (thighbone) 48 cm long would indicate a height of about 171.5 cm while a femur of 52 cm would indicate a height of about 179 cm.

a. What is the average rate of change of the height per length of femur for men?
b. Write a linear model for the height in terms of length of femur for men.
c. If a male femur has a length of 54 cm, how tall was the man according to the model?

C1–10. Graph and label the following inequality: $3x + 2y < 8$

For 11–12, solve the systems algebraically.

C1–11. $\begin{cases} x + 2y = 120 \\ 4x + 3y = 300 \end{cases}$

C1–12. $\begin{cases} x + 4y = -4 \\ 3x - 2y + 12 = 0 \end{cases}$

C1–13. Solve the system of equations graphically: $\begin{cases} x + 4 = 2y \\ 2x + y = -3 \end{cases}$

C1–14. Graph the system and label the solution set, S. $\begin{cases} 2x \leq 4y \\ -2x - y < 1 \end{cases}$

C1–15. Sound is formed by waves in air. The wavelength, λ, of sound varies inversely with the frequency, f. The note A below middle C has a frequency of 220 Hz and a wavelength of 157 cm. If the G note below that A has a frequency of 196 Hz, find the G note's wavelength to the nearest whole centimeter.

Polynomial Operations

WHAT YOU WILL LEARN

When compared with other functions, polynomials are easy to evaluate. Polynomials can be evaluated using only addition and multiplication. Because of this, polynomials are extremely useful in mathematics, so much so that scientists and engineers often approximate other functions using polynomials. This means it is important that you be able to work effectively with polynomials. In this chapter, you will learn to:

- Simplify expressions with exponent rules;
- Perform addition, subtraction, multiplication, and division of polynomials;
- Find powers of binomials;
- Factor completely using a variety of factoring techniques.

SECTIONS IN THIS CHAPTER

- Exponent Rules
- Four Basic Operations with Polynomials
- Powers of Binomials
- Factoring Techniques

Exponent Rules

When you first learned about exponents, they were positive whole numbers. This is how they were first defined centuries ago. An *exponent* tells the number of times another number, the *base*, is used as a factor. Thus $3^5 = 3 \cdot 3 \cdot 3 \cdot 3 \cdot 3$; five 3s are multiplied together.

POSITIVE WHOLE-NUMBER EXPONENTS

The following rules of exponents should be familiar from Algebra 1.

Multiplication rule

To multiply powers of the same base, keep the base and add the exponents.

$$b^m b^n = b^{m+n}$$

You add the exponents because you are counting the total number of factors of b.

EXAMPLE 2.1

Simplify: $2^3 \cdot 2^4$

SOLUTION

The bases are the same. Keep the base, and add the exponents.

$$2^3 \cdot 2^4 = 2^{3+4} = 2^7$$

With practice, this should become automatic. When in doubt, write it out.

$$2^3 \cdot 2^4 = (2 \cdot 2 \cdot 2) \cdot (2 \cdot 2 \cdot 2 \cdot 2) = 2^7$$

COMMON ERROR

Do *not* multiply the bases.

$$2^3 \cdot 2^4 \neq 4^7$$

Division rule

To divide powers of the same base, keep the base and *subtract* the exponents.

$$\frac{b^m}{b^n} = b^{m-n}$$

You subtract the exponents because dividing reduces the number of times the base is used as a factor.

EXAMPLE 2.2 Simplify: $\dfrac{3^9}{3^4}$

SOLUTION

Keep the base, and subtract the exponents. $\dfrac{3^9}{3^4} = 3^{9-4} = 3^5$

When in doubt, write it out. $\dfrac{3^9}{3^4} = \dfrac{\cancel{3} \cdot \cancel{3} \cdot \cancel{3} \cdot \cancel{3} \cdot 3 \cdot 3 \cdot 3 \cdot 3 \cdot 3}{\cancel{3} \cdot \cancel{3} \cdot \cancel{3} \cdot \cancel{3}} = 3^5$

Power rule

To raise a power to a power, *multiply* the exponents.

$$\left(b^m\right)^n = b^{mn}$$

To see why the exponents are multiplied, imagine writing out the power as a product.

$$\left(b^m\right)^n = \underbrace{b^m b^m \cdots b^m}_{n \text{ factors}} = b^{\overbrace{m+m+\cdots+m}^{n \text{ terms}}} = b^{mn}$$

EXAMPLE 2.3 Simplify: $(2^3)^4$

SOLUTION

Keep the base, and multiply the exponents. $(2^3)^4 = 2^{3 \cdot 4} = 2^{12}$

When in doubt, write it out. $(2^3)^4 = (2^3)\,(2^3)\,(2^3)\,(2^3) = 2^{3+3+3+3} = 2^{12}$

Distributive rule over multiplication

An exponent can be distributed over multiplication.

$$(ab)^n = a^n b^n$$

Distributive rule over division

An exponent can be distributed over division.

$$\left(\frac{a}{b}\right)^n = \frac{a^n}{b^n} \text{ (assuming } b \neq 0)$$

COMMON ERROR

Do not distribute exponents over addition and subtraction. This does not work.

$$(a + b)^n \neq a^n + b^n \text{ and } (a - b)^n \neq a^n - b^n$$

EXAMPLE 2.4

Which is the expression $3^4 \cdot 3^6$ equivalent to?

a. 3^{10}
b. 3^{24}
c. 9^{10}
d. 9^{24}

SOLUTION

By the multiplication rule, keep the common base and add the exponents: $3^4 \cdot 3^6 = 3^{4+6} = 3^{10}$, which is choice (a). When in doubt, write it out:

$3^4 \cdot 3^6 = (3 \cdot 3 \cdot 3 \cdot 3)(3 \cdot 3 \cdot 3 \cdot 3 \cdot 3 \cdot 3) = 3^{10}$. The exponent, 10, counts the total number of threes being multiplied.

Common mistakes students make are multiplying the exponents, choice (b), multiplying the bases, choice (c), or multiplying both, choice (d).

EXAMPLE 2.5

Simplify the expression $\dfrac{(2x^2)^3(4x^4)}{16x^7}$ where x is a nonzero number.

SOLUTION

Be careful to follow the order of operations. Evaluate exponents (in this case powers of exponents) *before* you multiply or divide.

$$\frac{(2x^2)^3(4x^4)}{16x^7}$$

$$\frac{((2)^3(x^2)^3)(4x^4)}{16x^7}$$ Use the distributive rule over multiplication

$$\frac{(8x^6)(4x^4)}{16x^7}$$ Use the power rule

$$\frac{32x^{10}}{16x^7}$$ Use the multiplication rule

$$2x^3$$ Use the division rule

Having a problem with powers of exponents? Write it out. $(2x^2)^3 = (2x^2)(2x^2)(2x^2) = 8x^6$

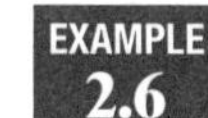

Are -2^6 and $(-2)^6$ the same?

SOLUTION

No, they are not the same because the order of operations is different. According to the order of operations, -2^6 means $-(2^6) = -64$. In this case, the exponent 6 is understood to be on the base 2 (not on -2) with the negative sign applied to the result. However in $(-2)^6$, the exponent is on the base -2. The result (since the exponent is even) is $(-2)^6 = +64$. It is important that you pay attention to this rule when using your calculator. Do not type -8^2 if what you mean is $(-8)^2$.

EXAMPLE 2.7

If $10^x = c$,
a. express 10^{x-3} in terms of c.
b. express c^3 in terms of x.

SOLUTION

a. Since the exponents are subtracted, two terms with the same base, 10, are being divided.

$$10^{x-3} = \frac{10^x}{10^3} = \frac{c}{1{,}000}$$

b. Substitute for c and then apply the power rule.

$$c^3 = \left(10^x\right)^3 = 10^{3x}$$

ZERO AS AN EXPONENT

Many beginning students expect that b^0 ought to equal 0. This is not true. By the division rule, we know $\frac{b^n}{b^n} = b^{n-n} = b^0$. However, reducing the fraction gives us 1, $\frac{b^n}{b^n} = 1$, provided $b \neq 0$. From this we get the following rule of exponents.

Zero power rule
Any nonzero number raised to the zero power equals 1.

$$b^0 = 1 \text{ for } b \neq 0$$

Note that 0^0 is undefined, much like division by 0.

EXAMPLE 2.8

Evaluate $3x^0 - (4x)^0$ for $x \neq 0$.

SOLUTION

Pay careful attention to the base of each exponent. In $3x^0$, the exponent applies only to the x (the base it is "touching"), not the 3. Thus $3x^0 = 3(1) = 3$. Because of the parentheses in $(4x)^0$, the exponent applies to both the 4 and the x, so $(4x)^0 = 1$.

$$3x^0 - (4x)^0 = 3 - 1 = 2$$

The basic rules of whole-number exponents are summarized below. These rules will be extended to cover rational exponents in Chapters 8 and 9.

EXPONENT RULES SUMMARY

For positive bases a and b, the following rules of exponents hold.

1. $b^m b^n = b^{m+n}$
2. $\dfrac{b^m}{b^n} = b^{m-n}$
3. $\left(b^m\right)^n = b^{mn}$
4. $b^0 = 1$
5. $(ab)^n = a^n b^n$
6. $\left(\dfrac{a}{b}\right)^n = \dfrac{a^n}{b^n}$

SECTION EXERCISES

In 1–3, simplify completely.

2–1. $(3x^4y^3)^2$

2–2. $(-2x^{4n}y^{7k})(3x^0y^{2k})$

2–3. $\dfrac{\left(2x^2y\right)^3\left(3xy^2\right)}{16x^7y^3}$

2–4. If $10^x = c$,

a. express $\dfrac{c}{10}$ as a power of 10 in terms of x.

b. express c^4 as a power of 10 in terms of x.

c. express 10^{6x} in terms of c.

2–5. Evaluate $(5x)^0 - 4x^0 + 3$ assuming $x \neq 0$.

Four Basic Operations with Polynomials

A *monomial* consists of a constant, called the *coefficient*, which may be multiplied by one or more variables. The variables may have positive integer exponents. Note that if the coefficient is 1, it is usually not written.

TABLE 2.1
WHAT IS A MONOMIAL?

Monomials	Not Monomials
-2, $3x$, $-5x^4$, x^2yz^3	$\frac{2}{x}$, $\sqrt{x}$, $3x^{1/3}y^{-2}$

A *polynomial* is either a monomial or a *sum* of monomials. The expressions 1, $3x + 5$, $2x^2 - 4x - 7$, and $x^3 - 2x^2y + 3xy^2 - 4y^3$ are all polynomials. A sum of two unlike monomials is a *binomial*; a sum of three unlike monomials is a *trinomial.* Although a polynomial is defined as a sum, it is often written using subtraction. The trinomial $2x^2 - 4x - 7$ could be written as $2x^2 + (-4x) + (-7)$. No matter which way it is written, the coefficients are 2, –4, and –7; each coefficient includes the sign to its left.

TABLE 2.2
TYPES OF POLYNOMIALS

Monomial	Binomial	Trinomial	Four-Term Polynomial
One term	Two terms	Three terms	Four terms
$3x^2$	$4x - 5$	$-2x^2 + 5x - 3$	$x^3 - 4x^2 + 5x + 7$

Polynomials can be combined using the four basic arithmetic operations: addition, subtraction, multiplication, and division.

ADDING POLYNOMIALS

Before adding polynomials, we need to review what *like terms* are. Like terms are monomials that differ, if at all, only in their coefficients. They have the same variables with the same exponents.

TABLE 2.3
LIKE AND UNLIKE TERMS

Like Terms	Unlike Terms
$-5x$ and $8x$	$7x^2$ and $2x^3$
$4x^2$ and $-2x^2$	$4x^5$ and $2y^5$
$3x^3y$ and $5x^3y$	$6x^2y$ and $4xy^2$

Polynomials are added by *combining like terms*. When like terms are combined, the coefficients are added and the variables and their exponents are left unchanged. So $5x + 2x = 7x$ and $3x^2 + (-5x^2) = -2x^2$. However, $2x^3 + 4x^2$ cannot be combined because they are not like terms (the exponents are different).

EXAMPLE 2.9

Add $x^3 + 5x - 4$ and $3x^3 - 2x - 6$.

SOLUTION

$x^3 + 5x - 4 + 3x^3 - 2x - 6$	Some people like to put parentheses around the original polynomials. For addition, this is optional and doesn't change anything.
$x^3 + 5x + (-4) + 3x^3 + (-2x) + (-6)$	Rewrite as a sum.
$\underbrace{x^3 + 3x^3} + \underbrace{5x + (-2x)} + \underbrace{(-4) + (-6)}$	Use the commutative property of addition to rearrange the terms so that like terms are together. (Most people do this mentally and do not bother to write it out.)
$4x^3 \quad + \quad 3x \quad + (-10)$	Combine the like terms.
$4x^3 + 3x - 10$	Final answer.

This example was written out in great detail to illustrate the algebra being used. Many people do the algebra mentally and go straight from the first line to the last.

EXAMPLE 2.10

Find the sum of $2x^4$, $5x^4 - 3x + 1$, and $-x^4 + 7$.

SOLUTION

$$\underline{2x^4} + \underline{5x^4} - 3x + \underline{\underline{1}} - \underline{x^4} + \underline{\underline{7}} = 6x^4 - 3x + 8$$

Underlining like terms can help you keep track of them.

SUBTRACTING POLYNOMIALS

You should already know that subtraction is the same as adding the opposite: $a - b = a + (-b)$. Some people like the mnemonic "same-change-change." Leave the first term the same, change addition to subtraction, and change the second term to its opposite. This is the same rule we use to subtract polynomials. Be very careful when negating a polynomial; you must change the sign of each term in the second polynomial. The most common error made when subtracting polynomials is to change the sign of only the first term.

EXAMPLE 2.11

Simplify: $4x^3 - 5x^2 + 7 - (5x^3 + 3x^2 - 8x - 9)$

SOLUTION

$$4x^3 - 5x^2 + 7 - (5x^3 + 3x^2 - 8x - 9)$$

$$4x^3 - 5x^2 + 7 + (-1)(5x^3 + 3x^2 - 8x - 9)$$

Change to adding the opposite. The opposite of a polynomial can be found by multiplying it by –1.

$$\underline{4x^3} \underline{\underline{- 5x^2}} \underline{\underline{\underline{+ 7}}} \underline{- 5x^3} \underline{\underline{- 3x^2}} + 8x \underline{\underline{\underline{+ 9}}}$$

Distribute the –1. Underline like terms.

$$-x^3 - 8x^2 + 8x + 16$$

Combine like terms.

SEE CALC TIP 2A

COMMON ERROR

Be careful to change each sign in the subtracted polynomial, not just the first term.

$$4x^3 - 5x^2 + 7 - (5x^3 + 3x^2 - 8x - 9) \neq 4x^3 - 5x^2 + 7 - 5x^3 + 3x^2 - 8x - 9$$

Sometimes words are used instead of symbols. Be careful when translating the words "subtract … from … ." Subtract 7 from 10 means 10 – 7, not 7 – 10. The *from* expression is *first* and the *subtract* expression is *second.*

EXAMPLE 2.12 Subtract $3x^2 + 5x - 4$ from $x^2 + 2x + 3$.

SOLUTION

$x^2 + 2x + 3 - (3x^2 + 5x - 4)$	The subtract expression is written after the subtraction sign. The parentheses around the polynomial being subtracted are *important*. They indicate that the whole polynomial is being subtracted, not just the first term.
$x^2 + 2x + 3 + (-1)\ (3x^2 + 5x - 4)$	Change to adding the opposite.
$x^2 + 2x + 3 - 3x^2 - 5x + 4$	Distribute the -1.
$-2x^2 - 3x + 7$	Combine like terms.

Again, the example was written out in more detail than is usually shown. Most people write out only these steps.

$$x^2 + 2x + 3 - (3x^2 + 5x - 4)$$
$$x^2 + 2x + 3 - 3x^2 - 5x + 4$$
$$-2x^2 - 3x + 7$$

Just make sure you use parentheses around the polynomial being subtracted and then distribute the subtraction to each term.

MULTIPLICATION OF POLYNOMIALS

Multiply polynomials by carefully using the distributive property.

1. Multiply each term of the first polynomial by each term of the second polynomial.
2. Combine like terms.

Note that the number of individual multiplications you need to do depends on the lengths of the polynomials being multiplied. Multiplying two binomials requires $2 \times 2 = 4$ multiplications. Multiplying a binomial by a trinomial needs $2 \times 3 = 6$ multiplications. Multiplying two trinomials takes $3 \times 3 = 9$ multiplications.

EXAMPLE 2.13 Find the product of $x - 3$ and $2x + 5$.

SOLUTION

$(x - 3)(2x + 5)$	Write as a product. Use parentheses!
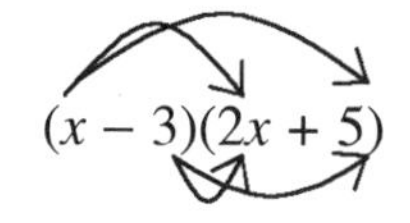	Distribute. There will be four separate multiplications. The arrows shown are optional but may help you keep track of what you need to multiply. Make sure you write all four products.
$x(2x) + x(5) - 3(2x) - 3(5)$	Remember that when multiplying monomials, you multiply the coefficients and add the exponents on like variables.
$2x^2 + 5x - 6x - 15$	
$2x^2 - x - 15$	Combine like terms.

Once more, extra details have been written. Most people would just write these steps.

$$(x - 3)(2x + 5)$$

$$2x^2 + 5x - 6x - 15$$

$$2x^2 - x - 15$$

Some students learn the mnemonic FOIL: first–outer–inner–last. This memory device applies only to multiplying a binomial by a binomial. To multiply any type of polynomials, just remember to distribute and combine.

EXAMPLE 2.14 Express $(3x - 2)(x^2 - 4x + 1)$ as a polynomial in simplest form.

SOLUTION

$$(3x - 2)(x^2 - 4x + 1) = 3x^3 - 12x^2 + 3x - 2x^2 + 8x - 2 = 3x^3 - 14x^2 + 11x - 2$$

Some people prefer to multiply vertically. This has the advantage of being very similar to how you multiply two- and three-digit (and more) numbers. Write the longer polynomial on top. Work right to left across the bottom polynomial. Keep like terms lined up in columns, squares above squares and so on. With this method, the example works out like this:

$$\begin{array}{rrrr} & x^2 & -4x & +1 \\ & & 3x & -2 \\ \hline & -2x^2 & +8x & -2 \\ 3x^3 & -12x^2 & +3x & \\ \hline 3x^3 & -14x^2 & +11x & -2 \end{array}$$

Step	Explanation
$-2x^2 + 8x - 2$	Multiply the first polynomial by –2.
$3x^3 - 12x^2 + 3x$	Multiply the first polynomial by $3x$.
$3x^3 - 14x^2 + 11x - 2$	Add (combine like terms).

EXAMPLE 2.15

Write $(a + b)^2$ as a polynomial.

SOLUTION

Remember: To square a quantity means to multiply it by itself.

$$(a + b)^2 = (a + b)(a + b) = a^2 + ab + ab + b^2 = a^2 + 2ab + b^2$$

EXAMPLE 2.16

Express $(3x + 1)(x – 2)(5 – x)$ as a polynomial in simplest form.

SOLUTION

Step	Explanation
$(3x + 1)(x - 2)(5 - x)$	First multiply the two binomials on the left.
$(3x^2 - 6x + x - 2)(5 - x)$	
$(3x^2 - 5x - 2)(5 - x)$	Combine like terms.
$15x^2 - 25x - 10 - 3x^3 + 5x^2 + 2x$	Multiply each term in the left trinomial by each term in the right binomial.
$-3x^3 + 20x^2 - 23x - 10$	Combine like terms, and rewrite in descending order of exponents.

EXAMPLE 2.17

Express $(x^2 – 2xy + y^2)(3x – 5y)$ as a polynomial in simplest form.

SOLUTION

Step	Explanation
$(x^2 - 2xy + y^2)(3x - 5y)$	Multiply each term on the left by each term on the right.
$3x^3 - 6x^2y + 3xy^2 - 5x^2y + 10xy^2 - 5y^3$	
$3x^3 - 11x^2y + 13xy^2 - 5y^3$	Combine like terms. When possible, write the x's in descending order and the y's in ascending order.

DIVISION OF A POLYNOMIAL BY A MONOMIAL

You may have noticed in the previous examples that adding, subtracting, or multiplying two polynomials always results in another polynomial. We say that polynomials are closed for these three operations. This is not the case for dividing two polynomials. A quotient of two polynomials may result in an expression that is not a polynomial.

To divide a polynomial by a monomial, divide each term in the polynomial by the monomial. When dividing two monomials, divide the coefficients and subtract exponents or like variables.

EXAMPLE 2.18 Divide $4x^3 - 6x^2 + 2x$ by $2x$.

SOLUTION

$\dfrac{4x^3 - 6x^2 + 2x}{2x}$	Write the quotient as a fraction.
$\dfrac{4x^3}{2x} - \dfrac{6x^2}{2x} + \dfrac{2x}{2x}$	Divide each term in the polynomial by the monomial.
$2x^2 - 3x + 1$	Simplify.

Note that we are assuming $x \neq 0$. If $x = 0$, we are dividing by 0, which is undefined. (Sometimes this will be stated explicitly in the problem but not always.)

Also remember that anything divided by itself (except 0) is 1.

COMMON ERROR

Many students simplify $\dfrac{4x^3 - 6x^2 + 2x}{2x}$ to $2x^2 - 3x$ because they believe that $\dfrac{2x}{2x}$ cancels, leaving nothing. However, $\dfrac{2x}{2x}$ simplifies to 1. This error is common not only when dividing but also when factoring (as discussed later in this chapter).

EXAMPLE 2.19 Simplify: $\dfrac{8x^3y^2 - 10x^2y^3}{4x^2y}$

SOLUTION

$$\frac{8x^3y^2 - 10x^2y^3}{4x^2y} = \frac{8x^3y^2}{4x^2y} - \frac{10x^2y^3}{4x^2y} = 2xy - \frac{5}{2}y^2$$

With practice, you will probably feel comfortable omitting the middle step.

POLYNOMIAL LONG DIVISION

Here is a blast from your past. Let's work out $\dfrac{2{,}982}{21}$ using long division. (Don't skip this! Study the example, and make sure you really remember how to do long division with whole numbers.)

$$21\overline{)2{,}982} \quad \rightarrow \quad \begin{array}{r} 1 \\ 21\overline{)2{,}982} \\ \underline{2{,}100} \\ 882 \end{array} \quad \rightarrow \quad \begin{array}{r} 14 \\ 21\overline{)2{,}982} \\ \underline{2{,}100} \\ 882 \\ \underline{840} \\ 42 \end{array} \quad \rightarrow \quad \begin{array}{r} 142 \\ 21\overline{)2{,}982} \\ \underline{2{,}100} \\ 882 \\ \underline{840} \\ 42 \\ \underline{42} \\ 0 \end{array}$$

We have $\dfrac{2{,}982}{21} = 142$. In this example, the remainder is 0. This means that 21 is a factor of 2,982, and we can write $2{,}982 = 21(142)$. If we change the problem to $\dfrac{2{,}987}{21}$, all the steps are exactly the same except we end up with a remainder of 5. This means 21 is not a factor of 2,987. We can express the answer as a mixed number by writing the remainder over the divisor: $\dfrac{2{,}987}{21} = 142 + \dfrac{5}{21}$, (In elementary school, you did not write an addition sign in your answer. However, that is what a mixed number is, a sum of a whole number and a fraction.) We can also write this as $2{,}987 = 21(142) + 5$.

What does this have to do with polynomial division? It turns out that polynomials behave a lot like integers. Let's look at $\dfrac{2x^3 + 9x^2 + 8x + 2}{2x + 1}$. Keep in mind that long division seems hard the first time you see it. You may need to read this example more than once. Like many things, it gets easier with practice. Here we go.

EXAMPLE 2.20 Find the quotient: $\dfrac{2x^3 + 9x^2 + 8x + 2}{2x + 1}$

SOLUTION

$$2x + 1\overline{)2x^3 + 9x^2 + 8x + 2}$$

Write the fraction in long division form. The numerator goes inside the division bar and the denominator goes outside.

$$\begin{array}{r} 1x^2 \qquad\qquad\quad \\ 2x + 1\overline{)2x^3 + 9x^2 + 8x + 2} \end{array}$$

Divide the first term inside by the first term outside, $\dfrac{2x^3}{2x} = 1x^2$. Write the answer on the bar.

$$\begin{array}{r} 1x^2 \qquad\qquad\quad \\ 2x + 1\overline{)2x^3 + 9x^2 + 8x + 2} \\ \underline{2x^3 + 1x^2 \qquad\qquad\quad} \end{array}$$

Multiply the outside, $2x + 1$, by x^2 (the answer from the previous step) to get $2x^3 + 1x^2$. Write it on the second line, lining up the like terms.

$$\begin{array}{r} 1x^2 \qquad\qquad\quad \\ 2x + 1\overline{)2x^3 + 9x^2 + 8x + 2} \\ \underline{-(2x^3 + 1x^2) \qquad\qquad\quad} \\ 8x^2 + 8x + 2 \end{array}$$

Subtract the two polynomials to get $8x^2 + 8x + 2$. The first terms should be a perfect match and subtract to 0. If not, the division above was not done correctly.

$$\begin{array}{r} 1x^2 + 4x \qquad \\ 2x + 1\overline{)2x^3 + 9x^2 + 8x + 2} \\ \underline{-(2x^3 + 1x^2) \qquad\qquad\quad} \\ 8x^2 + 8x + 2 \end{array}$$

Repeat the process using $8x^2 + 8x + 2$ divided by $2x + 1$. Divide the first terms, $\dfrac{8x^2}{2x} = 4x$ and add this to the answer bar.

$$\begin{array}{r} 1x^2 + 4x \qquad \\ 2x + 1\overline{)2x^3 + 9x^2 + 8x + 2} \\ \underline{-(2x^3 + 1x^2) \qquad\qquad\quad} \\ 8x^2 + 8x + 2 \\ \underline{8x^2 + 4x \qquad} \end{array}$$

Multiply the outside, $2x + 1$, by $4x$ (the answer from the previous step) to get $8x^2 + 4x$. Write it below, lining up the like terms.

$$\begin{array}{r} 1x^2 + 4x + 2 \\ 2x + 1\overline{)2x^3 + 9x^2 + 8x + 2} \\ \underline{-(2x^3 + 1x^2) \qquad\qquad\quad} \\ 8x^2 + 8x + 2 \\ \underline{-(8x^2 + 4x) \qquad} \\ 4x + 2 \end{array}$$

Subtract the two polynomials to get $4x + 2$.

$$\begin{array}{r|l} & 1x^2 + 4x + 2 \\ \hline 2x+1 & 2x^3 + 9x^2 + 8x + 2 \\ & -(2x^3 + 1x^2) \\ \hline & \quad 8x^2 + 8x + 2 \\ & \quad -(8x^2 + 4x) \\ \hline & \qquad 4x + 2 \end{array}$$

Repeat once more using $4x + 2$ divided by $2x + 1$. Divide the first terms, $\frac{4x}{2x} = 2$, and add this to the answer bar.

$$\begin{array}{r|l} & 1x^2 + 4x + 2 \\ \hline 2x+1 & 2x^3 + 9x^2 + 8x + 2 \\ & -(2x^3 + 1x^2) \\ \hline & \quad 8x^2 + 8x + 2 \\ & \quad -(8x^2 + 4x) \\ \hline & \qquad 4x + 2 \\ & \qquad -(4x + 2) \\ \hline & \qquad\qquad 0 \end{array}$$

Multiply the outside, $2x + 1$, by 2 (the answer from the previous step) to get $4x + 2$. Write it below, lining up the like terms. Subtract the two polynomials to get 0.

Because the remainder is 0, $2x + 1$ is a factor of $2x^3 + 9x^2 + 8x + 2$. We can write:

$$\frac{2x^3 + 9x^2 + 8x + 2}{2x + 1} = x^2 + 4x + 2$$

or

$$2x^3 + 9x^2 + 8x + 2 = (2x + 1)(x^2 + 4x + 2)$$

In the same way, if we try $\frac{2x^3 + 9x^2 + 8x + 7}{2x + 1}$ (everything is the same, except the constant in the numerator changed from 2 to 7), all the steps are exactly the same except we get a remainder of 5. This means $2x + 1$ is not a factor of $2x^3 + 9x^2 + 8x + 7$. Just as we did with integers, we can write the remainder over the divisor:

$$\frac{2x^3 + 9x^2 + 8x + 7}{2x + 1} = x^2 + 4x + 2 + \frac{5}{2x + 1}$$

The answer is in the form of a polynomial, $x^2 + 4x + 2$, and a *proper rational expression*, $\frac{5}{2x + 1}$, a ratio of two polynomials where the numerator is of lower degree than the denominator. This is the polynomial equivalent of a mixed number. Rational expressions are covered in Chapter 9.

You must be aware of a few things with polynomial division. When dividing integers, you never got negative numbers in the course of the division. This will not always be true when dividing polynomials. Because polynomial division can lead to sign errors, it is a good idea to change subtractions to addition of the opposite. Similarly, long division of integers does not involve fractions in the division process. Fractional coefficients can arise when dividing polynomials (although most Algebra 2 problems are chosen to avoid this). Finally, it would make no sense to leave out the 0 in 3,082 when dividing 3,082 by 23. Similarly, write missing terms with a 0 coefficient when dividing. For example, $3x^3 + 8x + 2$ should be written as $3x^3 + 0x^2 + 8x + 2$.

EXAMPLE 2.21 Find the quotient: $\dfrac{6x^3 - 19x^2 + 25}{3x - 5}$

SOLUTION

$$3x - 5\overline{\big)\,6x^3 - 19x^2 + 0x + 25}$$

Include $0x$ as a placeholder for the missing term.

$$\begin{array}{r} 2x^2 \qquad\qquad\qquad\quad \\ 3x - 5\overline{\big)\,6x^3 - 19x^2 + 0x + 25} \\ \underline{-\,6x^3 + 10x^2 \qquad\qquad} \\ -9x^2 + 0x + 25 \end{array}$$

Divide the first term inside by the first term outside, $\dfrac{6x^3}{3x} = 2x^2$.
Write the answer on the bar. Multiply $3x - 5$ by $2x^2$, and write the answer below. Subtract by changing to the opposite and adding.

$$\begin{array}{r} 2x^2 - 3x \qquad\qquad \\ 3x - 5\overline{\big)\,6x^3 - 19x^2 + 0x + 25} \\ \underline{-\,6x^3 + 10x^2 \qquad\qquad} \\ -9x^2 + 0x + 25 \\ \underline{+9x^2 - 15x \qquad} \\ -15x + 25 \end{array}$$

Divide the first term inside by the first term outside, $\dfrac{-9x^2}{3x} = -3x$.
Write the answer on the bar. Multiply $3x - 5$ by $-3x$, and write the answer below. Subtract by changing to the opposite and adding.

$$\begin{array}{r} 2x^2 - 3x - 5 \qquad \\ 3x - 5\overline{\big)\,6x^3 - 19x^2 + 0x + 25} \\ \underline{-\,6x^3 + 10x^2 \qquad\qquad} \\ -9x^2 + 0x + 25 \\ \underline{+9x^2 - 15x \qquad} \\ -15x + 25 \\ \underline{+15x - 25} \\ 0 \end{array}$$

Divide the first term inside by the first term outside, $\dfrac{-15x}{3x} = -5$.
Write the answer on the bar. Multiply $3x - 5$ by -5, and write the answer below. Subtract by changing to the opposite and adding.

Since the remainder is 0, $3x - 5$ is a factor of $6x^3 - 19x^2 + 25$:

$$\frac{6x^3 - 19x^2 + 25}{3x - 5} = 2x^2 - 3x - 5$$

SEE CALC TIP 2B

Divide: $8x^3 - 42x^2 + 21$ by $4x + 3$

SOLUTION

$$4x + 3\overline{)8x^3 - 42x^2 + 0x + 21}$$

Include $0x$ as a placeholder.

$$\begin{array}{r l} & 2x^2 \\ 4x + 3 \big) & \overline{8x^3 - 42x^2 + 0x + 21} \\ & \underline{-8x^3 - 6x^2} \\ & -48x^2 + 0x + 21 \end{array}$$

Divide the first term inside by the first term outside, $\dfrac{8x^3}{4x} = 2x^2$. Write the answer on the bar. Multiply $4x + 3$ by $2x^2$, and write the answer below. Subtract by changing to the opposite and adding.

$$\begin{array}{r l} & 2x^2 - 12x \\ 4x + 3 \big) & \overline{8x^3 - 42x^2 + 0x + 21} \\ & \underline{-8x^3 - 6x^2} \\ & -48x^2 + 0x + 21 \\ & \underline{48x^2 + 36x} \\ & 36x + 21 \end{array}$$

Divide the first term inside by the first term outside, $\dfrac{-48x^2}{4x} = -12x$. Write the answer on the bar. Multiply $4x + 3$ by $-12x$, and write the answer below. Subtract by changing to the opposite and adding.

$$\begin{array}{r l} & 2x^2 - 12x + 9 \\ 4x + 3 \big) & \overline{8x^3 - 42x^2 + 0x + 21} \\ & \underline{-8x^3 - 6x^2} \\ & -48x^2 + 0x + 21 \\ & \underline{48x^2 + 36x} \\ & 36x + 21 \\ & \underline{-36x - 27} \\ & -6 \end{array}$$

Divide the first term inside by the first term outside, $\dfrac{36x}{4x} = 9$. Write the answer on the bar. Multiply $4x + 3$ by 9, and write the answer below. Subtract by changing to the opposite and adding.

The remainder is not 0, so $4x + 3$ is not a factor of $8x^3 - 42x^2 + 21$. We write the quotient as a sum of a polynomial and a proper rational expression:

$$\frac{8x^3 - 42x^2 + 21}{4x + 3} = 2x^2 - 12x + 9 + \frac{-6}{4x + 3}$$

EXAMPLE 2.23 Express $\dfrac{x^4 + 2x^3 + 6x - 10}{x^2 + 4}$ as a sum of a polynomial and a proper rational expression.

SOLUTION

Start with $x^2 + 0x + 4\overline{)x^4 + 2x^3 + 0x^2 + 6x - 10}$. Missing terms should be filled in for both the numerator and denominator.

$$\begin{array}{r} x^2 + 2x - 4 \\ x^2 + 0x + 4\overline{)x^4 + 2x^3 + 0x^2 + 6x - 10} \\ \underline{-x^4 - 0x^3 - 4x^2 \qquad\qquad} \\ 2x^3 - 4x^2 + 6x - 10 \\ \underline{-2x^3 - 0x^2 - 8x \qquad} \\ -4x^2 - 2x - 10 \\ \underline{+4x^2 + 0x + 16} \\ -2x + 6 \end{array}$$

Now that the remainder, $-2x + 6$, is of a lower degree than the divisor, $x^2 + 4$, we are done dividing. Write the answer as the sum of a polynomial and a proper rational expression:

$$\frac{x^4 + 2x^3 + 6x - 10}{x^2 + 4} = x^2 + 2x - 4 + \frac{-2x + 6}{x^2 + 4}$$

SECTION EXERCISES

2–6. Subtract $4x^3 + 3x^2 - 5x + 9$ from $6x^3 - 3x^2 + 7$.

2–7. Express $(2x - 5)(x^2 - 4x + 2)$ as a polynomial in simplest form.

2–8. Simplify $\dfrac{25xy^5 - 10x^4y^3}{5xy^3}$.

2–9. Divide $3x^4 + 4x^3 - 19x^2 - 8x + 12$ by $3x - 2$.

2–10. Divide $x^4 - 3x^3 - x^2 - 12x - 20$ by $x^2 + 4$.

2–11. Divide $2x^3 - 11x^2 + 16x - 16$ by $x - 4$.

2–12. Express $2x^4 + 11x^3 + x^2 - 26x - 33$ divided by $x + 5$ as the sum of a polynomial and a proper rational expression.

Powers of Binomials

In the section on multiplying polynomials, we saw that $(a + b)^2 = a^2 + 2ab + b^2$. Multiplying that result by $(a + b)$ again gives $(a + b)^3 = a^3 + 3a^2b + 3ab^2 + b^3$ as shown below.

$$\begin{aligned}(a+b)^3 &= (a+b)(a+b)(a+b)\\ &= \left(a^2 + 2ab + b^2\right)(a+b)\\ &= a^3 + 2a^2b + ab^2 + a^2b + 2ab^2 + b^3\\ &= a^3 + 3a^2b + 3ab^2 + b^3\end{aligned}$$

In principle, we could keep multiplying by $(a + b)$ to find $(a + b)^n$ for any whole number n. However, this takes more and more work as the power gets larger. Fortunately, there is a shortcut.

POWERS OF BINOMIALS AND PASCAL'S TRIANGLE

The first few powers of $(a + b)$ are written out below. Terms have been rewritten to show coefficients of 1 and exponents of 0 and 1 to help see some patterns.

$$\begin{aligned}(a+b)^0 &= 1 &&= 1a^0b^0\\ (a+b)^1 &= a+b &&= 1a^1b^0 + 1a^0b^1\\ (a+b)^2 &= a^2 + 2ab + b^2 &&= 1a^2b^0 + 2a^1b^1 + 1a^0b^2\\ (a+b)^3 &= a^3 + 3a^2b + 3ab^2 + b^3 &&= 1a^3b^0 + 3a^2b^1 + 3a^1b^2 + 1a^0b^3\\ (a+b)^4 &= a^4 + 4a^3b + 6a^2b^2 + 4ab^3 + b^4 &&= 1a^4b^0 + 4a^3b^1 + 6a^2b^2 + 4a^1b^3 + 1a^0b^4\end{aligned}$$

You should be able to see that for any (whole number) exponent n:

1. $(a + b)^n$ has $n + 1$ terms. (To find the total number of terms, add 1 to the power.)
2. The first and last terms each have coefficients of 1.
3. The exponents on a start at n (the power) and decrease to 0.
4. The exponents on b start at 0 and increase to n (the power).

The only thing left to figure out is the coefficients. To help see the pattern, we write down just the coefficients for each power n as shown at right. When written in this way, the coefficients form a pattern known as Pascal's triangle. Each row begins and ends with 1. In between the 1s, each number is the sum of the two numbers diagonally above it in the preceding row as shown by the arrows between rows 3 and 4.

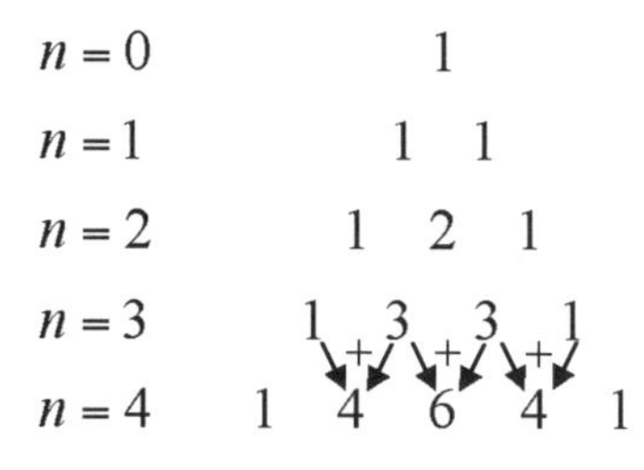

It is now easy to find each new line of the triangle. Using the coefficients from row $n = 4$, we find the coefficients for $n = 5$ are 1, 5, 10, 10, 5, and 1.

$n = 4$ 1   4   6   4   1

$n = 5$ 1   5   10   10   5   1

Now we are ready to expand $(a + b)^5$. The coefficients are found in the $n = 5$ row of Pascal's triangle. (This is actually the sixth row because the first row is $n = 0$.) Write down the coefficients with space between them for the variables.

$$(a + b)^5 = 1 \quad + 5 \quad + 10 \quad + 10 \quad + 5 \quad + 1$$

The exponents on a start at 5 and decrease to 0.

$$(a + b)^5 = 1a^5 \quad + 5a^4 \quad + 10a^3 \quad + 10a^2 \quad + 5a^1 \quad + 1a^0$$

The exponents on b start at 0 and increase to 5.

$$(a + b)^5 = 1a^5b^0 + 5a^4b^1 + 10a^3b^2 + 10a^2b^3 + 5a^1b^4 + 1a^0b^5$$

Remembering that $a^0 = b^0 = 1$ and omitting coefficients and exponents of 1, this simplifies to:

$$(a + b)^5 = a^5 + 5a^4b + 10a^3b^2 + 10a^2b^3 + 5ab^4 + b^5$$

At this point you may be thinking, "Well, that's all fine. What, though, if I want a power of something besides $(a + b)$? What if I want to expand $(2x - 3)^6$ or $(2x^2 - 3y)^4$?"

That is a good question.

Expand: $(2x - 3)^6$

SOLUTION

We need the $n = 6$ row of Pascal's triangle, which is 1 6 15 20 15 6 1. From that we have:

$$(a + b)^6 = 1a^6b^0 + 6a^5b + 15a^4b^2 + 20a^3b^3 + 15a^2b^4 + 6ab^5 + 1a^0b^6$$

Think of $(2x)$ as taking the place of a and (-3) (note the sign) as taking the place of b. Then we get:

$$\begin{aligned}(2x - 3)^6 = {} & 1(2x)^6(-3)^0 + 6(2x)^5(-3)^1 + 15(2x)^4(-3)^2 \\ & + 20(2x)^3(-3)^3 + 15(2x)^2(-3)^4 + 6(2x)^1(-3)^5 + 1(2x)^0(-3)^6\end{aligned}$$

Now simplify each term carefully. Remember, work with the exponents before doing the multiplication.

$$\begin{aligned}(2x - 3)^6 &= 1\left(64x^6\right)(1) + 6\left(32x^5\right)(-3) + 15\left(16x^4\right)(9) \\ &\quad +20\left(8x^3\right)(-27) + 15\left(4x^2\right)(81) + 6(2x)(-243) + 1(1)(729) \\ &= 64x^6 - 576x^5 + 2{,}160x^4 - 4{,}320x^3 + 4{,}860x^2 - 2{,}916x + 729\end{aligned}$$

EXAMPLE 2.25

Expand: $\left(2x^2 - 3y\right)^4$

SOLUTION

We need the $n = 4$ row of Pascal's triangle, which is 1 4 6 4 1. We have:

$$(a + b)^4 = 1a^4b^0 + 4a^3b + 6a^2b^2 + 4ab^3 + 1a^0b^4$$

Think of $(2x^2)$ as taking the place of a and $(-3y)$ as taking the place of b. Then we get:

$$\begin{aligned}\left(2x^2 - 3y\right)^4 = {} & 1\left(2x^2\right)^4(-3y)^0 + 4\left(2x^2\right)^3(-3y)^1 + 6\left(2x^2\right)^2(-3y)^2 \\ & + 4\left(2x^2\right)^1(-3y)^3 + 1\left(2x^2\right)^0(-3y)^4\end{aligned}$$

This simplifies to:

$$\begin{aligned}\left(2x^2 - 3y\right)^4 &= 1\left(16x^8\right)(1) + 4\left(8x^6\right)(-3y) + 6\left(4x^4\right)\left(9y^2\right) + 4\left(2x^2\right)\left(-27y^3\right) + 1(1)\left(81y^4\right) \\ &= 16x^8 - 96x^6y + 216x^4y^2 - 216x^2y^3 + 81y^4\end{aligned}$$

BINOMIAL THEOREM

The coefficients in Pascal's triangle can also be found using combinations, ${}_nC_r$, the number of ways r items can be chosen from n total items. The quantity ${}_nC_r$ can be found on most calculators, or you can use the formula ${}_nC_r = \frac{n!}{r!(n-r)!}$.

To use ${}_nC_r$ to find the coefficients, let n be the power and let r be one less than the position of the term (for the third term, $r = 3 - 1 = 2$). Here is a comparison of Pascal's triangle to ${}_nC_r$ for $n = 4$.

PASCAL'S TRIANGLE AND ${}_nC_r$

Pascal's triangle	1	4	6	4	1
${}_nC_r$	${}_4C_0 = 1$	${}_4C_1 = 4$	${}_4C_2 = 6$	${}_4C_3 = 4$	${}_4C_4 = 1$

The binomial theorem gives the formula for finding powers of binomials.

$$(a+b)^n = {}_nC_0a^nb^0 + {}_nC_1a^{n-1}b^1 + {}_nC_2a^{n-2}b^2 + \ldots + {}_nC_na^0b^n$$

A general expression for a term in the expansion, where n is the power, is:

$${}_nC_ra^{n-r}b^r$$

The first term has $r = 0$, the second term has $r = 1$, and so on. To find the kth term, let $r = k - 1$. There are $n + 1$ terms in the polynomial. Notice how the value of r matches the exponent on b. The exponents on a and b always add to the power, n.

EXAMPLE 2.26

Find the third term of $(x + 5)^8$

SOLUTION

Use ${}_nC_ra^{n-r}b^r$ and substitute in the appropriate values. For the third term, $r = 2$, one less than the position. The power is 8, so $n = 8$.

$${}_nC_ra^{n-r}b^r = {}_8C_2a^{8-2}b^2 = 28a^6b^2$$

Then substitute $a = x$ and $b = 5$.

$$28a^6b^2 = 28x^6(5)^2 = 28x^6(25) = 700x^6$$

You could use Pascal's triangle to do this problem, but you would need to write out nine rows to find the $n = 8$ row.

SEE CALC TIP 2C

Find the middle term of $(3x - 2)^6$.

SOLUTION

Use ${}_nC_r a^{n-r} b^r$, and substitute in the appropriate values. To find the position of the middle term, remember that the total number of terms is one more than the power. Since the power is 6, this expansion has 7 terms. The middle term of 7 terms is the 4th term. For the 4th term, $r = 3$, one less than the position. The power is 6, so $n = 6$.

$${}_nC_r a^{n-r} b^r = {}_6C_3 a^{6-3} b^3 = 20a^3b^3$$

Then substitute $a = 3x$ and $b = -2$.

$$20a^3b^3 = 20(3x)^3(-2)^3 = 20\left(27x^3\right)(-8) = -4{,}320x^3$$

SECTION EXERCISES

2–13. Expand $(3x - 2)^4$.

2–14. In the expansion of $(x + y)^{20}$, one of the terms is of the form Kx^8y^r.

- **a.** Find the value of r.
- **b.** Find the value of K.

2–15. Find the fourth term of the expansion of $(2x - y)^{12}$.

2–16. Find the last term of the expansion of $(2x + 3y)^6$.

Factoring Techniques

In an earlier section, we covered multiplying polynomials. You should be able to multiply out $(2x-1)(3x+4) = 6x^2 + 5x - 4$ fairly easily. You should have a procedure—distribute and combine—that you can use to multiply together any two polynomials.

Factoring is the opposite of multiplying. You can think of factoring as starting with just the answer to a multiplication problem and then trying to work backward to figure out what was multiplied. To factor $6x^2 + 5x - 4$ means to find two (or more) polynomials that multiply together to make $6x^2 + 5x - 4$. Factoring tends to be more difficult than multiplying. You probably figured out that $6x^2 + 5x - 4 = (2x-1)(3x+4)$, but how would you have known that if you had not just seen the previous example? This section covers several important factoring techniques. Some of them you may already be familiar with, but they are reviewed here because good factoring skills are essential in Algebra 2.

GREATEST COMMON FACTOR

The very first thing you should do with any polynomial factoring problem is check to see if there is a common factor (other than just 1) that can be factored out of all the terms in the polynomial.

EXAMPLE 2.28

Factor: $12x^4 - 24x^3 + 6x^2$

SOLUTION

First find the greatest common factor (GCF) of each of the three terms of this polynomial by looking at the coefficients and variables separately. The GCF of the coefficients 12, –24, and 6 is 6. The GCF for a variable that appears in all the terms of the polynomial is the one with the lowest exponent, in this case x^2. Together these make the GCF for the polynomial $6x^2$. Once we have found the GCF, we factor it out by dividing each term of the original polynomial by the GCF.

$$\begin{aligned} 12x^4 - 24x^3 + 6x^2 &= 6x^2\left(\frac{12x^4 - 24x^3 + 6x^2}{6x^2}\right) \\ &= 6x^2\left(\frac{12x^4}{6x^2} - \frac{24x^3}{6x^2} + \frac{6x^2}{6x^2}\right) \\ &= 6x^2\left(2x^2 - 4x + 1\right) \end{aligned}$$

Yet again, this example was written out in great detail. How much you actually write is up to you. Once you have figured out the GCF, you might feel you need only one of the two middle steps shown. You might even be able to write down the answer without any intermediate steps. (Practice helps you to do this.)

COMMON ERROR

Some students get so involved in the division that they forget to write the GCF in the answer. The correct answer is $6x^2(2x^2 - 4x + 1)$, not $(2x^2 - 4x + 1)$. The GCF gets factored out, but it does not just disappear.

COMMON ERROR

Remember (from dividing polynomials) that $\frac{6x^2}{6x^2} = 1$, not 0. A very common error among students is to omit the 1 and get the incorrect answer of $6x^2(2x^2 - 4x)$.

Because many people find factoring difficult, it is often worthwhile to check your answer by multiplying it. Someone who got $6x^2(2x^2 - 4x)$ as an answer to the last example would quickly see that something was missing when he or she multiplied it and got just $12x^4 - 24x^3$ instead of $12x^4 - 24x^3 + 6x^2$.

EXAMPLE 2.29

Factor: $42a^3b^2 - 30a^3b^4 + 12a^3b^5$

SOLUTION

The GCF of 42, –30, and 12 is 6. The lowest exponent on a is 3; the lowest exponent on b is 2. The GCF for the problem is $6a^3b^2$.

$$42a^3b^2 - 30a^3b^4 + 12a^3b^5 = 6a^3b^2\left(\frac{42a^3b^2}{6a^3b^2} - \frac{30a^3b^4}{6a^3b^2} + \frac{12a^3b^5}{6a^3b^2}\right)$$

$$= 6a^3b^2\left(7 - 5b^2 + 2b^3\right)$$

EXAMPLE 2.30

Factor: $3(x + 2)^2 + 4(x + 2)$

SOLUTION

In this example, the GCF is a binomial instead of a monomial. The factoring process is very similar but with an extra step at the end. After factoring out the common binomial, $x + 2$, distribute and combine to simplify the factor on the right.

$$3(x+2)^2 + 4(x+2) = (x+2)\left(\frac{3(x+2)^2}{(x+2)} + \frac{4(x+2)}{(x+2)}\right)$$
$$= (x+2)(3(x+2)+4)$$
$$= (x+2)(3x+6+4)$$
$$= (x+2)(3x+10)$$

Some students are tempted to multiply out the original expression. In this example, that would have led to extra, more difficult work.

FACTORING SPECIAL BINOMIALS

Most of the time when two binomials are multiplied, the result is (at least) a trinomial. However, the product of certain special binomials, called conjugates, is another binomial. A pair of conjugates is two binomials that differ only in the sign of the second term; $a + b$ and $a - b$ are conjugates. Note how the middle terms of the product are opposites and cancel each other.

$$(a+b)(a-b) = a^2 - \cancel{ab} + \cancel{ab} - b^2 = a^2 - b^2$$

What this means for factoring is that any binomial that is the difference of two perfect squares, $a^2 - b^2$, can be factored into two simpler binomials.

$$a^2 - b^2 = (a+b)(a-b)$$

EXAMPLE 2.31

Factor: $x^2 - 25$

SOLUTION

The expression x^2 is (obviously) a perfect square, and 25 is the square of 5. So the binomial can be written as the difference of two perfect squares:

$$x^2 - 25 = x^2 - 5^2 \text{ (Don't stop here, you are not done yet.)}$$

By thinking of x as a and 5 as b, we get:

$$x^2 - 25 = x^2 - 5^2 = (x+5)(x-5)$$

You can easily check that $(x+5)(x-5)$ multiplies out to $x^2 - 25$. Most people do not bother writing the middle step, $x^2 - 5^2$. However, that is what you need to be thinking when you factor a difference of squares.

Factor: $4x^2 - 9y^2$

SOLUTION

$$4x^2 - 9y^2 = (2x)^2 - (3y)^2 = (2x + 3y)(2x - 3y)$$

Again, writing out the middle step is optional. Skip it if you can mentally figure what is being squared.

Be aware that the method does not work for a sum of perfect squares.

EXAMPLE 2.33

Factor: $x^2 + 100$

SOLUTION

Although x^2 and 100 are both perfect squares, the fact that this is a sum instead of a difference means that this binomial does not factor with real numbers. It is *prime*.

> **REMEMBER**
> You can always factor a difference of two squares. You can never factor a sum of two squares using real numbers.

Are there other special binomials that can be factored? Yes, both the difference and sum of two perfect cubes can be factored. The formulas are a little more complicated, and the problems do not show up nearly as often as the difference of squares. However, you should be familiar with them.

$$a^3 - b^3 = (a - b)(a^2 + ab + b^2)$$

$$a^3 + b^3 = (a + b)(a^2 - ab + b^2)$$

The two factoring formulas for cubes are very similar. The terms are the same; the difference is the placement of the addition and subtraction signs. There is a mnemonic, SOAP, to help you place the signs properly. SOAP stands for same-opposite-always positive.

Same Opposite Always Positive

$$a^3 - b^3 = (a - b)(a^2 + ab + b^2)$$

Same Opposite Always Positive

$$a^3 + b^3 = (a + b)(a^2 - ab + b^2)$$

Factor: $x^3 - 64$

SOLUTION

Both x^3 and 64 are perfect cubes, so the problem can be done using the formula for the difference of cubes.

$$x^3 - 64 = x^3 - 4^3 = (x - 4)\left(x^2 + x(4) + 4^2\right) = (x - 4)\left(x^2 + 4x + 16\right)$$

COMMON ERROR

The formula is $a^3 - b^3 = (a - b)(a^2 + ab + b^2)$. However, some students get the formula confused with the formula for squaring a binomial and write $a^3 - b^3 = (a - b)(a^2 + 2ab + b^2)$. This is incorrect. There is no 2 in the middle term.

Factor: $1 + 27a^3$

SOLUTION

$$\begin{aligned} 1 + 27a^3 &= (1)^3 + (3a)^3 \\ &= (1 + 3a)\left((1)^2 - (1)(3a) + (3a)^2\right) \\ &= (1 + 3a)\left(1 - 3a + 9a^2\right) \end{aligned}$$

EXAMPLE 2.36

Factor: $a^3x^6 + 8y^3$

SOLUTION

$$\begin{aligned} a^3x^6 + 8y^3 &= \left(ax^2\right)^3 + (2y)^3 \\ &= \left(ax^2 + 2y\right)\left(\left(ax^2\right)^2 - \left(ax^2\right)(2y) + (2y)^2\right) \\ &= \left(ax^2 + 2y\right)\left(a^2x^4 - 2ax^2y + 4y^2\right) \end{aligned}$$

FACTORING BY GROUPING

Factoring polynomials with more than three terms can get very difficult. However, certain four-term polynomials can be factored with a technique called grouping. The idea is to group the four-term polynomial into two binomials, factor the GCF out of each binomial, and then factor a common binomial out of the resulting expression. The technique requires two steps: factor common monomials and then factor a common binomial.

Factor: $21x^3 - 6x^2 + 28x - 8$ by grouping

SOLUTION

$\underbrace{21x^3 - 6x^2} + \underbrace{28x - 8}$	Group into two binomials.
$3x^2(7x - 2) + 4(7x - 2)$	Factor the GCF out of each. Each grouping has a different GCF.
$(7x - 2)\left(3x^2 + 4\right)$	Factor out the common binomial.

EXAMPLE 2.38

Factor: $15x^3 + 12x^2 + 5x + 4$

SOLUTION

$\underbrace{15x^3 + 12x^2} + \underbrace{5x + 4}$	Group into two binomials.
$3x^2(5x + 4) + 1(5x + 4)$	Factor the GCF out of each. (If a grouping has no common factor, write 1 as its common factor.)
$(5x + 4)\left(3x^2 + 1\right)$	Factor out the common binomial.

EXAMPLE 2.39 Factor: $10x^3 + 2x^2 - 15x - 3$

SOLUTION

$\underbrace{10x^3 + 2x^2} - \underbrace{15x - 3}$	Group into two binomials.
$2x^2(5x + 1) - 3(5x + 1)$	Factor the GCF out of each. (If the third term is negative, factor out a negative common factor.)
$(5x + 1)(2x^2 - 3)$	Factor out the common binomial.

The success of the method depends on getting the same binomial after factoring out the two GCFs. If you do not get the same binomial, stop; the method is not working. This limitation means the method works for only certain four-term polynomials. Specifically, for the method to work, the product of the first and last terms of the polynomial must be the same as the product of the two middle terms. In Example 2.39, $(10x^3)(-3)$ and $(2x^2)(-15x)$ both equal $-30x^3$, so this factoring technique worked for that example.

FACTORING TRINOMIALS

In the beginning of this section, we mentioned that factoring tends to be more difficult than multiplying. So far, factoring out GCFs and factoring differences of squares and even cubes may seem fairly easy. Now it's time to get to the somewhat harder examples. You will factor trinomials of the form $ax^2 + bx + c$, where a, b, and c are numbers, usually integers in Algebra 2, and $a \neq 0$. So we want to factor expressions like $x^2 + 8x + 15$ and $3x^2 - 10x - 8$.

Be aware that not all trinomials can be factored using integers. Some can be factored only using irrational numbers (radicals). This is not usually done in Algebra 2. Others cannot be factored at all using real numbers. In this section, we will limit ourselves to ones that can be factored using integers.

The easiest trinomials to factor are ones that have a leading coefficient of 1 ($a = 1$). Let's start with those. We want to find two numbers, m and n, so that we can factor a trinomial of the form $x^2 + bx + c$ into $(x + m)(x + n)$. Working backward gives us:

$$\begin{aligned}(x + m)(x + n) = x^2 + nx + mx + mn &= x^2 + (m + n)x + mn \\ &= x^2 + \quad bx \quad + c\end{aligned}$$

This says we need $m + n = b$ and $mn = c$. In other words, to factor $x^2 + bx + c$, we need to find two numbers (if possible) whose product is c and whose sum is b. If we can find those numbers, we can factor the trinomial.

To factor $x^2 + bx + c$:

1. Find two numbers, m and n, that multiply to c and add to b.
2. Then factor $x^2 + bx + c = (x + m)(x + n)$.

EXAMPLE 2.40

Factor: $x^2 + 8x + 15$

SOLUTION

We need two numbers that multiply to 15 and add to 8. Start with the product. List the numbers that multiply to 15.

$$1 \times 15$$
$$3 \times 5$$
$$-1 \times -15$$
$$-3 \times -5$$

Of those four choices, the only pair that adds to 8 is 3 + 5:

$$x^2 + 8x + 15 = (x + 3)(x + 5)$$

If you know your multiplication and addition facts well, you will probably do most problems like this one (assuming the numbers involved are not too large) in your head. Remember when finding factors of c to consider both positive and negative numbers. Do not forget the simplest factors, $1 \times c$. Sometimes those are the ones you need.

Factor: $x^2 - 5x - 24$

SOLUTION

Pay attention to signs. In this problem, we need two numbers that multiply to –24. So we know we need one of each sign. Since they sum to –5, the one with the larger absolute value will be the negative. With a little thought (and making an organized list if necessary), we find $-8 \times 3 = -24$ and $-8 + 3 = -5$. So the numbers we want are –8 and 3.

$$x^2 - 5x - 24 = (x - 8)(x + 3)$$

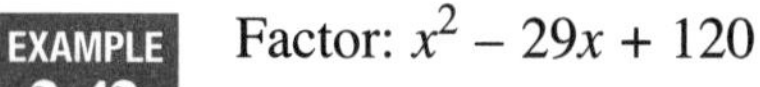

EXAMPLE 2.42

Factor: $x^2 - 29x + 120$

SOLUTION

We need two numbers that multiply to 120, so they must both be the same sign. Since they must add to –29, they must both be negative. Of all the possible pairs that multiply to 120, only –5 and –24 add to –29.

SEE CALC TIP 2D

$$x^2 - 29x + 120 = (x - 5)(x - 24)$$

If you find it difficult to list factors, see the "Graphing Calculator Tips" section at the back of the book for a technique to help you factor.

Factoring trinomials with $a = 1$ is not so bad. What if $a \neq 1$? There are two main techniques. The first is trial and error. The second is split the middle. Split the middle may seem more complicated, but trial and error can be extremely tedious. We will show one, fairly simple trial and error example to convince you that it is worth learning split the middle.

EXAMPLE 2.43

Factor: $8x^2 + 21x - 9$

SOLUTION

We will need two factors of the form $(px + m)(qx + n)$. In this problem, p and q must be factors of 8, and m and n must be factors of –9. Then somehow all four of them must go together to make 21. (If you are curious, we need $pn + qm = 21$, but that turns out not to be very helpful.)

Factors of 8 are 8×1 and 4×2. We do not consider negative factors for a, so we can ignore -8×-1 and -4×-2. Factors of –9 are -9×1, -3×3, and -1×9. For each pair of factors of 8, a pair of factors of –9 can be used in two different orders. For example, using 4×2 and -9×1 we could get either $(4x - 9)(2x + 1)$ or $(4x + 1)(2x - 9)$. There are a total of $2 \times 3 \times 2 = 12$ possible combinations of factors that will give $8x^2$ and –9. Only one of them gives $21x$ in the middle. The trial and error method involves multiplying out each pair until you find the one that works:

$$(4x - 9)(2x + 1) = 8x^2 - 14x - 9, \text{ NO}$$
$$(4x + 1)(2x - 9) = 8x^2 - 34x - 9, \text{ NO}$$

and so on until you try

$$(8x - 3)(x + 3) = 8x^2 + 21x - 9, \text{ YES!}$$

The amount of work involved in this method depends partly on how many factors a and c have and partly how good (or lucky) you are at guessing reasonable combinations to try. Practice helps some. However, many students find the method very frustrating.

The split the middle method rewrites the bx term into two terms and then factors by grouping. To rewrite the middle term, bx, find two numbers that multiply to ac (the product of a and c) and add to b (similar to the earlier method for $a = 1$ trinomials). This method is best learned by example.

EXAMPLE 2.44

Factor: $8x^2 + 21x - 9$ (the same expression as in Example 2.43)

SOLUTION

We start by turning our trinomial into a four-term polynomial. (This may not seem like "progress" at first, but be patient.)

$ac = 8(-9) = -72$	First multiply a and c.
$-3 \times 24 = -72$ and $-3 + 24 = 21$	Find two numbers that multiply to ac and add to b.
$8x^2 + 21x - 9$ $8x^2 - 3x + 24x - 9$	Use your two numbers to split the middle. Here $+21x$ splits into $-3x + 24x$

Now factor by grouping.

$\underbrace{8x^2 - 3x} + \underbrace{24x - 9}$	Group the binomials.
$x(8x - 3) + 3(8x - 3)$	Factor out the GCF from each group.
$(8x - 3)(x + 3)$	Factor out the common binomial.

You might be curious whether it matters which order we use when we split the middle. Switch the order of the middle terms and see.

$\underbrace{8x^2 + 24x} - \underbrace{3x - 9}$	Group the binomials.
$8x(x + 3) - 3(x + 3)$	Factor out the GCF from each group. If your third term is negative, include the negative sign in the GCF. Factor out -3 carefully, $\frac{-9}{-3} = +3$.
$(x + 3)(8x - 3)$	Factor out the common binomial.

Except for the order of the factors, this is the same as the previous answer. So no, it does not matter which order you use. However, if you have one positive and one negative number, you may make fewer sign errors if you write the negative term first when you split the middle.

EXAMPLE 2.45

Factor: $15x^2 - 38x + 24$

SOLUTION

We need two numbers that multiply to $15(24) = 360$ and add to -38. A little bit of experimenting (with help from your calculator) gives -20 and -18. So $-38x$ is split into $-20x + -18x$. Then factor by grouping.

$$\begin{aligned} 15x^2 - 38x + 24 &= 15x^2 - 20x - 18x + 24 \\ &= 5x(3x - 4) - 6(3x - 4) \\ &= (3x - 4)(5x - 6) \end{aligned}$$

SEE CALC TIP 2E

Factor: $12x^2 - 29x - 8$

SOLUTION

We need two numbers that multiply to $12(-8) = -96$ and add to -29. We find that -32 and $+3$ work. So $-29x$ is split into $-32x + 3x$. Then factor by grouping.

$$\begin{aligned} 12x^2 - 29x - 8 &= 12x^2 - 32x + 3x - 8 \\ &= 4x(3x - 8) + 1(3x - 8) \\ &= (3x - 8)(4x + 1) \end{aligned}$$

Split the middle factoring can be used to factor any trinomial that factors into two binomials.

Factor: $20x^4 + 7x^2 - 6$

SOLUTION

We need two numbers that multiply to $20(-6) = -120$ and add to $+7$. Try -8 and $+15$. Split $+7x^2$ into $-8x^2 + 15x^2$, and then factor by grouping.

$$\begin{aligned} 20x^4 + 7x^2 - 6 &= 20x^4 - 8x^2 + 15x^2 - 6 \\ &= 4x^2\left(5x^2 - 2\right) + 3\left(5x^2 - 2\right) \\ &= \left(5x^2 - 2\right)\left(4x^2 + 3\right) \end{aligned}$$

FACTORING COMPLETELY

Some factoring problems at first might seem to have more than one correct answer. For example, by factoring out the GCF, $4x^2 - 36 = 4(x^2 - 9)$. Using the difference of perfect squares, $4x^2 - 36 = (2x + 6)(2x - 6)$. Which one is right? The answer is that each is correct as far as it goes, but neither one is factored completely. In the first case, factoring out the GCF and then the difference of squares gives $4x^2 - 36 = 4(x^2 - 9) = 4(x + 3)(x - 3)$. In the second case, factoring the difference of squares and then the GCF (twice) gives $4x^2 - 36 = (2x + 6)(2x - 6) = 2(x + 3)(2)(x - 3) = 4(x + 3)(x - 3)$.

When a polynomial is factored completely, none of the remaining factors can be factored any further. To factor a polynomial completely may require more than one factoring step and may use more than one factoring technique.

1. First, factor out the GCF, if any. If the leading term is negative, factor out –1.
2. Try to factor the remaining polynomial. Count the number of terms to choose a technique. More than one technique may be used.

   - If it is a four-term polynomial, try grouping.
   - If it is a trinomial with $a = 1$, find two numbers that multiply to c and add to b. (Although it works, it is a little silly to use split the middle when $a = 1$.)
   - If it is a trinomial with $a > 1$, use split the middle.
   - If it is a binomial, see if it is a difference of perfect squares or a sum or difference of perfect cubes and factor appropriately.

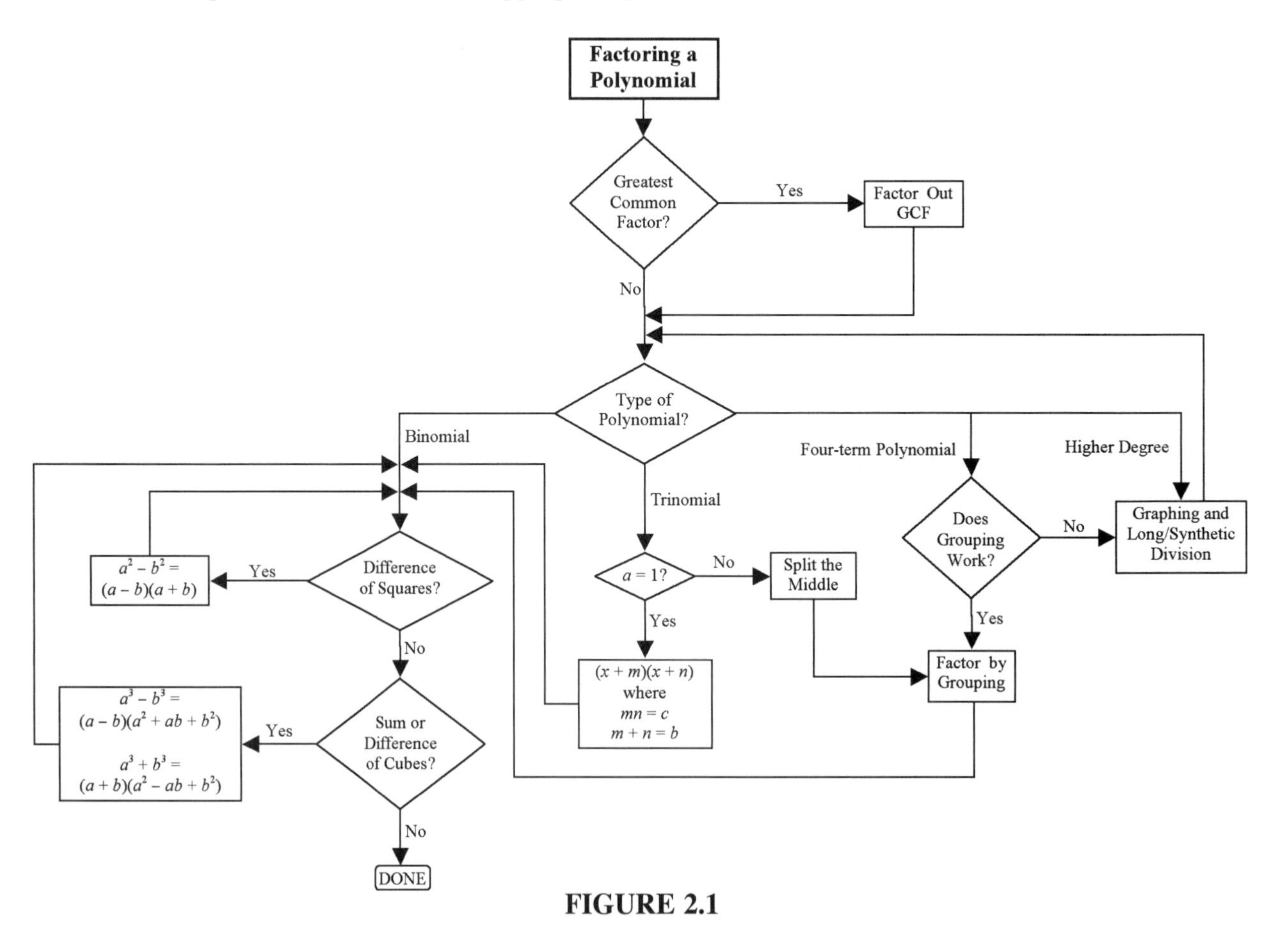

FIGURE 2.1

EXAMPLE 2.48

Factor: $5x^3 - 20x$

SOLUTION

$5x^3 - 20x$	
$5x(x^2 - 4)$	First factor out the GCF, $5x$.
$5x(x + 2)(x - 2)$	Then factor the difference of squares.

EXAMPLE 2.49

Factor: $x^4 - 16y^8$

SOLUTION

There is no GCF. (Always check. When there is one, factoring it out makes the rest of the problem easier.) It might not be obvious at first, but this binomial is a difference of perfect squares.

$$x^4 - 16y^8 = \left(x^2\right)^2 - \left(4y^4\right)^2$$
$$= \left(x^2 + 4y^4\right)\left(x^2 - 4y^4\right)$$

The left factor is a sum of squares and cannot be factored further. However, the factor on the right is another difference of squares and can be factored further.

$$\left(x^2 + 4y^4\right)\left(x^2 - 4y^4\right)$$
$$\left(x^2 + 4y^4\right)\left(x + 2y^2\right)\left(x - 2y^2\right)$$

None of these factors can be factored any more, so you are done.

Factor: $3x^2 - 9x - 84$

SOLUTION

Do not go right to split the middle without first checking for a GCF. In this case, there is one, 3. The resulting trinomial has $a = 1$ and factors more easily.

$$3x^2 - 9x - 84 = 3\left(x^2 - 3x - 28\right) = 3(x - 7)(x + 4)$$

Factor: $2x^2(x-1) - x(x-1) - 6(x-1)$

SOLUTION

A lot of students are tempted to distribute first to get rid of the parentheses. In this case, that will lead to a four-term polynomial that will not factor by grouping. Instead, note that each term contains $(x - 1)$. That is the GCF; factor it out.

$$2x^2(x-1) - x(x-1) - 6(x-1) = (x-1)\left(2x^2 - x - 6\right)$$

The resulting trinomial, $2x^2 - x - 6$, can be factored with split the middle. Two numbers that multiply to $2(-6)$ or -12 and add to -1 are -4 and $+3$.

$$\begin{aligned} 2x^2(x-1) - x(x-1) - 6(x-1) &= (x-1)\left(2x^2 - x - 6\right) \\ &= (x-1)\left(2x^2 - 4x + 3x - 6\right) \\ &= (x-1)(2x(x-2) + 3(x-2)) \\ &= (x-1)(x-2)(2x+3) \end{aligned}$$

EXAMPLE 2.52

Factor: $4x^5 - 14x^4 - 20x^3 + 70x^2$

SOLUTION

First factor out the CGF. Then try factoring by grouping.

$$\begin{aligned} 4x^5 - 14x^4 - 20x^3 + 70x^2 &= 2x^2\left(2x^3 - 7x^2 - 10x + 35\right) \\ &= 2x^2\left(x^2(2x-7) - 5(2x-7)\right) \\ &= 2x^2(2x-7)\left(x^2 - 5\right) \end{aligned}$$

EXAMPLE 2.53

Factor: $2y^3 - 54$

SOLUTION

First factor out the GCF. Then factor as a difference of cubes. Remember to apply SOAP.

$$\begin{aligned} 2y^3 - 54 &= 2\left(y^3 - 27\right) \\ &= 2\left(y^3 - (3)^3\right) \\ &= 2(y-3)\left(y^2 + 3y + 9\right) \end{aligned}$$

SECTION EXERCISES

In 17–32, factor the given expression completely.

2–17. $36a^4b^3 - 42a^3b^4 + 18a^2b^5$

2–18. $2(x + 3)^2 - 5(x + 3)$

2–19. $25x^2 - 9y^2$

2–20. $x^2 - 4xy - 12y^2$

2–21. $5x^2 - 80$

2–22. $2x^2 - 7x - 9$

2–23. $6x^2 + 27x - 15$

2–24. $x^3y - xy^3$

2–25. $x^4 - 81$

2–26. $4x^4 - x^2 - 3$

2–27. $8x^3 + 125$

2–28. $64x^3 - 27$

2–29. $125x^4y - xy^4$

2–30. $x^6 - 64$

2–31. $x^3 - 3x^2 + 4x - 12$

2–32. $3x^3 + x^2 - 12x - 4$

CHAPTER EXERCISES

C2–1. Simplify completely: $\dfrac{-8x^{10}y^4}{\left(2xy^2\right)^3}$

C2–2. If $10^x = c$, then express 10^{x+3} in terms of c.

C2–3. Find the sum of $x^5 + 2x^4 + 5$, x^3, and $3x^5 - x^3 + 4$.

C2–4. Express $(2x - 1)(x - 3)(4 - x)$ as a polynomial in simplest form.

C2–5. Find the quotient: $\dfrac{x^4 - 2x^3 + 5x^2 - 4x + 6}{x^2 + 2}$

C2–6. Divide $4x^4 + 18x^3 + 45x - 25$ by $2x - 1$.

C2–7. Express $x^3 - 8x^2 + 21x - 11$ divided by $x - 3$ as the sum of a polynomial and a proper rational expression.

C2–8. Expand the following and simplify: $(x - 3y)^3$

C2–9. Find the sixth term of the expansion of $(3x - 1)^7$

In 10–19, factor the given expressions completely.

C2–10. $8x^4 + 16x^3 - 4x^2$

C2–11. $3 - 27x^2$

C2–12. $2x^2 + 5x - 3$

C2–13. $15x^4 - 50x^3 - 40x^2$

C2–14. $x^4 + 5x^2 - 36$

C2–15. $x^4 + 8x$

C2–16. $2x^4 - 2000x$

C2–17. $4x^6 + 4$

C2–18. $2x^3 - 3x^2 + 10x - 15$

C2–19. $2x^5 + 4x^4 - 7x^3 - 14x^2$

3

Functions and Relations

WHAT YOU WILL LEARN

Functions are a cornerstone of mathematics. They are fundamental for understanding many mathematical relationships. Even before we learn about functions in algebra, we have experience with functions. When we press a button on a vending machine, we expect to get a particular item. If a different item is dispensed, we get frustrated. In a nutshell, this is the idea of a function: one input leads to one predictable output. In this chapter, you will learn to:

- Identify functions and relations;
- Use function notation;
- Determine domain and range;
- Recognize symmetry of functions;
- Perform operations on functions;
- Recognize and make transformations of functions;
- Find inverse functions.

SECTIONS IN THIS CHAPTER

- Relations and Functions
- Function Notation
- Domain and Range
- Symmetry
- Operations with Functions
- Transformations of Functions
- Inverse Functions

Relations and Functions

The ideas of relations and functions are extremely important in math. Although you may not have used the terminology, you have dealt with relations and functions for years both in your math classes and in everyday life.

A *relation* in math is something that can be represented by a set of ordered pairs. The first elements, usually referred to as x-values for convenience, form a set called the *domain*. The second elements, the y-values, form a set called the *range*.

For example, $\{(0, 1), (1, -2), (1, 3), (2, 7)\}$ represents a relation. The domain is the set $\{0, 1, 2\}$, and the range is the set $\{-2, 1, 3, 7\}$.

Relations do not have to be ordered pairs of numbers. An example from geometry is {(triangle, 3), (quadrilateral, 4), (pentagon, 5), …}, which represents the relation between the name of a polygon and its number of sides. A bag of jelly beans might have the relation {(red, cherry), (orange, tangerine), (yellow, lemon), (green, lime), (black, yucky)} to link the color of the jelly bean with its flavor. In algebra, most relations involve ordered pairs of numbers.

The first example above, $\{(0, 1), (1, -2), (1, 3), (2, 7)\}$, has an inconvenient feature. The x-values 0 and 2 are each related to only one y-value, but the x-value 1 is related to two different y-values. In other words, if you know $x = 2$, then y must be 7. However, if $x = 1$, then y could be either -2 or 3. This does not happen in the other two examples. If you say "pentagon," you know there are 5 sides. If you get a green jelly bean, you know it is lime flavored.

A *function* is a relation where no two ordered pairs have the same x-value but different y-values. In a function, each x-value, often thought of as the *input*, has only one y-value, or *output*. The geometry and jelly bean examples above are both functions; the first example with ordered number pairs is not.

Functions can be represented in many different ways besides a list of ordered pairs.

1. A table

A table makes really clear that a function is a set of ordered pairs. Note that no value can appear in the x-column twice but y-values may be repeated. In a function, no input can have more than one output but different inputs may have the same output. In other words, the x-values are unique, but the y-values may occur more than once.

x	y
-2	4
-1	3
0	4
1	7
2	12

2. A mapping diagram

These tend to be most common in algebra books (like this one). They show how each element of the domain is "mapped" to an element in the range. Note that only one arrow can originate from each domain element on the left.

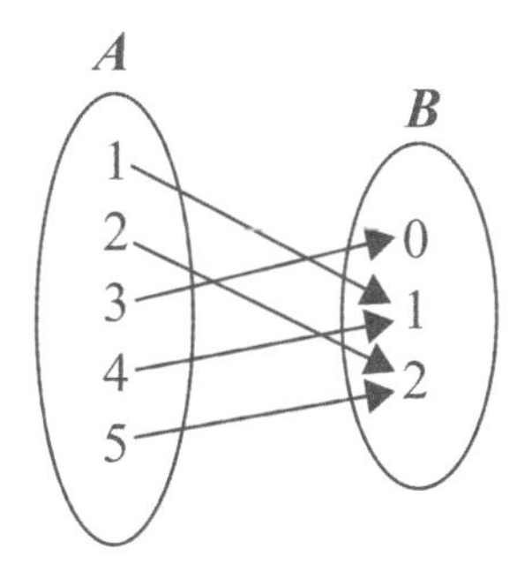

FIGURE 3.1

3. A graph

A graph has many advantages. It can represent an infinite number of ordered pairs. It helps show where a function has maximums and minimums and where a function is increasing and decreasing.

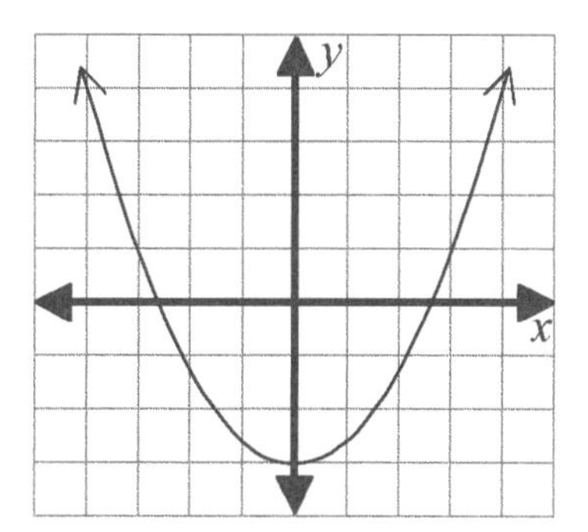

FIGURE 3.2

A quick visual way, called the vertical line test, will help you tell if a graph represents a function. If no vertical line intersects the graph more than once, the graph shows a function.

4. An equation or a formula

The equation $y = 10 - \sqrt{x}$, $x \geq 0$ represents a function. For each x-value in the domain, there is only one y-value. This will typically be the case for equations that are solved for y. Equations that are not solved for y are sometimes not functions. The equation $x^2 + y^2 = 25$ is the equation of a circle. It is not a function. The graph does not pass the vertical line test. For most values of x in the domain, there is more than one y-value. For example, when $x = 3$, y could be 4 or -4. Equations with even exponents on y are not functions.

5. A verbal rule or procedure

A written-out statement can indicate a function. An example is, "y is the sum of the digits of x, where x can be any whole number."

EXAMPLE 3.1 Classify the following relations as either a function or not a function. Justify your answer.

a. {(1, 2) (4, 5) (3, 2)}
b. {(1, 2) (1, 5) (3, 2)}
c. {(1, 4) (4, 1) (5, 3)}

SOLUTION

a. Yes, this is a function. Every input goes to only one output. (The *x*-values are unique. The duplicate *y*-values do not matter.)
b. No, this is not a function. The *x*-value 1 goes to 2 and 5.
c. Yes, this is a function. Every input goes to only one output. (The *x*-values are unique.)

EXAMPLE 3.2 Classify the following relations as either a function or not a function. Justify your answer.

a.

x	*y*
4	–2
1	–1
0	0
1	1
4	2

b.

x	*y*
0	3
1	5
2	7
3	9
4	11

c.

x	*y*
3	9
4	16
5	25
6	36
3	–9

SOLUTION

a. No, this is not a function. The *x*-value 4 goes to –2 or 2 and the *x*-value 1 goes to –1 and 1.
b. Yes, this is a function. Every input goes to only one output. (The *x*-values are unique.)
c. No, this is not a function. The *x*-value 3 goes to 9 and –9.

EXAMPLE 3.3 Classify the following relations as either a function or not a function. Justify your answer.

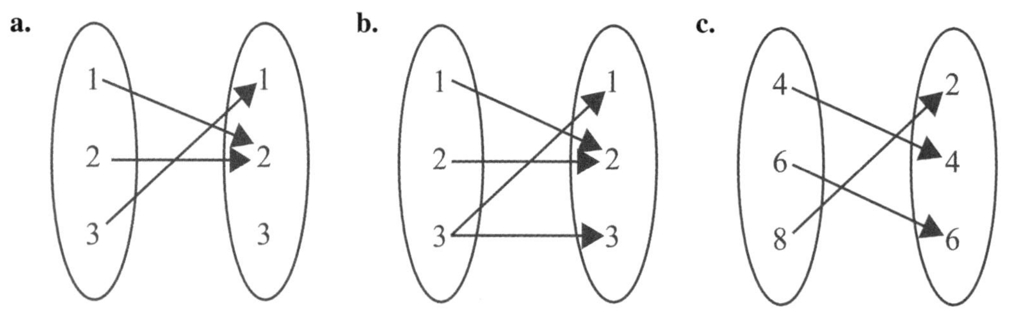

SOLUTION

a. Yes, this is a function. Every input goes to only one output. (The values on the left have only one arrow. Two arrows going into the same value on the right does not matter.)
b. No, this is not a function. The input 3 goes to 1 and 3. (Two arrows come out of 3 on the left.)
c. Yes, this is a function. Every input goes to only one output. (The values on the left each have only one arrow.)

Classify the following relations as either a function or not a function. Justify your answer.

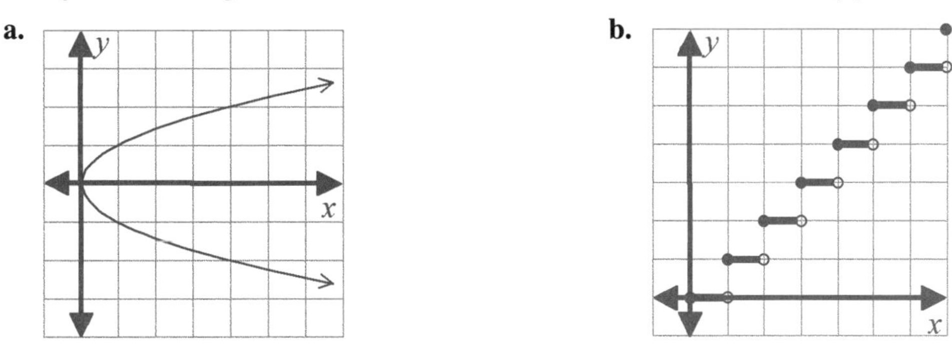

SOLUTION

a. No, this is not a function. The x-value 4 goes to -2 or 2. (It fails the vertical line test. Some vertical lines intersect the graph in more than one point.)

b. Yes, this is a function. Every input goes to only one output. (It passes the vertical line test. No vertical line intersects the graph more than once.)

EXAMPLE 3.5

Classify the following relations as either a function or not a function. Justify your answer.

a. $y = 2x + 3$

b. $x = y^2$

c. $x = 3$

SOLUTION

a. Yes, this is a function. Every x gives only one y. The graph is a diagonal line that passes the vertical line test.

b. No, this is not a function. If $x = 4$, then $y = -2$ or $+2$. The graph is a sideways parabola, which does not pass the vertical line test.

c. No, this is not a function. If $x = 3$, then y can be any number. The graph is a vertical line, which does not pass the vertical line test.

SECTION EXERCISES

3–1. Which relation is *not* a function?

(1) {(0, 1), (1, 2), (2, 3), (3, 4)}

(3) {(−1, 3), (0, 0), (1, 3)}

(2) {(2, prime), (3, prime), (4, composite), (5, prime)}

(4) {(2, −2), (1, −1), (0, 0), (1, 1), (2, 2)}

3–2. Which of the mappings diagrams shown below represent a function? Choose all that apply.

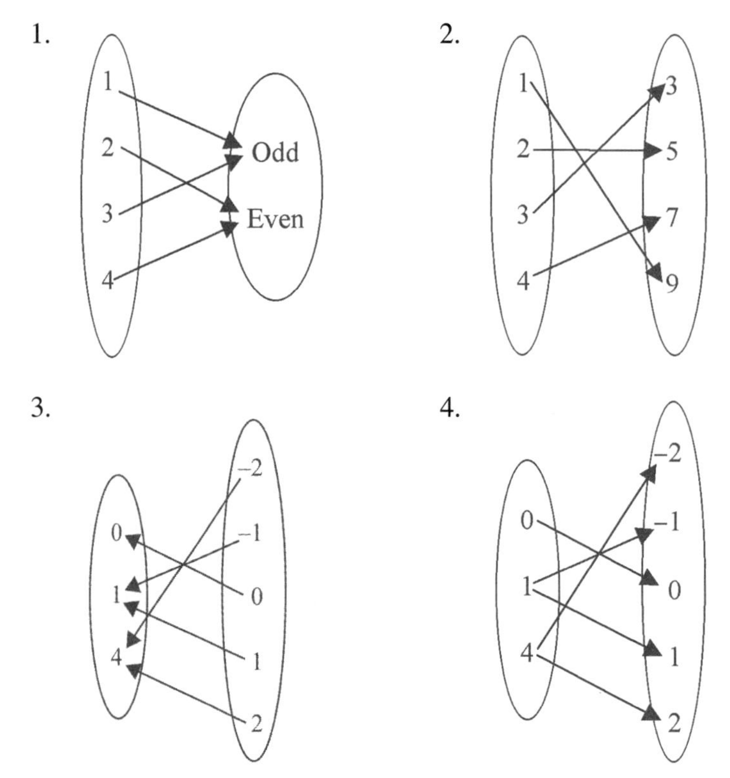

3–3. Which of the graphs shown below show(s) a function of x? Choose all that apply.

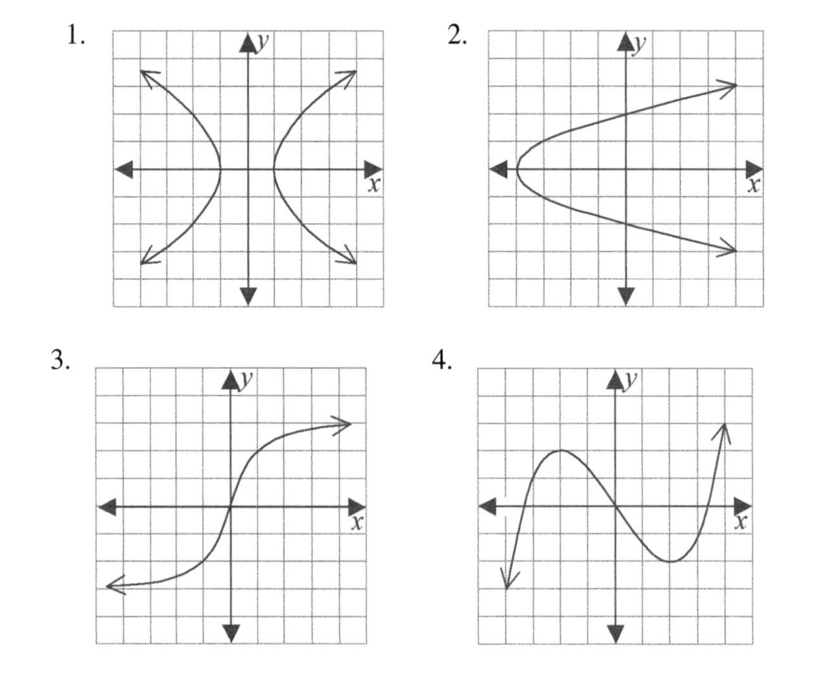

3–4. Which of the equations shown below represent(s) a function of x? Choose all that apply.

(1) $y = 5$
(2) $y = 3x + 5$
(3) $y = x^2 + 3x + 5$
(4) $y^2 = x^2 + 3x + 5$

Function Notation

Functions, like variables, are usually given single-letter names. The letters may be lower or upper case; lower is more common. The notation $y = f(x)$, which is read "y equals f of x," means that y is some function of x named f. Do not confuse the notation to mean a product; y is not f times x. Rather, y is the result when the function f is applied to the input value x.

EXAMPLE 3.6

The function f is defined by $f(x) = x^2 - 2x - 3$.

a. What does $f(5) = 12$ mean?
b. Evaluate $f(-2)$, and explain its meaning.
c. What does $y = f(x)$ mean?
d. For what values of x will $f(x) = 5$?
e. What is the y-intercept of the graph of $f(x)$?
f. What are the zeros of $f(x)$?

SOLUTION

a. In simple terms, $f(5) = 12$ means the function f turns 5 into 12. More precisely, when the input, x, is 5, the output is 12. The function f includes (5, 12) as one of its ordered pairs.

b. $f(-2)$ is read "f of -2" and means "the value of f when $x = -2$." To evaluate it, substitute -2 for each x and simplify: $f(-2) = (-2)^2 - 2(-2) - 3 = 5$. So $f(-2) = 5$.

c. In this problem, it is just a shorthand way to write $y = x^2 - 2x - 3$. In general, you can think of f as the name of the function, $f(x)$ as another name for y, and $f(7)$ as the specific value of y when $x = 7$.

d. Finding the values of x for which $f(x) = 5$ means to solve the equation $5 = x^2 - 2x - 3$.

$$5 = x^2 - 2x - 3 \rightarrow x^2 - 2x - 8 = 0$$
$$(x + 2)(x - 4) = 0$$
$$x = -2 \text{ or } x = 4$$

This means that either of the values $x = -2$ or $x = 4$ will result in the output, y, being equal to 5. Be very careful not to mix up the expression $f(5)$ and the equation $f(x) = 5$. The first means the value of y when $x = 5$; the second means to find the x-value(s) that will make $y = 5$.

e. The y-intercept of a graph is the point on the graph where $x = 0$. The y-intercept of the graph of $f(x)$ is $f(0) = (0)^2 - 2(0) - 3 = -3$.

f. *Zeros* (also called *roots*) of a function are the values of x for which the function equals zero. To find them, solve $f(x) = 0$.

$$x^2 - 2x - 3 = 0$$
$$(x + 1)(x - 3) = 0$$
$$x = -1 \text{ or } x = 3$$

The zeros of f are $x = -1$ and $x = 3$.

SEE CALC TIP 3A

Sometimes functions are evaluated for expressions. Even students that can evaluate functions for numerical values can find these problems challenging.

EXAMPLE 3.7 $f(x) = x^2 - 2x + 3$, find $f(x + 5)$ and simplify it.

SOLUTION

$$\begin{aligned} f(x+5) &= (x+5)^2 - 2(x+5) + 3 \\ &= x^2 + 10x + 25 - 2x - 10 + 3 \\ &= x^2 + 8x + 18 \end{aligned}$$

If this looks tricky to you, here is a helpful technique. Write the original function, $f(x) = x^2 - 2x + 3$ with boxes around each x, with all coefficients and exponents outside the boxes, $f\boxed{(x)} = \boxed{x}^2 - 2\boxed{x} + 3$. To evaluate $f(x + 5)$, replace the x in each box with $(x + 5)$ so it looks like this: $f\boxed{(x+5)} = \boxed{(x+5)}^2 - 2\boxed{(x+5)} + 3$. Now remove the boxes but leave the parentheses, $f(x + 5) = (x + 5)^2 - 2(x + 5) + 3$. Distribute carefully, and combine like terms.

SEE CALC TIP 3B

EXAMPLE 3.8 If $f(x) = \dfrac{x^2}{x + 4}$, find $f(x - 3)$ and simplify it.

SOLUTION

$$f\boxed{(x)} = \frac{\boxed{x}^2}{\boxed{x} + 4}$$

$$\begin{aligned} f\boxed{(x-3)} &= \frac{\boxed{(x-3)}^2}{\boxed{(x-3)} + 4} \\ &= \frac{(x-3)^2}{(x-3)+4} \\ &= \frac{x^2 - 6x + 9}{x + 1} \end{aligned}$$

EXAMPLE 3.9 If $f(x) = 2x^2 - x$, find $f(x+h)$ and simplify it.

SOLUTION

$$f\boxed{(x)} = 2\boxed{x}^2 - \boxed{x}$$

$$f\boxed{(x+h)} = 2\boxed{(x+h)}^2 - \boxed{(x+h)}$$

$$\begin{aligned} f(x+h) &= 2(x+h)^2 - (x+h) \\ &= 2(x+h)(x+h) - x - h \\ &= 2\left(x^2 + xh + xh + h^2\right) - x - h \\ &= 2\left(x^2 + 2xh + h^2\right) - x - h \\ &= 2x^2 + 4xh + 2h^2 - x - h \end{aligned}$$

None of the terms can be further combined.

SECTION EXERCISES

3–5. For the function $f(x) = \sqrt{x}$,

a. Evaluate $f(9)$.
b. Evaluate $f(225)$.
c. Evaluate $f(a)$. (Assume $a > 0$)
d. Evaluate $f\left(b^2\right)$. (Assume $b > 0$)
e. Evaluate $f(a + h)$. (Assume $a + h > 0$)
f. Solve $f(x) = 16$.

3–6. For the function $f(x) = -3x^2 - 2x + 1$,

a. Evaluate $f(1)$.
b. Evaluate $f(-4)$.
c. Evaluate $f(3a)$ and simplify.
d. Evaluate $f(z - 1)$ and simplify.
e. Find the value(s) of x for which $f(x) = -7$.

3–7. The graph of the function $g(x)$ is shown.

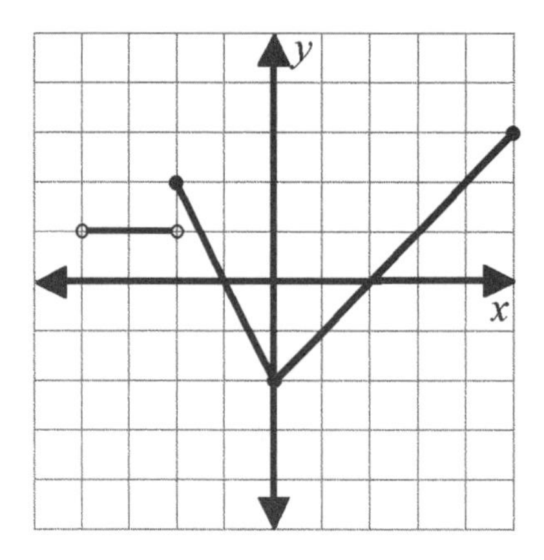

a. Find $g(3)$.
b. Find $g(-2)$.
c. Find $g(0.5)$.
d. Find the y-intercept of g.
e. For what value(s) of x will $g(x) = 2$?
f. Find the zeros of g.

3–8. The function h is shown in the table.

x	$h(x)$
–2	4
–1	3
0	2
1	0
2	1
3	2

a. Evaluate $h(2)$.
b. Solve $h(x) = 3$.
c. What is the y-intercept of the graph of h?
d. Find the zero(s) of h.

Domain and Range

Finding the domain and range of a function given as a set of ordered pairs, a table, or a map is easy. The domain is the set of all the first elements; the range is the set of all the second elements.

EXAMPLE 3.10

Find the domain and range of the function defined by the table below.

x	y
–2	4
–1	3
0	4
1	7
2	12

> **REMEMBER**
> **Domain:** All of the allowable inputs (usually x-values) for a function.
> **Range:** All of the possible outputs (usually y-values) for the function.

SOLUTION

The domain is {–2, –1, 0, 1, 2}, and the range is {3, 4, 7, 12}.

A function written as an equation should include the domain as part of the definition of the function. This is because the same equation can be used to represent different functions. For example, $y = \frac{1}{2}x + 3$ $\{x \mid x \text{ is a real number}\}$; $y = \frac{1}{2}x + 3$ $\{x \mid 0 \le x \le 6\}$, and $y = \frac{1}{2}x + 3$ $\{x \mid x \text{ is an integer}\}$ are three different functions because they each have different domains. Their graphs make the differences quite clear.

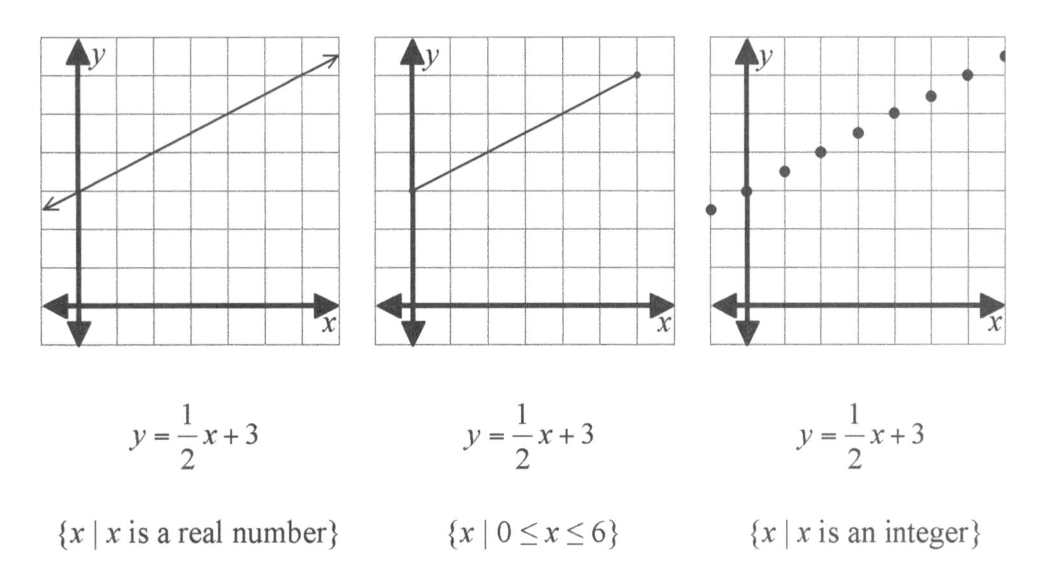

FIGURE 3.3

DOMAIN

If a domain is not given, use the following rules:

1. If the equation is a model where the variables have actual meaning, choose a domain that makes sense for the model.
2. If the equation has no context (not a word problem), assume the domain is the largest set of real numbers for which y will be defined and real. This is called the *natural domain* of the function. The following must be satisfied:

- The denominator must not equal 0. A zero denominator would make the function undefined.
- The radicand must be greater than or equal to 0. A negative radicand would give imaginary outputs (see Chapter 6).
- There is a third consideration if the function involves logarithms. It is discussed in Chapter 11.

EXAMPLE 3.11 A bus that seats 40 people has been assigned to take fans to the playoff game. Tickets for the bus cost \$3 each. The amount of money taken in, y, will be $y = 3x$, where x is the number of people who bought tickets. What is the domain of this function?

SOLUTION

The natural domain of the function would be all real numbers because $y = 3x$ is defined and real for all real x. However, people come in only whole numbers, so $x = -2$ and $x = 3.5$ would not be sensible domain values. In addition, at most 40 people can ride the bus.

The domain would be $\{x \mid x \text{ is a whole number and } 0 \leq x \leq 40\}$.

EXAMPLE 3.12 Find the natural domain of the function $y = \dfrac{3x}{2x - 4}$.

SOLUTION

We want all the values of x for which the function is defined and real. The following must be satisfied:

1. Denominator $\neq$ 0. $\quad 2x - 4 \neq 0$
   $x \neq 2$
2. Radicand $\geq$ 0. Since there is no radical in this equation, this adds no additional restrictions on x.

So the natural domain for this function is $\{x \mid x \neq 2\}$. Note that some people include $x \in \mathbb{R}$ (x is a real number) in the statement of the domain. We will consider that understood and specify only exceptions. This domain can also be written using interval notation as $(-\infty, 2) \cup (2, \infty)$. The open parentheses indicate that 2 is not included.

SEE CALC TIP 3C

EXAMPLE 3.13 Find the natural domain of the function $y = \sqrt{2x - 6}$.

SOLUTION

The following must be satisfied:

1. Denominator $\neq 0$. This is not a problem in this example.
2. Radicand ≥ 0.

$$2x - 6 \geq 0$$
$$2x \geq 6$$
$$x \geq 3$$

SEE CALC TIP 3D

The natural domain for this function is $\{x \mid x \geq 3\}$ or $[3, \infty)$.

EXAMPLE 3.14 Find the natural domain of the function $y = \dfrac{3}{\sqrt{2x - 6}}$.

SOLUTION

In this example, the variable appears in the denominator and in a radicand so both rules apply. The domain is the set that meets both requirements.

1. Denominator $\neq 0$.

$$\sqrt{2x - 6} \neq 0$$
$$2x \neq 6$$
$$x \neq 3$$

2. Radicand ≥ 0.

$$2x - 6 \geq 0$$
$$2x \geq 6$$
$$x \geq 3$$

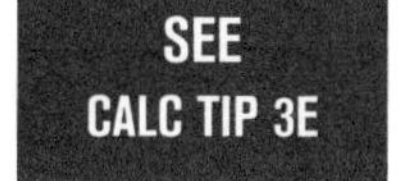

We have $x \geq 3$ and $x \neq 3$. Therefore, the natural domain for this function is $\{x \mid x > 3\}$ or $(3, \infty)$.

EXAMPLE 3.15 Find the natural domain of the function $y = \dfrac{\sqrt{5x + 15}}{x^2 - 16}$.

SOLUTION

1. Denominator $\neq 0$.

$$x^2 - 16 \neq 0$$
$$x^2 \neq 16$$
$$x \neq \pm 4$$

2. Radicand ≥ 0.

$$5x + 15 \geq 0$$
$$5x \geq -15$$
$$x \geq -3$$

We see that x can be any number greater than or equal to -3 *except* 4. This can be written $\{x \mid x \geq -3 \wedge x \neq 4\}$ or as $[-3, 4) \cup (4, \infty)$.

RANGE

Finding the range of a function defined by an equation can be difficult. Sometimes your knowledge of algebra can help. For the example $y = \sqrt{2x - 6}$, you know that the square root function gives only nonnegative answers (or you will know it after you read Chapter 8, "Radical Functions"). You also know that with the right input, any real nonnegative answer can be found. So the range for this example is $\{y \mid y \geq 0\}$.

A graph is another good way to find the range. The graph of the example $y = \dfrac{2x}{x^2 - 4}$ clearly suggests that the range is all real numbers.

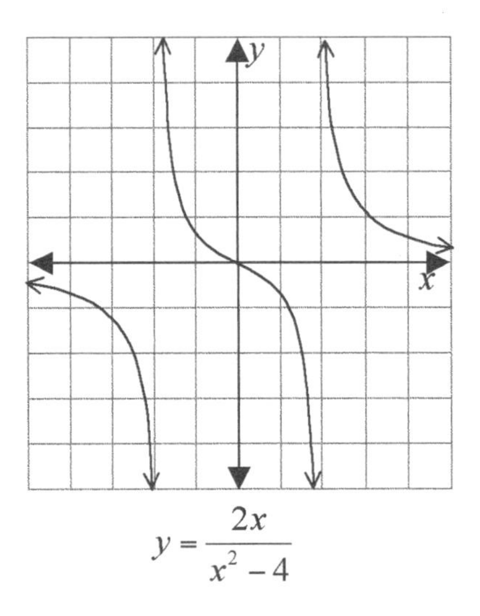

$y = \dfrac{2x}{x^2 - 4}$

FIGURE 3.4

Neither your knowledge of algebra nor a graph will work for all problems. The range of the function $y = \dfrac{12\sqrt{x}}{x^2 + 4}$ can be estimated from its graph; it looks to be approximately $[0, 2.4]$. To get a really accurate answer, though, you need either technology or calculus.

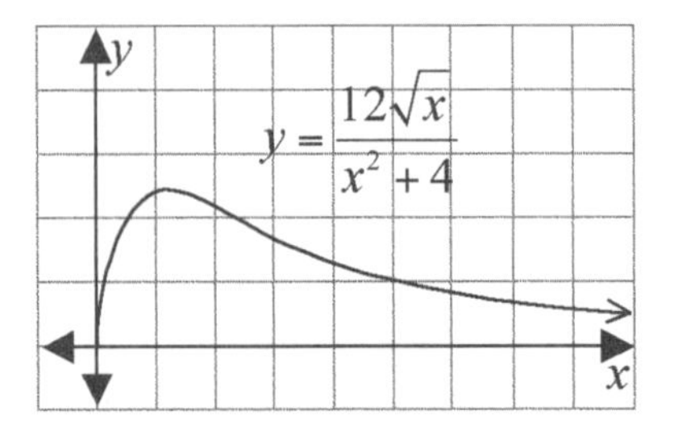

FIGURE 3.5

SECTION EXERCISES

3–9. Find the domain and range of the function defined by the table below.

x	y
0	7
1	4
2	3
3	4
4	7

3–10. Give both the domain and the range of the function in each graph in simplest interval notation.

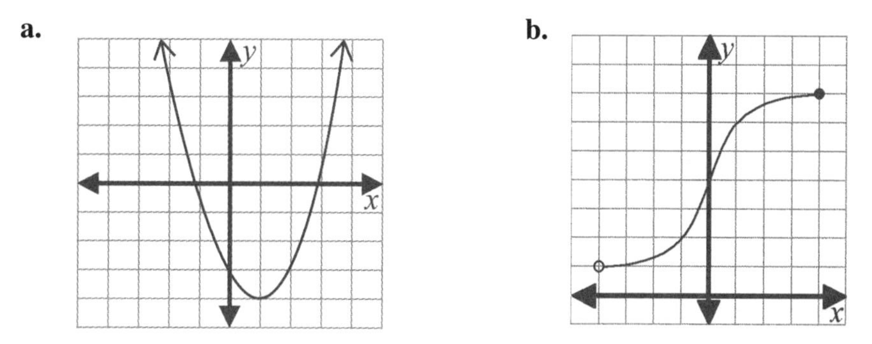

3–11. A radio station has 24 tickets to an upcoming concert. The station is giving them away in pairs to lucky callers. The number of tickets remaining, y, is given by the equation $y = 24 - 2x$, where x is the number of callers who have received tickets. What is the domain of this function in the context of the problem?

3–12. For $f(x) = \dfrac{\sqrt{x-2}}{x-3}$, tell if the following numbers are in the natural domain of $f(x)$. If *not*, explain why.

a. 1 **b.** 2 **c.** 3 **d.** 4

3–13. Find the natural domain of the function $y = \dfrac{2x-6}{x^2 - 4x - 5}$.

3–14. Find the natural domain of the function $f(x) = \dfrac{x}{\sqrt{x-1} - 3}$.

3-15. The function $f(x) = \dfrac{24}{x+1}$ has the domain $\{x \mid 1 \le x \le 5\}$. Find the range of the function.

Symmetry

The graphs of some functions show certain kinds of symmetry. Two types of symmetry are the most important.

1. y-axis symmetry
The graph of the function has line symmetry over the y-axis. Such functions are called *even functions.*

2. Origin symmetry
The graph of the function has point symmetry over the origin. These functions are called *odd functions.*

Note that with the rather boring exception of $f(x) = 0$, functions do not have x-axis symmetry.

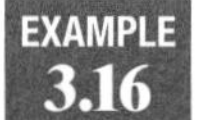

Determine what kind of symmetry, if any, each of the following functions has.

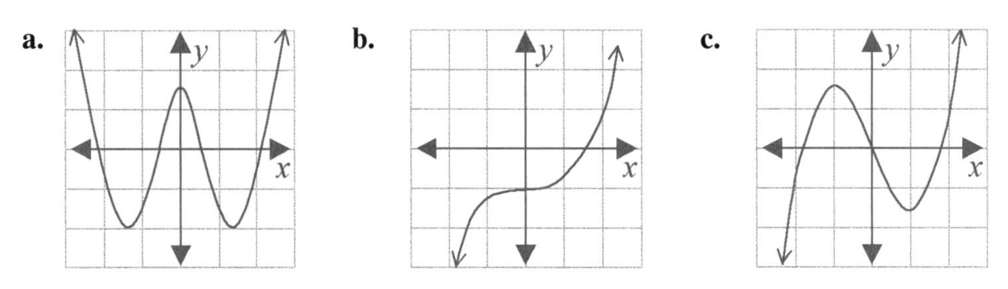

SOLUTION

a. This graph has y-axis symmetry and is an even function.
b. This graph has neither y-axis nor origin symmetry.
c. This graph has origin symmetry and is an odd function.

Most students find looking at the graph of a function to be the easiest way to check for symmetry. However, there is also an algebraic method based on the following definitions.

1. Even function
$f(-x) = f(x)$ for all x in the domain. This means that the y-value of the function at any given distance to the left of the y-axis is the same as the y-value at that same distance to the right of the axis. For example, in the even function graphed in Figure 3.6, $f(-1) = f(1)$, $f(-2) = f(2)$, and $f(-3) = f(3)$. In general, for any value of x in Figure 3.6, $f(-x) = f(x)$.

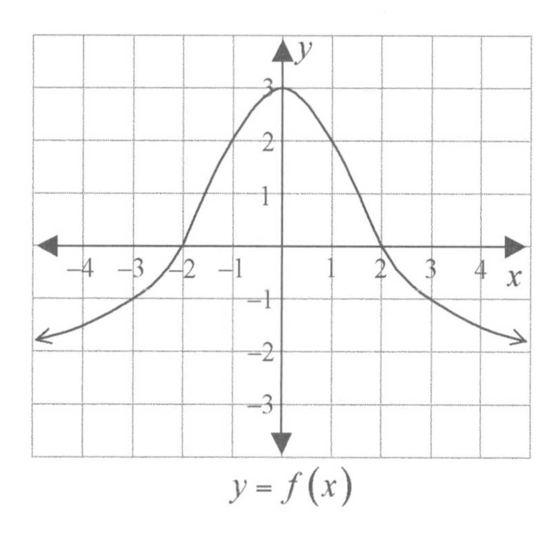

FIGURE 3.6

2. Odd function

$f(-x) = -f(x)$ for all x in the domain. This means that the y-value of the function at any given distance to the left of the y-axis is the opposite of the y-value at that same distance to the right of the axis. For example, in the odd function graphed in Figure 3.7, $f(-1) = -f(1)$, $f(-2) = -f(2)$, and $f(-3) = -f(3)$. In general, for any value of x in Figure 3.7, $f(-x) = -f(x)$.

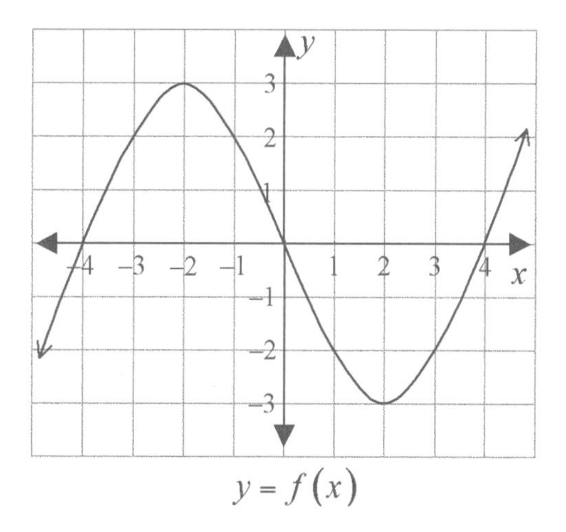

FIGURE 3.7

To check a function for symmetry, evaluate $f(-x)$. In other words, replace each x with $(-x)$ and simplify. (If this is tricky for you, remember the earlier tip about evaluating functions with boxes around the original x; the $(-x)$ goes into each box.)

1. If $f(-x)$ is the same as the original $f(x)$, $f(-x) = f(x)$, the function is even (has y-axis symmetry).
2. If $f(-x)$ is the opposite of the original $f(x)$, $f(-x) = -f(x)$, the function is odd (has origin symmetry).
3. If neither of the above is true, the function has neither kind of symmetry.

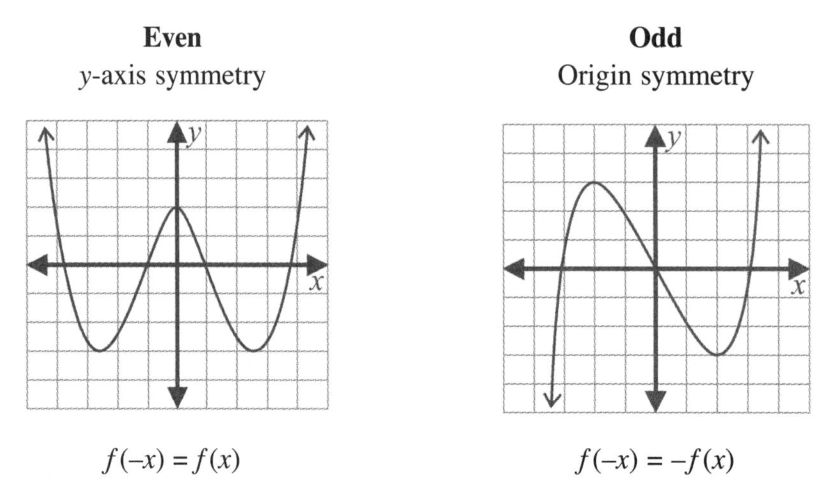

FIGURE 3.8

EXAMPLE 3.17

Determine what kind of symmetry, if any, each of the following functions has.

a. $f(x) = x^2 - 3$

b. $f(x) = 2x^3 + x$

c. $f(x) = x^2 - 2x$

d. $f(x) = \dfrac{x}{x^2 - 1}$

SOLUTION

Find $f(-x)$ and then compare it to $f(x)$.

a. $f(x) = x^2 - 3$

$$f(-x) = (-x)^2 - 3$$
$$f(-x) = x^2 - 3$$

$f(-x)$ is the same as $f(x)$, so $f(x)$ is an even function; it has y-axis symmetry.

b. $f(x) = 2x^3 + x$

$$f(-x) = 2(-x)^3 + (-x)$$
$$f(-x) = -2x^3 - x$$

$f(-x)$ is the opposite of $f(x)$, so $f(x)$ is an odd function; it has origin symmetry.

c. $f(x) = x^2 - 2x$

$$f(-x) = (-x)^2 - 2(-x)$$

$$f(-x) = x^2 + 2x$$

$f(-x)$ is neither the same nor the opposite of $f(x)$, so $f(x)$ is neither even nor odd.

d. $f(x) = \dfrac{x}{x^2 - 1}$

$$f(-x) = \frac{(-x)}{(-x)^2 - 1}$$

$$f(-x) = \frac{-x}{x^2 - 1}$$

$f(-x)$ is the opposite of $f(x)$, so $f(x)$ is an odd function; it has origin symmetry.

SECTION EXERCISES

3–16. What kind of symmetry do the following graphs show? If any graphs are functions, identify whether they are even, odd, or neither.

a. 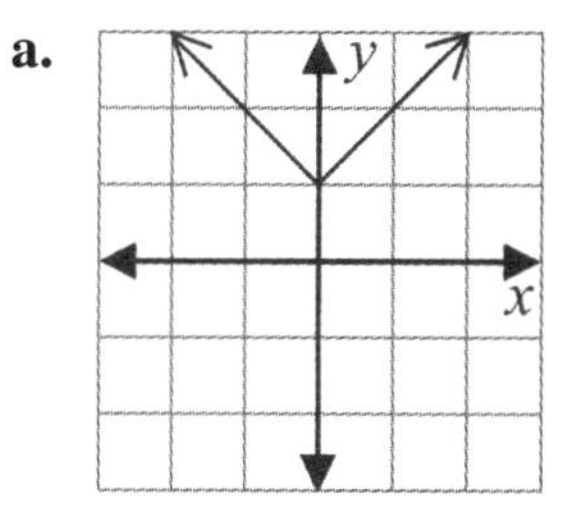

b. 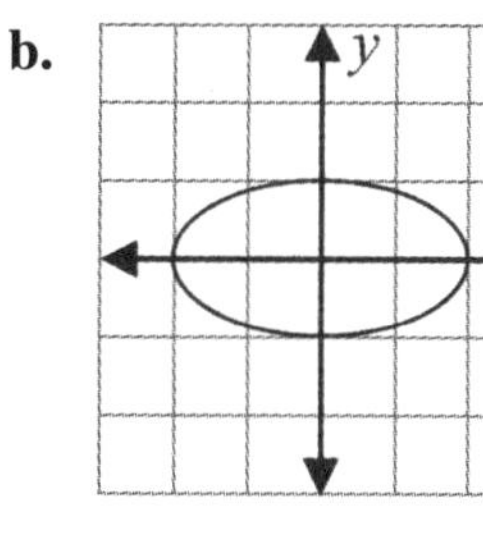

c. 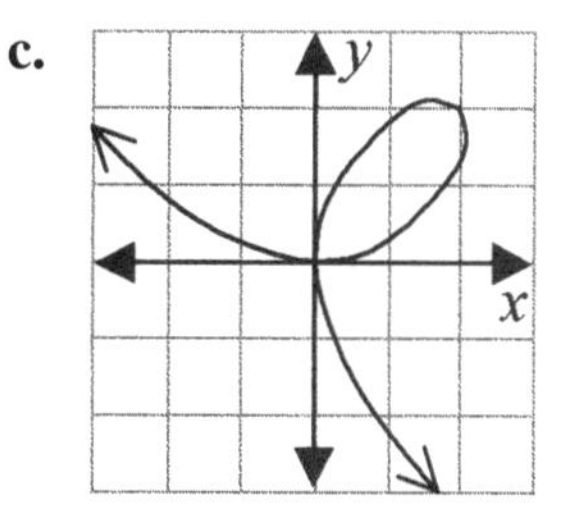

3–17. For each of the following functions, determine if it is even, odd, or neither.

a.

x	–4	–3	–2	–1	0	1	2	3	4
$f(x)$	–68	–30	–10	–2	0	2	10	30	68

b.

x	–4	–3	–2	–1	0	1	2	3	4
$f(x)$	0.063	0.125	0.25	0.5	1	2	4	8	16

3–18. For each of the following functions, determine if it is even, odd, or neither.

a. $f(x) = x^4 - bx^2 + c$

b. $f(x) = x^3 + bx$

Operations with Functions

Two or more functions can be added, subtracted, multiplied, or divided. They can also be combined together by a process called composition.

ALGEBRA OF FUNCTIONS

Two functions that have the same domain may be added, subtracted, multiplied, or divided to form new functions:

$$(f+g)(x) = f(x) + g(x)$$
$$(f-g)(x) = f(x) - g(x)$$
$$(fg)(x) = f(x)g(x)$$
$$\left(\frac{f}{g}\right)(x) = \frac{f(x)}{g(x)} \text{ where } g(x) \neq 0.$$

EXAMPLE 3.18

If $f(x) = x^2 - 4$ and $g(x) = 2x + 3$, write expressions for the following.

a. $(f+g)(x)$

b. $(f-g)(x)$

c. $(fg)(x)$

d. $\left(\frac{f}{g}\right)(x)$ and

e. $\left(\frac{g}{f}\right)(x)$

SOLUTION

a. $(f+g)(x) = x^2 - 4 + 2x + 3 = x^2 + 2x - 1$

b. $(f-g)(x) = x^2 - 4 - (2x + 3) = x^2 - 2x - 7$

c. $(fg)(x) = (x^2 - 4)(2x + 3) = 2x^3 + 3x^2 - 8x - 12$

d. $\left(\frac{f}{g}\right)(x) = \frac{x^2 - 4}{2x + 3},\ x \neq -\frac{3}{2}$

e. $\left(\frac{g}{f}\right)(x) = \frac{2x + 3}{x^2 - 4},\ x \neq \pm 2$

Note that for the two quotients, values of x that make the denominator equal zero must be excluded from the domain of the new function.

COMPOSITION OF FUNCTIONS

The function $h(x) = (x + 3)^2$ can be thought of as a combination of two simpler functions, $f(x) = x^2$ and $g(x) = x + 3$. However, h is not the sum, difference, product, or quotient of f and g. To get $h(x)$, we must apply the function f to the output of the function g: $h(x) = f(g(x))$. This new operation on functions is called the *composition of functions*. Compositions may be written with the symbol $\circ$:

So $(f \circ g)(x)$ is read "f of g of x" and means $f(g(x))$. Some books leave out the first set of parentheses and just write $f \circ g(x)$.

The expression $(f \circ g)(x)$ means we start with an input, x, and apply the function g to it, getting some unique output, $g(x)$. The function f is then applied to the output from g, giving the final, still unique, output, $f(g(x))$. This is shown schematically in Figure 3.9 below.

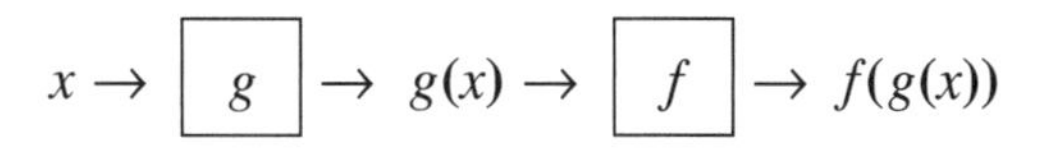

FIGURE 3.9

Using the functions defined previously, we have $(f \circ g)(x) = f(g(x)) = f(x + 3) = (x + 3)^2$, which is the function $h(x)$. Note that composition is not in general commutative. For these functions, the composition is $(g \circ f)(x) = g(f(x)) = g(x^2) = x^2 + 3$, which is not the same as $h(x)$.

EXAMPLE 3.19

Let $f(x) = x^2 - 3x + 2$ and $g(x) = \sqrt{x - 1}$. Evaluate each of the following.

a. $(f \circ g)(3)$

b. $(g \circ f)(3)$

c. $(f \circ g)(x)$

d. $(g \circ f)(x)$

SOLUTION

a.
$$\begin{aligned} (f \circ g)(3) &= f(g(3)) \\ g(3) &= \sqrt{3 - 1} = \sqrt{2} \\ (f \circ g)(3) &= f(\sqrt{2}) \\ &= (\sqrt{2})^2 - 3\sqrt{2} + 2 \\ &= 2 - 3\sqrt{2} + 2 \\ &= 4 - 3\sqrt{2} \end{aligned}$$

b. $(g \circ f)(3) = g(f(3))$

$$f(3) = (3)^2 - 3(3) + 2 = 2$$

$$(g \circ f)(3) = g(2)$$

$$= \sqrt{2 - 1}$$

$$= 1$$

SEE CALC TIP 3F

c. $(f \circ g)(x) = f(g(x)) = f(\sqrt{x - 1})$

$$(f \circ g)(x) = (\sqrt{x - 1})^2 - 3\sqrt{x - 1} + 2$$

$$= x - 1 - 3\sqrt{x - 1} + 2$$

$$= x - 3\sqrt{x - 1} + 1$$

d. $(g \circ f)(x) = g(f(x)) = g(x^2 - 3x + 2)$

$$(g \circ f)(x) = \sqrt{(x^2 - 3x + 2) - 1}$$

$$= \sqrt{x^2 - 3x + 1}$$

SECTION EXERCISES

3–19. If $f(x) = x^2 + 5$ and $g(x) = 3x - 2$, write expressions for the following.

a. $(f + g)(x)$

b. $(f - g)(x)$

c. $(fg)(x)$

d. $\left(\frac{f}{g}\right)(x)$

e. $\left(\frac{g}{f}\right)(x)$

3–20. On a certain remote island in the Indian Ocean lives a population of stink beetles. The stink beetles are eaten by the rose-breasted cuckoo birds that nest on the island. The cuckoos, in turn, are eaten by the bald rock cats that prowl the island. If the population of beetles is x, the population of cuckoo birds can be represented by $b(x) = \frac{x}{200}$. Further, if the bird population of the island is x, the cat population can be represented by $c(x) = 3\sqrt{x} - 2$.

a. If 20,000 stink beetles live on the island, what is the cuckoo population?

b. If 20,000 stink beetles live on the island, what is the cat population?

c. Write a composition of functions using $b(x)$ and $c(x)$ to give the population of bald rock cats as a function of the stink beetle population.

Transformations of Functions

The functions shown in Figure 3.10 are all different. However, after looking at their graphs it would seem that they are all related somehow (cousins maybe?). The graphs of $f(x) = -\sqrt{x}$ and $f(x) = \sqrt{-x}$ are reflections of the graph of $f(x) = \sqrt{x}$ over the coordinate axes. The graphs of $f(x) = \sqrt{x} + 2$ and $f(x) = \sqrt{x + 2}$ are translations, or shifts, of the graph of $f(x) = \sqrt{x}$. The graph of $f(x) = 2\sqrt{x}$ is a dilation, or stretch, of the graph of $f(x) = \sqrt{x}$

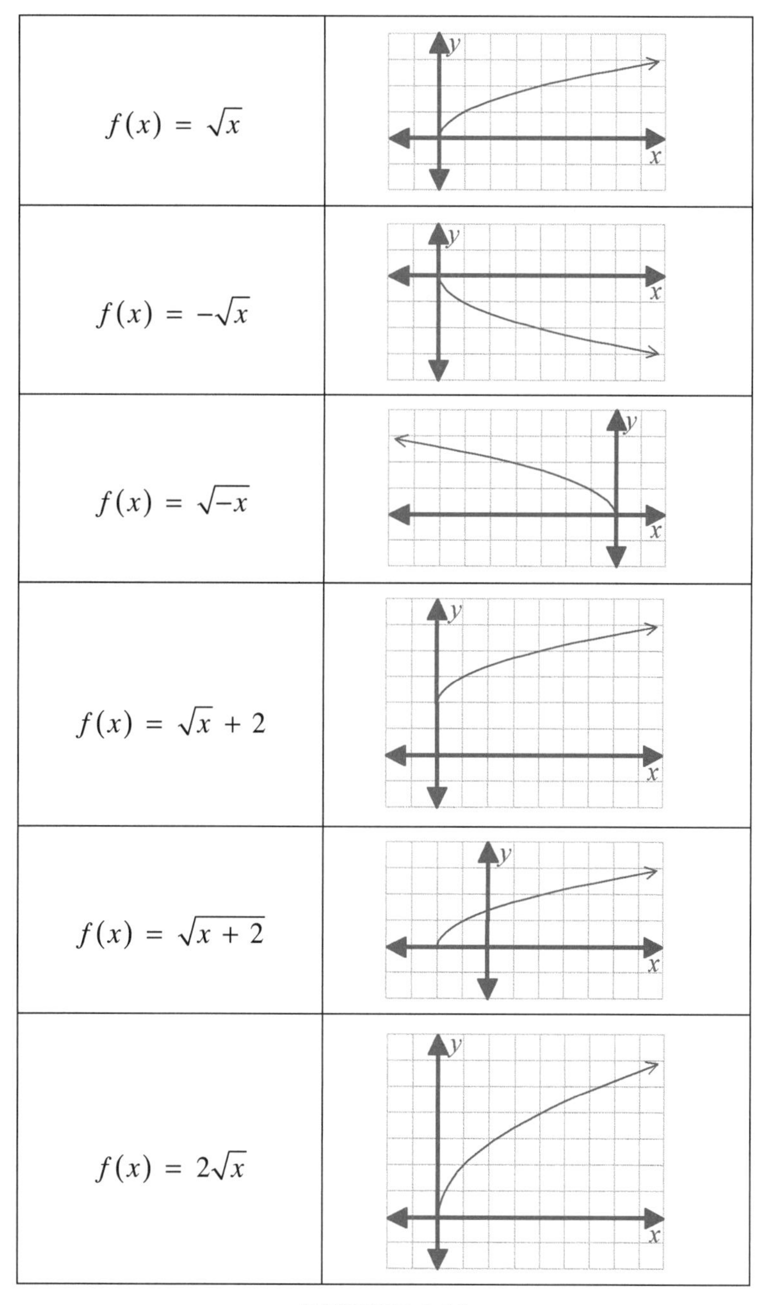

FIGURE 3.10

Reflections, translations, and dilations are basic transformations that can be done to any function to create a new function. Understanding these translations is very helpful when sketching graphs of functions and modeling with functions.

REFLECTIONS

You should be familiar with three basic reflections: reflections over each of the coordinate axes and a reflection in the origin. To reflect over the x-axis, negate the y. To reflect over the y-axis, negate the x. To reflect over the origin, negate both x and y. For any function $y = f(x)$, we have the following reflections shown in Figure 3.11.

Function	Transformation	Example	Graph
$y = f(x)$	Original function	$f(x) = \sqrt{4x - x^2}$	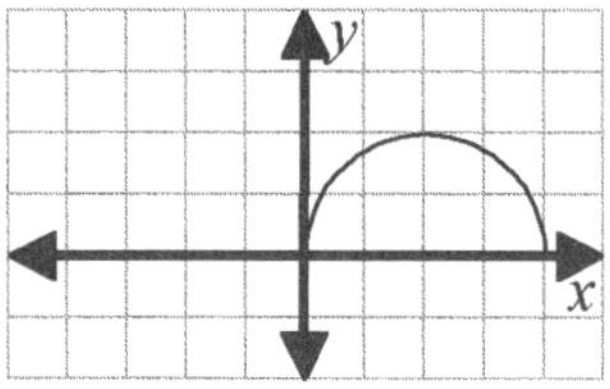
$-y = f(x)$ or $y = -f(x)$	Reflection over the x-axis	$f(x) = -\sqrt{4x - x^2}$	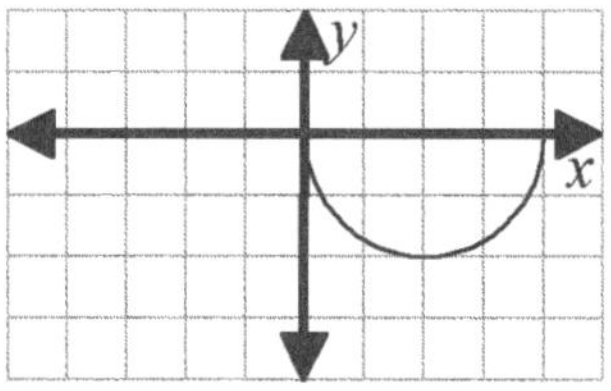
$y = f(-x)$	Reflection over the y-axis	$f(x) = \sqrt{4(-x) - (-x)^2} = \sqrt{-4x - x^2}$	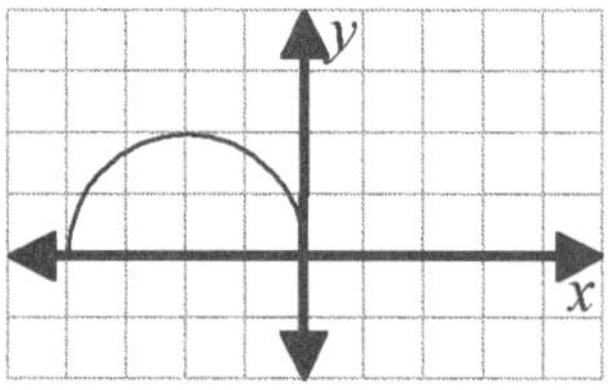
$-y = f(-x)$ or $y = -f(-x)$	Reflection over the origin	$f(x) = -\sqrt{4(-x) - (-x)^2} = -\sqrt{-4x - x^2}$	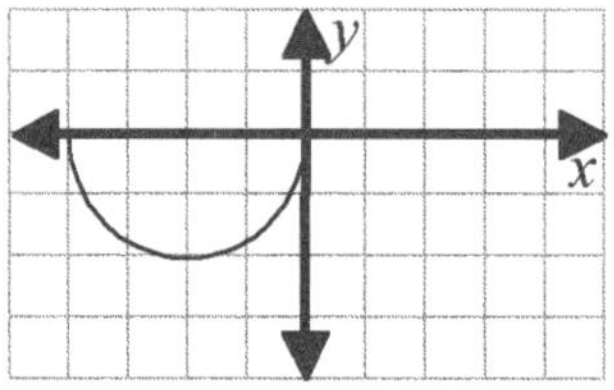

FIGURE 3.11

Consider the linear function $y = 2x + 1$.

a. Write an equation to reflect this function over the x-axis.

b. Write an equation to reflect this function over the y-axis.

c. Write an equation to reflect this function over the origin.

SOLUTION

a. To reflect over the x-axis, negate the y and then solve for y.

$-y = 2x + 1$

$y = -2x - 1$

b. To reflect over the y-axis, negate the x.

$y = 2(-x) + 1$

$y = -2x + 1$

c. To reflect over the origin, negate both x and y and then simplify.

$-y = 2(-x) + 1$

$y = 2x - 1$

Consider the quadratic function $y = (x - 2)^2$. Tell how the graph of each of the following will compare with the graph of $y = (x - 2)^2$.

a. $y = -(x - 2)^2$

b. $y = (-x - 2)^2$

c. $y = -(-x - 2)^2$

SOLUTION

a. When comparing $y = -(x - 2)^2$ to $y = (x - 2)^2$, you see a negative sign outside the parentheses. This negates the entire value of y, reflecting the graph over the x-axis.

b. When comparing $y = (-x - 2)^2$ to $y = (x - 2)^2$, you see a negative sign on only the x. This negates the value of x, reflecting the graph over the y-axis.

c. When comparing $y = -(-x - 2)^2$ to $y = (x - 2)^2$, you see negative signs both outside the parentheses and on the x. This negates the value of both x and y, reflecting the graph over the origin.

TRANSLATIONS (SHIFTS)

A function may be translated (shifted) vertically, horizontally, or in both directions. A combination of both a vertical and a horizontal translation results in a diagonal translation. Translations are done by subtracting (or adding) constants to the variables. Subtracting from x translates to the right; subtracting from y translates up.

Function	Transformation	Example	Graph
$y = f(x)$	Original function	$f(x) = \sqrt{4x - x^2}$	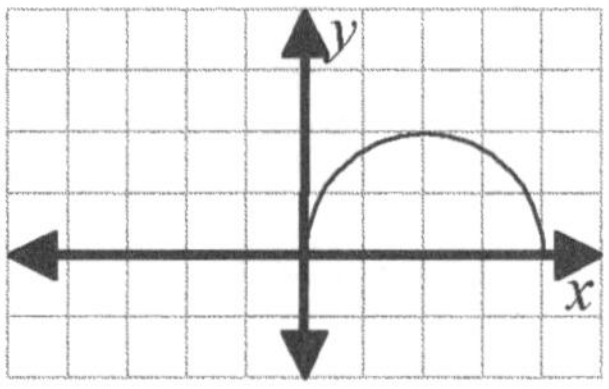
$y = f(x - h)$ $h > 0$	Horizontal translation "right h"	$f(x) = \sqrt{4(x - 2) - (x - 2)^2}$ Shift right 2	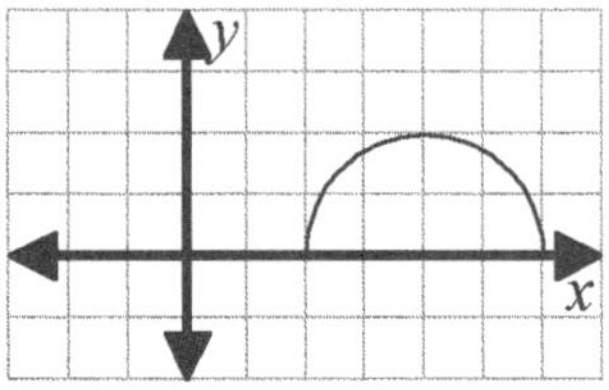
$y = f(x + h)$ $h > 0$	Horizontal translation "left h"	$f(x) = \sqrt{4(x + 3) - (x + 3)^2}$ Shift left 3	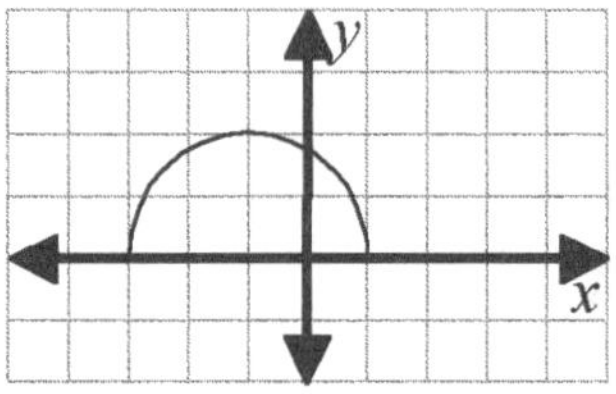
$y - k = f(x)$ or $y = f(x) + k$ $k > 0$	Vertical translation "up k"	$f(x) = \sqrt{4x - x^2} + 2$ Shift up 2	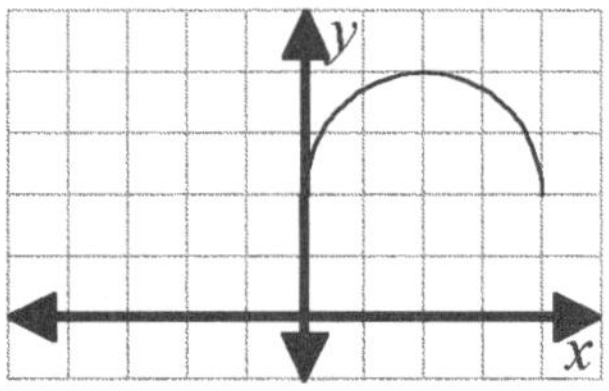
$y + k = f(x)$ or $y = f(x) - k$ $k > 0$	Vertical translation "down k"	$f(x) = \sqrt{4x - x^2} - 3$ Shift down 3	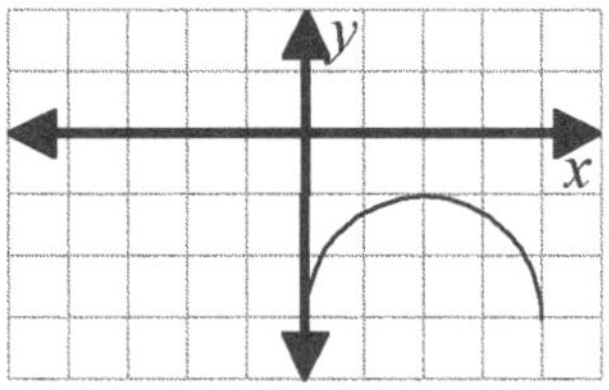
$y - k = f(x - h)$ or $y = f(x - h) + k$ $h > 0, k > 0$	Translation "right h and up k"	$f(x) = \sqrt{4(x - 3) - (x - 3)^2} + 2$ Shift right 3 and up 2	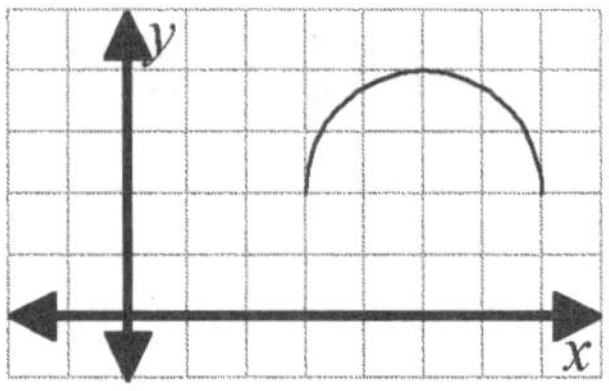

FIGURE 3.12

Most students have no trouble with the idea that $y = f(x) + 2$ translates the graph up 2 while $y = f(x) + (-3)$, usually written as $y = f(x) - 3$, translates the graph down 3. Positive is associated with up and negative with down in the vertical direction. However, the notion that $y = f(x - 2)$ translates the graph right 2 and $y = f(x + 3)$ translates the graph left 3 seems backward to many people. After all, in the horizontal direction, positive is associated with right and negative with left. It may help to know that when the subtraction or addition is done directly to the variable, x or y, both directions seem to be backward.

$$y = f(x - h) \text{ translates right } h$$
$$y - k = f(x) \text{ translates up } k$$

Only after the vertical translation is rewritten $y = f(x) + k$ does the sign seem to match with the direction of translation.

EXAMPLE 3.22 The function $y = f(x)$ is shown in the graph below along with three transformations of it labeled a, b, and c. Write equations in terms of $f(x)$ for each of the graphs a, b and c.

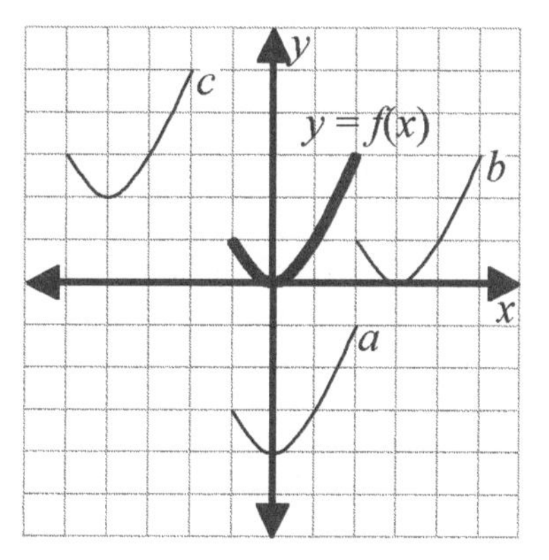

SOLUTION

a. The graph is shifted down 4 units. The new equation is $y = f(x) - 4$.
b. The graph is shifted right 3 units. The new equation is $y = f(x - 3)$.
c. The graph is shifted left 4 units and up 2 units. The new equation is $y = f(x + 4) + 2$.

EXAMPLE 3.23 Describe how the graph of $y = (x + 1)^2 + 2(x + 1) - 5$ compares with the graph of $y = x^2 + 2x$.

SOLUTION

Each individual x has 1 added to it, so the graph will shift left 1 unit. The entire equation has 5 subtracted from it, so the graph will shift down 5 units. The new graph will be a shift of 1 unit left and 5 units down.

DILATIONS (STRETCHES)

As you saw previously, adding or subtracting constants from the variables in a function produces translations of the graph. In very much the same way, multiplying or dividing the variables by constants produces dilations, a stretch or a shrink of the graph vertically, horizontally, or both.

Function	Transformation Description	Example	Graph
$y = f(x)$	Original function	$f(x) = \sqrt{4x - x^2}$	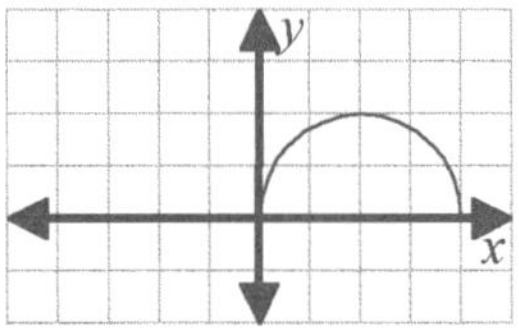
$y = f\left(\frac{x}{b}\right)$	Horizontal stretch by a factor of $\frac{1}{b}$, $b > 1$	$f(x) = \sqrt{4\left(\frac{x}{2}\right) - \left(\frac{x}{2}\right)^2}$ Horizontal stretch of $\frac{1}{2}$	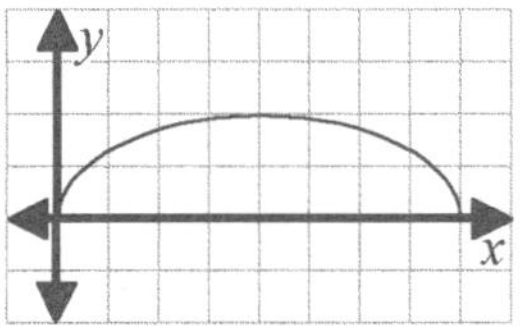
$y = f(bx)$	Horizontal shrink by a factor of $\frac{1}{b}$, $b > 1$	$f(x) = \sqrt{4(2x) - (2x)^2}$ Horizontal shrink of $\frac{1}{2}$	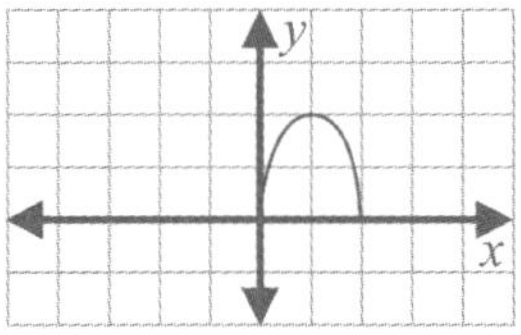
$\frac{y}{a} = f(x)$ or $y = af(x)$	Vertical stretch by a factor of a, $a > 1$	$f(x) = 3\sqrt{4x - x^2}$ Vertical stretch of 3	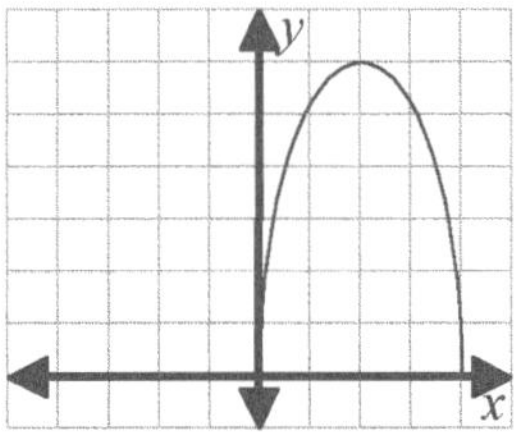
$\frac{y}{a} = f(x)$ or $y = af(x)$	Vertical shrink by a factor of a, $0 < a < 1$	$f(x) = \frac{1}{2}\sqrt{4x - x^2}$ Vertical shrink of $\frac{1}{2}$	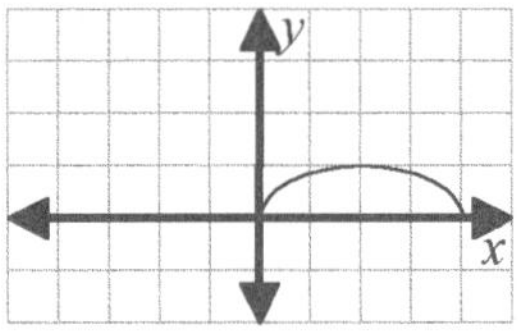
$\frac{y}{a} = f\left(\frac{x}{b}\right)$ or $y = af\left(\frac{x}{b}\right)$	Vertical dilation of k and horizontal dilation of b	$f(x) = 3\sqrt{4\left(\frac{x}{2}\right) - \left(\frac{x}{2}\right)^2}$ Vertical stretch of 3 and horizontal stretch of 2	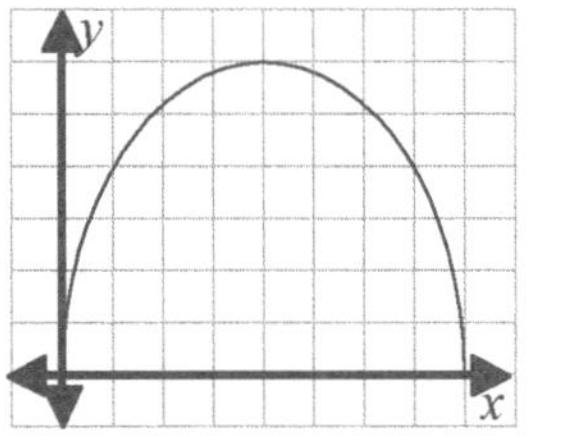

FIGURE 3.13

Notice how vertical dilations seem to make intuitive sense. Compared with the graph of $y = f(x)$, the graph of $y = 2f(x)$ is stretched vertically by a factor of 2 (becomes twice as tall), while the graph of $y = \frac{1}{2}f(x)$ shrinks by a factor of $\frac{1}{2}$ (becomes half as tall). Exactly like horizontal translations, though, horizontal dilations seem to work backward. The graph of $y = f(2x)$ is shrunk by a factor of $\frac{1}{2}$ (half as wide) compared with the graph of $y = f(x)$. The graph of $y = f\left(\frac{x}{2}\right)$ is stretched by a factor of 2 (twice as wide) compared with the graph of $y = f(x)$. The explanation is the same as for translations. If you multiply directly on either variable, the effect seems to be the opposite of your intuition. Only after solving for *y* does the vertical dilation seem to work the way most students think it should while the horizontal dilation remains "backward."

EXAMPLE 3.24 The function $y = f(x)$ is shown in the graph below along with two transformations of it labeled *a* and *b*. Write equations in terms of $f(x)$ for each of the graphs *a* and *b*.

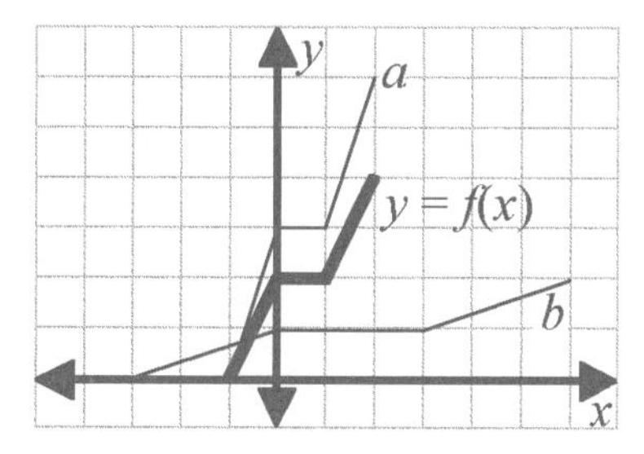

SOLUTION

a. This graph has been stretched vertically by a factor of $\frac{3}{2}$. Compare the *y*-intercepts to find this factor. The *y*-intercept of the original graph is 2, and the *y*-intercept of the new graph is 3. The *x*-values have not changed. The new equation is $y = \frac{3}{2}f(x)$.

b. This graph has been shrunk vertically by a factor of $\frac{1}{2}$. It has also been stretched horizontally by a factor of 3. The new equation is $y = \frac{1}{2}f\left(\frac{x}{3}\right)$.

Describe how the graph of each of the following functions compares with the graph of the original function $y = x^3 - 2x^2$.

a. $y = (x - 4)^3 - 2(x - 4)^2$

b. $y = 2\left(x^3 - 2x^2\right)$

c. $y = \left(\frac{x}{3}\right)^3 - 2\left(\frac{x}{3}\right)^2$

d. $y = -x^3 + 2x^2$

e. $y = -x^3 - 2x^2$

f. $y = x^3 - 2x^2 + 5$

SOLUTION

Compare each equation with the original function, $y = x^3 - 2x^2$.

a. In $y = (x - 4)^3 - 2(x - 4)^2$, each individual x has 4 subtracted from it. The graph of the original function is shifted 4 units to the right.

b. In $y = 2\left(x^3 - 2x^2\right)$, the original function has been multiplied by 2. The graph of the original function is stretched vertically by a factor of 2.

c. In $y = \left(\frac{x}{3}\right)^3 - 2\left(\frac{x}{3}\right)^2$, each individual x is divided by 3. This will stretch the graph of the original function by a factor of 3 horizontally.

d. In $y = -x^3 + 2x^2$, the original function has been multiplied by –1. This reflects the graph of the original function over the x-axis.

e. In $y = -x^3 - 2x^2$, although it may not be as obvious at first as the others, each individual x in the original equation has been negated: $y = (-x)^3 - 2(-x)^2 = -x^2 - 2x^2$. This graph is a reflection of the original graph over the y-axis.

f. In $y = x^3 - 2x^2 + 5$, 5 has been added to the function. This shifts the graph of the original function 5 units up.

EXAMPLE 3.26

Describe how the graphs of each of the following will compare with the graph of $y = f(x)$.

a. $y = f(x - 4)$
b. $y = f(-x)$
c. $y = 3f(x)$
d. $y = f(x) - 4$
e. $y = f(3x)$
f. $y = -f(-x)$
g. $y = f(x + 2) + 3$
h. $y = \frac{1}{2}f\left(\frac{1}{3}x\right)$

SOLUTION

a. $y = f(x - 4)$ means $y = f(x)$ is shifted right 4 units.
b. $y = f(-x)$ means $y = f(x)$ is reflected over the y-axis.
c. $y = 3f(x)$ means $y = f(x)$ is vertically stretched by a factor of 3.
d. $y = f(x) - 4$ means $y = f(x)$ is shifted down 4 units.
e. $y = f(3x)$ means $y = f(x)$ is horizontally shrunk by a factor of $\frac{1}{3}$.
f. $y = -f(-x)$ means $y = f(x)$ is reflected over the origin.
g. $y = f(x + 2) + 3$ means $y = f(x)$ is shifted left 2 units and up 3 units.
h. $y = \frac{1}{2}f\left(\frac{1}{3}x\right)$ means $y = f(x)$ is vertically shrunk by a factor of $\frac{1}{2}$ and horizontally stretched by a factor of 3.

EXAMPLE 3.27

The function $f(x)$, shown in the graph below, consists of two line segments.

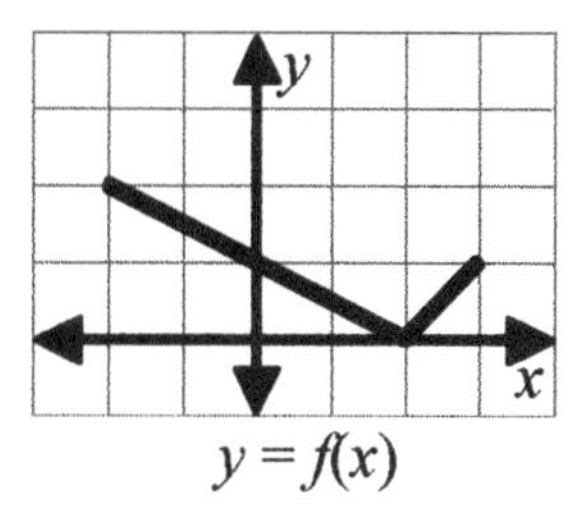

Sketch the graphs of each of the following:

a. $y = -f(x)$
b. $y = f(x - 6) - 4$
c. $y = f(2x)$
d. $y = 4f(x)$
e. $y = 3f\left(\frac{x}{2}\right)$

SOLUTION

a. The graph of $y = -f(x)$ is the reflection of the graph of $y = f(x)$ over the x-axis.

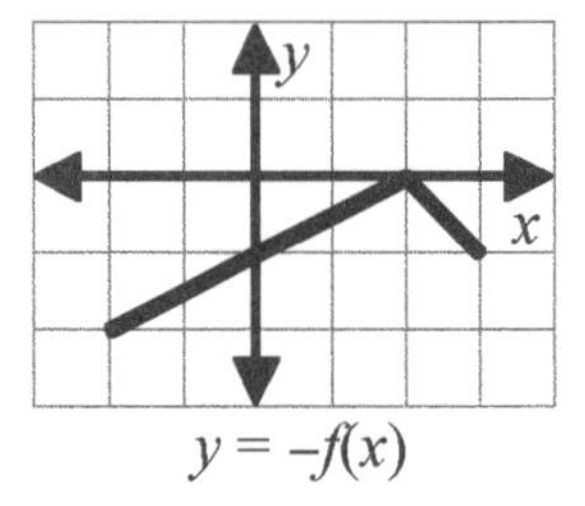

$y = -f(x)$

b. The graph of $y = f(x - 6) - 4$ is the result of the graph of $y = f(x)$ being shifted 6 units to the right and 4 units down.

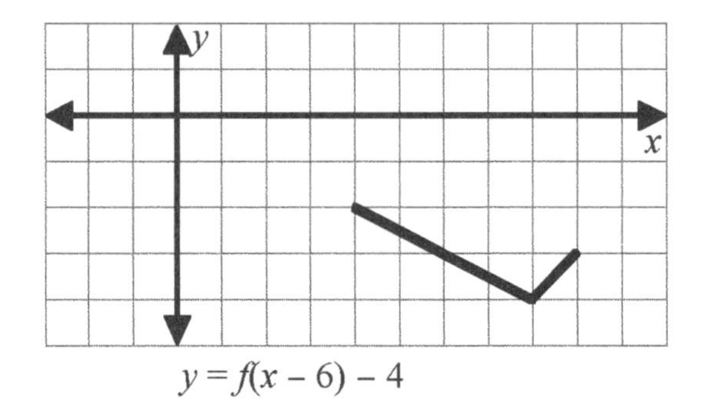

$y = f(x - 6) - 4$

c. The graph of $y = f(2x)$ is the horizontal dilation of the graph of $y = f(x)$ by a factor of $\frac{1}{2}$.

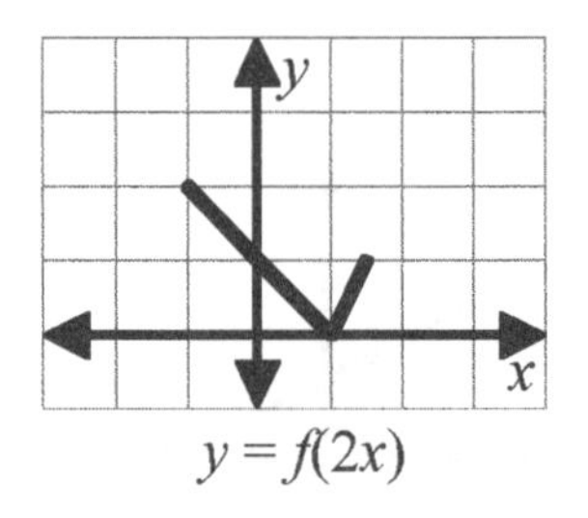

$y = f(2x)$

d. The graph of $y = 4f(x)$ is the vertical dilation of the graph of $y = f(x)$ by a factor of 4.

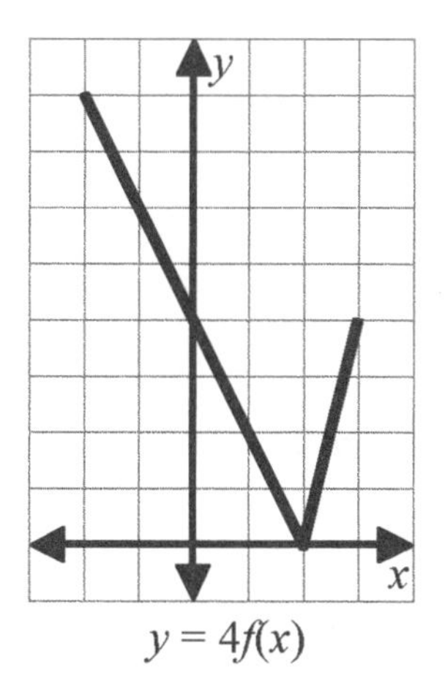

$y = 4f(x)$

e. The graph of $y = 3f\left(\frac{x}{2}\right)$ is the dilation of the graph of $y = f(x)$ horizontally by a factor of 2 and vertically by a factor of 3.

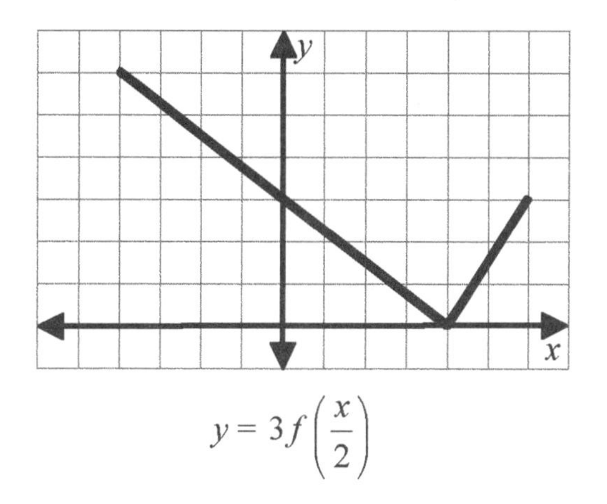

$y = 3f\left(\frac{x}{2}\right)$

SECTION EXERCISES

3–21. If $y = f(x)$, describe in words the result of each of the following transformations of the graph of $f(x)$.

a. $y = -f(x)$ **d.** $y = f(x - 5)$

b. $y = f(-x)$ **e.** $y = f(x + 4) - 6$

c. $y = f(x) + 3$ **f.** $y = f(2x)$

3–22. The graphs of the function $y = f(x)$ and four transformations are shown below.

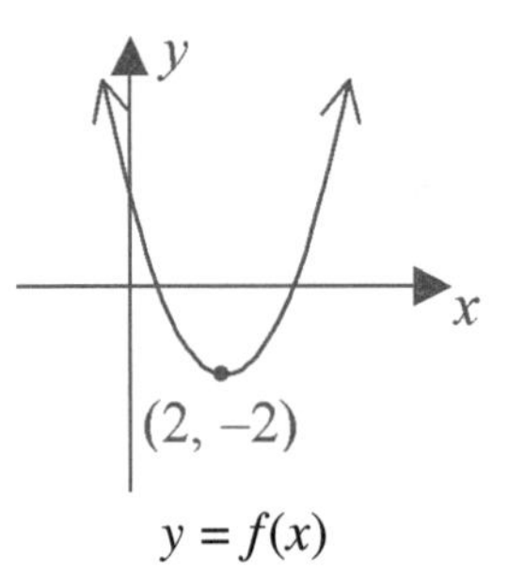

$y = f(x)$

Match the equations with the graphs of the transformations.

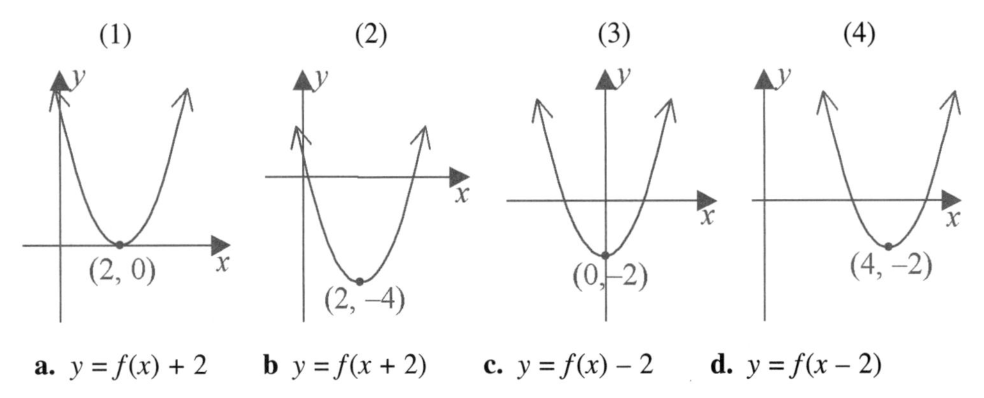

a. $y = f(x) + 2$ **b** $y = f(x + 2)$ **c.** $y = f(x) - 2$ **d.** $y = f(x - 2)$

3–23. The graphs of the function $y = f(x)$ and four transformations are shown below.

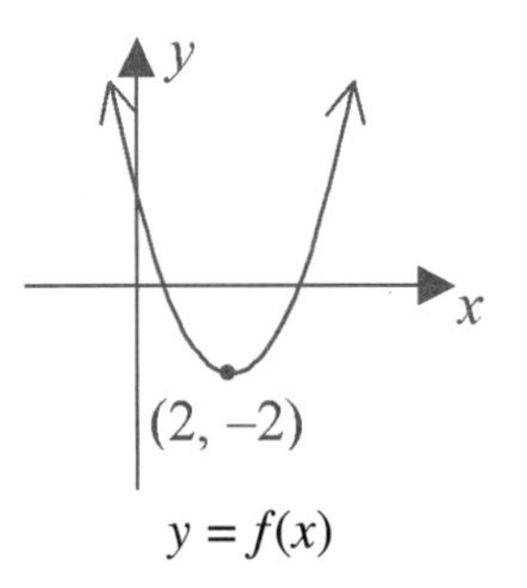

$y = f(x)$

Match the equations with the graphs of the transformations.

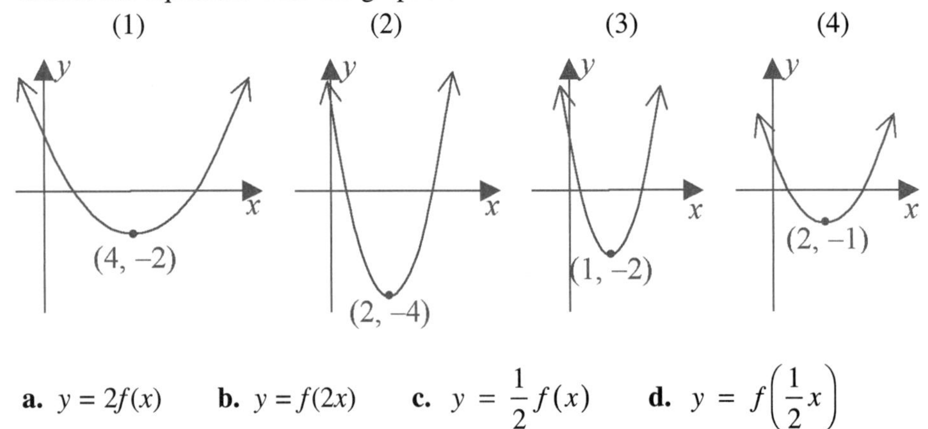

a. $y = 2f(x)$ **b.** $y = f(2x)$ **c.** $y = \frac{1}{2}f(x)$ **d.** $y = f\left(\frac{1}{2}x\right)$

3–24. The function $y = f(x)$ is graphed below. Copy $f(x)$ onto a larger grid (eight boxes in all directions from the origin) and then graph the following transformations. Label each graph.

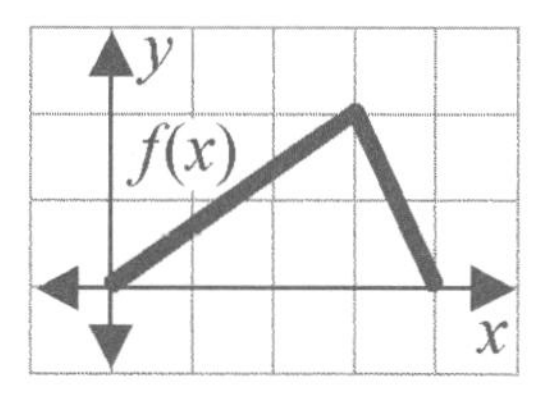

a. $y = f(-x)$
b. $y = -f(x)$
c. $y = f(x + 5)$
d. $y = f(x) + 3$
e. $y = f(x - 3) - 4$
f. $y = f\left(\frac{1}{2}x\right)$
g. $y = \frac{1}{2}f(x)$

3–25. Write an equation for the new graph of $y = (x - 4)\sqrt{x}$ after the following transformations.
a. A reflection over the y-axis.
b. A vertical stretch by a factor of 2.
c. A horizontal stretch by a factor of 3.

Inverse Functions

Each function has a relation associated with it called the inverse of the function. The inverse of a function may or may not be another function. Knowing whether the original function is one-to-one will tell you whether the inverse is also a function.

ONE-TO-ONE FUNCTIONS

A one-to-one function is a function where every y comes from only one x. In other words, no two ordered pairs can have the same y-value but different x-values. In graph form, a one-to-one function passes the *horizontal line test*—any horizontal line intersects the graph at most once.

Tell whether each of the following graphs represents a one-to-one function.

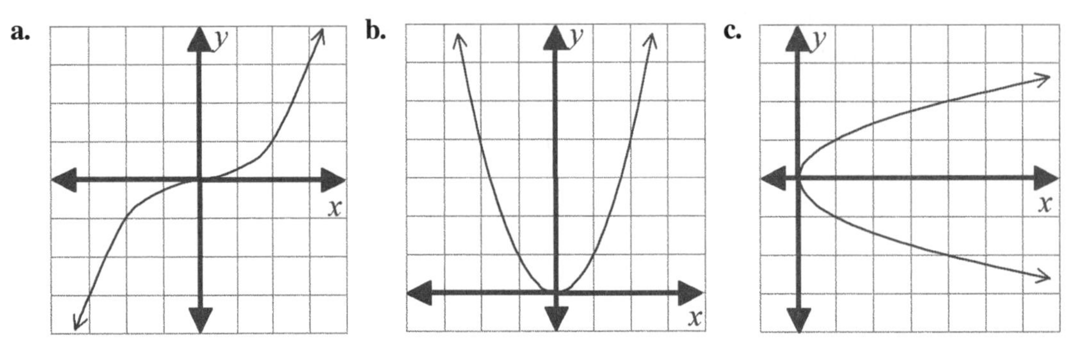

SOLUTION

a. Yes, this is a one-to-one function. It is a function because it passes the vertical line test; each x-value goes to only one y-value. It is one-to-one because it also passes the horizontal line test; each y-value comes from only one x-value.

b. This is a function by the vertical line test. However, it is not one-to-one; many horizontal lines will intersect the graph twice. This means many y-values come from two different x-values. For example, $y = 4$ comes from either $x = 2$ or $x = -2$.

c. Don't be hasty. The graph passes the horizontal line test. However, because it does not pass the vertical line test, it is not a function. So it certainly cannot be a one-to-one function. A one-to-one function must pass both the vertical and horizontal line tests.

INVERSE FUNCTIONS

Let's look at two functions, $f(x) = \sqrt{2x - 4}, x \geq 2$ and $g(x) = 0.5x^2 + 2, x \geq 0$.

Start with a number in the domain of $f(x)$, say $x = 10$, and substitute it into f:

$f(10) = \sqrt{2(10) - 4} = 4$. Now take that answer and substitute it into $g(x)$:

$g(4) = 0.5(4)^2 + 2 = 10$. We got back the original 10. Let's try a few more values of x. We will use each output from $f(x)$ as the input for $g(x)$.

x	$f(x)$	$\rightarrow$	x	$g(x)$
2	0	$\rightarrow$	0	2
3	$\sqrt{2}$	$\rightarrow$	$\sqrt{2}$	3
4	2	$\rightarrow$	2	4
5	$\sqrt{6}$	$\rightarrow$	$\sqrt{6}$	5

It appears that for any x-value, the function $g(x)$ undoes what $f(x)$ did and returns the original value x.

$$x \rightarrow \boxed{f} \rightarrow f(x) \rightarrow \boxed{g} \rightarrow x$$

The function $g(x)$ is called the *inverse* of the function $f(x)$. Note also that the function $f(x)$ is the inverse of $g(x)$; the two functions are inverses of each other.

The notation for the inverse of a function f is f^{-1}. So for $f(x) = \sqrt{2x - 4}$, $x \ge 2$, the inverse function is $f^{-1}(x) = 0.5x^2 + 2$, $x \ge 0$.

COMMON ERROR

Do not get the inverse symbol confused with an exponent.

$$f^{-1}(x) \text{ does } \textit{not} \text{ mean } \frac{1}{f(x)}.$$

PROPERTIES OF INVERSES

The following describes the main things you should know about inverses.

1. The inverse of any relation R is the new relation found by reversing the coordinates in each of the ordered pairs of R. For example, if R is the relation {(–1, 3), (0, 1), (1, 3), (2, 7)}, the inverse of R is {(3, –1), (1, 0), (3, 1), (7, 2)}. Just switch the x and the y in each ordered pair.
2. The inverse of a function will be another function only if the original function is one-to-one. In the example above, R is a function. Since R is not one-to-one, the inverse of R is not a function.
3. The graph of the inverse of a function is the reflection over the line $y = x$ of the graph of the original function.

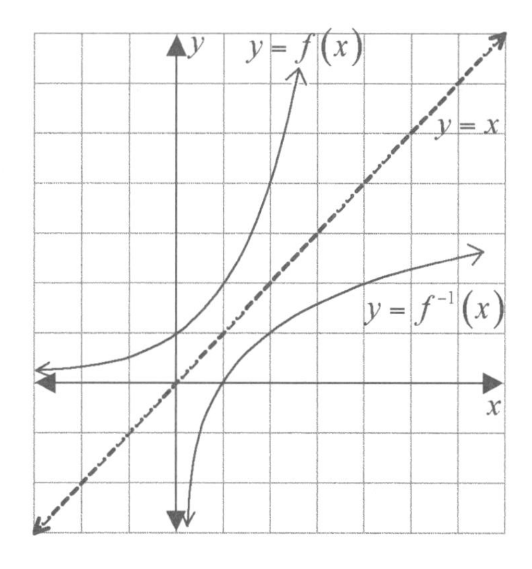

FIGURE 3.14

4. The domain of $f^{-1}(x)$ is the same as the range of $f(x)$ and the range of $f^{-1}(x)$ is the same as the domain of $f(x)$. In the previous example, R has domain $\{-1, 0, 1, 2\}$ and range $\{1, 3, 7\}$. The inverse of R has domain $\{1, 3, 7\}$ and range $\{-1, 0, 1, 2\}$.
5. If $f^{-1}(x)$ is the inverse of $f(x)$, then the following must both be true:
   - $f^{-1}(f(x)) = x$ for all x in the domain of $f(x)$.
   - $f\left(f^{-1}(x)\right) = x$ for all x in the domain of $f^{-1}(x)$ (the range of $f(x)$).

EXAMPLE 3.29 Verify algebraically that the functions $f(x) = \sqrt{2x - 4}$, $x \geq 2$ and $g(x) = 0.5x^2 + 2$, $x \geq 0$ are inverses of each other.

SOLUTION

Note that the tables shown previously are very convincing but do not prove the functions are inverses. Graphs of the functions are also quite convincing. The graph of $g(x)$ appears to be the reflection of the graph of $f(x)$ over the line $y = x$.

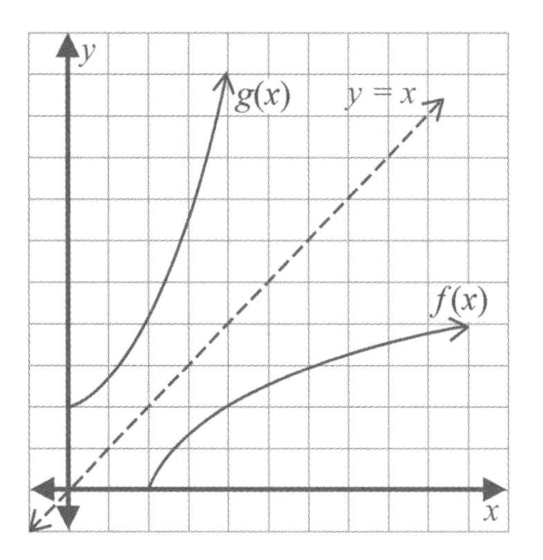

However, to prove that $f(x)$ and $g(x)$ are inverses, we must check that $g(f(x)) = x$ and $f(g(x)) = x$.

$$g(f(x)) = g\left(\sqrt{2x-4}\right) = 0.5\left(\sqrt{2x-4}\right)^2 + 2$$
$$= 0.5(2x - 4) + 2$$
$$= x - 2 + 2$$
$$= x$$

$$f(g(x)) = f\left(0.5x^2 + 2\right) = \sqrt{2\left(0.5x^2 + 2\right) - 4}$$
$$= \sqrt{x^2 + 4 - 4}$$
$$= \sqrt{x^2} = |x|$$
$$= x \text{ when } x \geq 0$$

Therefore, $f(x)$ and $g(x)$ are inverse functions, $g(x) = f^{-1}(x)$ and $f(x) = g^{-1}(x)$.

FINDING INVERSES

You should have one question left: how do we find the inverse of a given function? The key to the answer was given in the last section. To find the inverse of a function, switch the coordinates of each of the ordered pairs in the function. The details vary depending on how the function is defined.

1. If the function is defined by a set of ordered pairs or a table, switch the coordinates of each of the ordered pairs in the function. You are done.
2. If the function is given as a graph, find the inverse by reflecting the graph over the line $y = x$. This is done by switching the coordinates of points on the original graph and plotting the new points.
3. If the function is given as an equation, finding the inverse is done in two steps.
   - First: Replace y (or $f(x)$) with x and replace x (all of them!) with y.
   - Second: Solve for y. This new y is $f^{-1}(x)$.

Find the inverse of the function $f(x)$ shown in the table below.

x	$f(x)$
–2	–5
–1	2
0	3
1	4
2	11

SOLUTION

To find the inverse of a function represented by a table, simply switch the coordinates of each ordered pair represented in the table. So (–2, –5) becomes (–5, –2) and so on. Note that the heading of the column on the right has changed.

x	$f^{-1}(x)$
–5	–2
2	–1
3	0
4	1
11	2

Graph the inverse of the function shown in the graph below.

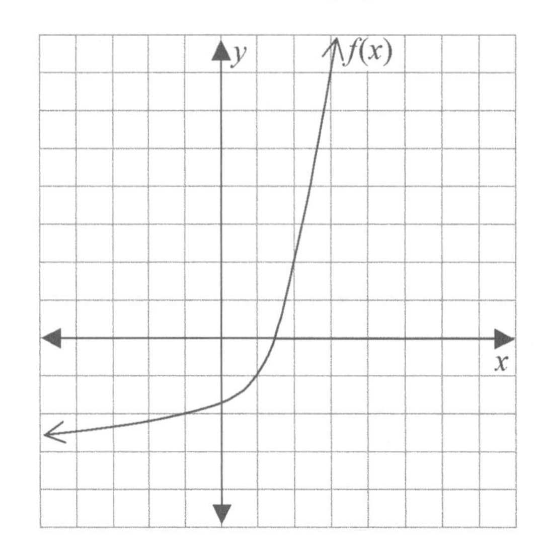

SOLUTION

Select some representative points on the graph of $f(x)$, and switch the coordinates of each.

$f(x)$	→	$f^{-1}(x)$
(–1, –2)	→	(–2, –1)
(1, –1)	→	(–1, 1)
(2, 2)	→	(2, 2)
(3, 7)	→	(7, 3)

Graph the points found for $f^{-1}(x)$ and connect them with an appropriate graph. In this case, the graph should be a smooth curve. Some functions, though, have corners.

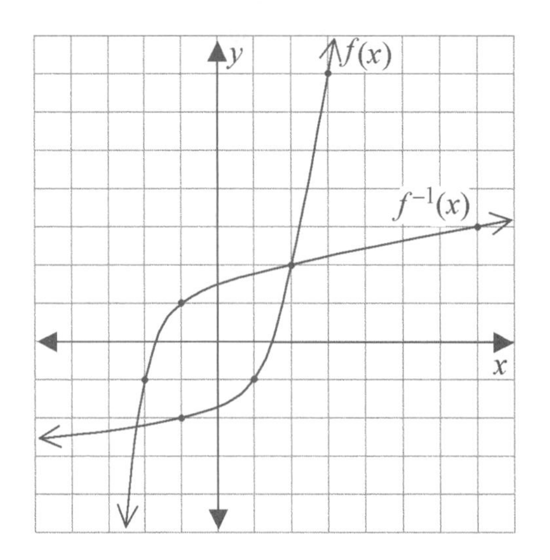

EXAMPLE 3.32 Find the inverse of the function $f(x) = 3x - 6$.

SOLUTION

First rename $f(x)$ as y. Then switch x and y, and solve for (the new) y.

$$f(x) = 3x - 6 \rightarrow y = 3x - 6$$

Inverse:
$$x = 3y - 6$$
$$3y = x + 6$$
$$y = \frac{x+6}{3} \text{ or } y = \frac{1}{3}x + 2$$

Finally, rename y as $f^{-1}(x)$: $f^{-1}(x) = \frac{x+6}{3}$ or $f^{-1}(x) = \frac{1}{3}x + 2$.

Either form of the inverse shown above is acceptable. Note that the first form helps clarify the inverse relation between $f(x)$ and $f^{-1}(x)$. The original function $f(x) = 3x - 6$ says "multiply by 3 and then subtract 6." The inverse, $f^{-1}(x) = \frac{x+6}{3}$, says we can undo f if we "add 6 and then divide by 3." Algebraically, this is equivalent to "multiply by $\frac{1}{3}$ and then add 2." However, it is not as obvious that it undoes "multiply by 3 and then subtract 6."

EXAMPLE 3.33 Find the inverse of the function $f(x) = \frac{2x}{x-5}$, $x \neq 5$.

SOLUTION

Rename $f(x)$ as y, switch x and y, and solve for y.

$$f(x) = \frac{2x}{x-5} \rightarrow y = \frac{2x}{x-5}$$

Inverse:
$$x = \frac{2y}{y-5}$$
$$x(y-5) = 2y$$
$$xy - 5x = 2y$$
$$xy - 2y = 5x$$
$$y(x-2) = 5x$$
$$y = \frac{5x}{x-2}$$

SEE CALC TIP 3G

The inverse of f is $f^{-1}(x) = \frac{5x}{x-2}$.

Do not let examples like this one tempt you into looking for shortcuts like "just switch the 2 and the 5." Shortcuts that might work for one specific type of function may not work for another. Do the algebra. It's good practice.

Find the inverse of the function $f(x) = 2x^2 - 8$, $x \geq 0$.

SOLUTION

First note the domain. On its natural domain, $(-\infty, \infty)$, the function $f(x) = 2x^2 - 8$ is not one-to-one and so its inverse would not be a function. By restricting the domain to $x \geq 0$, that problem is avoided. For $x \geq 0$, $f(x)$ is one-to-one and so its inverse will be a function.

$$f(x) = 2x^2 - 8,\ x \geq 0 \rightarrow y = 2x^2 - 8,\ x \geq 0$$

Inverse:
$$x = 2y^2 - 8,\ y \geq 0$$

$$y^2 = \frac{x+8}{2}$$

$$y = \sqrt{\frac{x+8}{2}}$$

SEE CALC TIP 3H

Note that only the positive square root is kept because of the restriction $y \geq 0$.

The inverse of $f(x)$ is $f^{-1}(x) = \sqrt{\dfrac{x+8}{2}}$.

SECTION EXERCISES

3–26. A function is shown in the table. Create a table for the inverse of the function.

x	$f(x)$
0	10
1	5
2	2
3	1

3–27. Find the inverse of each function.

a. $f(x) = \frac{1}{3}(x + 2)$

b. $f(x) = \frac{2}{3}x - 12$

c. $f(x) = (2x - 4)^2; x \geq 2$

d. $f(x) = 3 - \sqrt{x + 9}; x \geq -9$

e. $f(x) = \frac{3x}{x + 2}; x \neq -2$

f. $f(x) = (36 - 2x)^2; x \geq 18$

3–28. Draw the graph of the inverse of the function shown below.

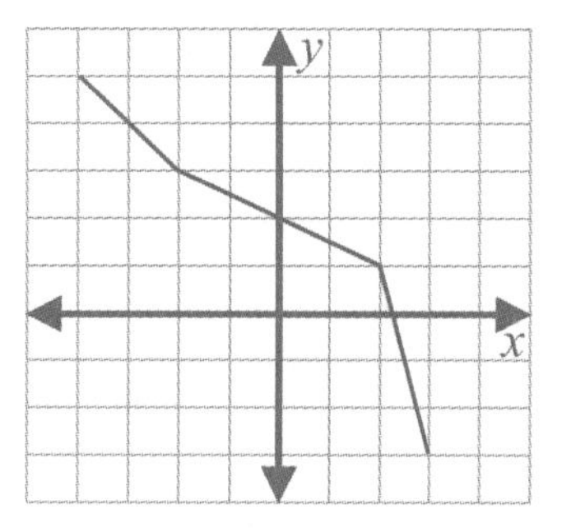

CHAPTER EXERCISES

C3–1. Below are tables of values for four different relationships.

(1)

x	y
–2	4
–1	1
0	0
1	1
2	4

(2)

x	y
9	–3
4	–2
0	0
4	2
9	3

(3)

x	y
1	8
2	6
3	4
4	2
5	0

(4)

x	y
0.0	1
0.25	0
0.5	–1
0.75	0
1.0	1

a. In which relationship(s) is y a function of x? (Choose all that apply.)
b. In which relationship(s) is y a one-to-one function of x? (Choose all that apply.)

C3–2. Suppose $y = f(x)$ for some function $f(x)$.

a. What does $f(9) = 3$ mean?
b. Write an ordered pair that corresponds to the statement $f(9) = 3$.
c. If $f(9) = 3$, write an ordered pair that would be on the graph of the *inverse* of $f(x)$.

C3–3. The function $f(x)$ is defined by $f(x) = \dfrac{4x + 8}{x - 1}$.

a. Tell what $f(5) = 7$ means.
b. Evaluate $f(-3)$.
c. For what value of x will $f(x) = 10$?
d. What is the y-intercept of the graph of $f(x)$?
e. What is the zero of $f(x)$?
f. What is the domain of the function $f(x)$?
g. Evaluate $f(x - 2)$ and simplify.

C3–4. Give both the domain and the range of the function in each graph in simplest interval notation.

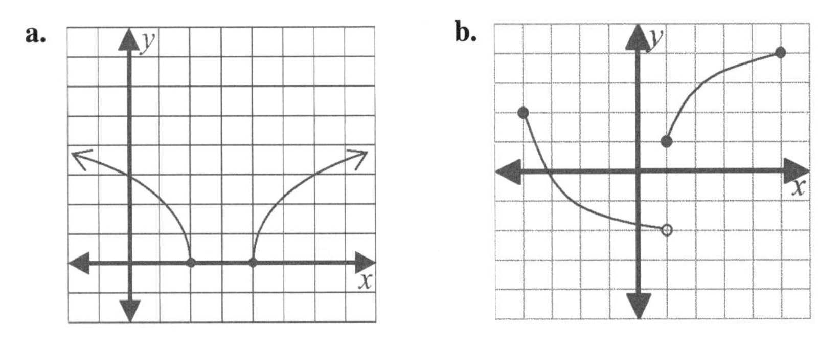

C3–5. Find the natural domain of the function $y = \sqrt{32 - 4x}$.

C3–6. What is the domain of the function $f(x) = \dfrac{\sqrt{x + 1}}{x}$ in the real numbers expressed in interval notation?

C3–7. The function $f(x)$ is defined as $f(x) = x^2 - 4x$ with domain $\{x| -1 \le x < 6\}$. What is the range of $f(x)$?

C3–8. What kind of symmetry do the following graphs show? If any graphs are functions, identify them as even, odd, or neither.

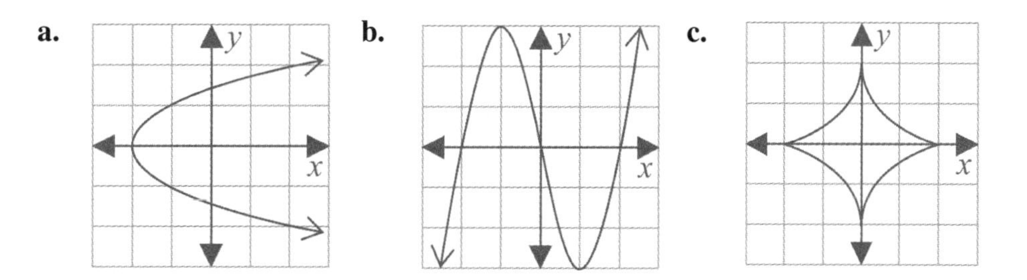

C3–9. For the following function, determine if it is even, odd, or neither.

x	–4	–3	–2	–1	0	1	2	3	4
$f(x)$	0.751	1.2	2.4	6	12	6	2.4	1.2	0.751

C3–10. For each of the following functions, determine if it is even, odd, or neither.

a. $f(x) = 3x^4 - 5x^2 + 1$

b. $f(x) = x\sqrt{x^2 - a^2}$

c. $f(x) = \dfrac{ax^2 + 1}{x^3 + bx}$

C3–11. If $f(x) = 2x - 3$ and $g(x) = x^2 - 9$,

a. Evaluate $(f \circ g)(4)$.

b. Evaluate $(g \circ f)(4)$.

c. Evaluate $(f \circ g)(x)$ and simplify.

d. Evaluate $(g \circ f)(x)$ and simplify.

C3–12. If $y = f(x)$, describe in words the result of each of the following transformations of the graph of $f(x)$.

a. $y = -f(-x)$

b. $y = f(x + 2)$

c. $y = 3f(x)$

C3–13. The graphs of the function $y = f(x)$ and four transformations are shown below. Match the equations with the graphs of the transformations.

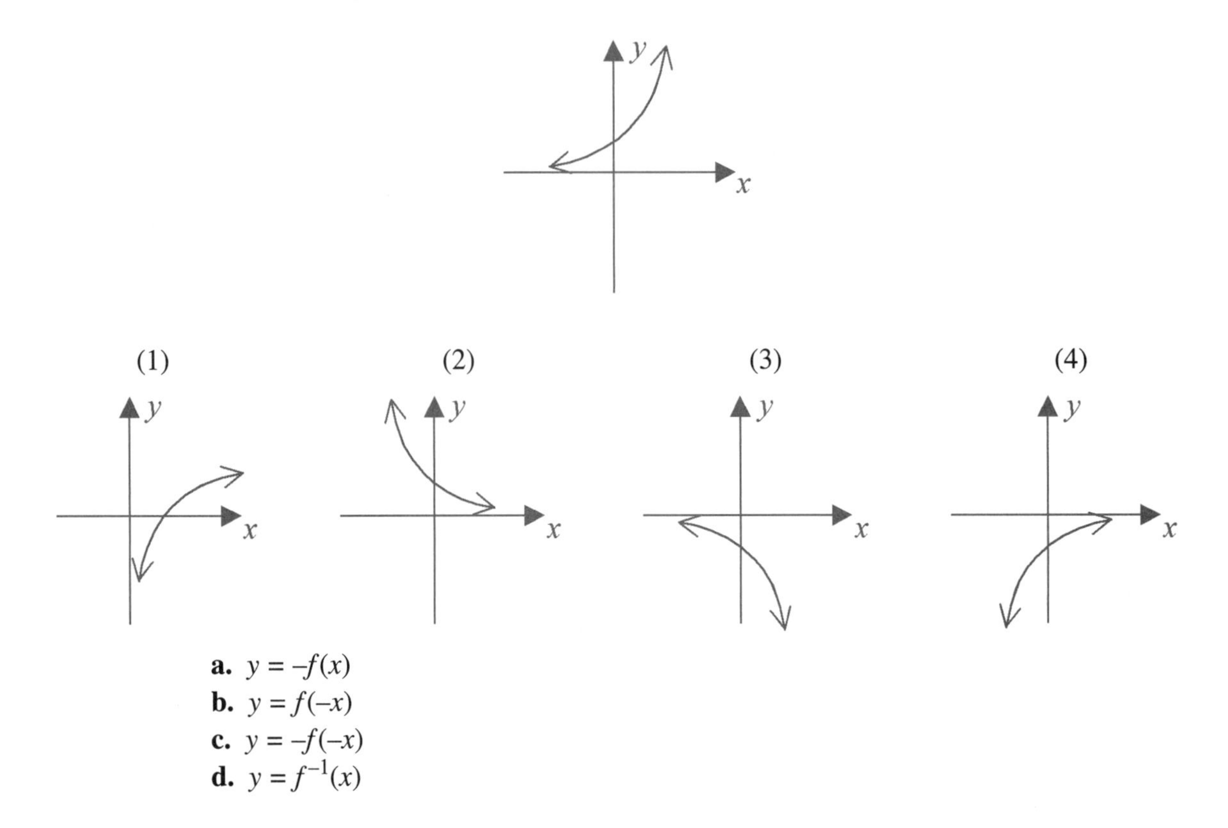

a. $y = -f(x)$
b. $y = f(-x)$
c. $y = -f(-x)$
d. $y = f^{-1}(x)$

C3–14. A function $y = f(x)$ is graphed below. Copy $f(x)$ onto a larger grid and then graph the following transformations. Label each graph.

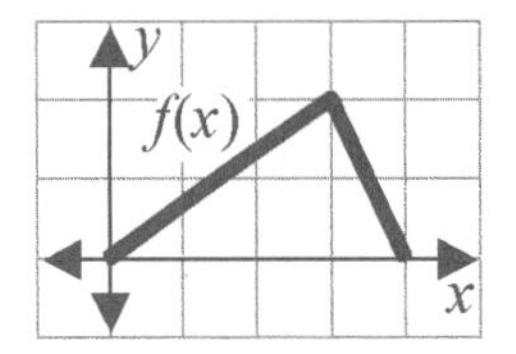

a. $y = -f(-x)$
b. $y = f(x) - 3$
c. $y = f(x - 4)$
d. $y = 3f(x)$
e. $y = f(2x)$

C3–15. Write an equation for the new graph of $y = \dfrac{2x}{x^2 + 1}$ after the following transformations.

a. A reflection over the x-axis.

b. A translation 4 units left and 3 units up.

C3–16. If the point (2, –5) is on the graph of $y = f(x)$, what point will be on the graph of the inverse function?

C3–17. Find the inverse of each function.

a. $y = \dfrac{3}{x + 2}; x \neq -2$

b. $f(x) = \dfrac{2\sqrt{x} - 12}{3}, x \geq 0$

C3–18. Draw the graph of the inverse of the function shown below.

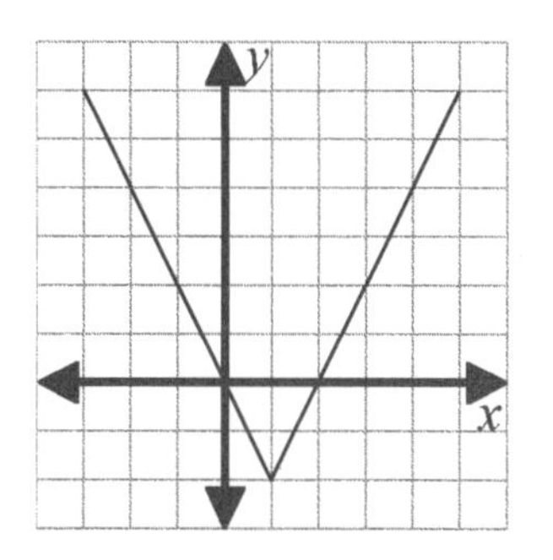

C3–19. A group of students decides to rent a stretch limousine to go to the prom. The cost per person depends on how many people ride and is shown in the table below.

Number of people	1	2	3	4	5	6	7	8
Cost/person ($)	240.00	120.00	80.00	60.00	48.00	40.00	34.50	30.00

a. Is cost per person a function of the number of people? Give a reason to support your answer.

b. Write the domain of the relation (based only on the data in the table).

c. Write the range of the relation.

C3–20. The function $g(x)$ is graphed below.

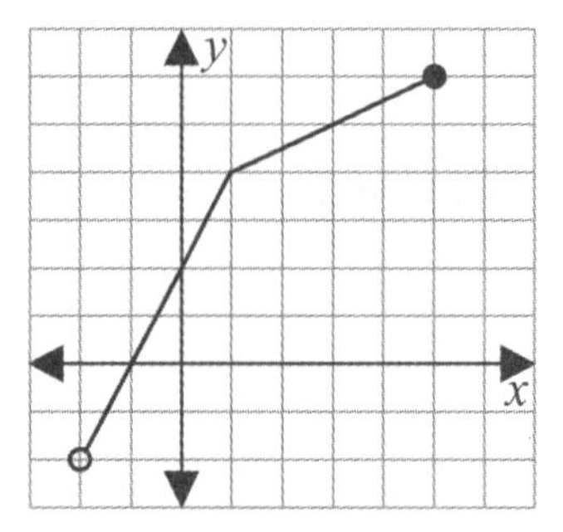

Use the graph to answer the following.

a. Evaluate $g(3)$.

b. Solve $g(x) = 1$.

c. Find the y-intercept of $g(x)$.

d. Find the domain of $g(x)$.

e. Find the range of $g(x)$.

f. Determine if $g(x)$ is one-to-one, and justify your answer.

g. Evaluate $g^{-1}(g(4))$.

Absolute Value Functions

WHAT YOU WILL LEARN

Absolute value functions are a particular type of piecewise defined functions. These are special functions that are made of parts of simpler functions "cut and pasted" together. In this chapter, you will learn to:

- Solve absolute value equations and inequalities, both algebraically and graphically;
- Graph absolute value functions;
- Interpret, write, and graph piecewise defined functions;
- Interpret and graph step functions.

SECTIONS IN THIS CHAPTER

Absolute Value Models

Often in mathematics, knowing the distance between a value of x and a particular number, such as 2, is useful. Since distance is never negative, the smallest possible distance is 0. In our example, this occurs exactly when $x = 2$. As x moves away from 2 in either direction, the distance between x and 2 increases linearly.

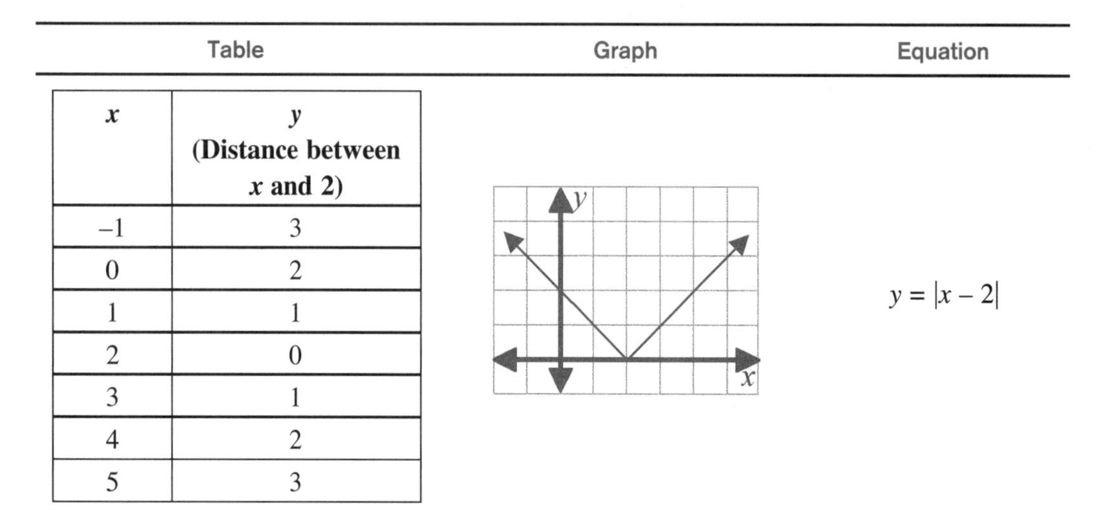

x	y (Distance between x and 2)
–1	3
0	2
1	1
2	0
3	1
4	2
5	3

FIGURE 4.1

Figure 4.1 shows an example of an *absolute value function.* A characteristic feature of an absolute value graph is its V shape. The V can be wide or narrow, right side up or upside down.

The absolute value of a number a is written $|a|$ and is defined as the distance from a to the origin (zero) on a number line. From the number line below, you can see that $|5| = 5$ because the number 5 is five units away from 0 and $|-3| = 3$ because –3 is three units away from 0. Also, $|0| = 0$ because 0 is zero units away from itself.

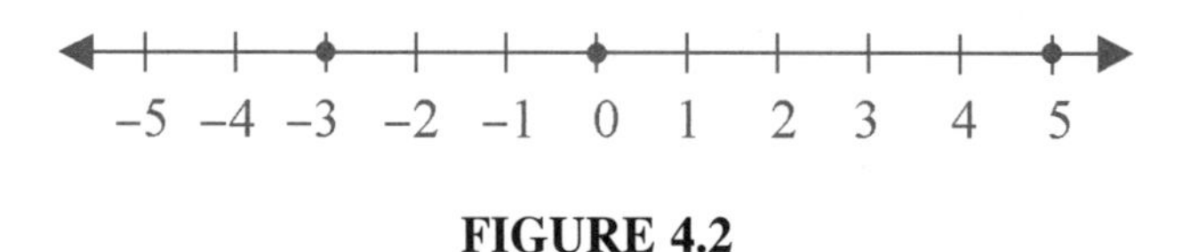

FIGURE 4.2

From the simple examples above, it should be clear that if a is already greater than or equal to 0, the absolute value function will do nothing to it: $|a| = a$ if $a \geq 0$. If a is less than 0, the absolute value function will make a positive by negating it: $|a| = -a$ if $a < 0$. This second part confuses some people at first. Remember, the negative sign in $-a$ does *not* say "a is a negative number." It means the opposite of a or a multiplied by –1. So if a itself is negative, the opposite of a, which is $-a$, will be positive.

Putting this information together provides the algebraic definition of absolute value.

ALGEBRAIC DEFINITION OF ABSOLUTE VALUE

$$|a| = \begin{cases} a & \text{for } a \geq 0 \\ -a & \text{for } a < 0 \end{cases}$$

Many people take a mental shortcut: absolute value means "make it positive." This will often get them through simple problems but can lead to misconceptions. The most common misuse of this shortcut is to conclude $|-a| = a$. This will be true only if a is originally nonnegative. If $a < 0$, then $|-a| = a$ is not true. For example, if $a = -2$, then $|-a| = |-(-2)| = |2| = 2$, which is not the same as a.

Another important thing to remember about absolute value signs is they act as grouping symbols (like parentheses). In general, operations inside absolute value signs should be performed before the absolute value is taken. This is especially important with addition and subtraction. The examples below illustrate this topic.

EXAMPLE 4.1

a. Evaluate $|3 + (-5)|$
b. Evaluate $|-6 - 4|$

SOLUTION

a. $|3 + (-5)| = |-2| = 2$
b. $|-6 - 4| = |-10| = 10$

COMMON ERROR

Do *not* distribute absolute value over addition or subtraction.

$$|3 + (-5)| \neq |3| + |-5| = 3 + 5 = 8$$
$$|-6 - 4| \neq |-6| - |4| = 6 - 4 = 2$$

COMMON ERROR

Another error, related to the "make it positive" shortcut, is believing that absolute value changes subtraction to addition, as in $|7 - 5| = 7 + 5$. This is *wrong*; do not fall into this trap.

Some important properties of absolute value are given below.

ABSOLUTE VALUE PROPERTIES

1. $|a| \ge 0$
2. $|-a| = |a|$
3. $|ab| = |a| \bullet |b|$
4. $\left|\frac{a}{b}\right| = \frac{|a|}{|b|}$

$|a + b| \neq |a| + |b|$

$|a - b| \neq |a| - |b|$

Solving Absolute Value Equations Algebraically

An *absolute value equation* in one variable is an equation that contains a variable expression in an absolute value symbol. A common absolute value equation is of the form $|ax + b| = c$, where a, b, and c are integers. The equation could be more complicated. For example, there could be a quadratic or other nonlinear expression in the absolute value symbol. There could be variables outside the absolute value symbol. There could even be more than one absolute value expression. Although the algebraic solution method shown here will work on all these equations, you may want to consider a graphical solution (shown later) for more complicated ones.

To solve an absolute value equation with one absolute value expression algebraically, remember that the absolute value will do one of two things. If the expression inside the absolute value symbols is already positive or zero, the absolute value will do nothing and can be ignored. If the expression inside the absolute value symbols is negative, the absolute value will negate the expression (multiply it by –1), making it positive. We can use this information to rewrite an absolute value equation as two separate equations without absolute value symbols. Then we can solve each and check the answers in the original equation.

EXAMPLE 4.2

Solve for x: $|3x - 4| = 20$

SOLUTION

Rewrite as two non-absolute value equations, and solve each equation. The equation on the left assumes the expression is already positive and the absolute value bars are not needed. The equation on the right assumes the expression is negative and the absolute value bars have the effect of multiplying the expression by –1.

If $3x - 4 \geq 0$, then	If $3x - 4 < 0$, then
$3x - 4 = 20$	$-(3x - 4) = 20$
$3x = 24$	$-3x + 4 = 20$
$x = 8$	$-3x = 16$
	$x = -\dfrac{16}{3}$

Check the candidate solutions in the original equation.

$$|3(8) - 4| = 20 \qquad \left|3\left(-\frac{16}{3}\right) - 4\right| = 20$$

$$|20| = 20 \checkmark \qquad |-20| = 20 \checkmark$$

Both candidates check, so the solutions are $x = 8$ and $x = -\dfrac{16}{3}$.

It is important to check the candidate answers in the original equation because sometimes rewriting the equation as two separate equations introduces extraneous (wrong) answers. These extraneous answers occur when the candidate solution contradicts the assumption. Consider what happens if we change Example 4.2 to $|3x - 4| = -20$. The solution process is very similar:

If $3x - 4 \geq 0$, then	If $3x - 4 < 0$, then
$3x - 4 = -20$	$-(3x - 4) = -20$
$3x = -16$	$-3x + 4 = -20$
$x = -\dfrac{16}{3}$	$-3x = -24$
	$x = 8$

Neither of these solutions checks in the equation $|3x - 4| = -20$. This equation has no solution. (This should not have been a surprise since an absolute value can never be negative.) Observe that the candidate solution $x = -\dfrac{16}{3}$ was found after assuming $3x - 4 \geq 0$. However, if $x = -\dfrac{16}{3}$, this assumption is false; $3\left(-\dfrac{16}{3}\right) - 4 = -20 \not\geq 0$.

Since the candidate solution contradicts the assumption it came from, the candidate does not solve the original equation. Similarly, the candidate solution $x = 8$ came from assuming $3x - 4 < 0$, which is not true when $x = 8$.

Most people do not write the assumptions "if expression ≥ 0" and "if expression < 0" when rewriting the absolute value equation as two separate equations. (Neither will we in the rest of the examples.) Not writing the assumption makes it absolutely necessary that you check candidate answers in the original equation. If you don't check candidate answers, you could accept incorrect solutions.

Solve for x: $|2x - 1| - 2 = 3x$

SOLUTION

Rewrite as two non-absolute value equations, and solve each equation.

$$2x - 1 - 2 = 3x$$
$$2x - 3 = 3x$$
$$x = -3$$

Check:

$$|2(-3) - 1| - 2 = 3(-3)$$
$$|-7| - 2 = -9$$
$$5 = -9 \text{ No.}$$

$$-(2x - 1) - 2 = 3x$$
$$-2x + 1 - 2 = 3x$$
$$-1 = 5x$$
$$x = -\frac{1}{5}$$

Check:

$$\left|2\left(-\frac{1}{5}\right) - 1\right| - 2 = 3\left(-\frac{1}{5}\right)$$
$$\left|-\frac{7}{5}\right| - 2 = -\frac{3}{5}$$
$$-\frac{3}{5} = -\frac{3}{5} \checkmark$$

Note: Only the terms inside the absolute value signs are negated, not the –2 outside.

Read the check on the left carefully. When $x = -3$, the expression inside the absolute value symbols equals –7. However, that candidate solution came from assuming that the expression was positive so that the absolute value signs could be ignored. This contradiction results in an extraneous solution on that side.

The only solution is $x = -\frac{1}{5}$.

An equation that contains two absolute value expressions can be solved the same way except that you may need up to four separate equations instead of two.

Solve for x: $|3x + 5| + |x - 3| = 12$

SOLUTION

Rewrite as non-absolute value equations, and solve each equation. Because there are two absolute value expressions, your solution must contain all four possible combinations of negative and positive expressions. For clarity, we have shown the different assumptions made but most people usually do not.

	If $3x + 5 \geq 0$, then	If $3x + 5 < 0$, then
If $x - 3 \geq 0$, then	$3x + 5 + x - 3 = 12$ $x = 2.5$	$-(3x + 5) + x - 3 = 12$ $x = -10$
If $x - 3 < 0$, then	$3x + 5 + -(x - 3) = 12$ $x = 2$	$-(3x + 5) + -(x - 3) = 12$ $x = -3.5$

Now check all four candidates in the original equation. You can verify that two of them ($x = 2.5$ and $x = -10$) do not check. The solution is $x = -3.5$ or $x = 2$.

Because of the work involved in solving four separate equations, the example above might be more easily solved graphically.

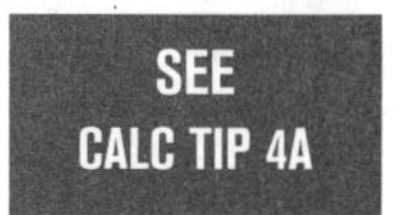

SECTION EXERCISES

For 1–6, solve the following equations algebraically.

4–1. $|2x + 5| = 13$

4–2. $|x - 3| = -3$

4–3. $|3x - 7| = 5$

4–4. $|2x - 9| - 4x = 3$

4–5. $|x - 3| - 5 = 2x$

4–6. $|1 + 3x| - x = x - 1$

Solving Absolute Value Inequalities Algebraically

Several different algebraic methods can be used to solve an absolute value inequality. The method shown below has the advantage that it can be used for other kinds of inequalities, such as quadratic inequalities (see Chapter 5).

To solve an absolute value inequality, follow these steps:

1. First solve for the corresponding equation.
2. Place the solutions on a number line. Use open circles for strict inequalities ($<$ or $>$) and closed circles for inclusive inequalities ($\leq$ or $\geq$).
3. Check a value from each interval to determine if the inequality is true on that interval.
4. Write the solution in the appropriate notation.

EXAMPLE 4.5 Solve for x: $|2x - 3| < 7$

SOLUTION

Solve for the corresponding equation. (Write and solve two equations that represent the absolute value relationship.)

$2x - 3 = 7$	$-(2x - 3) = 7$
$2x = 10$	$-2x + 3 = 7$
$x = 5$	$-2x = 4$
	$x = -2$

Both check.

Place the solutions on a number line. Use open circles at each solution because < is a strict inequality. Then check a number from each interval in the original inequality.

$-2 \qquad 5$

Check $x = -3$	Check $x = 0$	Check $x = 6$
$2(-3) - 3| < 7$	$|2(0) - 3| < 7$	$|2(6) - 3| < 7$
$9 < 7$ No	$3 < 7$ Yes	$9 < 7$ No

Shade the appropriate intervals on the number line to show the solution. Then write the solution.

The solution is $-2 < x < 5$ or, in interval notation, $(-2, 5)$.

EXAMPLE 4.6 Solve for x: $|3x - 8| \geq x + 4$

SOLUTION

Solve for the corresponding equation. (Write and solve two equations that represent the absolute value relationship.)

$3x - 8 = x + 4$	$-(3x - 8) = x + 4$
$x = 6$	$x = 1$

Both check.

Use closed circles for ≥ on the number line.

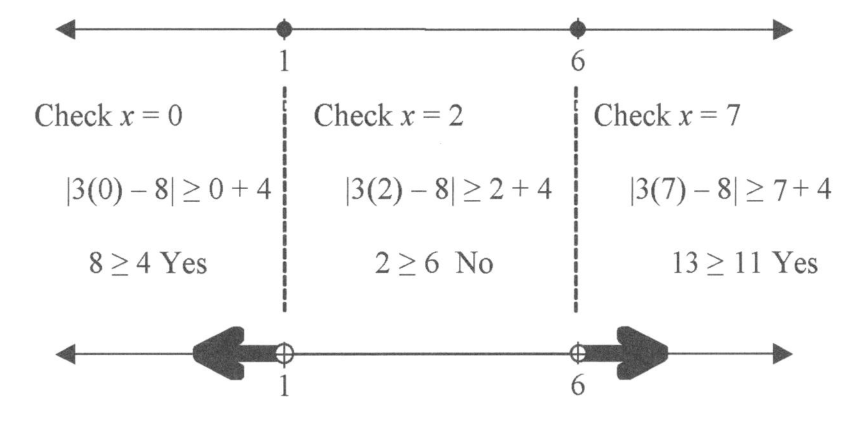

The solution is $x \leq 1$ or $x \geq 6$, which can also be written $(-\infty, 1] \cup [6, \infty)$.

Absolute value can be used to model simple intervals. Some tropical fish do best if their water temperature is 78°F but can tolerate temperatures within 3°F of that. The absolute value inequality $|T - 78| \leq 3$ represents all the temperatures, T, that these fish will tolerate.

Simple absolute value inequalities of the form $|x - m| < d$ or $|x - m| > d$ can be easily sketched on a number line as shown below. With an absolute value inequality, the order of subtraction does not matter. So the same graphs may be used for inequalities involving $|m - x|$. Inequalities involving $|x + m|$ should be rewritten $|x - (-m)|$. In the inequalities below, m is the middle of an interval and d represents the distance between m and the endpoints of the interval. Note that although m may be any number, d must be positive. If the inequalities are inclusive (≥ or ≤), the endpoints should be solid instead of open.

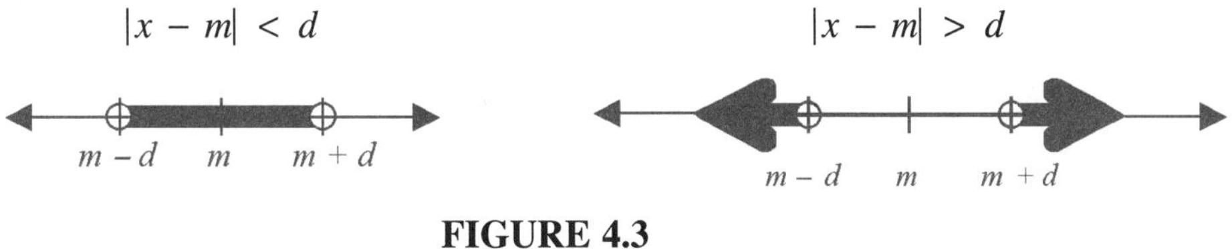

FIGURE 4.3

To graph the temperatures that the tropical fish can tolerate, we note that 78°F is the middle of the interval and the fish can stand temperatures up to 3°F either side of that.

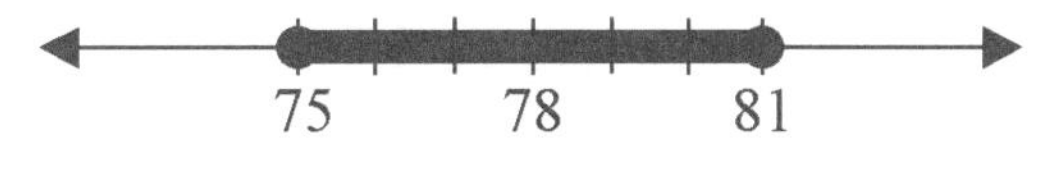

FIGURE 4.4

SECTION EXERCISES

For 7–12, solve the following inequalities algebraically.

4–7. $|3 + x| \leq 5$

4–8. $|x| < -2$

4–9. $|3 - x| > 4$

4–10. $|3x + 1| + x \leq 5$

4–11. $|2x - 11| > x + 7$

4–12. $\left|\frac{1}{2}x - 3\right| < x$

4–13. Sue's Sweet Shoppe keeps a big jar of candy at the door. Any customer who can guess the true number of candies in the jar to within 10 wins the whole jar. If 678 candies are in the jar, which inequality describes the winning guesses G?
(1) $|G - 10| \leq 678$
(2) $|G - 678| \leq 10$
(3) $G \leq 678 \pm 10$
(4) $|G| \leq 678 \pm 10$

4–14. The environmental system on the starship *Enterprise* is supposed to keep the temperature within 2°C of 20°C at all times. If the temperature is represented by T, write a simple absolute value inequality to represent the requirement stated above.

Graphing Absolute Value Functions

The simplest absolute value function is $y = |x|$. Its graph is shown below and looks like a V with its vertex at the origin. More precisely, it is the union of two half lines (or rays).

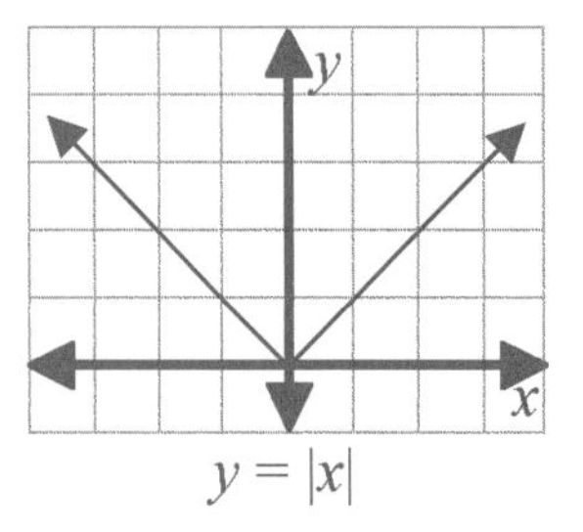

FIGURE 4.5

This makes sense if you remember the algebraic definition of absolute value.

$$|x| = \begin{cases} x & \text{for } x \geq 0 \\ -x & \text{for } x < 0 \end{cases}$$

This means that when $x \geq 0$, $y = |x|$ is the same as $y = x$, a line of slope 1 through the origin. When $x < 0$, $y = |x|$ is the same as $y = -x$, a line of slope -1 through the origin.

The transformations you learned in Chapter 3 can all be applied to the absolute value function.

EXAMPLE 4.7

Sketch the graph of $y = -|x| + 2$.

SOLUTION

The graph of $y = -f(x)$ is the reflection of the graph of $y = f(x)$ over the x-axis (see the section on transformations in Chapter 3). The graph of $y = f(x) + k$ is the translation of the graph of $y = f(x)$ up k units. These mean that $y = -|x| + 2$ will reflect the graph of $y = |x|$ over the x-axis and then translate it up 2 units. (Make sure you do it in the right order. If the translation were first and then the reflection, the equation would be written $y = -(|x| + 2)$ and would not give the same graph.)

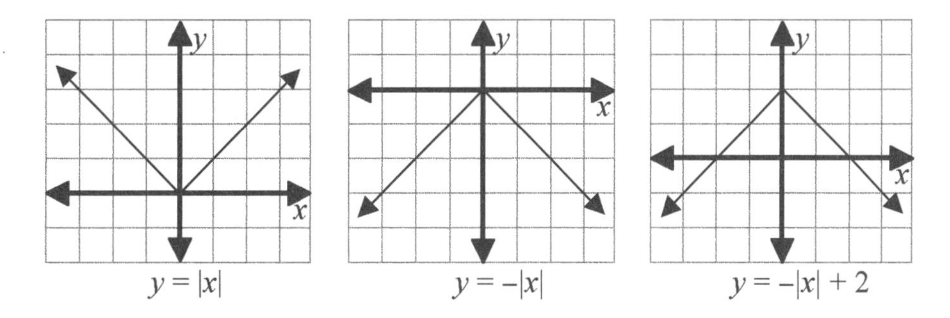

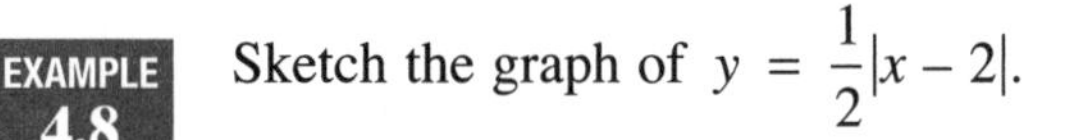

EXAMPLE 4.8

Sketch the graph of $y = \frac{1}{2}|x - 2|$.

SOLUTION

The graph of $y = af(x)$ is a vertical dilation of the graph of $y = f(x)$ by a factor of a. The graph of $y = f(x - h)$ is the translation of the graph of $y = f(x)$ right h units. This means $y = \frac{1}{2}|x - 2|$ will combine a vertical dilation (or shrink) of the graph of $y = |x|$ by a factor of $\frac{1}{2}$ and a translation right 2 units. (Remember, horizontal translations go in the opposite direction from the apparent sign in the equation. $f(x - 2)$ translates to the right; $f(x + 2)$ translates to the left.) The vertical dilation is done before the translation. In this example, though, either order of the transformations gives the same graph.

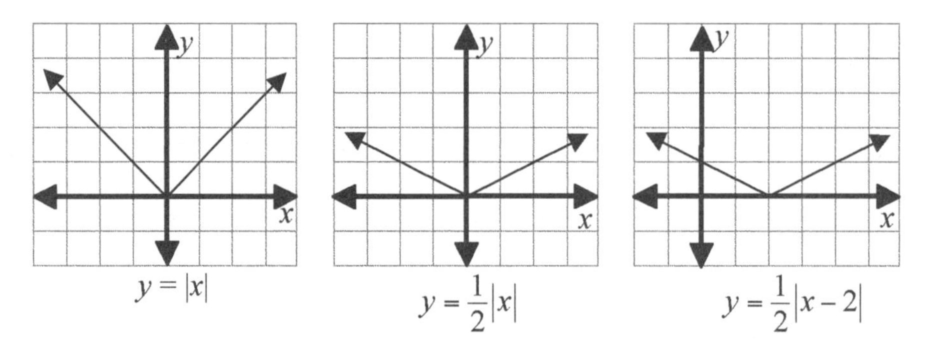

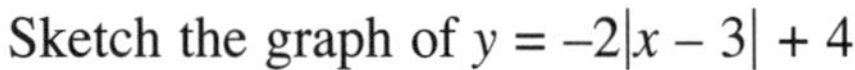
Sketch the graph of $y = -2|x - 3| + 4$

SOLUTION

Compared with the graph of $y = |x|$, the graph of $y = -2|x - 3| + 4$ will be upside down (reflected over the x-axis), vertically dilated by a factor of 2, and translated right 3 and up 4 so that its vertex is at (3, 4). These transformations are shown below.

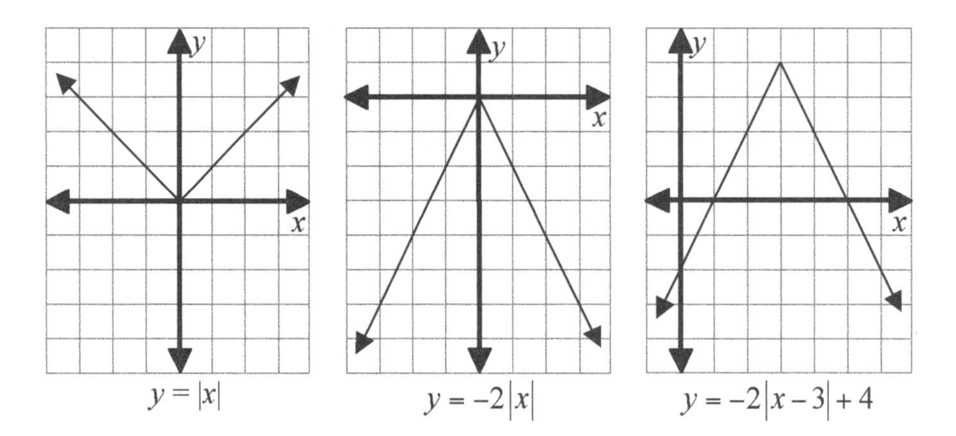

Be careful when using transformations to graph something like $y = |-3x + 6|$. Some students are tempted to say the graph will be reflected over the y-axis, horizontally dilated by a factor of $\frac{1}{3}$, and translated 6 units to the left. It is not. Although the equation can be graphed with these transformations, the order of the transformations is tricky and not intuitive. It is simpler to apply transformations if the coefficient of x is factored out before applying the transformation rules.

Sketch the graph of $y = |-3x + 6|$.

SOLUTION

Factor out the coefficient of x, and simplify using absolute value properties.

$$\begin{aligned} y &= |-3x + 6| \\ &= |-3(x - 2)| \\ &= |-3||x - 2| \\ &= 3|x - 2| \end{aligned}$$

Now we can easily see that the graph is vertically stretched by a factor of 3 and translated 2 units to the right.

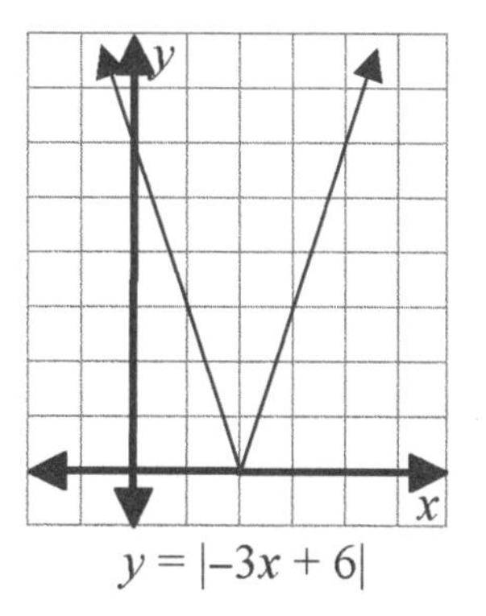

$y = |-3x + 6|$

SECTION EXERCISES

For 15–18, graph the following.

4–15. $y = |x + 1| - 2$

4–16. $y = -2|x| + 3$

4–17. $y = 2|x - 3| + 1$

4–18. $y = |0.5x + 2| - 2$

Solving Absolute Value Equations and Inequalities Graphically

Graphing is a powerful and important method for solving equations and inequalities. Not all equations can be solved algebraically. With today's graphing calculators or computer graphing programs, though, most equations can be solved (at least to extremely close approximations) quickly and easily using graphical methods. Graphs help us see the solutions more clearly. Graphs also do not give extraneous zeros as sometimes happens when absolute value equations are solved algebraically.

To solve an equation of the form $f(x) = g(x)$ graphically:

1. Graph $y = f(x)$ and $y = g(x)$ on the same set of axes.
2. Find the x-values of the point(s) where the graphs intersect.

Solve graphically: $|2x - 4| = x + 1$

SOLUTION

Graph both $y = |2x - 4|$ and $y = x + 1$. You should be able to graph the line $y = x + 1$ quickly using the slope–intercept method. For the absolute value function, you can use the transformations covered in the preceding section. (If you do, remember to factor out the 2 first to get $y = 2|x - 2|$.) Instead you can make a table of values, either by hand or with a calculator. The graphs are shown below.

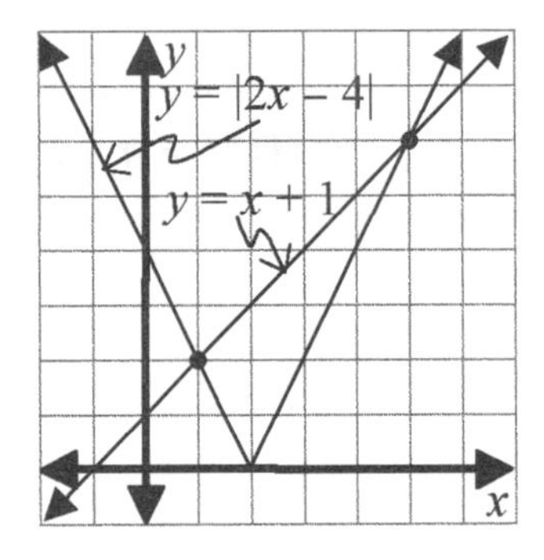

The solutions are the x-values of the points where the two graphs intersect, $x = 1$ and $x = 5$.

The solution set is $\{1, 5\}$.

Note that the original equation involves only the variable x. We introduced y into the problem so we could graph it on coordinate axes. However, the solutions to the problem are just x-values, not ordered pairs.

To solve an inequality graphically, we need to know how to interpret the inequality symbols on a graph.

TABLE 4.1
INEQUALITY SYMBOLS ON A GRAPH

Symbol	Meaning
$<$	Below
$\leq$	On or below
$>$	Above
$\geq$	On or above

To solve an inequality of the form $f(x) < g(x)$ graphically:

1. Graph $y = f(x)$ and $y = g(x)$ on the same set of axes.
2. Find all the x-values where the graph of $f(x)$ is *below* the graph of $g(x)$. (For inequalities other than $<$, substitute the appropriate word(s) from Table 4.1.) These x-values will usually be in one or more intervals created by the points of intersection of the graphs.

EXAMPLE 4.12 Solve graphically: $|2x + 3| \leq 7$

SOLUTION

Graph both $y = |2x + 3|$ and $y = 7$. We want the x-values where the graph of $y = |2x + 3|$ is *on or below* the graph of $y = 7$. The two graphs intersect at $x = -5$ and $x = 2$. Because the problem involves less than *or equal*, these points will be included (solid circles). The absolute value graph is *below* (less than) the horizontal line for all x-values between $x = -5$ and $x = 2$. This portion of the graph is bold on the diagram.

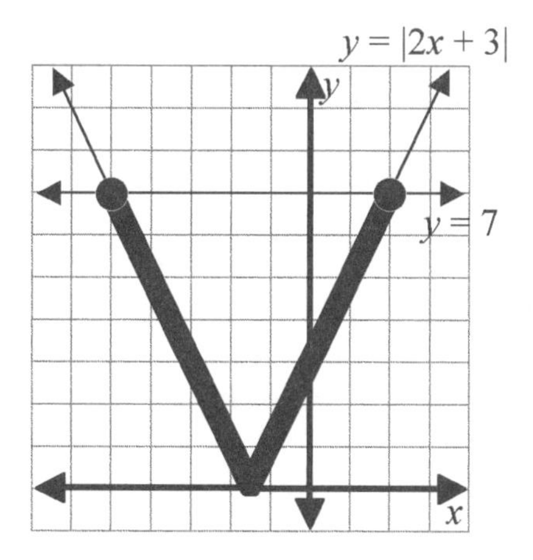

The solution set consists of those x-values where the absolute value graph is on or below the horizontal line, the part that is bold on the graph. The solution is $-5 \leq x \leq 2$.

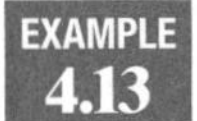

Solve graphically: $|3x - 1| > 7 - x$

SOLUTION

In this problem, we want to find the x-values where the absolute value function (the V graph) is *above* the straight line. The two graphs intersect at $x = -3$ and $x = 2$. Because of the strict inequality, these points will not be included in the solution, so they have been indicated with open circles. The absolute value function is above the line to the left of $x = -3$ and to the right of $x = 2$.

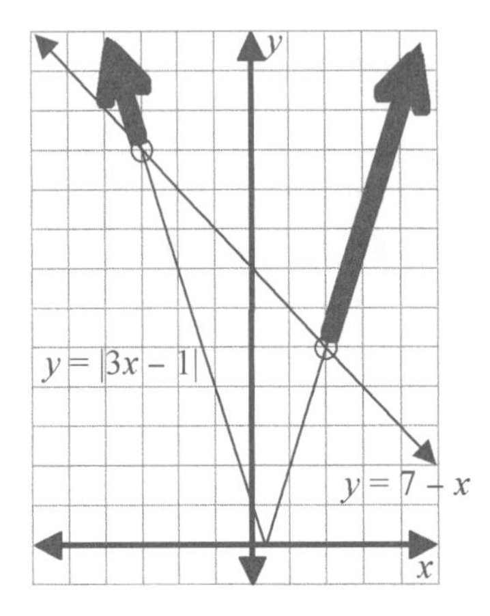

The solution is $x < -3$ or $x > 2$.

SECTION EXERCISES

For 19–22, solve the following graphically.

4–19. $|x + 3| = 4$

4–20. $|2x - 3| = |x - 2| + 3$

4–21. $|2 + 3x| > 4$

4–22. $|4x - 3| \le 9$

Piecewise Defined Functions

The absolute value function $y = |x|$ is a simple example of a *piecewise defined function*. A piecewise defined function is a function whose definition, or equation, is different on different intervals of x. In the case of the absolute value function, we have:

$$y = |x| = \begin{cases} x & \text{for } x \geq 0 \\ -x & \text{for } x < 0 \end{cases}$$

This means that when $x \geq 0$, $y = x$. For values of $x < 0$, though, $y = -x$. The equation is different for negative and positive values of x. Both parts of the function are lines, but they are not the same line.

Piecewise functions can have any number of pieces. The different pieces do not all have to be lines. Consecutive pieces do not even have to intersect at their endpoints. The only thing the pieces cannot do is overlap each other.

EVALUATING PIECEWISE DEFINED FUNCTIONS

EXAMPLE 4.14

Consider the following function: $y = \begin{cases} 5x - 1 & \text{if } x < 2 \\ 3x + 1 & \text{if } x \geq 2 \end{cases}$

a. Find y when $x = -3$.
b. Find y when $x = 4$.
c. Find y when $x = 2$.

SOLUTION

a. $x = -3$ is in the interval $x < 2$, so use $y = 5x - 1$ to find y.
$y = 5(-3) - 1 = -16$

b. $x = 4$ is in the interval $x \geq 2$, so use $y = 3x + 1$ to find y.
$y = 3(4) + 1 = 13$

c. $x = 2$ is in the interval $x \geq 2$, so use $y = 3x + 1$ to find y.
$y = 3(2) + 1 = 7$

GRAPHING PIECEWISE DEFINED FUNCTIONS

EXAMPLE 4.15

Graph the function: $y = \begin{cases} 3 & \text{if } x < 1 \\ -2 & \text{if } x \geq 1 \end{cases}$

SOLUTION

To graph this function using tables, construct two tables, one for each part of the function. The point $x = 1$, the x-value where the equations change, should be in both tables. Sometimes students mistakenly believe that $x < 1$ means that the function stops at $x = 0$. However, $x < 1$ indicates that the graph continues until $x = 1$ but is open at

$x = 1$. Include any endpoints of intervals in your tables, but indicate if the graph is open (○) or closed (●) at each endpoint.

x	$y = 3$
–2	3
–1	3
0	3
1	3 ○

x	$y = -2$
1	–2 ●
2	–2
3	–2
4	–2

Use these tables to sketch the graph.

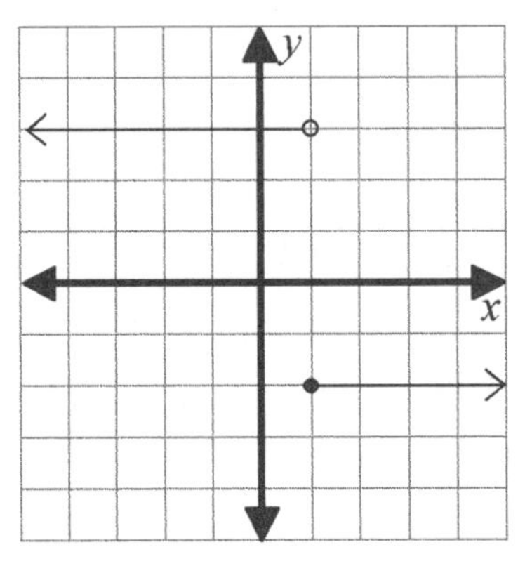

Graph the function: $y = \begin{cases} -x & \text{if } x \leq -1 \\ 2x & \text{if } x > -1 \end{cases}$

SOLUTION

Make a set of tables for each interval of x in the function. Include the endpoint in your table, and note if the endpoint is open or closed.

x	$y = -x$
–4	4
–3	3
–2	2
–1	1 ●

x	$y = 2x$
–1	–2 ○
0	0
1	2
2	4

Use these tables to sketch the graph.

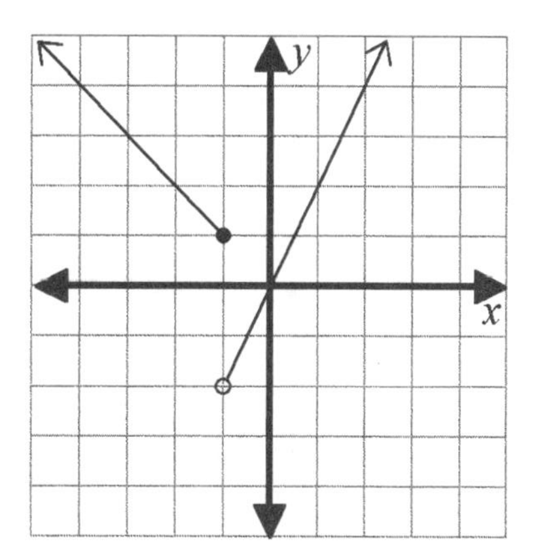

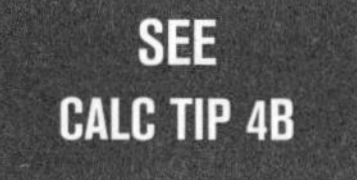

EXAMPLE 4.17

Graph the function: $y = \begin{cases} -x + 2 & \text{if } x < 0 \\ 2 & \text{if } 0 \leq x < 3 \\ 2x - 4 & \text{if } x \geq 3 \end{cases}$

SOLUTION

This function has three pieces.

1. When $x < 0$, $y = -x + 2$. This is a part of a line with slope -1 and y-intercept 2. This part of the graph extends indefinitely to the left but stops at $x = 0$ (the y-axis), where the function changes definition.
2. For $0 \leq x < 3$, $y = 2$. This part of the graph is a horizontal line segment from (0, 2) to (3, 2).
3. For $x \geq 3$, $y = 2x - 4$. This is part of another line, beginning at (3, 2) and extending indefinitely up and to the right with a slope of 2.

Here are the tables.

x	$y = -x + 2$
–3	5
–2	4
–1	3
0	2 ○

x	$y = 2$
0	2 ●
1	2
2	2
3	2 ○

x	$y = 2x - 4$
3	2 ●
4	4
5	6
6	8

If open and closed pieces meet at the same point, the graph is continuously drawn without a large open or closed circle.

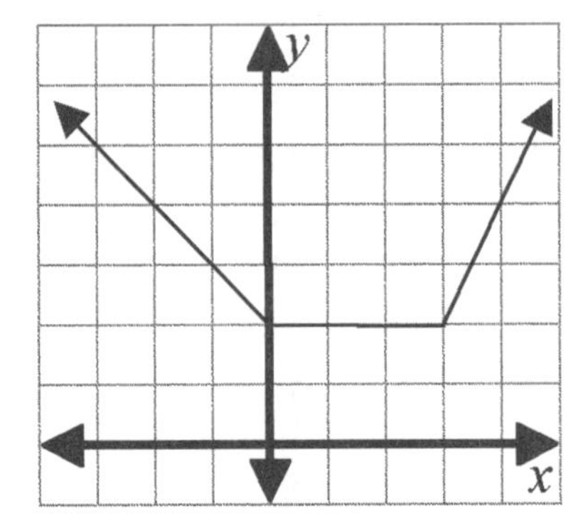

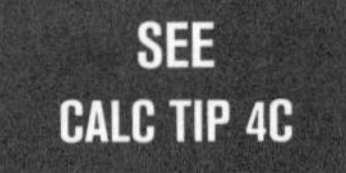

Piecewise functions have many real-life applications. A common example is pricing for items bought in quantity. A company might charge \$25 apiece for up to 10 items but only \$23 apiece if you order more than 10. Another example is income tax. Many state and the federal governments have different tax rates depending on how much income you earn.

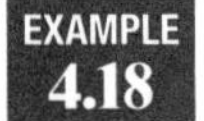

Suppose a state taxes income as follows: a rate of 5% on income up to \$20,000, a rate of 10% on income over \$20,000 and up to \$50,000, and finally a top rate of 20% on income over \$50,000. (Note: these numbers are for example purposes only. As of 2012, the highest state tax rate on personal income was 11%.)

Write the marginal tax rate, r, in percent, as a piecewise function of income, x, in dollars and graph it.

SOLUTION

According to the problem, the tax rate is 5% for $x \leq \$20{,}000$, 10% for $\$20{,}000 < x \leq \$50{,}000$, and 20% for $x > \$50{,}000$. This can be written as a piecewise function:

$$r = \begin{cases} 5\% & \text{if } 0 \leq x \leq \$20{,}000 \\ 10\% & \text{if } \$20{,}000 < x \leq \$50{,}000 \\ 20\% & \text{if } x > \$50{,}000 \end{cases}$$

The graph of this function is shown on the next page. Note that consecutive pieces of the function do not meet at their endpoints; the function has *discontinuities* (breaks in its graph) at $x = \$20{,}000$ and $x = \$50{,}000$. Note also the open and closed circles to show what the tax rate is at each of those discontinuities.

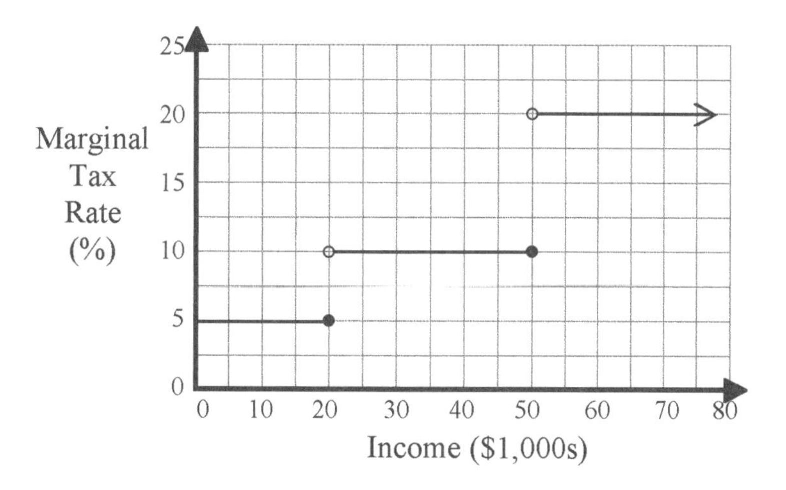

The previous example showed just tax rates. The actual dollar amount of taxes owed, T, can be expressed using the point-slope form of the equation of a line.

$$T = \begin{cases} 0.05x & \text{if } 0 \le x \le \$20{,}000 \\ \$1{,}000 + 0.10(x - \$20{,}000) & \text{if } \$20{,}000 < x \le \$50{,}000 \\ \$4{,}000 + 0.20(x - \$50{,}000) & \text{if } x > \$50{,}000 \end{cases}$$

Sketch the graph of actual taxes owed, T.

SOLUTION

This graph consists of two line segments and a ray. The first segment starts at the origin and has a slope of 0.05, ending at x = $20,000. The next segment starts at ($20,000, $1,000) and has a slope of 0.1, ending at x = $50,000. Finally, the ray starts at ($50,000, $4,000) and has a slope of 0.2.

Here are the tables with x and T expressed in 1,000's of dollars.

x	$T = 0.05x$
0	0
10	0.5
20	1 ●

x	$T = 1{,}000 + 0.10(x - 20{,}000)$
20	1 ○
30	2
40	3
50	4 ●

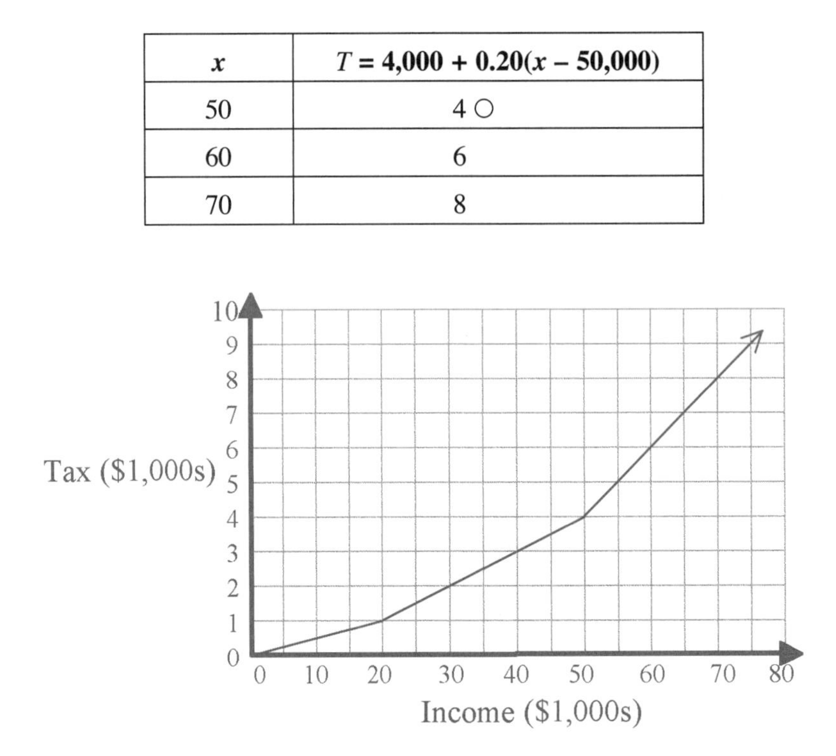

x	$T = 4{,}000 + 0.20(x - 50{,}000)$
50	4 O
60	6
70	8

SECTION EXERCISES

4–23. $y = \begin{cases} -x + 4 & \text{if } x < 3 \\ 2x - 4 & \text{if } x \geq 3 \end{cases}$

a. Find y when $x = 0$.
b. Find y when $x = 5$.
c. Find y when $x = 3$.

4–24. Graph the function: $y = \begin{cases} 1 & \text{if } x \leq 2 \\ x - 1 & \text{if } x > 2 \end{cases}$

4–25. Graph the function: $y = \begin{cases} 0 & \text{if } x < 0 \\ 2x & \text{if } 0 \leq x < 3 \\ 6 & \text{if } x \geq 3 \end{cases}$

4–26. A solid-waste disposal facility charges $15.00 to dispose of up to 300 pounds of waste and then $0.05 per pound for each additional pound.

a. Write this as a piecewise function
b. Sketch the graph.

Step Functions

A *step function* is a piecewise function where each piece is a constant. Its graph is a series of horizontal line segments. It often looks like a series of steps (or a staircase), which is how the function gets its name.

EXAMPLE 4.20 A parking lot charges customers \$2.00 for the first hour or any part of an hour and then \$0.50 for each additional hour with a \$5.00 maximum for the day. No one may park more than 12 hours. Graph the parking fee as a function of the number of hours parked.

SOLUTION

This parking lot charges \$2.00 to park for an hour. Even if a customer stays only 10 minutes, the cost is \$2.00. If the customer stays past that first hour, even just a minute, the cost increases by \$0.50. The same is true if the customer stays past the second hour and so on up to a maximum of \$5.00. The graph is shown below.

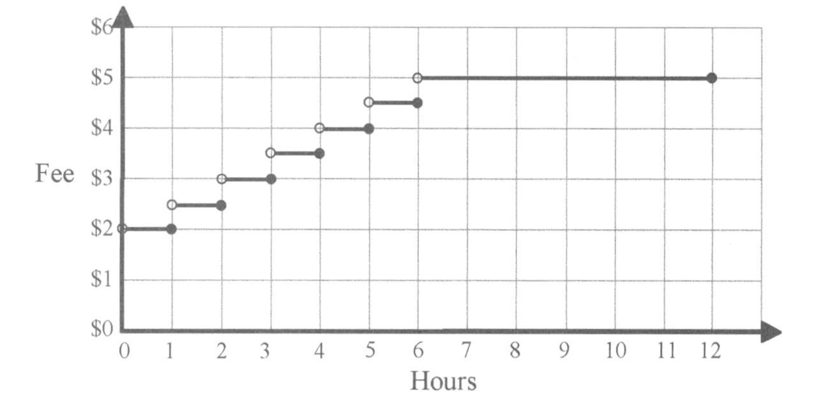

Two special step functions are the *floor function* (sometimes called the *greatest integer function*) and the *ceiling function*. The floor function is usually written as $y = \lfloor x \rfloor$ and represents the greatest integer less than or equal to x. The ceiling function is usually written as $y = \lceil x \rceil$ and means the smallest integer greater than or equal to x. (Note: Some people use $[x]$ or $[\![x]\!]$ to represent the floor function and $]x[$ or $]\!]x[\![$ to represent the ceiling function.)

TABLE 4.2
EXAMPLES OF FLOOR AND CEILING FUNCTIONS

Function	Value	Reason
$\lfloor 3.7 \rfloor$	3	3 is the greatest integer less than or equal to 3.7
$\lfloor 4 \rfloor$	4	4 is the greatest integer less than or equal to 4.
$\lceil 3.7 \rceil$	4	4 is the smallest integer greater than or equal to 3.7
$\lceil 3.7 \rceil$	4	4 is the smallest integer greater than or equal to 4

Evaluate the following

a. $\lfloor 6.5 \rfloor$

b. $\lceil 2.3 \rceil$

c. $\lfloor -2.1 \rfloor$

d. $\lceil -5.6 \rceil$

SOLUTION

a. $\lfloor 6.5 \rfloor = 6$
The floor function goes down to the next integer.

b. $\lceil 2.3 \rceil = 3$
The ceiling function goes up to the next integer.

c. $\lfloor -2.1 \rfloor = -3$
The floor function goes down to the next integer.

d. $\lceil -5.6 \rceil = -5$
The ceiling function goes up to the next integer.

GRAPHS OF FLOOR AND CEILING FUNCTIONS

The graphs of the floor and ceiling functions look similar. However, they are actually the same only at integer values of x, as shown in Figures 4.6 and 4.7.

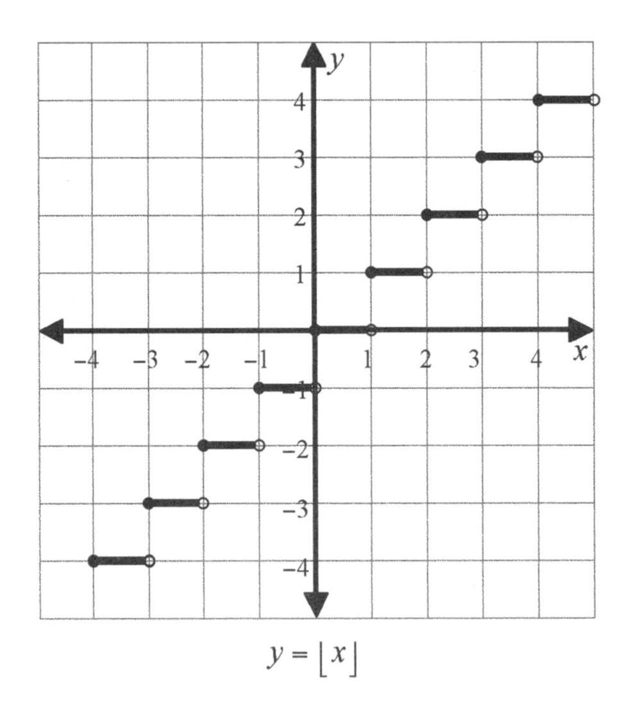

FIGURE 4.6

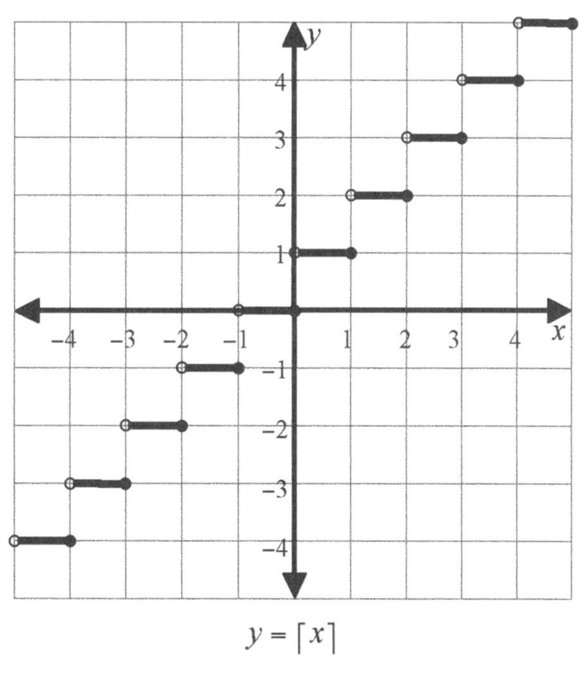

$y = \lceil x \rceil$

FIGURE 4.7

SECTION EXERCISES

For 27–28, sketch the graph over the interval $-2 \le x \le 2$.

4–27. $y = 2\lceil x \rceil$

4–28. $f(x) = -2\lfloor x \rfloor$

CHAPTER EXERCISES

For 1–2, solve the following equations algebraically.

C4–1. $\frac{2}{3}|4 - x| = 6$

C4–2. $|5 - 2x| = 3x - 20$

For 3–4, solve the following inequalities algebraically.

C4–3. $|2x + 5| \geq 7$

C4–4. $\left|\frac{1}{2}x - 3\right| < 2$

C4–5. A poll is taken to determine potential support for Agatha Bluenose in a run for president. Which of the following inequalities shows that the proportion P of people who support Agatha is within 3% of 42%?

(1) $|P - 3| \leq 42$
(2) $|P - 3| \geq 42$
(3) $|P - 42| \leq 3$
(4) $|P - 42| \geq 3$

C4–6. ACME widgets are supposed to be 15 cm in diameter. Widgets are rejected if their diameter differs from 15 cm by more than 0.25 cm. Write a simple absolute value inequality to represent the rejected diameters, D, of ACME widgets.

For 7–8, graph the following.

C4–7. $y = -|x - 2| + 4$

C4–8. $y = |-2x + 4| - 4$

For 9–10, solve the following graphically.

C4–9. $|2x + 5| - 7 = x$

C4–10. $|3 - 2x| \geq 7$

C4–11. Bert's Rent-a-Car will rent customers a compact car for \$50 a day or for \$200 a week. Draw a graph showing the cost to rent a car from Bert's Rent-a-Car for times ranging from 1 to 14 days.

C4–12. Let f be the following function: $f(x) = \begin{cases} 2x & 0 \leq x \leq 2 \\ 6 - x & 2 < x \leq 4 \end{cases}$

a. Sketch the graph of $f(x)$.
b. Sketch the graph of $y = -f(x)$.
c. Sketch the graph of $y = f(-x)$.
d. Sketch the graph of $y = -f(-x)$.

C4–13. From April 3, 1988 until February 2, 1991, the first-class postage rate in the United States was given by $P = 0.05 + 0.20\lceil x \rceil$, where P was the postage required to mail a letter weighing x ounces.

a. What was the least amount for which a first-class letter could be mailed?
b. What was the additional cost for each extra ounce?
c. Graph the function for the domain $0 \leq x \leq 5$ ounces.

Quadratic Functions

WHAT YOU WILL LEARN

If you have ever watched the water come out of a water fountain, then you have seen a parabola, which is a quadratic function. Quadratic functions are useful for modeling a wide variety of everyday events. In this chapter, you will learn to:

- Solve quadratic equations and inequalities;
- Graph quadratic functions;
- Find and interpret the average rate of change;
- Write quadratic equations to satisfy a variety of given information;
- Solve linear-quadratic systems of equations.

SECTIONS IN THIS CHAPTER

- Quadratic Models
- Solving Quadratic Equations
- The Discriminant and the Nature of Zeros
- Graphing Quadratic Functions
- Average Rate of Change
- Transformations and Vertex Form of a Parabola
- Solving Quadratic Inequalities
- Writing a Quadratic Equation
- Solving Quadratic-Linear Systems of Equations

Quadratic Models

In your imagination only, suppose you dropped a golf ball from the roof of the Empire State Building. (Actually doing this is frowned upon.) If we ignore the effect of air resistance, the ball would drop 16 feet in the first second. Then because gravity would constantly speed it up, the ball would fall 32 feet farther in each succeeding second than it had in the second before. The height of the ball can be modeled as shown in Figure 5.1.

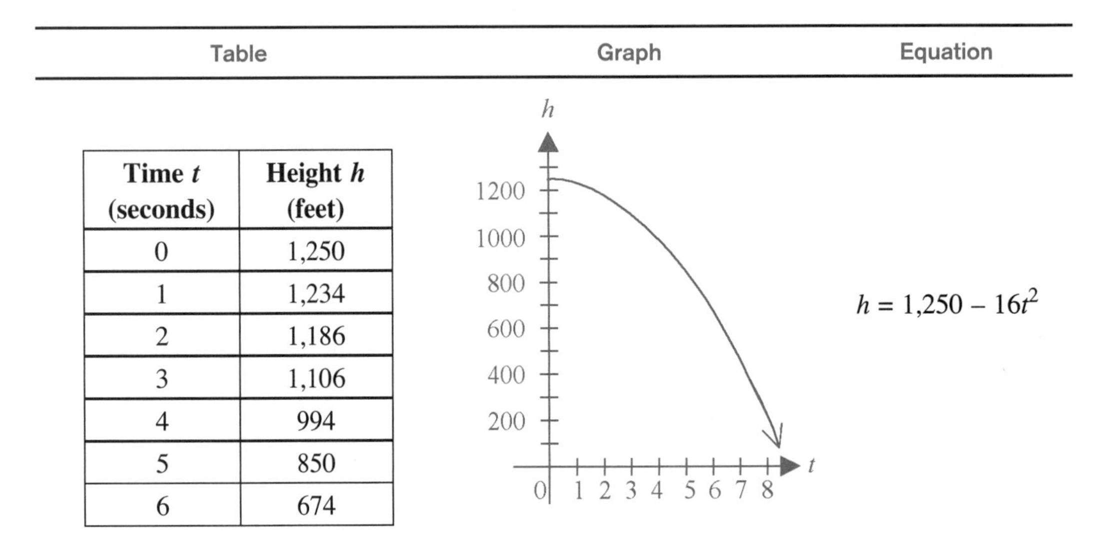

Time t (seconds)	Height h (feet)
0	1,250
1	1,234
2	1,186
3	1,106
4	994
5	850
6	674

FIGURE 5.1

This is an example of a *quadratic model*. The equation contains a t^2-term. The graph forms a curve rather than a straight line. Quadratic models and their related equations are the topic of this chapter.

Solving Quadratic Equations

A *quadratic equation* in one variable is an equation that can be written in the form $ax^2 + bx + c = 0$, where a, b, and c are real numbers and $a \neq 0$. This form is called the *general form* (some people call it the *standard form*) for a quadratic equation. Unlike a linear equation that typically has exactly one real solution, quadratic equations have two solutions. Four different algebraic methods can be used to solve quadratic equations.

EXTRACTING SQUARE ROOTS

A quadratic equation where $b = 0$ has no linear (x) term and is often called a *pure quadratic equation*. These are the easiest to solve.

To solve a pure quadratic equation by taking square roots:

1. Isolate the x^2-term (get it by itself).
2. Take the square root of each side. Use a $\pm$ symbol to show there are two answers.

EXAMPLE 5.1

Solve: $3x^2 - 7 = 8$

SOLUTION

$3x^2 - 7 = 8$	Solve for x^2.
$3x^2 = 15$	Add 7 to each side.
$x^2 = 5$	Divide each side by 3.
$\sqrt{x^2} = \sqrt{5}$	Take the square root of each side.
$\|x\| = \sqrt{5}$	$\sqrt{x^2} = \|x\|$. This is why there is both a positive and a negative solution shown in the next step.
$x = \sqrt{5}$ or $x = -\sqrt{5}$	Solve for x.

The solution set is $\{-\sqrt{5}, \sqrt{5}\}$.

Note several important things:

1. The method works only for pure quadratic equations (or equations in pure quadratic form; see the next example). *Do not use this method for equations that contain a linear* (b) *term.*
2. The two solutions are often written together as $x = \pm\sqrt{5}$ or $\{\pm\sqrt{5}\}$. This is a shorthand form that says there are two solutions, one positive and one negative.
3. Most people skip the absolute value step shown previously and go straight from $x^2 = k$ to $x = \pm\sqrt{k}$. This will be done in all future examples.
4. If $x^2 = k$ and $k < 0$, the equation will have no real solutions. (There are still two solutions, but they are imaginary. This is covered in the next chapter.)

EXAMPLE 5.2

Solve: $2(x - 3)^2 = 12$

SOLUTION

Treat $(x - 3)$ as a single quantity and solve for it. Then solve for x by adding 3 to each side.

$2(x - 3)^2 = 12$	
$(x - 3)^2 = 6$	Divide each side by 2.
$x - 3 = \pm\sqrt{6}$	Take the square root of each side. Remember to include ±.
$x = 3 \pm \sqrt{6}$	Add 3 to each side.

Be careful how you interpret the solution. The answers are *not* $x = 3$ and $x = \pm\sqrt{6}$. The answers are $x = 3 + \sqrt{6} \approx 5.45$ and $x = 3 - \sqrt{6} \approx 0.55$.

EXAMPLE 5.3

Find the time, to the nearest tenth, at which the golf ball dropped from the Empire State Building hits the ground. Use $h = 1{,}250 - 16t^2$, given at the beginning of the chapter, as a model for dropping a ball from the Empire State Building.

SOLUTION

When the ball hits the ground, $h = 0$. So solve $0 = 1{,}250 - 16t^2$.

$0 = 1{,}250 - 16t^2$	
$-1{,}250 = -16t^2$	Subtract 1,250 from each side.
$78.125 = t^2$	Divide each side by –16.
$\sqrt{78.125} = t$	Take the square root of each side. Disregard the negative root; time should be a positive value.
$t \approx 8.8$	Round to the nearest tenth.

The ball hits the ground in about 8.8 seconds.

FACTORING

A quadratic equation that contains a linear (x) term, sometimes called a *complete quadratic equation*, cannot be solved directly by taking square roots. If such an equation can be factored, it can be solved by applying the *zero product property* (ZPP). The ZPP says that if the product of two (or more) quantities is zero, at least one of the quantities must equal 0. Note that since the ZPP works only when the product is zero, quadratic equations must first be put in the general form before applying this method.

To solve a quadratic equation by factoring:

1. Write the equation in general form: $ax^2 + bx + c = 0$.
2. Factor.
3. Set each factor equal to 0 and solve (apply ZPP).

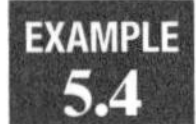

Solve: $2x^2 = 12 - 5x$

SOLUTION

$2x^2 = 12 - 5x$	
$2x^2 + 5x - 12 = 0$	Rewrite in general form.
$2x^2 - 3x + 8x - 12 = 0$	Factor. Use split the middle technique.
$x(2x - 3) + 4(2x - 3) = 0$	
$(2x - 3)(x + 4) = 0$	
$2x - 3 = 0$ or $x + 4 = 0$	Set each factor equal to 0.
$x = 1.5$ or $x = -4$	Solve for x.

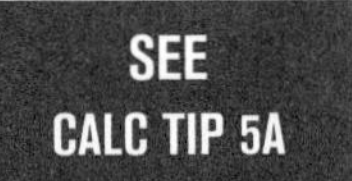

The solution set is $\{-4, 1.5\}$.

The usefulness of this method depends partly on how hard the equation is to factor. If it factors easily, this method is a good choice. Remember, though, that not all quadratic equations factor using integers. Some that can be factored are so difficult that using a different solution method may be easier.

COMPLETING THE SQUARE

Completing the square is a technique for rewriting a binomial of the form $x^2 + bx$ into a perfect square trinomial, one that can be factored into two equal binomials. For example, completing the square twice transforms the equation $x^2 + y^2 - 8x + 6y - 11 = 0$ into $(x - 4)^2 + (y + 3)^2 = 36$, which you may recognize as the equation of a circle. Completing the square may be the least commonly used method for actually solving quadratic equations, but it is a helpful skill for rewriting certain quadratic relations into equivalent forms. It can also be used to derive the quadratic formula (shown in the next section).

The expression $x^2 + bx$ can be made into a perfect square trinomial by adding $\left(\frac{b}{2}\right)^2$ to it. The resulting trinomial, $x^2 + bx + \left(\frac{b}{2}\right)^2$, factors into $\left(x + \frac{b}{2}\right)^2$. Of course, if you are solving an equation, you must add $\left(\frac{b}{2}\right)^2$ to each side to keep the equation balanced. These steps must all be done in the right order. Study the following directions and Example 5.5 carefully.

To solve a quadratic equation by completing the square:

1. Write the equation in the form $x^2 + bx = c$. Move the variables to one side and constant to the other side.
2. Complete the square by adding $\left(\frac{b}{2}\right)^2$ to both sides.
3. Factor the left side into $\left(x + \frac{b}{2}\right)^2$.
4. Solve by taking square roots, and then get x by itself by subtracting $\frac{b}{2}$.
5. Express the final answer in the requested form (simplest radical form or decimal approximations).

EXAMPLE 5.5 Solve $x^2 - 12x - 5 = 0$ by completing the square.

SOLUTION

$x^2 - 12x - 5 = 0$

$x^2 - 12x = 5$ — Rewrite the equation in the form $x^2 + bx = c$ by adding 5 to each side.

$x^2 - 12x + 36 = 5 + 36$ — Complete the square by adding $\left(\frac{b}{2}\right)^2$ to each side.

Since $b = -12$, $\left(\frac{b}{2}\right)^2 = \left(\frac{-12}{2}\right)^2 = (-6)^2 = 36$.

$x^2 - 12x + 36 = 41$ — Simplify the right side.

$(x - 6)^2 = 41$ — Factor the trinomial into a perfect square binomial. The constant inside the binomial will always be $\frac{b}{2}$. In this problem, $b = -12$. So the constant in the binomial is $\frac{b}{2} = \frac{-12}{2} = -6$, and the binomial is $x - 6$.

$(x - 6) = \pm\sqrt{41}$ — Take the square root of each side. Remember to include $\pm$.

$x = 6 \pm \sqrt{41}$ — Add 6 to each side to solve for x.

The solutions, in simplest radical form, are $\{6 \pm \sqrt{41}\}$.

EXAMPLE 5.6 Solve $4x^2 - 24x - 15 = 0$ by completing the square.

SOLUTION

$4x^2 - 24x - 15 = 0$

$4x^2 - 24x = 15$

$x^2 - 6x = \frac{15}{4}$ — Rewrite the equation in the form $x^2 + bx = c$ by adding 15 to each side and dividing each term by 4.

$x^2 - 6x + 9 = \frac{15}{4} + 9$ — Complete the square by adding $\left(\frac{b}{2}\right)^2$ to each side.

Since $b = -6$, $\left(\frac{b}{2}\right)^2 = \left(\frac{-6}{2}\right)^2 = (-3)^2 = 9$.

$x^2 - 6x + 9 = \frac{51}{4}$ Simplify the right side.

$(x - 3)^2 = \frac{51}{4}$ Factor the trinomial into a perfect square binomial.

$(x - 3) = \pm\sqrt{\frac{51}{4}}$ Take the square root of each side.

$x = 3 \pm \sqrt{\frac{51}{4}}$ Add 3 to each side to solve for x.

$x = 3 \pm \frac{\sqrt{51}}{2} = 3 \pm \frac{1}{2}\sqrt{51}$

The solutions, in simplest radical form, are $\left\{3 \pm \frac{1}{2}\sqrt{51}\right\}$.

QUADRATIC FORMULA

Although it is not always the quickest or easiest method to solve a quadratic equation, the quadratic formula is important. Unlike taking square roots or factoring, it can be used to solve *any* quadratic equation. The formula can be derived by completing the square for the equation $ax^2 + bx + c = 0$. All Algebra 2 students should have the formula memorized. If you find this formula hard to remember, many funny videos on the Internet can help you remember it. Search for "quadratic formula song video" and learn it to music.

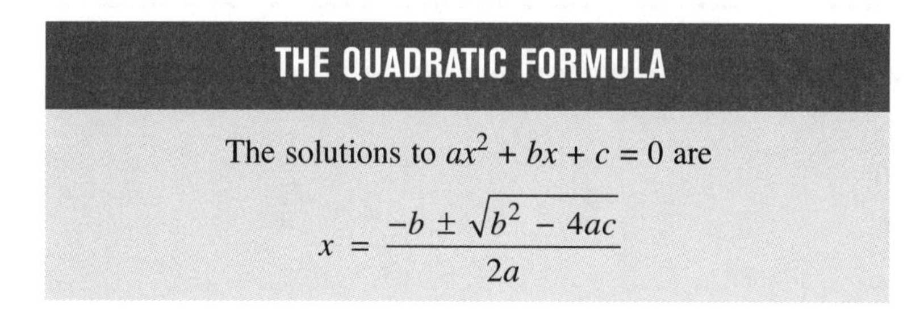

THE QUADRATIC FORMULA

The solutions to $ax^2 + bx + c = 0$ are

$$x = \frac{-b \pm \sqrt{b^2 - 4ac}}{2a}$$

To solve a quadratic equation using the quadratic formula:

1. Put the equation in the general form: $ax^2 + bx + c = 0$.
2. Find the values of a, b, and c.
3. Substitute the values of a, b, and c into the quadratic formula $x = \frac{-b \pm \sqrt{b^2 - 4ac}}{2a}$ and simplify.
4. Express the answers in the desired form.

EXAMPLE 5.7

Solve: $3x^2 = 5(x + 5)$ and write the solutions in simplest radical form.

SOLUTION

$3x^2 = 5(x + 5)$
$3x^2 = 5x + 25$
$3x^2 - 5x - 25 = 0$

Write the equation in general form by distributing the 5 and solving for 0.

$a = 3, b = -5, c = -25$

Identify the values for a, b, and c.

$$x = \frac{-b \pm \sqrt{b^2 - 4ac}}{2a}$$

Substitute the values for a, b, and c into the quadratic formula.

$$x = \frac{-(-5) \pm \sqrt{(-5)^2 - 4(3)(-25)}}{2(3)}$$

$$x = \frac{5 \pm \sqrt{325}}{6}$$

Simplify.

$$x = \frac{5 \pm 5\sqrt{13}}{6}$$

Simplify the radical, $325 = (25)(13)$.

The solutions, in simplest radical form, are $\left\{\frac{5}{6} \pm \frac{5\sqrt{13}}{6}\right\}$.

SECTION EXERCISES

For 1–7, solve algebraically. Express irrational answers in simplest radical form.

5–1. $x^2 - 100 = 0$

5–2. $\frac{x}{6} = \frac{10}{x}$

5–3. $\frac{2}{3}x^2 + 7 = 15$

5–4. $\frac{2x + 9}{9} = \frac{11}{2x - 9}$

5–5. $-3(x + 5)^2 = -12$

5–6. $x^2 + 5x = 40 - x$

5–7. $3x^2 = 13x + 10$

5–8. Solve $4x^2 - 8x - 10 = 0$ by completing the square.

5–9. Solve $3x^2 + 13x + 9 = 0$ using the quadratic formula.

5–10. The base of a rectangle is represented by $x + 5$, and the height is represented by $x - 5$. The area of the rectangle is 56. Find the dimensions of the rectangle by solving algebraically.

The Discriminant and the Nature of Zeros

The quantity $b^2 - 4ac$ inside the radical in the quadratic formula is called the *discriminant*. It is useful for telling what type of zeros a particular quadratic equation will have. Remember that *zeros* (also called *roots*) of a quadratic equation are the solutions to $ax^2 + bx + c = 0$ and are the x-values where the graph of $y = ax^2 + bx + c$ crosses or touches the x-axis. Classifying the zeros gives information about the graph of $y = ax^2 + bx + c$ and also about the expected solutions to $ax^2 + bx + c = 0$. If a, b, and c are all rational numbers, there are four possible types of solutions, which are summarized in Table 5.1.

TABLE 5.1
THE EFFECT OF THE DISCRIMINANT ON A QUADRATIC EQUATION

Discriminant $b^2 - 4ac$	Nature of Zeros	Behavior of the Graph	Graph ($a > 0$)
Positive and a perfect square	2 different real, rational zeros	Crosses the x-axis twice at rational values	r_1 and r_2 are both rational
Positive but not a perfect square	2 different real, irrational zeros	Crosses the x-axis twice at irrational values	r_1 and r_2 are both irrational
Equals 0	2 equal (seems like one), real, rational zeros	Touches the x-axis once (tangent to the x-axis)	r is rational
Negative	No real zeros (see Chapter 6)	Never intersects the x-axis	

EXAMPLE 5.8

Will the equation $5x^2 - 6x + 2 = 0$ have real zeros?

SOLUTION

In this equation, $a = 5$, $b = -6$, and $c = 2$. The discriminant equals:

$$\begin{aligned} b^2 - 4ac &= (-6)^2 - 4(5)(2) \\ &= 36 - 40 \\ &= -4 \end{aligned}$$

Since the discriminant is less than 0, the zeros to this equation will not be real.

EXAMPLE 5.9

For what value(s) of k will the equation $2x^2 + kx + 50 = 0$ have equal zeros?

SOLUTION

For the zeros of a quadratic equation to be equal, the discriminant must be 0.

$a = 2, b = k, c = 50$	Identify the values for a, b, and c.
$b^2 - 4ac = 0$ $k^2 - 4(2)(50) = 0$	Substitute the values into the formula for the discriminant and set it equal to 0.
$k^2 - 400 = 0$ $k^2 = 400$ $k = \pm 20$	Solve for k.

The zeros to the equation will be equal if k is either 20 or -20.

SECTION EXERCISES

5–11. Choose all answers that satisfy the given condition. A quadratic equation with rational coefficients has discriminant D.

a. The equation will have distinct rational roots if D equals
(1) -8 (2) 0 (3) 8 (4) 16

b. The graph of the equation will be tangent to the x-axis if D equals
(1) -8 (2) 0 (3) 8 (4) 16

c. The graph of the equation will not intersect the x-axis if D equals
(1) -8 (2) 0 (3) 8 (4) 16

5–12. For what value of a will the parabola $y = ax^2 - 8x + 36$ have two real, equal roots?

5–13. For what values of c will the quadratic equation $2x^2 + 5x + c = 0$ have imaginary roots?

5–14. For the quadratic equation $2x^2 + kx + 8 = 0$,

a. What value(s) of k would make the roots of the equation real and equal?

b. What positive values of k would make the roots of the equation real?

c. What is the largest integer value of k for which the roots of the equation would be imaginary?

Graphing Quadratic Functions

An equation of the form $y = ax^2 + bx + c$ $(a \neq 0)$ is called a *quadratic function.* Its graph will always be in the shape of a *parabola*. The parabola $y = x^2 - 4x - 3$ is graphed in Figure 5.2. with some of its important features labeled.

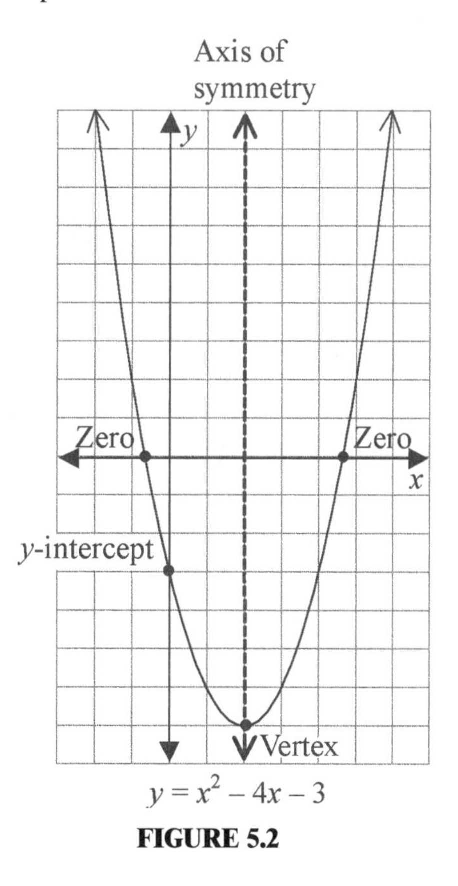

FIGURE 5.2

Table 5.2 describes features about the graph of a parabola, $y = ax^2 + bx + c$ $(a \neq 0)$.

TABLE 5.2
FEATURES OF THE GRAPH OF A PARABOLA

Feature	Details
Axis of symmetry	The parabola has a vertical axis of symmetry. The axis of symmetry is the vertical line that cuts the parabola in half. The equation for the axis of symmetry is given by $x = \frac{-b}{2a}$. In Figure 5.2, the axis of symmetry is $x = \frac{-(-4)}{2(1)}$ or $x = 2$. Be sure you include the "$x =$" in your axis of symmetry; this is how we know it is a vertical line.
Vertex	The parabola has a vertex (or turning point) on the axis of symmetry. Since the vertex is on the axis, the x-coordinate of the vertex is $x = \frac{-b}{2a}$. The y-coordinate is found by substituting the x-value into the equation of the parabola. In Figure 5.2, $x = 2$ and $y = (2)^2 - 4(2) - 3 = -7$, so the vertex is at $(2, -7)$.
Zeros	A parabola may have up to two real zeros, which can be found by solving $ax^2 + bx + c = 0$. In Figure 5.2, the quadratic formula can be used to solve $x^2 - 4x - 3 = 0$, giving $x \approx -0.65$ and $x \approx 4.65$.
y-intercept	The y-intercept is where the parabola crosses the y-axis and is found from the equation by letting $x = 0$ and finding the y-value. For an equation in the form $y = ax^2 + bx + c$, the y-intercept is c. In Figure 5.2, the y-intercept is $c = -3$.
Minimum or maximum value; Increasing and decreasing	If $a > 0$ (positive), the minimum value of the function will be the y-value of the vertex. When the graph is read left to right, the function will be decreasing on the interval to the left of the vertex and increasing on the interval to the right of the vertex. If $a < 0$ (negative), the maximum value of the function will be the y-value of the vertex with the function increasing to the left of the vertex and decreasing to the right. In Figure 5.2, $a > 0$, the vertex is $(2, -7)$. So the function has a minimum value of -7. The function decreases on the interval $(-\infty, 2)$ and increases on the interval $(2, \infty)$.
End behavior	The left and right sides of the parabola both increase without bound (go to ∞) when $a > 0$ or both decrease without bound (go to $-\infty$) when $a < 0$.

Feature	Details
Concavity	A parabola that opens up ($a > 0$) is said to be *concave up*. A parabola that opens down ($a < 0$) is said to be *concave down*. Loosely speaking, a graph (not just a parabola) is concave up if it forms a cup (or part of a cup) that "holds water." It is concave down if it forms part of a cup that "dumps water." Concave up: Holds water. Concave down: Dumps water

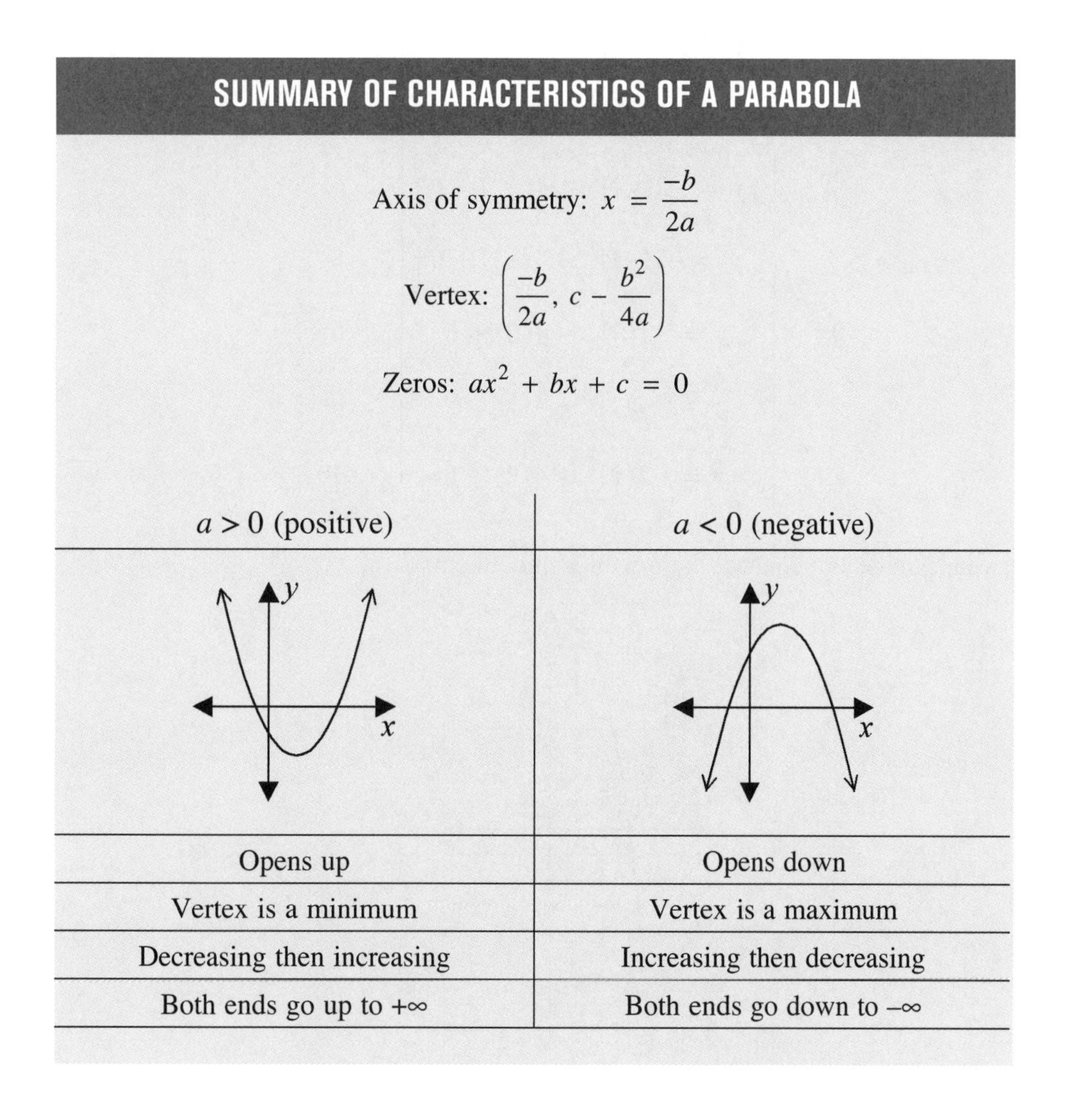

SUMMARY OF CHARACTERISTICS OF A PARABOLA

Axis of symmetry: $x = \dfrac{-b}{2a}$

Vertex: $\left(\dfrac{-b}{2a}, c - \dfrac{b^2}{4a}\right)$

Zeros: $ax^2 + bx + c = 0$

$a > 0$ (positive)	$a < 0$ (negative)
Opens up	Opens down
Vertex is a minimum	Vertex is a maximum
Decreasing then increasing	Increasing then decreasing
Both ends go up to $+\infty$	Both ends go down to $-\infty$

EXAMPLE 5.10 Sketch the graph of $y = -2x^2 - 4x + 6$. Identify the axis of symmetry, the zeros, the vertex, and the y-intercept. Describe where the graph is increasing and decreasing. Describe the end behavior.

SOLUTION

Sketching graphs of quadratic functions given in general form, $y = ax^2 + bx + c$, is usually done by making a table of values and then plotting the points. The x-value in the center of the table should be the value of the axis of symmetry; this point will be the vertex of the parabola. As a rule of thumb, try to graph three points on either side of the vertex.

For this example, the axis of symmetry is $x = \frac{-b}{2a} = \frac{-(-4)}{2(-2)} = \frac{4}{-4} = -1$.

Use this value as the center of your table, and include three points on each side.

x	$y = -2x^2 - 4x + 6$	y
−4	$y = -2(-4)^2 - 4(-4) + 6$	−10
−3	$y = -2(-3)^2 - 4(-3) + 6$	0
−2	$y = -2(-2)^2 - 4(-2) + 6$	6
−1	$y = -2(-1)^2 - 4(-1) + 6$	8
0	$y = -2(0)^2 - 4(0) + 6$	6
1	$y = -2(1)^2 - 4(1) + 6$	0
2	$y = -2(2)^2 - 4(2) + 6$	−10

Plot your points, and connect them with a smooth curve.

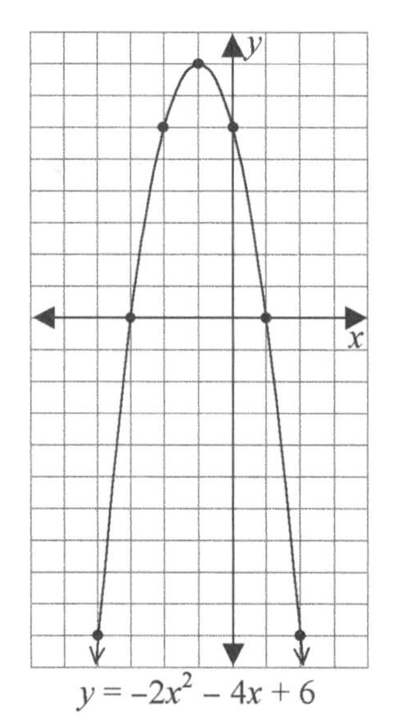

Identify the following from your graph and table.

Axis of symmetry	$x = -1$
Zeros	$x = -3$, $x = 1$
Vertex	$(-1, 8)$
y-intercept	6
Increasing and decreasing	Increasing on $x < -1$ and decreasing on $x > -1$
End behavior	Both ends go to $-\infty$

SEE CALC TIP 5B

EXAMPLE 5.11

A parabola has the equation $y = 2x^2 + 5x - 12$.

a. Does the parabola open up or down?
b. What is the y-intercept of the parabola?
c. What is the equation of the axis of symmetry of the parabola?
d. What are the coordinates of the vertex of the parabola? Is it a maximum or a minimum?
e. What are the zeros of the parabola?
f. Sketch the parabola.

SOLUTION

a. For $y = 2x^2 + 5x - 12$, $a = 2$. Since a is positive, the parabola opens up.
b. For $y = 2x^2 + 5x - 12$, $c = -12$. The y-intercept is -12.
c. The formula for the axis of symmetry, is $x = \frac{-b}{2a}$. In $y = 2x^2 + 5x - 12$, $a = 2$ and $b = 5$. So the axis of symmetry is $x = \frac{-5}{2(2)} = -\frac{5}{4} = -1.25$.

d. The vertex is on the axis of symmetry, so the x-value of the vertex is -1.25. Find the y-value of the vertex by evaluating the function at $x = -1.25$; $y = 2(-1.25)^2 + 5(-1.25) - 12$ gives $y = -15.125$. The vertex is $(-1.25, -15.125)$.
e. To find the zeros, set $y = 0$ and solve for x.

$$0 = 2x^2 + 5x - 12$$
$$0 = 2x^2 + 8x - 3x - 12$$
$$0 = 2x(x + 4) - 3(x + 4)$$
$$0 = (x + 4)(2x - 3)$$
$$x + 4 = 0 \text{ or } 2x - 3 = 0$$
$$x = -4 \text{ or } x = 1.5$$

The zeros are $x = -4$ and $x = 1.5$.

f. Make a table of values. Some students are bothered because the table and the resulting points plotted don't appear to show symmetry. This is because the axis of symmetry, $x = -1.25$, does not fall on a whole number or a number ending in 0.5.

x	y
−5	13
−4	0
−3	−9
−2	−14
−1	−15
0	−12
1	−5
2	6
3	21

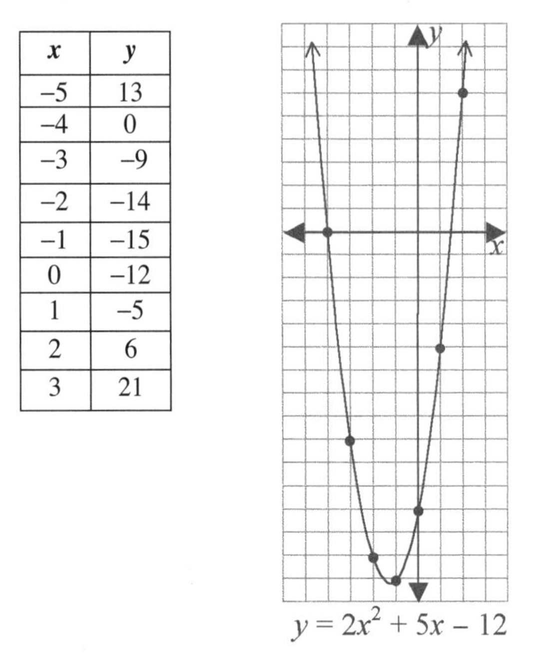

$y = 2x^2 + 5x - 12$

The techniques used to answer questions about graphs of parabolas are the same skills used to solve word problems modeled with quadratic functions.

EXAMPLE 5.12 After a baseball is hit by a bat, it follows a path given by $y = 3 + 0.4x + ax^2$, where y is the vertical height of the ball and x is the horizontal distance of the ball from the batter, both in feet.

a. How high off the ground was the ball when it was hit?
b. What should be the sign of a?
c. Suppose $a = -0.002$. What is the greatest height reached by the ball?
d. If the ball was not caught, how far from the batter, to the nearest tenth of a foot, did the ball first hit the ground if $a = -0.002$.

SOLUTION

a. When the ball was hit, $x = 0$. Substituting $x = 0$ into the equation gives $y = 3$. The ball was 3 feet off the ground when it was hit.
b. Since the ball will eventually fall to the ground, a should be negative. A negative a causes the graph of the parabola to open down (concave down).
c. Find the maximum height by finding the y-value of the vertex. First find the x-value of the vertex by using the formula for the axis of symmetry, $x = \frac{-b}{2a}$. For $y = 3 + 0.4x - 0.002x^2$, the x-value of the vertex is $x = \frac{-0.4}{2(-0.002)} = 100$.

Evaluate $x = 100$ in the function to find the y-value. So

$y = 3 + 0.4(100) - 0.002(100)^2 = 23$. The greatest height of the ball is 23 feet.

d. The ball hits the ground when $y = 0$.

$0 = 3 + 0.4x - 0.002x^2$	Set $y = 0$. Use the quadratic formula to solve for x.
$a = -0.002, b = 0.4, c = 3$	Identify a, b, and c. Be careful, the equation is not written in descending order.
$x = \dfrac{-b \pm \sqrt{b^2 - 4ac}}{2a}$ $x = \dfrac{-0.4 \pm \sqrt{(0.4)^2 - 4(-0.002)(3)}}{2(-0.002)}$	Substitute values into the quadratic formula and evaluate.
$x = -7.2$ or $x = 207.2$	Round to the nearest tenth.
$x = 207.2$	Choose only the positive value to match the question.

The ball hits the ground about 207.2 feet away from the batter.

SEE CALC TIP 5D

SECTION EXERCISES

5–15. Let $y = -\frac{1}{2}x^2 + 2x + 6$.

a. Sketch the graph. (Suggestion: Graph only even values of x.)
b. Identify the axis of symmetry.
c. Identify the zeros.
d. Write the coordinates of the vertex.
e. Identify the y-intercept.
f. Describe where the graph is increasing and decreasing.
g. Describe the end behavior of the graph.

5–16. Let $y = \frac{1}{2}x^2 - \frac{1}{2}x - 3$.

a. Does the parabola open up or down?
b. What is the y-intercept of the parabola?
c. What is the equation of the axis of symmetry of the parabola?
d. What are the coordinates of the vertex of the parabola? Is it a maximum or a minimum?
e. What are the zeros of the parabola?
f. Sketch the parabola.

5–17. A golfer hits a tee shot so that the ball's height, h, in feet is given by $h(t) = -16t^2 + 72t$, where t is time in seconds since the ball was hit.

a. When did the ball reach its maximum height?
b. What was the maximum height reached by the ball?
c. When will the ball hit the ground?

5–18. Doofus has a tree fort in a tree 60 feet away from his regular house. Doofus wants to watch TV in his tree fort, so he runs a long extension cord out from his bedroom window and up to his tree fort. Between Doofus's bedroom window and the tree fort, the extension cord hangs in the shape of a parabola with the equation $y = ax^2 - x + 16$, where y is the height of the cord above the ground and x is the horizontal distance away from Doofus's house. The cord reaches its lowest point 24 feet horizontally from Doofus's house.

a. Find the value of a.
b. Using the value you found in part a, how high is Doofus's tree fort?
c. Doofus's friend Rufus would like to ride his four-wheeler around Doofus's yard. When on the four-wheeler, the top of Rufus's head is 5 feet off the ground. Can Rufus safely ride his four-wheeler around the yard?

5–19. Chuck Rock is standing on a cliff above a level plain on the moon when an alien taps him on the shoulder. Startled, Chuck jumps upward and, unfortunately, outward off the edge of the cliff. The equation for Chuck's height above the rocky plain below is given by $y = 42 + 27t - 5.4t^2$, where y is Chuck's height in feet and t is the time in seconds since Chuck jumped.

a. How high was the cliff?
b. When did Chuck reach the highest point of his jump?
c. How high above the plain did Chuck get?
d. To the nearest tenth of a second, how long did it take Chuck to hit the ground?
e. How much time did Chuck spend more than 42 feet above the ground?

Average Rate of Change

The slope of a line, $m = \frac{\Delta y}{\Delta x}$, tells the rate of change of y with respect to x (see Chapter 1). On a line with slope m, each time x increases by one unit, y changes by m units, increasing if m is positive and decreasing if m is negative. For a line, this rate of change is constant. This means we can use any two points on the line to calculate the slope; we'll get the same answer no matter which points we use. It also means that the rate of change of y is the same everywhere along the line. The y-value will change by exactly the same amount when x changes from 0 to 1 as when x changes from 1 to 2 or from 1,000 to 1,001.

The slope of a nonlinear function such as a parabola is not constant like that of a line. The parabola $y = x^2$ is graphed in Figure 5.3. The rate of change of y with respect to x is clearly not constant. When x changes from 0 to 1, y changes by 1. When x changes from 1 to 2, though, y changes by 3.

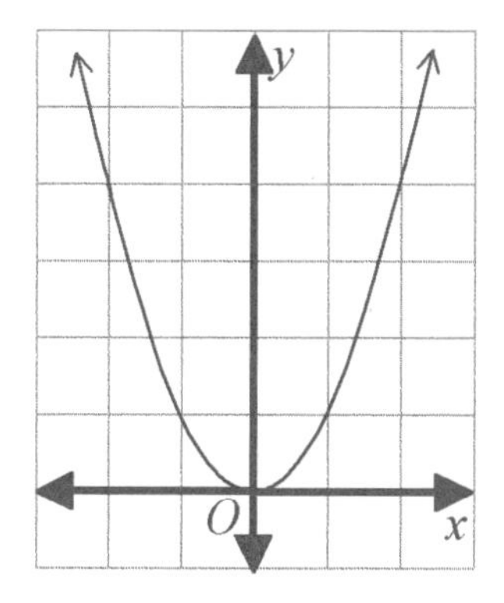

FIGURE 5.3

When talking about the rate of change of a nonlinear function, we distinguish between two different rates of change, the *instantaneous rate of change* at a particular point and the *average rate of change* over some interval. Think of traveling by car a distance of 90 miles in two hours. During the trip, the car's speedometer measures instantaneous speed, the speed the car is going at each particular instant. The instantaneous speed of the car may change often during the trip. However, average speed for the trip is given by:

$$\text{average speed} = \frac{\text{distance}}{\text{time}} = \frac{90 \text{ miles}}{2 \text{ hours}} = 45 \text{ mph}$$

The graph in Figure 5.4 illustrates one possible trip. The instantaneous speed was 30 mph at time $x = 0.5$ hours, 65 mph at time 1 hour, and 40 mph at time 1.5 hours.

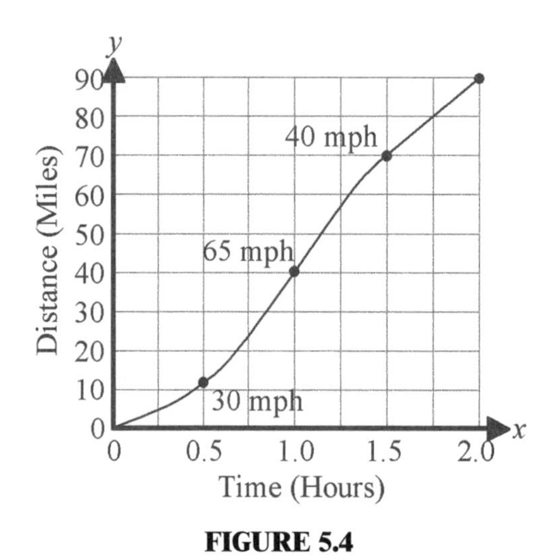

FIGURE 5.4

The average speed for the trip was 45 mph. How is that represented on the graph?

Suppose the car had actually traveled exactly 45 mph for the whole two hours. Then the graph of your trip would have been the straight line shown in Figure 5.5. This line is called a *secant line*. A secant line connects any two points on a graph, in this case the points (0, 0) and (2, 90). The slope of the secant line, $m = \dfrac{90 - 0}{2 - 0} = 45$, is the average speed for the trip or, in more mathematical terms, the average rate of change of y with respect to x over the interval $x = 0$ to $x = 2$.

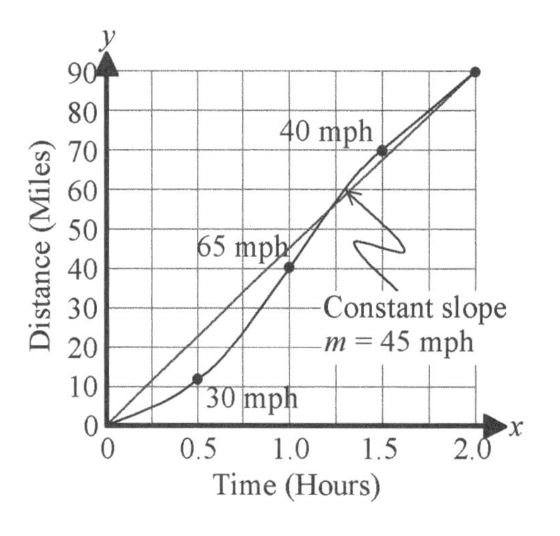

FIGURE 5.5

We can calculate an average rate of change between any two points on the function. For example, the average speed from $t = 0.5$ to $t = 1.5$ was approximately $\dfrac{70 - 12}{1.5 - 0.5} = 58$ mph. (It is approximate because we had to estimate the y-values from the graph.)

Let's summarize. If a function $f(x)$ is defined on some interval $x = a$ to $x = b$, then the *average rate of change* of $f(x)$ on the interval $[a, b]$ is the *slope of the secant line* from $(a, f(a))$ to $(b, f(b))$.

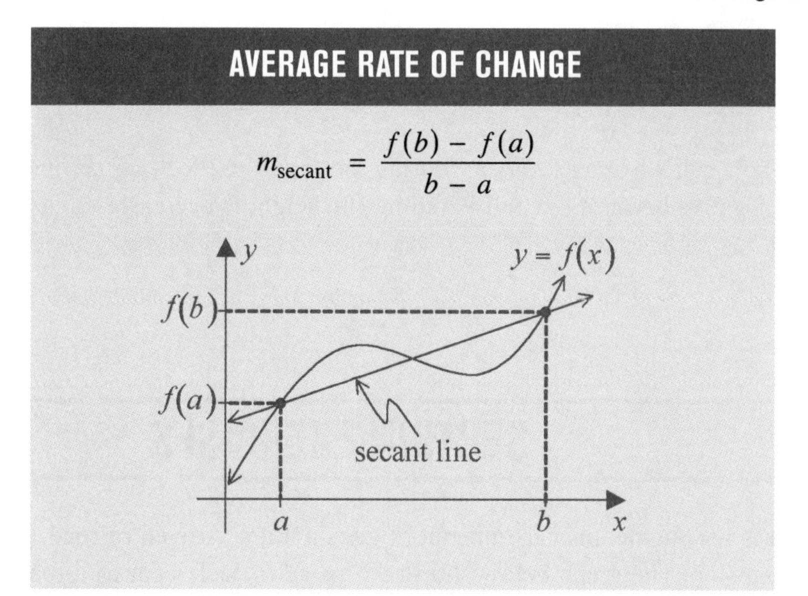

The table below shows the temperature each hour during a recent morning.

Time (A.M.)	6:00	7:00	8:00	9:00	10:00	11:00	12:00
Temperature (°F)	28	34	39	41	42	44	44

What was the average rate of change in the temperature from 6:00 A.M. to 10:00 A.M.?

SOLUTION

The average rate of change is $\dfrac{42 - 28}{10 - 6} = \dfrac{14}{4} = 3.5\,°\text{F}$ per hour. From 6:00 A.M. to 10:00 A.M., the temperature increased by 3.5°F per hour.

When interpreting average rate of change, state that the value is increasing for positive change and decreasing for negative change. Be sure to include the units for the rate.

EXAMPLE 5.14

If a golf ball is dropped from the top of the Empire State Building, its height, h, in feet as a function of time, t, in seconds since it was dropped is given by $h = 1{,}250 - 16t^2$. Find the average rate of change of the golf ball's height on the interval $t = 2$ seconds to $t = 5$ seconds, and explain what it means in this problem.

SOLUTION

First find the value of h at $t = 2$ seconds and $t = 5$ seconds.

$$h(2) = 1{,}250 - 16(2)^2 = 1{,}186$$

$$h(5) = 1{,}250 - 16(5)^2 = 850$$

Use h in place of y and t in place of x:

$$\frac{\Delta h}{\Delta t} = \frac{850 - 1{,}186}{5 - 2} = \frac{-336}{3} = -112 \text{ ft/s}$$

This is the golf ball's average velocity during the time $t = 2$ to $t = 5$ seconds. The velocity is negative because the ball is falling (its height is decreasing at a rate of 112 ft/s).

SECTION EXERCISES

5–20. On Jack's way home, his car runs out of gas on a long stretch of road. He lets it coast to a stop. The graph below shows the speed of Jack's car as a function of time in seconds since the engine quit.

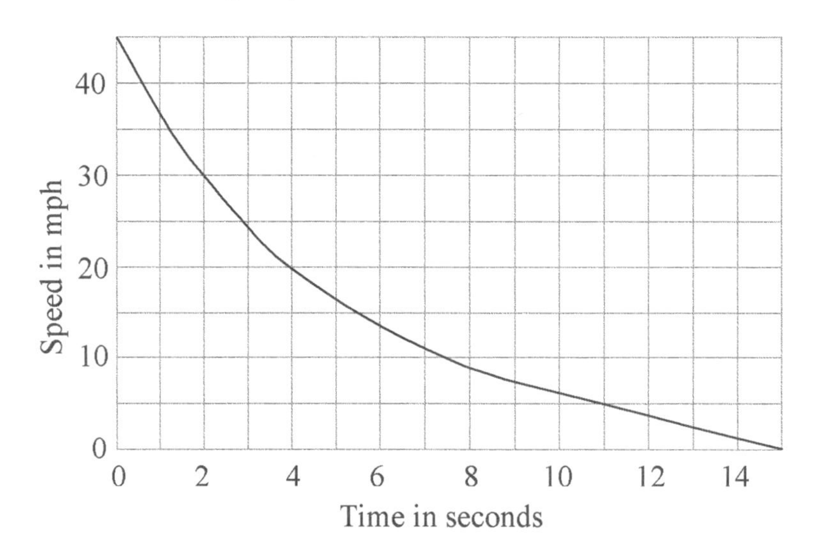

What was the average rate of change of the car's velocity from $t = 2$ to $t = 11$ seconds?

5–21. The stopping distances in feet for a car are given in the table below as a function of the car's speed in mph.

Speed (mph)	20	30	40	50	60
Stopping Distance (ft)	25	55	105	188	300

Find the average rate of change of stopping distance with respect to speed over the interval from 30 to 50 mph. Include units, and tell briefly but precisely what the answer means in this problem.

Transformations and Vertex Form of a Parabola

The simplest quadratic function is the function $y = x^2$, whose graph is shown in Figure 5.6.

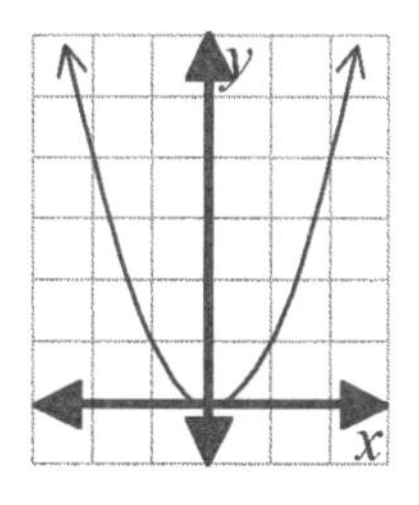

FIGURE 5.6

All other parabolas of the form $y = ax^2 + bx + c$ can be obtained from this one using reflections, dilations, and translations. (This would be a good time to review the section on transformations in Chapter 3.)

REFLECTION

If we want a parabola that opens down, we can reflect it over the x-axis. To do this, we make the transformation $y = f(x) \rightarrow y = -f(x)$. Our new equation is $y = -x^2$, graphed in Figure 5.7.

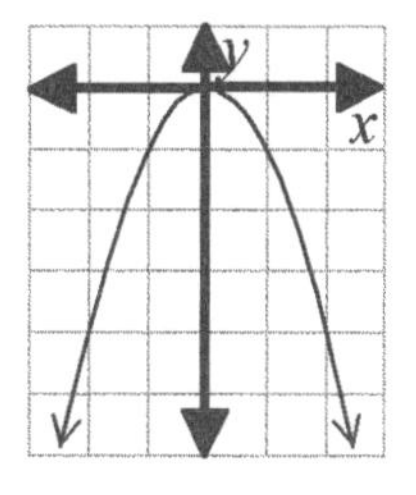

FIGURE 5.7

Because $y = x^2$ has y-axis symmetry, its graph is unchanged by a reflection over the y-axis. Also, a reflection over the origin is the same graph as a reflection over the x-axis.

DILATION

The graph of $y = af(x)$ is a vertical dilation of the graph of $y = f(x)$ by a factor of a. If $a > 1$, the graph will be stretched vertically. If $0 < a < 1$, the graph will shrink vertically. The graphs of $y = 2x^2$ and $y = 0.5x^2$ are shown in Figure 5.8.

Vertically stretching a parabola with $a > 1$ produces a graph that appears narrower than the original graph. Vertically shrinking a parabola with $0 < a < 1$ produces a graph that appears wider than the original graph. Both of these effects make it appear that a vertical dilation could be represented as a horizontal dilation, which is true.

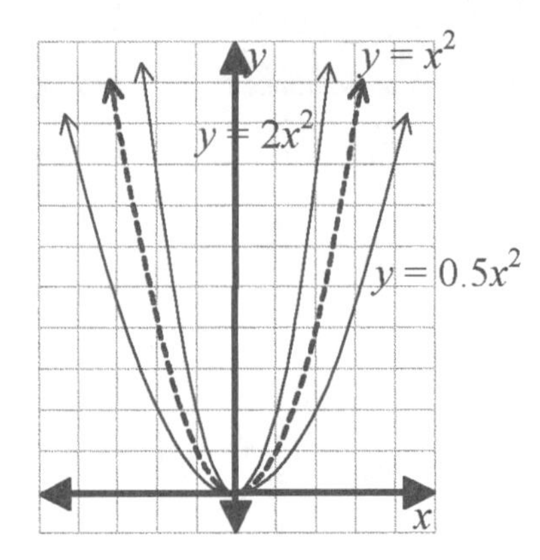

FIGURE 5.8

A horizontal dilation of $y = f(x)$ by a factor b is given by $y = f\left(\frac{x}{b}\right)$. Applying that to the function $y = x^2$ produces $y = \left(\frac{x}{b}\right)^2 = \frac{x^2}{b^2} = \frac{1}{b^2}x^2$. From this we see that for $y = x^2$, a horizontal dilation by b is equivalent to a vertical dilation by $a = \frac{1}{b^2}$. So we really only need to do one or the other. Making a vertical dilation is easier than making a horizontal dilation.

If $a < 0$, the transformation $y = ax^2$ includes both a vertical dilation of $y = x^2$ by a factor of $|a|$ and a reflection over the x-axis.

TRANSLATIONS

The parabola $y = x^2$ has its vertex at the origin, (0, 0). To translate the vertex to another point (h, k), we need to translate the whole parabola "right h and up k." This is done by changing $y = f(x)$ to $y = f(x - h) + k$. When applied to the function $y = x^2$, this gives $y = (x - h)^2 + k$. The graph of $y = (x - 3)^2 + 2$ is shown in Figure 5.9.

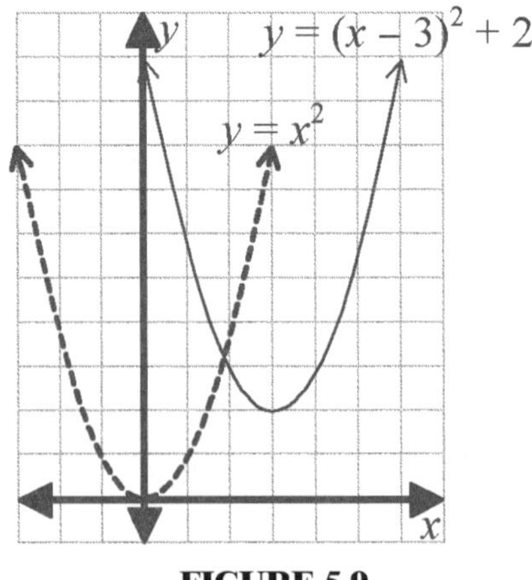

FIGURE 5.9

VERTEX FORM OF A PARABOLA

Combining reflections, dilations, and translations of the function $y = x^2$ gives the following equation of a parabola, often called the *vertex form.*

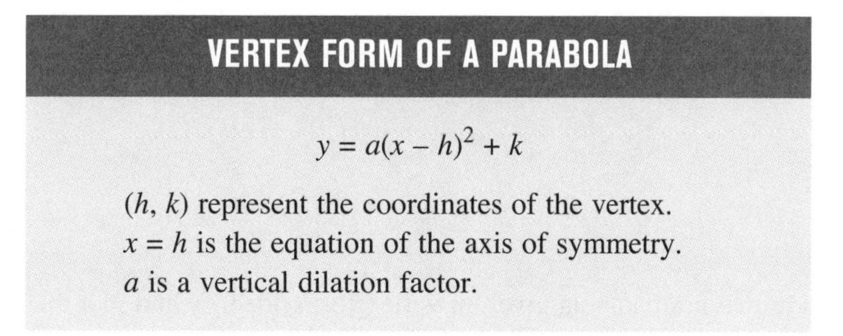

VERTEX FORM OF A PARABOLA

$$y = a(x - h)^2 + k$$

(h, k) represent the coordinates of the vertex.
$x = h$ is the equation of the axis of symmetry.
a is a vertical dilation factor.

In the vertex form of the equation of a parabola, the dilation factor, a, will affect the graph of the parabola in the following ways:

- If $|a| > 1$, the parabola will appear narrower than $y = x^2$.
- If $0 < |a| < 1$, the parabola will appear wider than $y = x^2$.
- If $a < 0$, the parabola will be reflected vertically; it will open down.

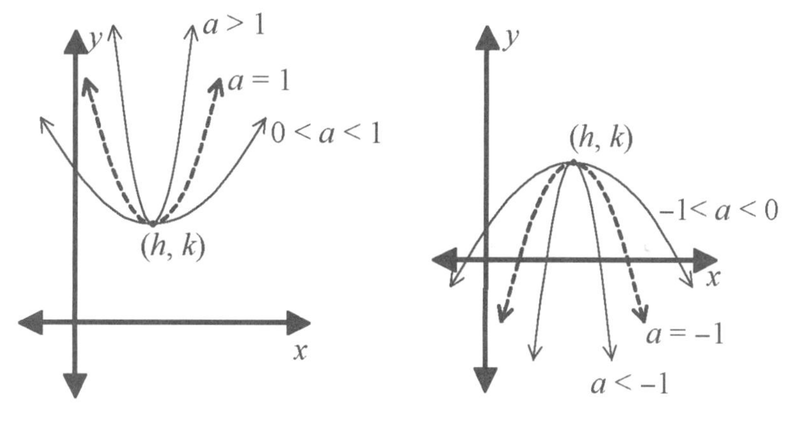

FIGURE 5.10

Using the vertex form of the equation of a parabola has several advantages over the general form, $y = ax^2 + bx + c$. First (and most obvious) is that the vertex form lets you easily identify the vertex of the parabola, one of the parabola's most important points. Finding the zeros of a parabola in vertex form is also easier; you don't need to factor or to use the quadratic formula. Finally, using the vertex form of a parabola makes graphing quadratic functions much easier.

EXAMPLE 5.15

Describe how to transform $y = x^2$ to get the graph of $y = 2(x + 3)^2 - 1$.

SOLUTION

$y = 2(x + 3)^2 - 1$ is the graph of $y = x^2$ after a vertical stretch by a factor of 2 and a translation left 3 units and down 1 unit.

EXAMPLE 5.16

Graph: $y = (x - 2)^2 - 3$

SOLUTION

To sketch a graph of a parabola given in vertex form, identify and plot the point at the vertex. For $a = 1$, sketch points at integral distances to the left and the right of the vertex with y-values increasing by 1, 3, 5, ...

For $y = (x - 2)^2 - 3$, we have $a = +1$, $h = +2$, $k = -3$. Positive signs have been written for emphasis. Remember, the value of h is always the opposite of the value of the constant in the binomial.

The vertex (h, k) is $(2, -3)$. Plot $(2, -3)$ on the graph. Then plot points to the left and right, going over 1 block in the x-direction each time but going up (because a is $+1$) 1 block, then 3 blocks, then 5 blocks, and so on.

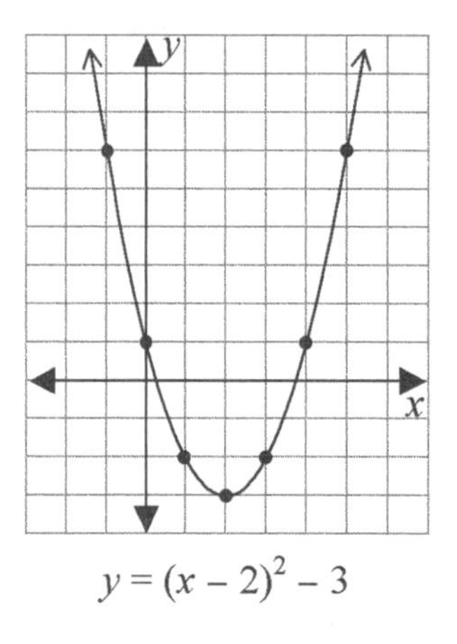

$y = (x - 2)^2 - 3$

EXAMPLE 5.17

Graph: $y = -2(x + 3)^2 + 9$

SOLUTION

In this example, we have $a = -2$, $h = -3$, $k = 9$. Again, remember the value of h is always the opposite of the value of the constant in the binomial. (It may help to rewrite the equation as $y = -2(x - (-3))^2 + 9$.)

The vertex (h, k) is $(-3, 9)$. Plot it on the graph. Then plot points to the left and right, going over 1 block in the x-direction each time. Because $a < 0$, we will be going down instead of up as we move away from the vertex. Because the graph is dilated vertically by a factor of 2, the y-values will decrease by 2 times 1, 3, 5, . . . In other words, we will go down by 2 blocks, then by 6 blocks, then by 10 blocks, and so on.

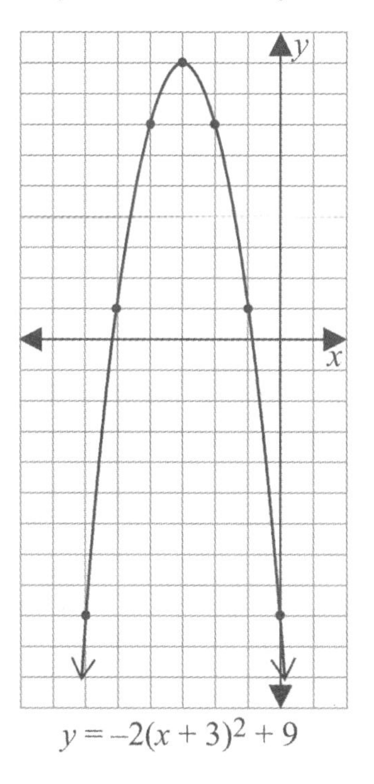

$y = -2(x + 3)^2 + 9$

EXAMPLE 5.18 Graph: $y = \frac{1}{2}(x - 3)^2 + 4$

SOLUTION

Graph the vertex at (3, 4). For each block to the left and right of the vertex, count up $\frac{1}{2}$ times 1, 3, 5, ... blocks. So count up $\frac{1}{2}$, then up $\frac{3}{2}$, and then up $\frac{5}{2}$ blocks, always moving just one block horizontally. Plot the points and connect them with a smooth curve.

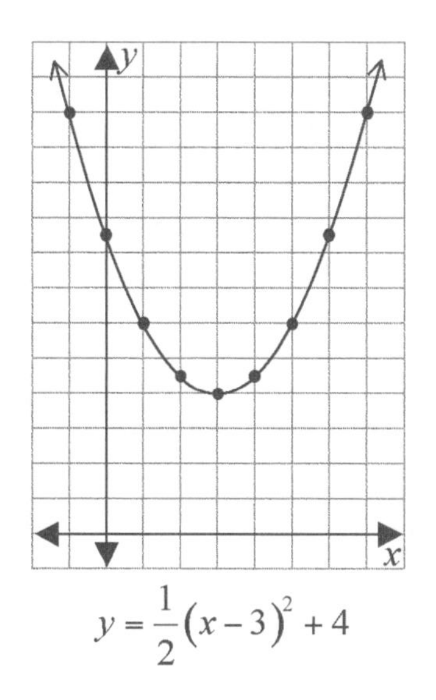

$y = \frac{1}{2}(x - 3)^2 + 4$

COMMON ERROR

Counting up $\frac{1}{2}$ block and over 1 block horizontally is not the same as counting up 1 block and over 2 blocks horizontally. Do not get graphing the y differences confused with graphing slope for a line.

A precise graph of a parabola is not always necessary. In some problems, a rough sketch may be adequate.

EXAMPLE 5.19

A center fielder throws the ball toward home plate. The ball follows the path shown by the equation $y = -0.002(x - 110)^2 + 32$, where y is the vertical height of the ball and x is the horizontal distance of the ball from the fielder, both in feet. Find the vertex and the zeros (to the nearest tenth) of this function, and explain what they mean in the context of this problem. Then make a rough sketch of the player's throw.

SOLUTION

The vertex is at (110, 32). This means the ball reached a height of 32 feet above the ground when it was 110 feet from the outfielder.

To find the zeros, set $y = 0$ and solve for x.

$$0 = -0.002(x - 110)^2 + 32$$

$0.002(x - 110)^2 = 32$	Add $0.002(x - 110)^2$ to both sides to move the term to the left side.
$(x - 110)^2 = 16{,}000$	Divide each side by 0.002.
$x - 110 \approx \pm 126.5$	Take the square root of each side.
$x \approx 110 \pm 126.5$	Add 110 to each side to get x by itself.
$x \approx -16.5 \lor x \approx 236.5$	Round to the nearest tenth.

The positive zero says the ball would travel 236.5 feet from the fielder before hitting the ground if no other player caught it first. The negative zero is behind the fielder and has no real meaning in this example.

With the vertex and both zeros, we can make a rough sketch of the graph as below. Recall that the vertex is on the axis of symmetry, which always passes halfway between the two zeros.

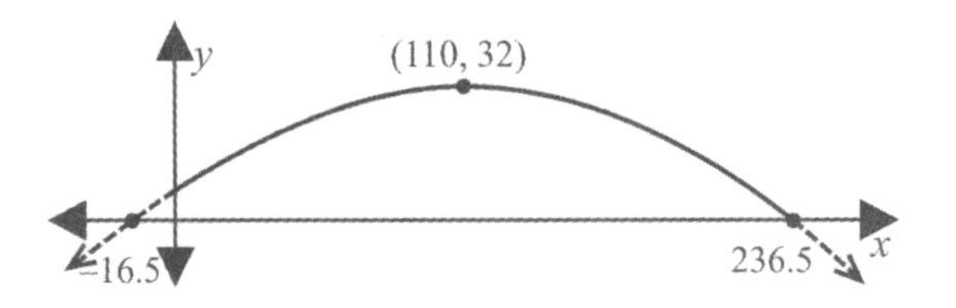

Mathematically, the ends of the parabola continue indefinitely down on both the right and left. However, the only part of the parabola that is meaningful in this problem is the part between the y-intercept and the positive zero. This represents the in-air path of the ball from where it left the fielder's hand to where it first hit the ground, assuming it wasn't caught.

Note that the scales on the two axes are not the same. To have roughly equal scales, the vertex would need to be about half as high. How important that is depends on the purpose of the graph.

CHANGING BETWEEN GENERAL AND VERTEX FORMS

The same parabola may be written in either general form, $y = ax^2 + bx + c$, or vertex form, $y = a(x - h)^2 + k$. Converting from one form to the other is sometimes helpful.

The vertex form, $y = a(x - h)^2 + k$, of the parabola can be obtained by finding the vertex of the parabola from its general form, $y = ax^2 + bx + c$. First find the axis of symmetry, $x = -\frac{b}{2a}$. This gives the x-coordinate of the vertex. Then substitute this value into the equation to find the y-coordinate of the vertex. The vertex is at $\left(-\frac{b}{2a}, c - \frac{b^2}{4a}\right)$. This gives the values of h and k. The vertex form can then be written as

$$\begin{aligned} y &= a(x - h)^2 + k \\ &= a\left(x - \frac{-b}{2a}\right)^2 + c - \frac{b^2}{4a} \end{aligned}$$

You need to remember the formula for the axis of symmetry, $x = -\frac{b}{2a}$, to find h. It is probably not worth memorizing the formula for k. After finding h, just substitute it for x in the general equation to find k. Note that the value of a does not change. A given parabola will have the same value of a when written in vertex form as when written in general form.

EXAMPLE 5.20

Rewrite $y = x^2 - 6x + 7$ in the vertex form.

SOLUTION

Find the coordinates of the vertex.

$$x = -\frac{b}{2a} = -\frac{(-6)}{2(1)} = 3$$

$$y = 3^2 - 6(3) + 7 = -2$$

The vertex is at $(3, -2)$. From the original equation, we can see that the value of a is 1.

In vertex form, the equation is $y = (x - 3)^2 - 2$.

EXAMPLE 5.21

Rewrite $y = 2x^2 - 8x + 9$ in the vertex form.

SOLUTION

Find the vertex.

SEE CALC TIP 5E

$$x = -\frac{b}{2a} = -\frac{(-8)}{2(2)} = 2$$

$$y = 2(2)^2 - 8(2) + 9 = 1$$

The vertex is (2, 1) and $a = 2$. The vertex form of the equation is $y = 2(x - 2)^2 + 1$.

Changing the vertex form of a quadratic function into the general form is really easy. Just distribute and combine.

EXAMPLE 5.22

Rewrite $y = -2(x + 3)^2 - 9$ in the general form.

SOLUTION

$$y = -2(x + 3)^2 - 9$$

$$y = -2\left(x^2 + 6x + 9\right) - 9 \qquad \text{First square the binomial.}$$

$$y = -2x^2 - 12x - 18 - 9 \qquad \text{Multiply the trinomial by } -2.$$

$$y = -2x^2 - 12x - 27 \qquad \text{Combine the constants.}$$

COMMON ERROR

Some students multiply the constant into the binomial before squaring.

$y = -2(x+3)^2 - 9$ is not the same as $y = (-2x-6)^2 - 9$. Order of operations requires the binomial be squared before multiplying by –2.

SECTION EXERCISES

5–22. For the parabola $y = -2(x - 5)^2 - 3$:

a. Find the coordinates of the vertex.

b. Write the equation of the axis of symmetry.

5–23. Describe how to transform $y = x^2$ to get the graph of $y = -\frac{1}{2}(x - 5)^2 + 4$.

5–24. Graph: $y = -\frac{1}{2}(x - 4)^2 + 6$

5–25. A basketball player stands at the origin and shoots at the basket, the center of which is 23 feet away and 10 feet off the floor. The equation of flight of the ball is $y = \frac{-x^2}{16} + \frac{3}{2}x + 7$, where y is the height of the ball and x is the horizontal distance of the ball from the player (both in feet).

a. How high off the floor was the ball when it left the player's hands?

b. How high does the ball reach? How far (horizontally) from the player is it at the highest point?

c. Put the equation into vertex form $y = a(x - h)^2 + k$.

d. Does the ball go through the hoop? Justify your answer.

5–26. Rewrite $y = x^2 - 8x + 23$ in vertex form.

5–27. Rewrite $y = -3x^2 - 54x - 256$ in vertex form.

5–28. Rewrite $y = -4(x - 5)^2 + 3$ in general form.

Solving Quadratic Inequalities

If you have read Chapter 4, you will recognize the following as similar to the method we used to solve absolute value inequalities.

To solve a quadratic inequality, follow these steps:

1. Find the zeros of the associated quadratic equation (just replace the inequality sign with an equal sign and solve).
2. Graph the zeros on a number line. Use open or closed circles as appropriate depending on the inequality symbol. This will (usually) divide the number line into three intervals.
3. Check each of the intervals to see if it satisfies the original inequality. These intervals may be checked graphically or by checking a value in the original inequality algebraically.
4. Write the solution set with the appropriate notation.

EXAMPLE 5.23 Solve $x^2 - 2x \geq 8$, and write its solution in interval notation.

SOLUTION

First solve for the corresponding equality.

$$x^2 - 2x = 8$$
$$x^2 - 2x - 8 = 0$$
$$(x + 2)(x - 4) = 0$$
$$x = -2 \text{ or } x = 4$$

Now place the zeros on a number line. Use closed circles at each solution because ≥ includes the zeros. Then check a number from each interval in the original inequality.

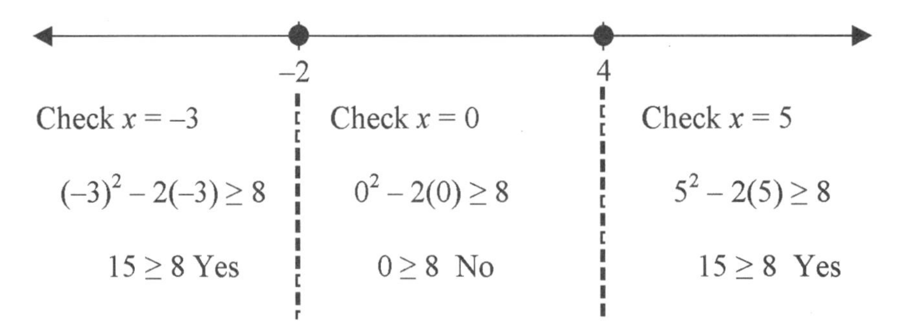

Shade the intervals that check, and write the solution set in appropriate notation.

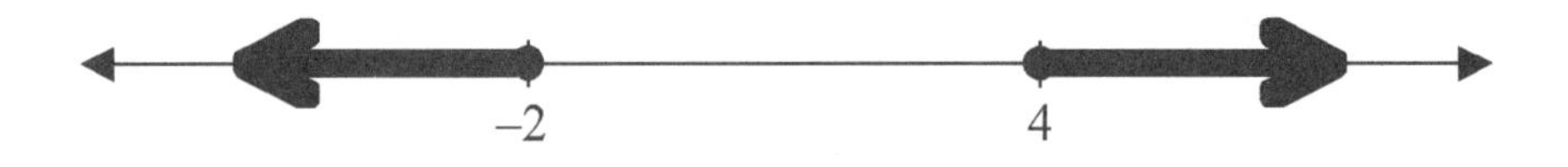

The solution set is $x \leq -2$ or $x \geq 4$, which can be written in interval notation as $(-\infty, -2] \cup [4, \infty)$.

An alternative to checking individual points in the intervals is to look at a sketch of the graph of $y = x^2 - 2x - 8$. You know the zeros are −2 and 4 (because you solved the equation) and that it opens up (because $a = 1$ is positive). Because the inequality is "greater than or equal," you want x-values where the graph of the parabola is on or above the x-axis. These are to the left of −2 and to the right of 4.

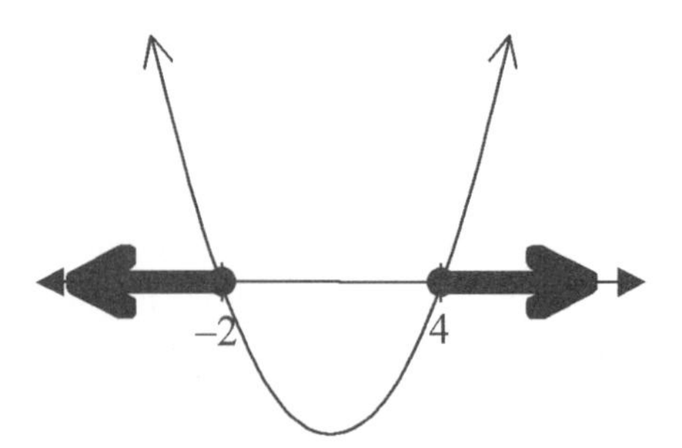

By looking at this sketch, you should be able to write that the solution set is $(-\infty, -2] \cup [4, \infty)$.

EXAMPLE 5.24 Solve: $(x - 3)^2 - 16 < 0$

SOLUTION

Solve for the corresponding equality first.

$$
\begin{aligned}
(x - 3)^2 - 16 &= 0 \\
(x - 3)^2 &= 16 \\
x - 3 &= \pm\sqrt{16} \\
x - 3 &= \pm 4 \\
x &= 3 \pm 4 \\
x &= -1 \text{ or } x = 7
\end{aligned}
$$

Now place the zeros on a number line. Use open circles at each solution because < does not include the zeros. Then check a number from each interval in the original inequality.

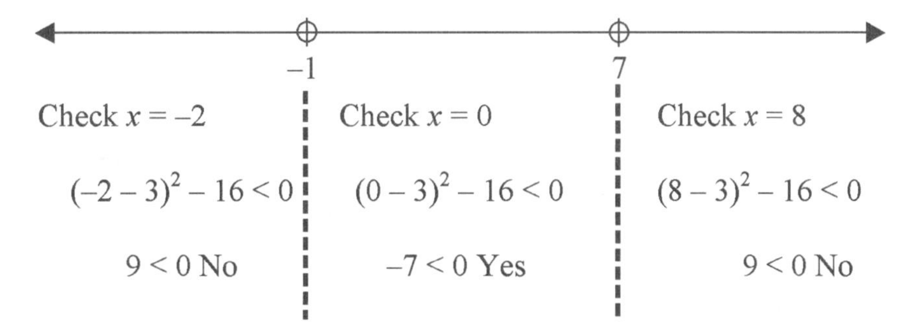

Shade the intervals that check, and write the solution set in appropriate notation.

The solution set is $-1 < x < 7$.

SECTION EXERCISES

For 29–34, solve for x *either algebraically or graphically.*

5–29. $x^2 - 16 < 0$

5–30. $x^2 - 10 \geq 0$

5–31. $x^2 + 4x + 5 > 0$

5–32. $x^2 - 16x + 64 \leq 0$

5–33. $x^2 - 3x + 4 \leq 0$

5–34. $15 - 2x - x^2 < 0$

5–35. A batter hits a baseball so that the ball's height, *h,* in feet is given by $h(t) = -16t^2 + 60t + 4$, where *t* is time in seconds since the ball was hit. For what interval of time was the ball more than 40 feet off the ground?

Writing a Quadratic Equation

At times, you may want to write a quadratic equation that satisfies certain information, perhaps from a word problem or a graph.

A QUADRATIC EQUATION FROM RATIONAL ZEROS

If you want to write a quadratic equation and you know the values of its zeros, the easiest way to find an equation is to work backward from the zeros to the factors and then to the equation. Remember how we found the zeros from the factors? We set each factor equal to 0 and solved for *x*. To work backward, we will undo operations until we have an equation equal to 0.

EXAMPLE 5.25

Write a quadratic equation having zeros $x = -2$ and $x = \frac{7}{3}$.

SOLUTION

$x = -2$	$x = \frac{7}{3}$	Write the solutions.
	$3x = 7$	Clear fractions if necessary. (Multiply the second solution by 3.)
$x + 2 = 0$	$3x - 7 = 0$	Add or subtract to get two factors equal to 0.

The two factors are $(x + 2)$ and $(3x - 7)$.

$(x + 2)(3x - 7) = 0$	Multiply the factors, and set them equal to 0 to write the equation.
$3x^2 - x - 14 = 0$	Distribute and combine to rewrite the equation in general form.

A possible equation is $3x^2 - x - 14 = 0$.

Note that the answer is not unique. The equation can be multiplied by any nonzero constant to get another equation with the same zeros. For example, multiplying by -2 gives $-6x^2 + 2x + 28 = 0$, which has the same zeros.

Just knowing the zeros of a quadratic equation is not enough information to determine the equation; there are an infinite number of equations with those zeros. With one more piece of information, a point on the graph or a y-intercept, you can determine the exact equation.

EXAMPLE 5.26

Write the quadratic function having zeros $x = -2$ and $x = \frac{7}{3}$ and a y-intercept of 7.

SOLUTION

The zeros are the same as those in the previous example, so we know we need an equation of the form $y = a(x + 2)(3x - 7)$. A y-intercept of 7 means $y = 7$ when $x = 0$; substitute these values of x and y into the equation to find a.

$$\begin{aligned} 7 &= a\big((0) + 2\big)\big(3(0) - 7\big) \\ 7 &= a(2)(-7) \\ 7 &= -14a \\ a &= -\frac{1}{2} \end{aligned}$$

Write the value for a in the equation. Then distribute and combine to write the equation in general form

$$\begin{aligned} y &= -\frac{1}{2}(x + 2)(3x - 7) \\ &= -\frac{1}{2}(3x^2 - x - 14) \\ &= -\frac{3}{2}x^2 + \frac{1}{2}x + 7 \end{aligned}$$

The quadratic function is $y = -\frac{3}{2}x^2 + \frac{1}{2}x + 7$.

A QUADRATIC EQUATION FROM IRRATIONAL ZEROS

If the zeros are irrational (or imaginary, see Chapter 6), the above procedure can still be used. However, it is easier to isolate the radical term and then square both sides.

Write a quadratic equation having zeros $x = \dfrac{-1 \pm 2\sqrt{5}}{3}$

SOLUTION

$$x = \frac{-1 \pm 2\sqrt{5}}{3}$$

$3x = -1 \pm 2\sqrt{5}$	Multiply by 3 to clear the fraction.
$3x + 1 = \pm 2\sqrt{5}$	Add 1 to isolate the radical term.
$(3x + 1)^2 = \left(\pm 2\sqrt{5}\right)^2$	Square both sides.
$9x^2 + 6x + 1 = 20$	Square the binomial on the left. On the right, note that squaring eliminates the ± sign.
$9x^2 + 6x - 19 = 0$	Subtract 20 to write in general form.

One equation having the given roots is $9x^2 + 6x - 19 = 0$.

EXAMPLE 5.28

Write the quadratic function having zeros $x = \dfrac{-1 \pm 2\sqrt{5}}{3}$ and passing through the point (1, 8).

SOLUTION

The zeros are the same as those in the equation from the previous example, so we need an equation of the form $y = a\left(9x^2 + 6x - 19\right)$. Use the fact that the graph passes through the point (1, 8) to find the value of a.

$$8 = a\left(9(1)^2 + 6(1) - 19\right)$$
$$8 = -4a$$
$$a = -2$$

Write the value for a in the equation. Then distribute to write the equation in general form

$$y = -2\left(9x^2 + 6x - 19\right)$$
$$= -18x^2 - 12x + 38$$

The desired quadratic function is $y = -18x^2 - 12x + 38$.

A QUADRATIC EQUATION FROM ITS VERTEX AND A POINT

If you know the vertex for a parabola, it is easier to use the vertex form of the quadratic equation, $y = a(x - h)^2 + k$. Again, you need one more point on the graph to find a unique equation.

EXAMPLE 5.29

Write the quadratic function with a vertex at (–3, 7) that passes through the point (–1, 8).

SOLUTION

The vertex is given as (–3, 7) so the vertex form of the equation is:

$$y = a(x - (-3))^2 + 7$$
$$= a(x + 3)^2 + 7$$

Use the fact that the graph of the function passes through (–1, 8) to find the value of a.

$$8 = a(-1 + 3)^2 + 7$$
$$8 = a(2)^2 + 7$$
$$8 = 4a + 7$$
$$1 = 4a$$
$$\frac{1}{4} = a$$

The desired function is $y = \frac{1}{4}(x + 3)^2 + 7$. This can be multiplied out if you prefer the general form.

COMMON ERROR

When solving for a, students sometimes change $8 = a(2)^2 + 7$ to $8 = a4 + 7$ and then conclude that $8 = a11$ and $a = \frac{8}{11}$. Avoid this error by always writing the coefficient on the left side of the variable.

SECTION EXERCISES

5–36. Find the equation of the parabola having vertex (–2, 6) and passing through the point (4, –3).

5–37. Write the general form of the quadratic function having zeros $x = 5$ and $x = -\frac{2}{3}$ and a y-intercept of 20.

5–38. Write the general form of the quadratic function having zeros $x = \frac{-7 \pm 3\sqrt{2}}{5}$ and passing through the point (–2, 9).

5–39. A young ballplayer hits a pop fly. He hits the ball when it is 2 feet above the ground. The ball reaches its highest point 20 feet horizontally from the batter and 10 feet off the ground. Assume the ball's path is in the shape of a parabola. Find the vertex form for the equation for the path of the ball with the origin at ground level below where the ball was hit.

Solving Quadratic-Linear Systems of Equations

A system of one quadratic and one linear equation is called a *quadratic-linear system* (a Q-L system). Such systems can be solve algebraically, using substitution, or graphically.

To solve a Q-L system algebraically:

1. Solve one equation for one variable in terms of the other.
2. Substitute this expression into the other equation and solve.

EXAMPLE 5.30

Solve: $\begin{cases} 3x - 2y = 4 \\ y = \frac{1}{2}x^2 + x - 3 \end{cases}$

SOLUTION

$$3x - 2y = 4$$
$$y = \frac{1}{2}x^2 + x - 3$$

The bottom equation is solved for y, so substitute the entire expression for y into the top equation.

$$3x - 2\left(\frac{1}{2}x^2 + x - 3\right) = 4$$

Remember to use parentheses.

$$3x - x^2 - 2x + 6 = 4$$

Distribute –2 carefully.

$$x^2 - x - 2 = 0$$

Combine like terms, and write the equation in general form. If necessary, multiply every term by –1 so the quadratic term is positive (this makes factoring easier).

$$(x + 1)(x - 2) = 0$$
$$x = -1 \text{ or } x = 2$$

Factor, set each factor equal to 0, and solve for x.

When $x = -1$, $y = \frac{1}{2}(-1)^2 + (-1) - 3 = -3.5$

Find the corresponding value of y for each value of x from one of the original equations.

When $x = 2$, $y = \frac{1}{2}(2)^2 + 2 - 3 = 1$

$(-1, -3.5), (2, 1)$

Write the solutions as ordered pairs, matching each x with its corresponding value of y.

The solution set is $\{(-1, -3.5), (2, 1)\}$.

To solve a Q-L system graphically:

1. Graph both equations on the same set of axes.
2. Find the point(s) of intersection.
3. Write the solution(s).

SEE CALC TIP 5F

EXAMPLE 5.31

Solve graphically: $\begin{cases} 3x - 2y = 4 \\ y = \frac{1}{2}x^2 + x - 3 \end{cases}$

SOLUTION

Solve the linear equation for y:

$$\begin{aligned} 3x - 2y &= 4 \\ -2y &= -3x + 4 \\ y &= \frac{-3x}{-2} + \frac{4}{-2} \\ y &= \frac{3}{2}x - 2 \end{aligned}$$

Graph $\begin{cases} y = \dfrac{3}{2}x - 2 \\ y = \dfrac{1}{2}x^2 + x - 3 \end{cases}$ on the same set of axes.

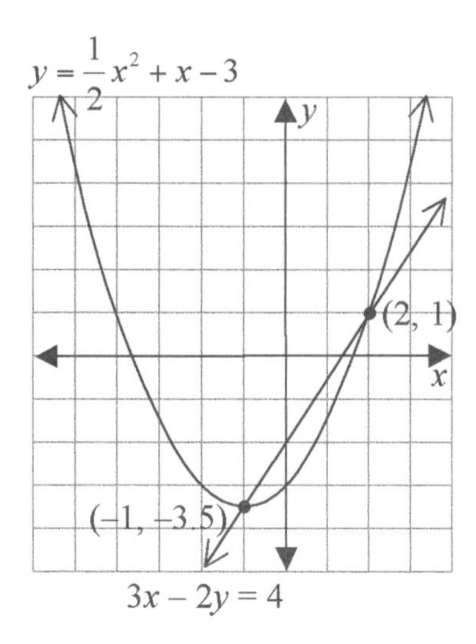

Label the equations, and write the ordered pairs of the points of intersection.

The solution set is {(–1, –3.5), (2, 1)}.

SECTION EXERCISES

For 40–41, solve the following systems of equations algebraically.

5–40. $\begin{cases} y = x^2 + 2x + 1 \\ y = x + 3 \end{cases}$

5–41. $\begin{cases} y - 2x = -3 \\ y = -2x^2 + 2x + 5 \end{cases}$

Solve the following system of equations graphically.

5–42. $\begin{cases} y = -3x + 6 \\ y = -x^2 + x + 6 \end{cases}$

CHAPTER EXERCISES

For 1–2, solve algebraically. Express irrational answers in simplest radical form.

C5–1. $x^2 + 5x - 6 = 0$

C5–2. $3x^2 = x$

C5–3. $6x^2 + 5x = 4$

C5–4. $\dfrac{x}{5} = \dfrac{x + 6}{x - 2}$

C5–5. Solve $25x^2 + 50x - 1 = 0$ by completing the square.

C5–6. Solve $5x^2 - 7x - 1 = 0$ using the quadratic formula.

C5–7. Choose all answers that satisfy the given condition. A quadratic equation with rational coefficients has discriminant D.

a. The roots of the equation will be imaginary if D equals
(1) –8 (2) 0 (3) 8 (4) 16

b. The graph of the equation will intersect the x-axis twice if D equals
(1) –8 (2) 0 (3) 8 (4) 16

c. The roots of the equation will be real and equal if D equals
(1) –8 (2) 0 (3) 8 (4) 16

C5–8. For what values of c will the parabola $y = 5x^2 + 12x + c$ have two distinct real roots?

C5–9. For $y = 2x^2 - 4x - 16$:

a. Sketch the graph.
b. Identify the axis of symmetry.
c. Identify the zeros.
d. Identify the vertex.
e. Identify the y-intercept.
f. Describe where the graph is increasing and decreasing.
g. Describe the end behavior of the graph.

C5–10. For $y = -2x^2 - 2x + 12$:

a. Does the parabola open up or down?
b. What is the y-intercept of the parabola?
c. What is the equation of the axis of symmetry of the parabola?
d. What are the coordinates of the vertex of the parabola? Is it a maximum or a minimum?
e. What are the zeros of the parabola?
f. Sketch the parabola.

C5–11. A soccer goalie kicks the ball downfield. The path of the ball can be represented by $y = -0.04x^2 + 2x + 2$, where y is the height of the ball and x is the horizontal distance from where the goalie kicked it, both in feet.

a. How high off the ground was the ball when it was kicked?
b. How high did the ball reach?
c. How far away was the ball when it reached its maximum?
d. How far from the kicker did the ball first hit the ground (round to the nearest foot)?

C5–12. An object's position as a function of time is given by $x(t) = -t^2 + 6t + 5$, $t \geq 0$ (x in meters, t in seconds).

a. What is the object's average velocity for the first 4 seconds? (Average velocity = $\Delta x/\Delta t$).

b. What is the object's average velocity from $t = 1$ to $t = 9$ seconds?

c. Write an equation for the secant line that intersects the graph of $x(t)$ at the points where $t = 1$ and $t = 9$.

C5–13. Describe how to transform $y = x^2$ to get the graph of $y = 3(x + 2)^2 - 7$.

C5–14. Graph $y = 2(x + 3)^2 - 5$.

C5–15. Rewrite $y = x^2 + 10x + 24$ in vertex form.

C5–16. Rewrite $y = 2x^2 - 4x - 4$ in vertex form.

For 17–18, solve algebraically or graphically.

C5–17. $x^2 - 5x \leq 24$

C5–18. $5 - 4x - x^2 < 0$

C5–19. Using the graph of $y = f(x)$ below, write the solution set to:

a. $f(x) \geq 0$

b. $f(x) < 5$

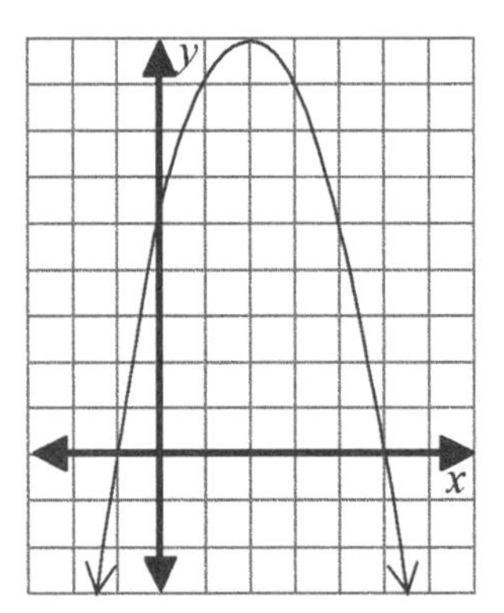

C5–20. Find the equation of the parabola having vertex (10, 15) and a y-intercept of 315.

C5–21. Write the factored form of a quadratic function having zeros $x = -7$ and $x = \frac{5}{2}$ and passing through the point (1, –6).

C5–22. Write the general form of the quadratic function having zeros $x = \frac{7 \pm 2\sqrt{11}}{2}$ and a y-intercept of –10.

Solve the following system of equations algebraically.

C5–23. $\begin{cases} y = 0.5x - 1 \\ y = -0.25x^2 + x + 5 \end{cases}$

Solve the following system of equations graphically.

C5–24. $\begin{cases} y = x^2 - 2x - 8 \\ y - x = 2 \end{cases}$

6

Complex Numbers

WHAT YOU WILL LEARN

Throughout your whole life, you have been slowly extending your ideas about what can be a number. When you were young, you may have been taught that you could not divide 2 by 5 because 5 does not "go into" 2. Later, you learned about fractions. You may also have heard that you can't subtract 5 from 2 because the 2 "isn't big enough" to take 5 away from it. Then you learned about negative numbers. Later still, you learned that 5 really does have a square root; it's just irrational. As recently as middle school or even Algebra I, you were probably taught that you couldn't take a square root of a negative number because nothing multiplied by itself can equal a negative. Well, guess what? It's time to extend our number system once more to include imaginary and complex numbers. Despite their name, imaginary numbers are useful for solving some very real problems in mathematics. In this chapter, you will learn to:

- Write square roots of negative numbers as imaginary numbers;
- Compute powers of i;
- Perform arithmetic on complex numbers;
- Solve any quadratic equation;
- Write a quadratic equation from its complex zeros;
- Graph complex numbers.

SECTIONS IN THIS CHAPTER

Imaginary Numbers

The quadratic equation $x^2 - 1 = 0$ is easily solved: $x^2 = 1$ so $x = \pm\sqrt{1} = \pm 1$. However, the equation $x^2 + 1 = 0$ presents a problem. We get $x^2 = -1$ so $x = \pm\sqrt{-1}$. However, there is no real number that when squared gives -1. The problem has no real solution. In the mid-sixteenth century, mathematicians invented a new kind of number to solve this problem. Over time, these numbers came to be called *imaginary numbers* and are represented using the *imaginary unit*, i, which is defined as follows:

$$i = \sqrt{-1} \text{ or } i^2 = -1$$

By using i, we can take square roots of all negative numbers. For $k > 0$:

$$\sqrt{-k} = \sqrt{-1}\sqrt{k} = i\sqrt{k}$$

EXAMPLE 6.1

Rewrite the following in terms of i.

a. $\sqrt{-16}$

b. $\sqrt{-17}$

c. $\sqrt{-18}$

SOLUTION

SEE CALC TIP 6A

a.
$$\begin{aligned}\sqrt{-16} &= \sqrt{-1}\sqrt{16}\\ &= i(4)\\ &= 4i\end{aligned}$$

b.
$$\begin{aligned}\sqrt{-17} &= \sqrt{-1}\sqrt{17}\\ &= i\sqrt{17}\end{aligned}$$

c.
$$\begin{aligned}\sqrt{-18} &= \sqrt{-1}\sqrt{18}\\ &= i\sqrt{18}\\ &= i\left(3\sqrt{2}\right)\\ &= 3i\sqrt{2}\end{aligned}$$

Note that i is usually written after rational numbers but before irrational numbers. (This is done mostly for clarity. It minimizes the chance of confusing $\sqrt{2}i$ with $\sqrt{2i}$.)

Like real numbers, imaginary numbers have opposites. For example, $-\sqrt{-9} = -3i$.

COMMON ERROR

Be careful not to mix up $\sqrt{-9}$ and $-\sqrt{9}$.

SECTION EXERCISES

For 1–4, express each number in terms of i *and simplify.*

6–1. $\sqrt{-81}$

6–2. $-\sqrt{-24}$

6–3. $\frac{2}{3}\sqrt{-144}$

6–4. $2\sqrt{-75}$

Powers of *i*

If we accept that i obeys the rules of exponents and remember the definition $i = \sqrt{-1}$, we can look for a pattern in the powers of i.

TABLE 6.1
PATTERNS IN POWERS OF *i*

Power of *i*		Result
$i^0 =$		1
$i^1 =$		i
$i^2 =$	$(\sqrt{-1})^2 =$	-1
$i^3 =$	$i^2 i = -1i =$	$-i$
$i^4 =$	$i^2 i^2 = (-1)(-1) =$	1
$i^5 =$	$i^4 i^1 = 1i =$	i
$i^6 =$	$i^4 i^2 = 1(-1) =$	-1
$i^7 =$	$i^4 i^3 = 1(-i) =$	$-i$
$i^8 =$	$i^4 i^4 = 1(1) =$	1

The powers of i are cyclic. They keep repeating: 1, i, -1, $-i$ and then back to 1. This can be represented on the "clock" shown in Figure 6.1 where the numbers inside the circle show the first four exponents on i (starting at 0) and the numbers outside show the results.

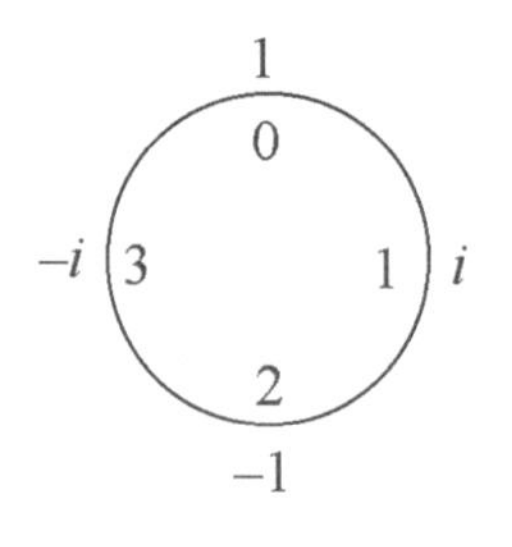

FIGURE 6.1

For small powers of i, you can start at the exponent 1 and simply count around the circle until you get to the desired power. For example, for i^{11}, you start counting at the exponent 1. When you get to 11, you are at 3 on the circle. So $i^{11} = -i$.

This is no fun for calculating large powers of i, but there is a shortcut. Since $i^4 = 1$, remove all the groups of i^4 by dividing the exponent by 4. Keep just the remainder, and find that remainder inside the circle to get the answer.

Evaluate: i^{11}

SOLUTION

First divide the power by 4 and find the remainder. The number 4 divides into 11 twice with the remainder 3. Then find the result of i to the power of the remainder. So $i^{11} = i^3 = -i$, which matches the answer we found by counting around the circle.

SEE CALC TIP 6B

Evaluate: i^{126}

SOLUTION

$$4\overline{)126}^{\;31r2} \qquad \text{So } i^{126} = i^2 = -1.$$

SEE CALC TIP 6B

SECTION EXERCISES

For 5–8, simplify each of the following.

6–5. i^{24}

6–6. $2i^9 \cdot 3i^5$

6–7. i^{999}

6–8. $i + i^2 + i^3 + i^4 + \ldots + i^{100}$

Operations with Complex Numbers

Numbers of the form bi, where b is a nonzero real number, are called *pure imaginary numbers*. The set of all numbers of the form $a + bi$, where a and b are both real numbers, is called the set of *complex numbers*. Note that the real numbers, $b = 0$, and the pure imaginary numbers, $a = 0$ and $b \neq 0$, are both subsets of the complex numbers.

ADDING AND SUBTRACTING COMPLEX NUMBERS

We add and subtract complex numbers the same way we add and subtract binomials, by combining like terms. Real numbers are combined with real numbers, and imaginary numbers are combined with imaginary numbers.

Find the sum of $3 + 7i$ and $5 - 3i$.

SOLUTION

$(3 + 7i) + (5 - 3i) = (3 + 5) + (7i - 3i) = 8 + 4i$

For addition, using the parentheses around the complex numbers is optional. This example shows how the like terms are grouped; usually students add the like terms mentally.

Subtract $2 - 6i$ from $-1 + 3i$.

SOLUTION

$(-1 + 3i) - (2 - 6i) = -1 + 3i - 2 + 6i = -3 + 9i$

For subtraction, parentheses around the first number are optional. However, those around the second number are required to show that the subtraction applies to the whole binomial, reminding you to distribute the negative sign.

COMMON ERROR

Subtract $2 - 6i$ from $-1 + 3i$ means $-1 + 3i$ is first and $2 - 6i$ is second (after the subtraction sign). Remember: "From is First and Subtract is Second."

MULTIPLYING COMPLEX NUMBERS

Multiplying complex numbers is the same as multiplying monomials and binomials except you must remember that $i^2 = -1$ and simplify the answer.

Simplify: $3i(2 - 5i)$

SOLUTION

$$3i(2 - 5i) = 6i - 15i^2 = 6i - 15(-1) = 15 + 6i$$

EXAMPLE 6.7

Find the product: $\left(4 - i\sqrt{3}\right)\left(5 + 2i\sqrt{3}\right)$

SOLUTION

$$\begin{aligned}\left(4 - i\sqrt{3}\right)\left(5 + 2i\sqrt{3}\right) &= 20 + 8i\sqrt{3} - 5i\sqrt{3} - 2i^2\left(\sqrt{3}\right)^2 \\ &= 20 + 3i\sqrt{3} - 2(-1)(3) \\ &= 20 + 3i\sqrt{3} + 6 \\ &= 26 + 3i\sqrt{3}\end{aligned}$$

EXAMPLE 6.8

Simplify: $(3 + 2i)(3 - 2i)$

SOLUTION

$$\begin{aligned}(3 + 2i)(3 - 2i) &= 9 - 6i + 6i - 4i^2 \\ &= 9 - 4(-1) \\ &= 9 + 4 \\ &= 13\end{aligned}$$

SEE CALC TIP 6C

In Example 6.8, two complex numbers are multiplied together to give a real answer. The numbers $a + bi$ and $a - bi$ are called *complex conjugates*. They have the same real part, a, but opposite imaginary parts, bi and $-bi$. Their product will always be a real number. All the imaginary parts cancel out.

$$(a + bi)(a - bi) = a^2 - \cancel{abi} + \cancel{abi} - b^2i^2 = a^2 - b^2(-1) = a^2 + b^2$$

DIVIDING COMPLEX NUMBERS

When a complex number is properly written in $a + bi$ form, there is no i in a denominator. This presents a small problem when dividing complex numbers; we need to express the final answer with real denominators. There are two ways to do this, depending on whether you are dividing by a pure imaginary number or by a complex number.

When dividing by a pure imaginary, write the quotient as a fraction, multiply both the numerator and denominator by i, and simplify. Remember that $i^2 = -1$.

EXAMPLE 6.9

Simplify: $\dfrac{2}{3i}$

SOLUTION

$$\frac{2}{3i} = \frac{2}{3i} \cdot \frac{i}{i} = \frac{2i}{3i^2} = \frac{2i}{3(-1)} = -\frac{2}{3}i$$

EXAMPLE 6.10

Express the quotient $\dfrac{2 - 3i\sqrt{6}}{2i\sqrt{6}}$ in $a + bi$ form.

SOLUTION

Many people prefer denominators to be *rationalized.* Not only should there be no i in a denominator but there should also be no radicals. Both the i and the $\sqrt{6}$ in the denominator can be taken care of at the same time by multiplying both the numerator and denominator by $i\sqrt{6}$. (Rationalizing denominators is covered more thoroughly in Chapter 8.)

$$\left(\frac{2 - 3i\sqrt{6}}{2i\sqrt{6}}\right)\left(\frac{i\sqrt{6}}{i\sqrt{6}}\right) = \frac{2i\sqrt{6} - 3i^2\left(\sqrt{6}\right)^2}{2i^2\left(\sqrt{6}\right)^2}$$

$$= \frac{2i\sqrt{6} - 3(-1)(6)}{2(-1)(6)}$$

$$= \frac{2i\sqrt{6} + 18}{-12}$$

$$= \frac{18}{-12} + \frac{2i\sqrt{6}}{-12}$$

$$= -\frac{3}{2} - \frac{1}{6}i\sqrt{6}$$

Recall that the product of two complex conjugates is a real number. We can use this fact to simplify quotients with binomial denominators. To divide by a complex binomial, $a + bi$, write the quotient as a fraction, multiply both the numerator and denominator by the conjugate of the denominator, $a - bi$, and simplify.

EXAMPLE 6.11

Express the quotient $\dfrac{3-12i}{2-4i}$ in simplest $a + bi$ form.

SOLUTION

Multiply both the numerator and denominator by the conjugate of the denominator, $2 + 4i$.

$$\frac{(3-12i)}{(2-4i)}\cdot\frac{(2+4i)}{(2+4i)} = \frac{6+12i-24i-48i^2}{4+8i-8i-16i^2}$$
$$= \frac{6+12i-24i+48}{4+16}$$
$$= \frac{54-12i}{20}$$
$$= \frac{27}{10}-\frac{3}{5}i$$

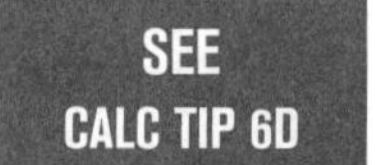
SEE CALC TIP 6D

COMMON ERROR

Do *not* reduce the real and imaginary parts separately!

$$\frac{8-12i}{2-4i} \neq \frac{{}^{4}\cancel{8}-{}^{3}\cancel{12i}}{\cancel{2}-\cancel{4i}} \neq 4-3i$$

POWERS OF COMPLEX NUMBERS

In the section on powers of binomials (Chapter 2), we learned the binomial theorem and Pascal's triangle. You may want to review that section at this time. The binomial theorem shows the patterns for a power on a binomial.

$$(a+b)^n = {}_nC_0a^nb^0 + {}_nC_1a^{n-1}b^1 + {}_nC_2a^{n-2}b^2 + \ldots + {}_nC_na^0b^n$$

Expanding powers of complex numbers combines the binomial theorem with rules you have learned for simplifying complex numbers.

EXAMPLE 6.12

Simplify: $(2 + 3i)^5$

SOLUTION

When we learned to expand polynomials, we did this example.

$$(a+b)^5 = 1a^5b^0 + 5a^4b^1 + 10a^3b^2 + 10a^2b^3 + 5a^1b^4 + 1a^0b^5$$

To simplify $(2 + 3i)^5$, let $a = (2)$ and $b = (3i)$. Then simplify each term (exponents before multiplication), evaluate powers of i, and add like terms.

$$\begin{aligned}(2+3i)^5 &= 1(2)^5(3i)^0 + 5(2)^4(3i)^1 + 10(2)^3(3i)^2 + 10(2)^2(3i)^3 + 5(2)^1(3i)^4 + 1(2)^0(3i)^5 \\ &= 1(32)(1) + 5(16)(3i) + 10(8)\left(9i^2\right) + 10(4)\left(27i^3\right) + 5(2)\left(81i^4\right) + 1(1)\left(243i^5\right) \\ &= 32 + 240i - 720 - 1{,}080i + 810 + 243i \\ &= 122 - 597i\end{aligned}$$

SEE CALC TIP 6E

EXAMPLE 6.13 Simplify: $(2 - i)^6$

SOLUTION

$$(a+b)^6 = 1a^6b^0 + 6a^5b + 15a^4b^2 + 20a^3b^3 + 15a^2b^4 + 6ab^5 + 1a^0b^6$$

Think of (2) as taking the place of a and $(-i)$ (note the sign) as taking the place of b. Then we get:

$$\begin{aligned}(2-i)^6 = {} & 1(2)^6(-i)^0 + 6(2)^5(-i)^1 + 15(2)^4(-i)^2 \\ & + 20(2)^3(-i)^3 + 15(2)^2(-i)^4 + 6(2)^1(-i)^5 + 1(2)^0(-i)^6\end{aligned}$$

Now simplify each term carefully (remember, exponents before multiplication).

$$\begin{aligned}(2-i)^6 &= 1(64)(1) + 6(32)(-i) + 15(16)(-1) \\ &\qquad +20(8)(i) + 15(4)(1) + 6(2)(-i) + 1(1)(-1) \\ &= 64 - 192i - 240 + 160i + 60 - 12i - 1 \\ &= -117 - 44i\end{aligned}$$

SECTION EXERCISES

For 9–11, write each number in terms of i, *perform the operation, and simplify the answer.*

6–9. $\sqrt{-64} + \sqrt{-36}$

6–10. $\sqrt{-12}\sqrt{-3}$

6–11. $\dfrac{\sqrt{-72}}{3\sqrt{2}}$

For 12–13, perform the operation and express the answer in a + bi *form.*

6–12. $(4 + 3i) + (3 - 7i)$

6–13. $\left(-5 + \sqrt{-32}\right) - \left(2 - \sqrt{-18}\right)$

For 14–15, express each product in simplest a + bi *form.*

6–14. $1 + 4i(2 - 3i)$

6–15. $\left(3 + 2i\sqrt{3}\right)\left(2 + i\sqrt{3}\right)$

For 16–17, each of the following involves the product of a pair of complex conjugates. Evaluate each and simplify.

6–16. $(2 + i)(2 - i)$

6–17. $\left(4 - 2i\sqrt{5}\right)\left(4 + 2i\sqrt{5}\right)$

For 18–19, express each answer in simplest a + bi *form.*

6–18. $\dfrac{5 - 3i}{1 - i}$

6–19. $\dfrac{7 - 2i}{4i}$

6–20. Divide $2 - 3i$ by its conjugate.

6–21. For the number $4 - 3i$:

a. Find its additive inverse.
b. Find its complex conjugate.
c. Find its multiplicative inverse.

6–22. For $f(x) = x^5 + 5x$, find $f(2i)$.

6–23. Evaluate $z^2 + \dfrac{10i}{z}$ for $z = 1 - 2i$.

For 24–25, evaluate and express in a + bi *form.*

6–24. $(2 - i)^3$

6–25. $(1 + i)^6$

Solving Quadratic Equations with Complex Zeros

We can now solve all possible quadratic equations if we allow the solutions to include imaginary numbers. Pure quadratic equations can still be solved by taking square roots; all other quadratic equations can be solved using the quadratic formula.

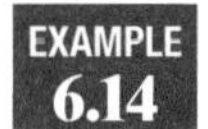

Solve for x: $2x^2 + 15 = 7$

SOLUTION

$2x^2 + 15 = 7$	Get x by itself.
$2x^2 = -8$	Subtract 15.
$x^2 = -4$	Divide by 2.
$x = \pm\sqrt{-4}$	Take the square root of each side.
$x = \pm 2i$	Rewrite the square root of a negative number as an imaginary number. Simplify the radical.

The solutions are $x = \pm 2i$.

EXAMPLE 6.15

Solve for x in simplest $a + bi$ form: $4x^2 - 20x + 61 = 0$

SOLUTION

Remember the quadratic formula: $x = \dfrac{-b \pm \sqrt{b^2 - 4ac}}{2a}$. Use this formula for all quadratic equations with a linear term, bx, that has complex zeros.

$4x^2 - 20x + 61 = 0$	Use the quadratic formula.
$a = 4, b = -20, c = 61$	Identify values for a, b, and c.
$x = \dfrac{-(-20) \pm \sqrt{(-20)^2 - 4(4)(61)}}{2(4)}$	Substitute values into the quadratic formula.
$= \dfrac{20 \pm \sqrt{-576}}{8}$	Simplify.
$= \dfrac{20 \pm 24i}{8}$	Write the square root of negative number in terms of i. Simplify the radical.
$= \dfrac{20}{8} \pm \dfrac{24i}{8}$	Divide each numerator term by the denominator.
$= \dfrac{5}{2} \pm 3i$	Reduce.

The solutions are $x = \dfrac{5}{2} \pm 3i$.

COMMON ERROR

Many students successfully use the quadratic formula only to make a mistake reducing the final fraction. If you think $\frac{20 \pm 24i}{8}$ reduces to $\frac{\overset{5}{\cancel{20}} \pm 24i}{\underset{2}{\cancel{8}}} = \frac{5}{2} \pm 24i$ or to $\frac{20 \pm \overset{3}{\cancel{24}}i}{\cancel{8}} = 20 \pm 3i$, you should write out the single fraction as two separate fractions before reducing:

$$\frac{20 \pm 24i}{8} = \frac{20}{8} \pm \frac{24i}{8} = \frac{5}{2} \pm 3i$$

Note that for a quadratic equation with all real coefficients, if the zeros are complex, they will form a complex conjugate pair. One zero will be of the form $a + bi$ and the other zero of the form $a - bi$.

SECTION EXERCISES

For 26–29, solve each equation and express the answers in terms of i.

6–26. $\frac{x}{6} = \frac{-2}{3x}$

6–27. $(2x - 3)(2x + 3) + 90 = 0$

6–28. $5x + \sqrt{-8} = 4\sqrt{-18}$

6–29. If $\frac{x^2}{6} + \frac{y^2}{8} = 1$, find the values of y if $x = 3$.

For 30–33, solve the following equations. Express the answer in simplest a + bi *form.*

6–30. $x^2 - 6x + 13 = 0$

6–31. $9x^2 - 6x + 5 = 0$

6–32. $\frac{x}{4} = \frac{x - 5}{x - 3}$

6–33. $2x^2 + 3 = 4x$

Writing Quadratic Factors from Complex Zeros

When given a complex conjugate pair, it is possible to find a quadratic equation that has those zeros. The method is the same one we used in Chapter 5 to find an equation for a pair of irrational conjugate zeros. First isolate the imaginary term, and then square both sides.

EXAMPLE 6.16 Find a quadratic equation having the zeros $\pm 3i$.

SOLUTION

$x = \pm 3i$

$x^2 = (\pm 3i)^2$	Square both sides.
$x^2 = 9i^2$	Squaring removes the ± symbol.
$x^2 = -9$	Remember $i^2 = -1$.
$x^2 + 9 = 0$	Add 9 to set equation equal to 0.

A quadratic equation with the desired zeros is $x^2 + 9 = 0$.

EXAMPLE 6.17 Find a quadratic equation having zeros $5 \pm 4i$.

SOLUTION

$x = 5 \pm 4i$

$x - 5 = \pm 4i$	Subtract 5 to get the imaginary term by itself.
$(x - 5)^2 = (\pm 4i)^2$	Square both sides.
$(x - 5)(x - 5) = 16i^2$	Squaring removes the ± symbol. Write out the binomials to be accurate.
$x^2 - 10x + 25 = -16$	Remember that $i^2 = -1$.
$x^2 - 10x + 41 = 0$	Add 16 to set equation equal to 0.

A quadratic equation with the desired zeros is $x^2 - 10x + 41 = 0$.

EXAMPLE 6.18 Find a quadratic equation having the zeros $4 \pm 2i\sqrt{3}$.

SOLUTION

$x = 4 \pm 2i\sqrt{3}$	
$x - 4 = \pm 2i\sqrt{3}$	Subtract 4 to get the imaginary term by itself.
$(x - 4)^2 = \left(\pm 2i\sqrt{3}\right)^2$	Square both sides.
$x^2 - 8x + 16 = (4)i^2(3)$	Squaring removes the ± symbol.
$x^2 - 8x + 16 = -12$	Remember that $i^2 = -1$.
$x^2 - 8x + 28 = 0$	Add 12 to set the equation equal to 0.

A quadratic equation with the desired zeros is $x^2 - 8x + 28 = 0$.

SECTION EXERCISES

6–34. The zeros of a quadratic equation with integral coefficients are $\pm 5i$. Write one such equation.

6–35. One zero of a quadratic equation with integral coefficients is $2 - i\sqrt{6}$.

a. Write the other zero.

b. Write a quadratic equation with these zeros.

The Complex Plane

All real numbers can be represented on a number line. Imaginary and complex numbers, not being real, cannot be shown on a number line. Instead, we use a set of coordinate axes called the *complex plane*. The horizontal axis is the real axis; the vertical axis is the imaginary axis. On the complex plane, the number $a + bi$ is graphed at the point with coordinates (a, b).

Graph the following numbers on the complex plane.

a. 3

b. $-2i$

c. $-1 + 4i$

SOLUTION

a. Since 3 is a real number, it is graphed at 3 on the real (horizontal) axis. You can think of it as $3 + 0i$, so it is graphed at the point (3, 0).

b. Since $-2i$ is a pure imaginary number, it is graphed at -2 on the imaginary (vertical) axis. Think of it as $0 - 2i$. It is graphed at the point (0, -2).

c. Since $-1 + 4i$ is a complex number, it is graphed at the point (-1, 4).

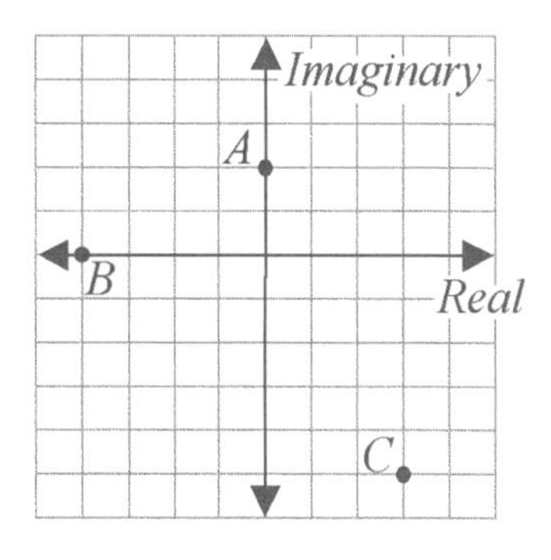

EXAMPLE 6.20

Write the complex number represented by the points labeled *A*, *B*, and *C* on the graph to the right.

SOLUTION

Point *A* is at (0, 2). It can be written as $0 + 2i$, which is usually simplified to just $2i$. Point *A* is on the imaginary axis and so represents a pure imaginary number.

Point *B* is at (-4, 0). It can be written as $-4 + 0i$ but is usually simplified to just -4. Point *B* is on the real axis and represents a real number.

Point *C* has the coordinates (3, -5) and represents the complex number $3 - 5i$.

SECTION EXERCISES

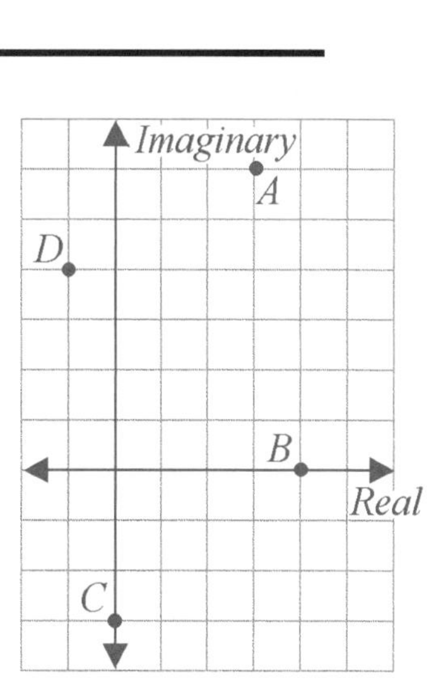

6–36. Write, in $a + bi$ form, the complex numbers represented by the points *A*, *B*, *C*, and *D* in the diagram.

6–37. Graph and label the complex numbers.

a. $3i$

b. -2

c. $-4 + 2i$

d. $4 - i$

CHAPTER EXERCISES

C6–1. Express $2\sqrt{\frac{-45}{64}}$ in terms of i and simplify.

For 2–3, write each number in terms of i*, perform the operation, and simplify the answer.*

C6–2. $2\sqrt{-45} - \sqrt{-80}$

C6–3. $\dfrac{\sqrt{-300}}{\sqrt{-25}}$

C6–4. Simplify $i^3 - i^2 + i - 1$.

C6–5. From the sum of $4 + 5i$ and $2 + 3i$, subtract $5 - 7i$.

For 6–7, express each product in simplest a + bi *form.*

C6–6. $(1 + 4i)(2 - 3i)$

C6–7. $\left(\sqrt{3} + 2i\right)\left(\sqrt{3} - 2i\right)$

For 8–9, express each answer in simplest a + bi *form.*

C6–8. $\dfrac{1 + 3i}{2 + 4i}$

C6–9. $\dfrac{5i}{2 - i}$

C6–10. Evaluate $2x^2 - 4x + 5$ for $x = 1 + 3i$. Express the answer in simplest $a + bi$ form.

C6–11. Evaluate $(3 - 2i)^4$ and express in $a + bi$ form.

C6–12. Solve $x^2 + 64 = 0$. Express the answers in terms of i.

C6–13. If $3x^2 - 2y^2 = 72$, find the value(s) of y when $x = 4$.

For 14–16, solve the following equations. Express the answers in simplest a + bi *form.*

C6–14. $4x^2 + 16x + 17 = 0$

C6–15. $\dfrac{x}{4x - 10} = \dfrac{2}{x}$

C6–16. $4x^2 + 19 = 16x$

C6–17. The zeros of a quadratic equation with integral coefficients are $4 \pm 3i$. Write one such equation.

C6–18. Write, in $a + bi$ form, the complex numbers represented by the points C, A, T, and S in the diagram.

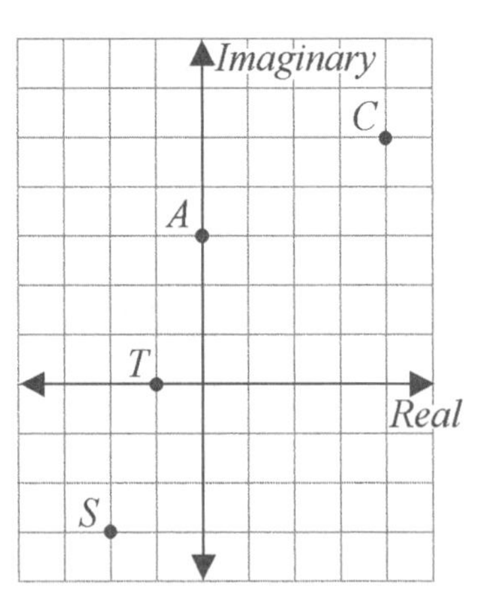

C6–19. Graph and label the complex numbers.

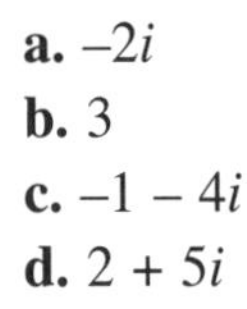

a. $-2i$
b. 3
c. $-1 - 4i$
d. $2 + 5i$

Polynomial Functions

WHAT YOU WILL LEARN

Linear functions and quadratic functions are two examples of a more general class of functions called polynomial functions. As the name suggests, these functions take the form of a polynomial in one variable. In addition to linear and quadratic functions, we can have cubics, quartics, and quintics. The possibilities are quite literally infinite; we can have polynomial functions of any degree. In this chapter, you will learn to:

- Graph polynomial functions;
- Rewrite polynomials in factored form;
- Write polynomial functions having given zeros;
- Solve polynomial equations algebraically and graphically;
- Solve polynomial inequalities algebraically and graphically.

SECTIONS IN THIS CHAPTER

Polynomial Models

Suppose you have a rectangular piece of cardboard that measures 18 inches by 24 inches. You cut a square of side x out of each corner and fold up the resulting flaps to form an open-topped box as shown in Figure 7.1.

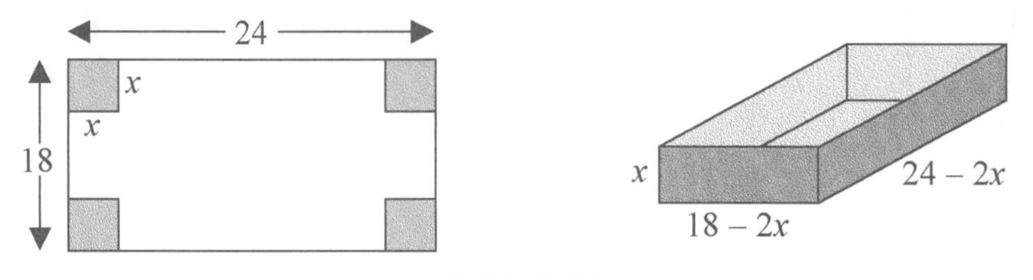

FIGURE 7.1

The volume of your box is $V = x(24 - 2x)(18 - 2x)$ or $V = 4x^3 - 84x^2 + 432x$. This is an example of a polynomial function.

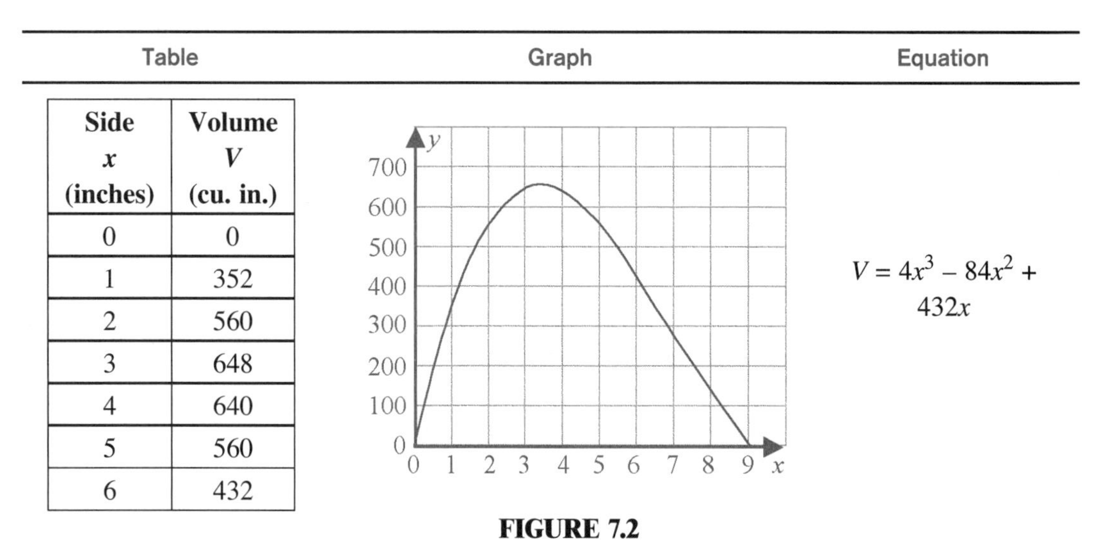

Side x (inches)	Volume V (cu. in.)
0	0
1	352
2	560
3	648
4	640
5	560
6	432

$V = 4x^3 - 84x^2 + 432x$

FIGURE 7.2

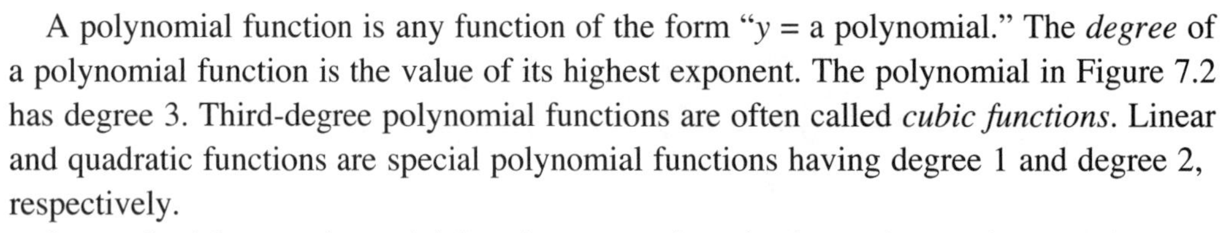

A polynomial function is any function of the form "y = a polynomial." The *degree* of a polynomial function is the value of its highest exponent. The polynomial in Figure 7.2 has degree 3. Third-degree polynomial functions are often called *cubic functions.* Linear and quadratic functions are special polynomial functions having degree 1 and degree 2, respectively.

In standard form, polynomial functions are written in *descending order,* meaning the highest exponent comes first and the rest of the terms follow in decreasing order of exponents. Thus a polynomial of degree n would be written $y = a_nx^n + a_{n-1}x^{n-1} + \ldots + a_2x^2 + a_1x + a_0$, where a_n, a_{n-1}, . . . a_2, a_1, and a_0 are real numbers, the coefficients, and $a_n \neq 0$. The first coefficient, a_n, is called the *leading coefficient* and the last coefficient, a_0, is the y-intercept. The subscript on a indicates its position in the polynomial. For example, a_3 represents the value of the coefficient of the x^3 term. (Remember, the subscript is not an exponent. It simply indicates the position of the coefficient.)

Characteristics of Polynomial Graphs

Polynomial functions and their graphs all share two important characteristics. Their domain is all real numbers, and their graphs are continuous, smooth curves. This means there are no breaks in the graph, such as those sometimes found in piecewise functions (Chapter 4) or rational functions (Chapter 9) and no sharp corners, such as in absolute value functions (Chapter 4).

To learn about other characteristics of polynomial functions and their graphs, let's look at some representative polynomial functions of increasing degree. We are particularly interested in three things: the number of zeros the polynomial can have, the number of turning points (places where it changes from increasing to decreasing or vice versa) it can have, and its end behavior. By end behavior we mean what happens to the y-value of the polynomial and its graph as x approaches infinity (∞) and as x approaches negative infinity ($-\infty$).

POLYNOMIAL FUNCTIONS OF DEGREE 0

A polynomial function of degree 0 is a constant function, $y = a_0$ where $a_0 \neq 0$ (the zero function, $y = 0$, is a special case and does not have degree 0). The graph of a constant function is a horizontal line (Figure 7.3). Constant functions have no zeros and no turning points. Their graphs neither increase nor decrease; the end behavior is the same as the rest of the graph. In short, they are not very interesting.

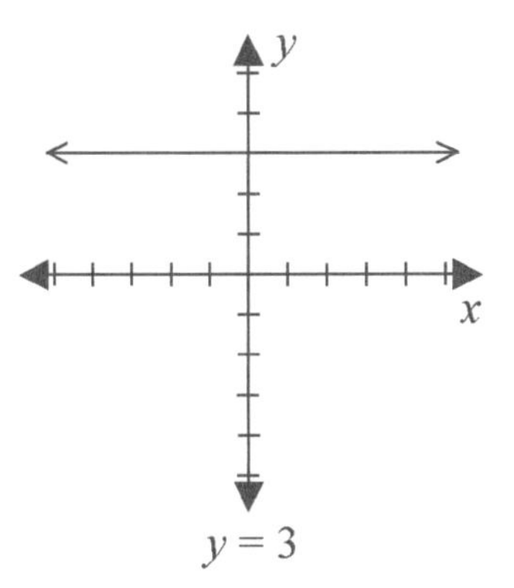

FIGURE 7.3

POLYNOMIAL FUNCTIONS OF DEGREE 1

Polynomial functions of degree 1 are linear functions. They have equations of the form $y = a_1x + a_0$ where $a_1 \neq 0$. Figure 7.4 shows two representative examples. All first-degree polynomial functions have exactly one zero. Depending on the sign of the leading coefficient, a_1, they are either always increasing or always decreasing, which means they have no turning point. If $a_1 > 0$, the function is always increasing. This means the right side of the function will rise to ∞ and the left side will fall to $-\infty$. If $a_1 < 0$, the situation is exactly the opposite. The function is always decreasing, with the right side falling to $-\infty$ and the left side rising to ∞. Either way, note that the two ends go in opposite directions, one up and one down.

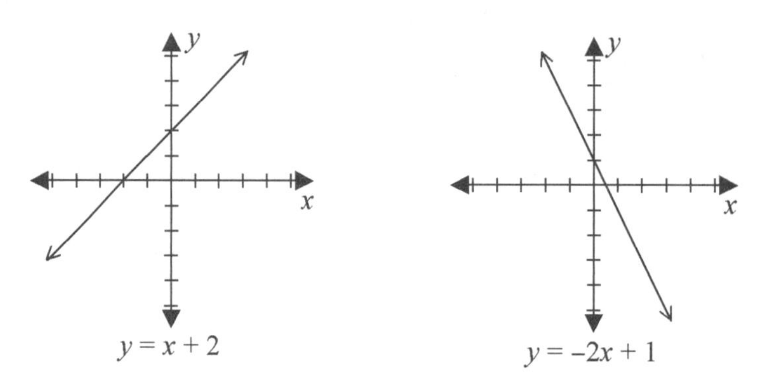

FIGURE 7.4

POLYNOMIAL FUNCTIONS OF DEGREE 2

A polynomial function of degree 2, also called a quadratic function, has the form $y = a_2x^2 + a_1x + a_0$ where $a_2 \neq 0$. Three representative quadratic functions are shown in Figure 7.5. A quadratic function may have two, one, or no distinct real zeros. It always has exactly one turning point, which is either the maximum or the minimum value of the function. The ends of the graph both go in the same direction. That direction depends on the sign of the leading coefficient, a_2. If $a_2 > 0$, both sides rise to ∞. If $a_2 < 0$, both sides fall to $-\infty$.

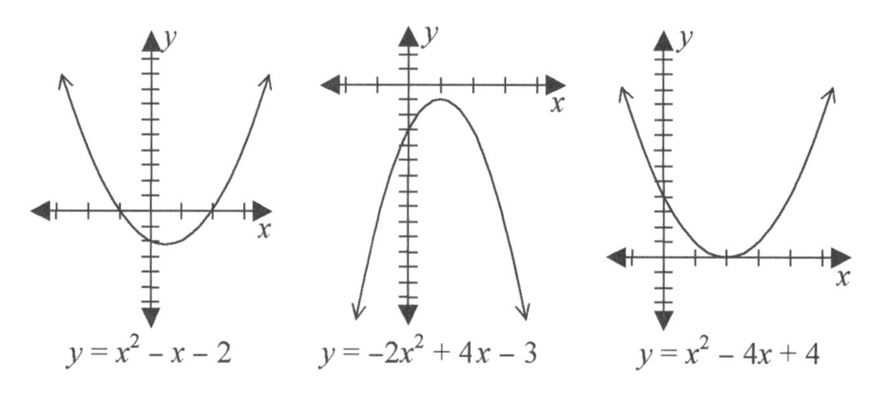

FIGURE 7.5

POLYNOMIAL FUNCTIONS OF DEGREE 3

Polynomial functions of degree 3 have equations of the form $y = a_3x^3 + a_2x^2 + a_1x + a_0$ where $a_3 \neq 0$. They are called cubic functions. Three examples are shown in Figure 7.6. From the graphs, it appears that cubic functions can have three, two, or one real zeros (but not none) and either two or no turning points. Each turning point is a relative (or local) maximum or minimum. They are neither the highest nor the lowest point on the entire graph. However, they are higher or lower than any of the nearby points on either side. The y-values of the highest and lowest points anywhere on a graph are called the absolute (or global) maximum or minimum. Cubic functions have neither an absolute maximum nor an absolute minimum.

The end behavior of polynomials of degree 3 depends on the leading coefficient in the same way as the end behavior of first-degree polynomials. The two ends go in opposite directions. If $a_3 > 0$, the right side rises to ∞ and the left side falls to $-\infty$. If $a_3 < 0$, the behavior is reversed, with the right side falling to $-\infty$ and the left side rising to ∞. The fact that the two sides go in opposite directions means it is not possible for a third-degree polynomial function to have no real zero; it must cross the x-axis at least once. It is also impossible for a third-degree polynomial to have exactly one turning point.

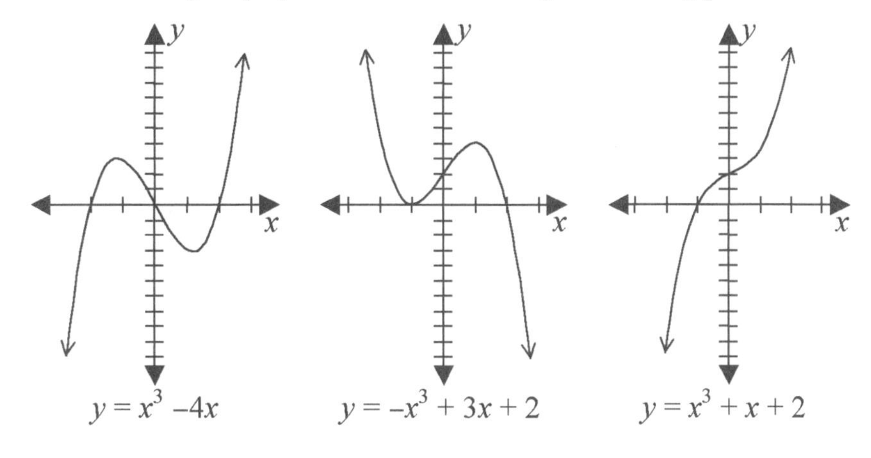

FIGURE 7.6

POLYNOMIAL FUNCTIONS OF DEGREE 4

Let's look at one more specific case before making some generalizations about polynomial functions. Fourth-degree polynomial functions, sometimes called quartic functions, have the form $y = a_4x^4 + a_3x^3 + a_2x^2 + a_1x + a_0$ where $a_4 \neq 0$. Some representative examples are shown in Figure 7.7. They may have anywhere from zero to four distinct real zeros and either one or three turning points. Their end behavior depends on the leading coefficient a_4 in the same way as the end behavior of second-degree polynomial functions. Both ends go in the same direction, rising to ∞ if $a_4 > 0$ and falling to $-\infty$ if $a_4 < 0$. Because the ends go in the same direction, it is possible for the function to have no real zeros. However, it must have an odd number of turning points.

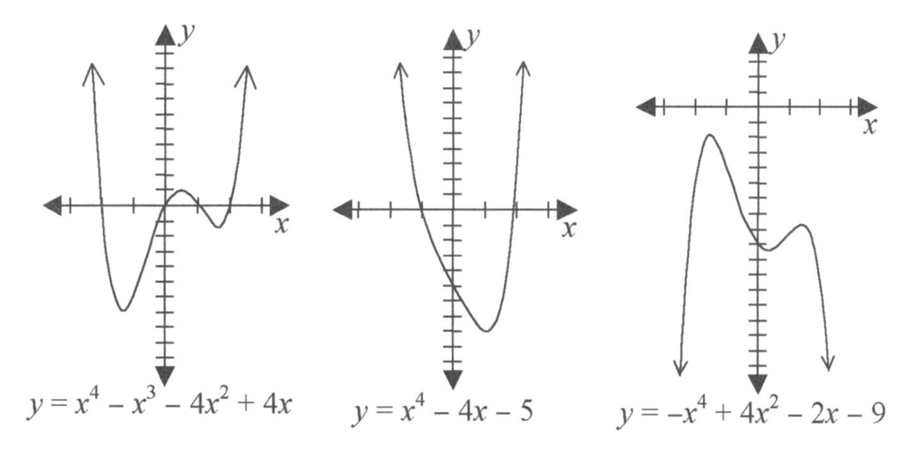

FIGURE 7.7

POLYNOMIAL FUNCTIONS OF DEGREE *n*

From the various previous examples, we can make some generalizations about polynomial functions and their graphs.

It appears that for any polynomial function, the number of real zeros will be no more than the degree of the polynomial. In addition, if the polynomial is of odd degree, there must be at least one real zero. For example, a fifth-degree polynomial function could have any number of real zeros from 1 to 5 while a sixth-degree polynomial function could have any number of real zeros from 0 to 6.

The number of turning points on the graph of a polynomial function is at most one less than the degree of the polynomial. A polynomial of odd degree will have an even number of turning points (or possibly none), while a polynomial of even degree will have an odd number of turning points, meaning at least one. Thus the graph of a fifth-degree polynomial function could have 0, 2, or 4 turning points, while the graph of a sixth-degree polynomial function could have 1, 3, or 5 turning points.

For all polynomial functions of degree 1 or more, which means all except constant functions, the end behavior is such that each side of the graph either rises to ∞ or falls to $-\infty$. The right side will rise to ∞ if the leading coefficient is positive and will fall to $-\infty$ if the leading coefficient is negative. For functions of odd degree, the two sides, left and right, will go in opposite directions. For functions of even degree, the two sides will go in the same direction. The end behavior of polynomial functions of degree 1 or greater is shown graphically in Figure 7.8.

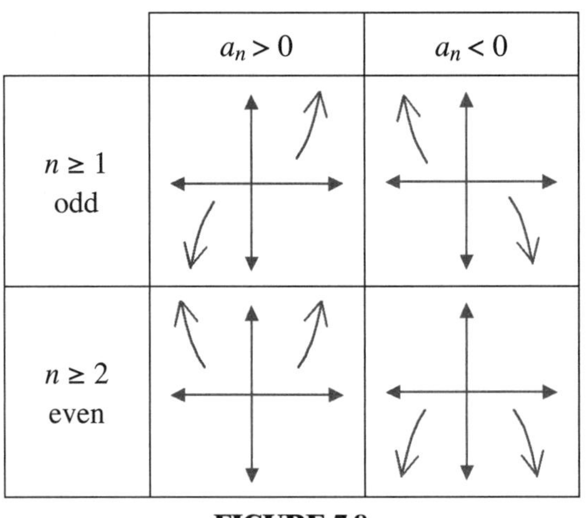

FIGURE 7.8

PROPERTIES OF POLYNOMIAL FUNCTION OF DEGREE $n \geq 1$

The following summarizes the properties of polynomial functions of degree $n \geq 1$.

1. The function has a domain of all real numbers.
2. The graph of the function will be continuous (with no breaks in the graph).
3. The function will have at most n real zeros. If n is odd, the function will have at least one real zero.

4. The graph of the function will have at most $n - 1$ turning points. If n is even, the function will have an odd number of turning points, meaning at least one. If n is odd, the function will have an even number of turning points, meaning possibly none.
5. Each side of the graph will either rise to ∞ or fall to $-\infty$ depending on the sign of the leading coefficient and the degree of the function.
   - Positive leading coefficient: right side rises to ∞
   - Negative leading coefficient: right side falls to $-\infty$
   - Odd degree: right and left sides go in opposite directions
   - Even degree: right and left sides go in the same direction

EXAMPLE 7.1

Consider the polynomial function $y = 5 - 3x^2 - 2x^3$.

a. Write the function in standard form.
b. What is the degree of the function?
c. What is the leading coefficient? List all the coefficients.
d. How many distinct real zeros could the function have?
e. How many turning points could the function have?
f. What is the end behavior of the graph of this function?

SOLUTION

a. Standard form means the exponents are in descending order:
$y = -2x^3 - 3x^2 + 5$

b. The degree is the highest exponent, 3.

c. The leading coefficient is the first one. In this case, $a_3 = -2$. The other coefficients are $a_2 = -3$, $a_1 = 0$, and $a_0 = 5$. Note that the coefficient a_1 is not "nothing." It is 0 because there is not an x-term (linear term) in the equation.

d. We cannot usually tell exactly how many real zeros a polynomial function has just by looking at the equation. Since the degree is $n = 3$, we know the function can have at most 3 distinct real zeros. Also, since the degree is odd, the function must have at least one real zero. This function could have 1, 2, or 3 distinct real zeros.

e. As with zeros, we cannot usually tell exactly how many turning points a polynomial function has just by looking at the equation. The greatest number of turning points the function can have is one less than its degree, in this case 2. Because the degree is odd, there will be an even number of turning points. This function will have either 0 or 2 turning points.

f. The leading coefficient of the function is negative, so the right side will fall to $-\infty$. The degree of the function is odd, so the two sides will go in opposite directions. This means the left side will rise to ∞.

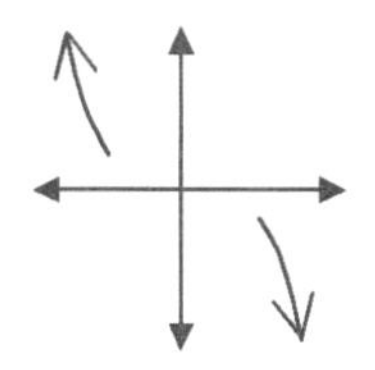

Our conclusions for parts d, e, and f are verified by the graph of the function shown below.

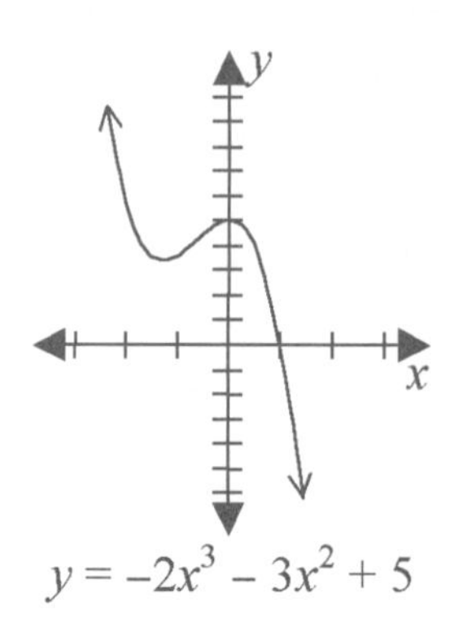

$y = -2x^3 - 3x^2 + 5$

y-INTERCEPTS OF POLYNOMIAL FUNCTIONS

Since the domain of a polynomial function is all real numbers, it must have a y-intercept. As for any function, the y-intercept can be found by substituting 0 for x and evaluating. Note that when the function is written in standard form, $y = a_n x^n + a_{n-1}x^{n-1} + \ldots + a_2x^2 + a_1x + a_0$, the y-intercept will always be a_0. This is consistent with what you know about lines and parabolas. For a linear function written in the form $y = mx + b$ (or $y = a_1x + a_0$), the y-intercept is b (or a_0). For a parabola written in the form $y = ax^2 + bx + c$ (or $y = a_2x^2 + a_1x + a_0$), the y-intercept is c (or a_0). Keep in mind that polynomial functions are not always written in standard form. In many cases, finding the y-intercept is simplest by setting $x = 0$ and evaluating.

EXAMPLE 7.2

Find the y-intercept for $y = 4 - 3x + 2x^2 - x^3$.

SOLUTION

Set $x = 0$ to find y.

$$y = 4 - 3(0) + 2(0)^2 - (0)^3$$
$$y = 4$$

The y-intercept is 4.

EXAMPLE 7.3 Find the y-intercept and the degree: $y = -(x-4)\left(3-x^2\right)^2(2x-1)^3$

SOLUTION

To find the y-intercept, set $x = 0$.

$$y = -(x-4)\left(3-x^2\right)^2(2x-1)^3$$
$$y = -((0)-4)\left(3-(0)^2\right)^2(2(0)-1)^3$$
$$y = -(-4)(3)^2(-1)^3$$
$$y = -36$$

The y-intercept is -36.

To find the degree, total the exponents on each variable or factor. If both the variable and the factor have exponents, multiply them to get that subtotal.

$$y = -(x-4)^1\left(3-x^2\right)^2(2x-1)^3 \qquad 1 + 2\cdot 2 + 3 = 8$$

The exponents total to 8. The polynomial function has degree 8.

SYMMETRY OF POLYNOMIALS

Even and odd functions were introduced in Chapter 3. Determining these types of symmetry for polynomials is especially easy.

Here, in Figure 7.9, is a quick review of even and odd functions.

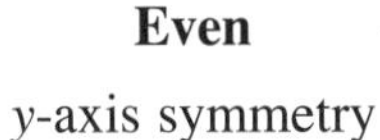

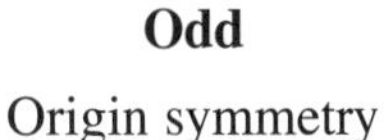

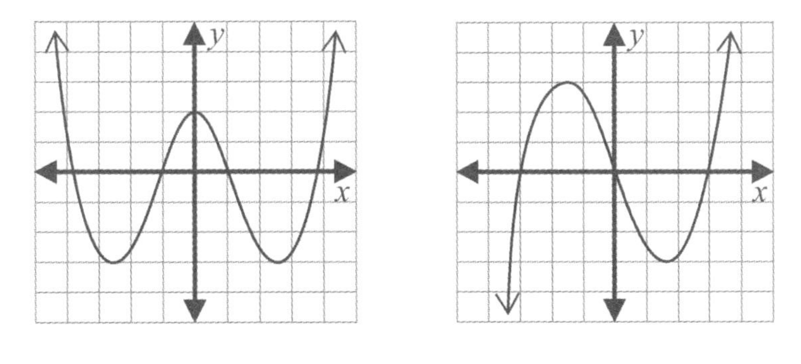

$f(-x) = f(x)$

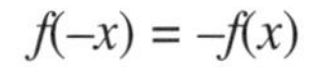

FIGURE 7.9

EXAMPLE 7.4

Determine if the following polynomials are even or odd functions or if they are neither.

a. $f(x) = 3x^4 - 5x^2 - 9$

b. $f(x) = 2x^3 - 6x$

c. $f(x) = x^3 + 2x^2$

d. $f(x) = x^3 - 3$

SOLUTION

Find $f(-x)$ and compare it to $f(x)$. If they are the same, the function is even. If they are opposites, the function is odd. If it is mixed, the function is neither.

a. $f(x) = 3x^4 - 5x^2 - 9$

$$\begin{aligned} f(-x) &= 3(-x)^4 - 5(-x)^2 - 9 \\ &= 3x^4 - 5x^2 - 9 \\ &= f(x) \end{aligned}$$

Even

b. $f(x) = 2x^3 - 6x$

$$\begin{aligned} f(-x) &= 2(-x)^3 - 6(-x) \\ &= -2x^3 + 6x \\ &= -f(x) \end{aligned}$$

Odd

c. $f(x) = x^3 + 2x^2$

$$\begin{aligned} f(-x) &= (-x)^3 + 2(-x)^2 \\ &= -x^3 + 2x^2 \\ &\neq f(x) \\ &\neq -f(x) \end{aligned}$$

Neither

d. $f(x) = x^3 - 3$

$$\begin{aligned} f(-x) &= (-x)^3 - 3 \\ &= -x^3 - 3 \\ &\neq f(x) \\ &\neq -f(x) \end{aligned}$$

Neither

Perhaps you noticed the pattern above. Polynomials that are *even functions* have only even exponents on the variable, including the constant term, which has an unwritten x^0. Polynomials that are *odd functions* have only odd exponents. Any polynomial that has a mix of odd and even exponents in its terms is neither even nor odd.

SECTION EXERCISES

7–1. For the polynomial function $y = -2x^4 + 5x^2 - x - 3$:

a. What is the degree of the function?

b. What are the values of a_n and a_0? What is the value of a_3?

c. What is the largest number of real zeros the function could have? (Do not use your calculator.)

d. What is the largest number of turning points it could have? (Do not use your calculator.)

e. What is the y-intercept of the function? (Do not use your calculator.)

f. Describe the end behavior of the graph. (Do not use your calculator.)

7–2. Find the y-intercept and degree of the function:

$$y = \frac{1}{6}(x + 3)^2 (x + 1)(x - 2)^2$$

7–3. Determine if the following functions are even, odd, or neither.

a. $y = -3x^4 + 5x^2 - 7$

b. $y = (x - 2)^3$

c. $y = 2x^3 - 6x$

7–4. If $f(x) = x^5 + x^3 + x$ and $f(a) = 10$, find the value of $f(-a)$.

Real Zeros of Polynomial Functions

We have seen that a polynomial function of degree n can have at most n real zeros. How can we locate these zeros? The principle is simple. To find the zeros of a polynomial (or any) function, set $y = 0$ and solve for x. The difficulty can arise in solving for x.

One of the first methods you learned for solving quadratic equations was factoring. Set the equation equal to 0, factor, and then apply the zero product property (ZPP). This property says if the product of two or more quantities is zero, then at least one of the quantities must equal 0.

Find the zeros: $f(x) = 2x^2 + 3x - 35$

SOLUTION

Set $f(x) = 0$, factor (using split the middle), and apply the ZPP.

$$\begin{aligned} 2x^2 + 3x - 35 &= 0 \\ 2x^2 - 7x + 10x - 35 &= 0 \\ x(2x - 7) + 5(2x - 7) &= 0 \\ (2x - 7)(x + 5) &= 0 \end{aligned}$$

$2x - 7 = 0$ or $x + 5 = 0$

$x = \frac{7}{2}$ or $x = -5$

The zeros are -5 and $\frac{7}{2}$.

This same method can be used to find the zeros of higher-degree polynomial equations.

EXAMPLE 7.6

Find the zeros: $f(x) = 2x^3 - 6x^2 - 8x + 24$

SOLUTION

Set $f(x) = 0$, factor out the GCF, factor by grouping, and then apply the ZPP.

$$\begin{aligned} 2x^3 - 6x^2 - 8x + 24 &= 0 \\ 2\left(x^3 - 3x^2 - 4x + 12\right) &= 0 \\ 2\left(x^2(x - 3) - 4(x - 3)\right) &= 0 \\ 2\left(x^2 - 4\right)(x - 3) &= 0 \\ 2(x + 2)(x - 2)(x - 3) &= 0 \end{aligned}$$

$x + 2 = 0$ or $x - 2 = 0$ or $x - 3 = 0$

$x = -2$ or $x = 2$ or $x = 3$

The zeros are ± 2 and 3.

EXAMPLE 7.7 Find the zeros: $f(x) = -3x^5 + 18x^3 - 15x$

SOLUTION

Set $f(x) = 0$, factor, and apply the ZPP.

$$-3x^5 + 18x^3 - 15x = 0$$
$$-3x\left(x^4 - 6x^2 + 5\right) = 0$$
$$-3x\left(x^2 - 1\right)\left(x^2 - 5\right) = 0$$
$$3x(x + 1)(x - 1)\left(x^2 - 5\right) = 0$$

$-3x = 0$ or $x + 1 = 0$ or $x - 1 = 0$ or $x^2 - 5 = 0$
$x = 0$ or $x = -1$ or $x = 1$ or $x = \pm\sqrt{5}$

The zeros are 0, ± 1, and $\pm\sqrt{5}$.

Unfortunately, not all polynomial equations factor using integers. Even some that do may be very difficult to factor with just paper and pencil. Fortunately, technology provides us with some powerful tools for finding zeros of functions. With a graphing calculator or computer graphing program, you can easily graph polynomial functions and find their zeros, or at least very good approximations, graphically.

EXAMPLE 7.8 Find the zeros: $f(x) = 4x^3 - 7x^2 - 5x + 6$

SOLUTION

Since factor by grouping doesn't work for this example, we don't know how to factor this (yet). With the help of technology, though, we can graph it.

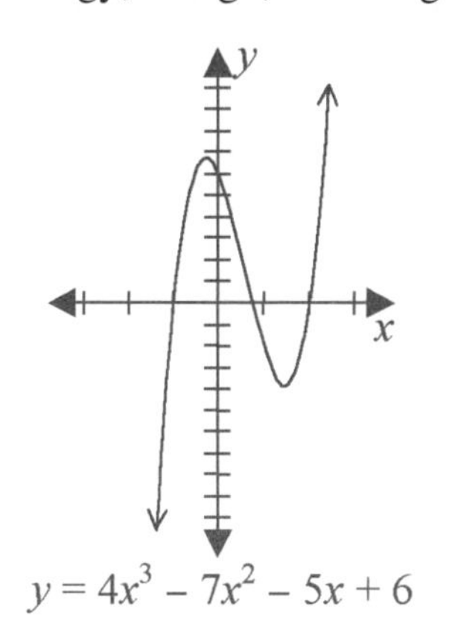

$y = 4x^3 - 7x^2 - 5x + 6$

From the graph, it appears that there are zeros at $x = -1$ and $x = 2$ and a third zero somewhere around $x = 0.8$. A graphing calculator confirms the first two and finds that the third zero is actually 0.75 or $\frac{3}{4}$. All three can be checked by substituting into the equation. All give $f(x) = 0$.

THE FACTOR THEOREM

You have probably noticed that as a result of the zero product property, if $(x - r)$ is a factor of a polynomial equation, then $x = r$ is a zero of the equation. This can be seen in Example 7.6, where $(x + 2)$, $(x - 2)$, and $(x - 3)$ were all factors of the polynomial and $x = -2$, $x = 2$, and $x = 3$ were all zeros.

Although it is not as obvious, the converse is also true. If $x = r$ is a zero of a polynomial function, then $(x - r)$ must be a factor. This is true even if r is not an integer. In Example 7.5, the function $f(x) = 2x^2 + 3x - 35$ was factored to $f(x) = (2x - 7)(x + 5)$ from which the zeros were found to be $x = \frac{7}{2}$ and $x = -5$. Clearly, the zero $x = -5$ has a corresponding factor of $(x - (-5))$ or $(x + 5)$. It may not be as clear that the zero $x = \frac{7}{2}$ has a corresponding factor of $\left(x - \frac{7}{2}\right)$. By factoring 2 out of the first factor above, we can rewrite the function as $f(x) = 2\left(x - \frac{7}{2}\right)(x + 5)$. In this form, we see that each zero, r, has a corresponding factor, $(x - r)$.

This works even for irrational zeros. In Example 7.7, by writing the first x as $(x - 0)$ and by treating the factor $(x^2 - 5)$ as a difference of squares, we can write the function in the form $f(x) = -3x^5 + 18x^3 - 15x = -3(x - 0)(x + 1)(x - 1)\left(x + \sqrt{5}\right)\left(x - \sqrt{5}\right)$. Again, each real zero, r, has a corresponding factor, $(x - r)$.

This relationship between factors and zeros of polynomial functions is called the factor theorem.

FACTOR THEOREM

A polynomial function $f(x)$ has a factor $(x - r)$ if and only if r is a zero of the function.

We can use the factor theorem to help factor polynomials and also to create polynomials that have given zeros.

EXAMPLE 7.9

Factor the polynomial function: $f(x) = 4x^3 - 7x^2 - 5x + 6$

SOLUTION

In Example 7.8, we used technology to find the zeros of this function: -1, $\frac{3}{4}$, and 2. By the factor theorem, $(x + 1)$, $\left(x - \frac{3}{4}\right)$, and $(x - 2)$ should all be factors. However, $(x + 1)\left(x - \frac{3}{4}\right)(x - 2) = x^3 - \frac{7}{4}x^2 - \frac{5}{4}x + \frac{3}{2}$. To get our original function f, we need to multiply by an additional factor of 4: $f(x) = 4(x+1)\left(x - \frac{3}{4}\right)(x-2)$. Note that multiplying by 4 is just a vertical stretch of the graph. It does not change the zeros.

If you prefer your factors to have all integer coefficients, you can distribute the 4 through the $\left(x - \frac{3}{4}\right)$ factor to get $f(x) = (x+1)(4x-3)(x-2)$.

The function $f(x) = 4x^3 - 7x^2 - 5x + 6$ can be factored into either $f(x) = 4(x+1)\left(x - \frac{3}{4}\right)(x-2)$ or $f(x) = (x+1)(4x-3)(x-2)$.

EXAMPLE 7.10 Write a polynomial function of degree 3 having zeros at $x = -2$, $x = 1$, and $x = \frac{7}{4}$ and a y-intercept of 42.

SOLUTION

We know the three zeros. By the factor theorem, the function will have factors $(x + 2)$, $(x - 1)$, and $\left(x - \frac{7}{4}\right)$. The third factor has a fraction in it, which many people would prefer to avoid. It may be multiplied by 4 without changing the zero to get $(4x - 7)$. So a polynomial function having the given zeros is $y = (x + 2)(x - 1)(4x - 7)$. However, this has a y-intercept (found by putting in $x = 0$) of $(2)(-1)(-7) = 14$. We want a y-intercept of 42. How can we fix this without changing our zeros?

We cannot just add 28 or, even worse, 42, on the end of the function. These are vertical translations of the graph, which will change the zeros. However, we can multiply this function by any nonzero constant. This will cause a vertical dilation of the graph and will not change the zeros. So any polynomial of the form $y = a(x + 2)(x - 1)(4x - 7)$ for any nonzero a will have the given zeros. We can find a by using the fact that when $x = 0$, $y = 42$.

$$
\begin{aligned}
42 &= a(0+2)(0-1)(4(0)-7) \\
42 &= 14a \\
3 &= a
\end{aligned}
$$

So the cubic polynomial having zeros $x = -2$, $x = 1$, and $x = \frac{7}{4}$ and a y-intercept of 42 is $y = 3(x + 2)(x - 1)(4x - 7)$.

Note that multiplying this out is optional. Most people do not do it for polynomials above degree 2 (quadratics).

Many polynomial functions can be factored and their zeros found using a combination of technology, the factor theorem, and polynomial long division. (Review Chapter 2 if you need to.)

EXAMPLE 7.11 Factor and find the zeros of the polynomial function: $y = 6x^3 - 23x^2 - 13x + 70$

SOLUTION

The only technique we know so far for factoring a quadrinomial is grouping, and that won't work with this polynomial. Unless we can factor it, we have no algebraic way to find the zeros. (Actually, there is a cubic formula. It makes the quadratic formula look like child's play. Just for fun, search for "cubic formula" sometime on the Internet. Just don't plan to memorize it or use it a lot.) So let's check out the graph of the function and see if that helps.

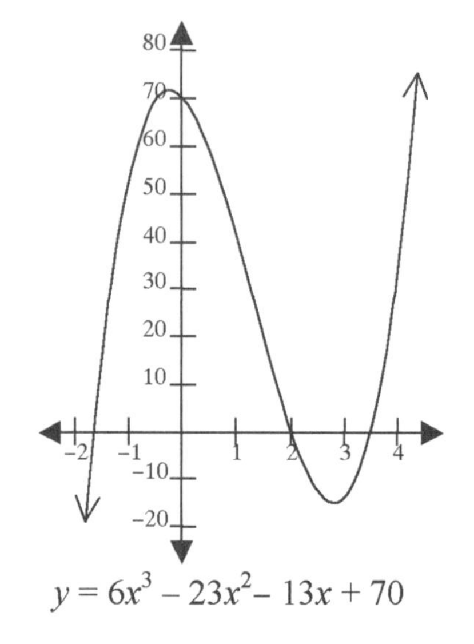

$y = 6x^3 - 23x^2 - 13x + 70$

The graph suggests that $x = 2$ is an integer zero of the function. We can check that this is true: $6(2)^3 - 23(2)^2 - 13(2) + 70 = 0$. So now we have one zero. The graph shows two others. Unfortunately, they are not integers and might not even be rational. How can we find them?

We can use the factor theorem. Because $x = 2$ is a zero, $x - 2$ must be a factor. We can factor $(x - 2)$ out of the polynomial using long division (Chapter 2).

$$\begin{array}{r} 6x^2 - 11x - 35 \\ x-2\overline{\smash{)}\,6x^3 - 23x^2 - 13x + 70} \\ \underline{-6x^3 + 12x^2} \qquad\qquad\quad \\ -11x^2 - 13x + 70 \\ \underline{-11x^2 - 22x} \qquad \\ -35x + 70 \\ \underline{+35x - 70} \\ 0 \end{array}$$

Note that we had to get a 0 remainder because we know $(x - 2)$ is a factor. If our remainder isn't 0, we made a mistake.

So now we have factored the original polynomial into a binomial and trinomial, $y = (x - 2)(6x^2 - 11x - 35)$. We can factor the trinomial using split the middle and get $6x^2 - 11x - 35 = (3x + 5)(2x - 7)$.

SEE CALC TIP 7A

Factored polynomial: $y = (x - 2)(3x + 5)(2x - 7)$

Zeros: $x = 2,\ x = -\frac{5}{3},\ x = \frac{7}{2}$

SYNTHETIC DIVISION

We are sure you agree that polynomial division is lots of fun. However, if we are going to do it often, it would be nice to have a shortcut. Synthetic division is a quick, easy way to divide a polynomial by a factor of the form $x - r$, exactly the sort of factor we get from finding a zero of a polynomial graphically.

Here are long division and synthetic division shown for the same problem.

$$\begin{array}{r r}
 & 6x^2 - 11x - 35 \\
x - 2 \,\big) & 6x^3 - 23x^2 - 13x + 70 \\
 & \underline{-6x^3 + 12x^2 \qquad\qquad\quad} \\
 & -11x^2 - 13x + 70 \\
 & \underline{-11x^2 - 22x \qquad} \\
 & -35x + 70 \\
 & \underline{+ 35x - 70} \\
 & 0
\end{array}$$

$$\begin{array}{c|rrrr}
2 & 6 & -23 & -13 & 70 \\
 & & 12 & -22 & -70 \\
\hline
 & 6 & -11 & -35 & 0
\end{array}$$

Clearly, synthetic division is much shorter, with a lot less writing. There are some important differences. In synthetic division, only coefficients are written and the exponent for each term is implied by the position of the coefficient. Keeping the numbers lined up precisely is important. The quotient is written at the bottom instead of the top. Can you compare the answer on the left, $6x^2 - 11x - 35$, with the values on the right, 6 –11 –35 0? The 0 at the end represents the remainder and confirms that 2 is a zero. It is not used when rewriting the polynomial answer. Another important difference is that the number on the outside in the synthetic division is the value of the zero; it is the opposite of the constant in the factor. Here are the steps to create the synthetic division on the right.

In the initial setup, draw a right angle with an extra-long base. Write the zero (not the factor), 2, on the outside and just the coefficients of the dividend, 6, –23, –13, and 70, on the inside. Then do the following:

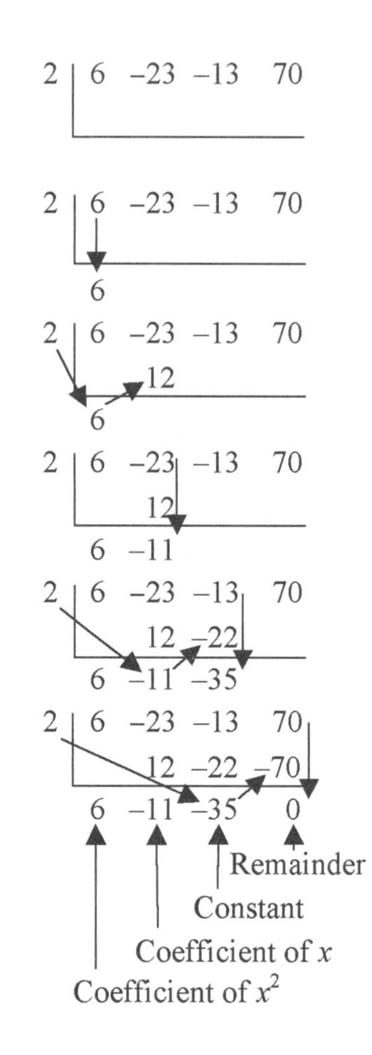

Copy the 6 below the horizontal line.

Multiply the 2 by the 6 and place the answer, 12, below the –23 and above the horizontal line.

Add the –23 and the 12. Then place the sum, –11, below the line.

Repeat the last two steps, this time using –11.

Repeat one more time with –35.

This division shows that $6x^3 - 23x^2 - 13x + 70$ factors into $(x - 2)\left(6x^2 - 11x - 35\right)$.

EXAMPLE 7.12 The polynomial function $y = 2x^3 - 3x^2 - 27$ has a zero at $x = 3$. Factor $x - 3$ out of the polynomial, and rewrite the function in factored form.

SOLUTION

We write the zero (not the factor), 3, on the outside and just the coefficients of the dividend, 2, –3, 0, and –27, on the inside. We must include a 0 coefficient for any missing term, in this case the x-term.

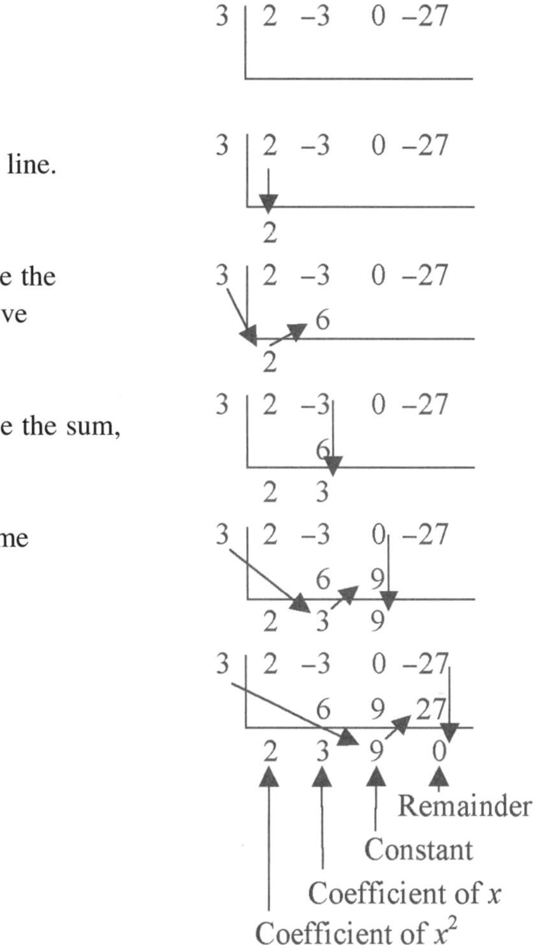

Copy the 2 below the horizontal line.

Multiply the 3 by the 2 and place the answer, 6, below the –3 and above the horizontal line.

Add the –3 and the 6. Then place the sum, 3, below the line.

Repeat the last two steps, this time using the new 3.

Repeat one more time with 9.

The function $y = 2x^3 - 3x^2 - 27$ can be written in factored form as $y = (x - 3)(2x^2 + 3x + 9)$.

EXAMPLE 7.13 The polynomial $y = x^4 - 2x^3 - 7x^2 + 8x + 12$ has zeros at $x = -1$ and $x = 3$. Use synthetic division to factor the polynomial and find the other zeros.

SOLUTION

The order of division does not matter. We chose to divide out the –1 first.

–1	1	–2	–7	8	12
		–1	3	4	–12
	1	–3	–4	12	0

Now it is time to divide out the other zero, 3. Do not go back and divide it into the original polynomial. Divide it into the answer you just got. This is easily done by extending the table we already have. Do not include the last column, which is the remainder from the first division.

–1	1	–2	–7	8	12
		–1	3	4	–12
3	1	–3	–4	12	0
		3	0	–12	
	1	0	–4	0	

The last row represents the polynomial $x^2 + 0x - 4$, written as $x^2 - 4$. Our answer so far is $y = (x + 1)(x - 3)\left(x^2 - 4\right)$. We can factor $x^2 - 4$ as a difference of squares into $(x - 2)(x + 2)$ and get $y = (x + 1)(x - 3)(x - 2)(x + 2)$. By the factor theorem, the previously unknown zeros are $x = -2$ and $x = 2$.

In factored form, the polynomial is $y = (x + 1)(x - 3)(x - 2)(x + 2)$. Its zeros are $x = -2$, $x = -1$, $x = 2$, and $x = 3$.

The polynomial $y = 2x^4 - 5x^3 + 6x^2 + 7x - 30$ has zeros at $x = -1.5$ and $x = 2$. Use synthetic division to factor it.

SOLUTION

Divide out 2 first, just because it seems easier than using a decimal.

2	2	–5	6	7	–30
		4	–2	8	30
	2	–1	4	15	0

Extend the previous chart, and divide out –1.5.

2	2	–5	6	7	–30
		4	–2	8	30
–1.5	2	–1	4	15	0
		–3	6	–15	
	2	–4	10	0	

Our answer so far is $y = (x + 1.5)(x - 2)\left(2x^2 - 4x + 10\right)$. We can factor a 2 out of the quadratic factor and get $y = 2(x + 1.5)(x - 2)\left(x^2 - 2x + 5\right)$. Some people leave it at this.

However, if the decimal inside the first factor bugs you, you can distribute the 2 into that factor. The final answer is $y = (2x + 3)(x - 2)\left(x^2 - 2x + 5\right)$.

SECTION EXERCISES

7–5. Find the zeros: $f(x) = x^4 + 5x^3 - 9x^2 - 45x$

7–6. Find all the zeros of the polynomial $y = 3x^4 + 4x^3 - 19x^2 - 8x + 12$ and write the function in factored form.

7–7. Find a polynomial function that has zeros $x = -2$, $x = 1$, and $x = \frac{7}{2}$ and y-intercept of 28.

7–8. Find all the zeros of the polynomial $y = 2x^4 - 11x^3 + 6x^2 + 22x + 8$ and write the function in factored form.

7–9. The polynomial function $y = x^3 + 6x^2 + 3x - 20$ has a zero at $x = -4$. Rewrite the function in factored form.

7–10. The polynomial $y = 2x^4 + 5x^3 - 13x^2 - 22x + 24$ has zeros at $x = -3$ and $x = 2$. Use synthetic division to factor it and find the other zeros.

Total Zeros of Polynomial Functions

We have already seen that a polynomial function of degree n will have at most n real distinct zeros. One of the most famous theorems in algebra takes that a step further.

FUNDAMENTAL THEOREM OF ALGEBRA

A polynomial of degree $n \geq 1$ will always have exactly n total zeros.

To get an idea of the reasonableness of this theorem, consider the polynomial $y = a_n x^n + a_{n-1}x^{n-1} + \ldots + a_2x^2 + a_1x + a_0$. Imagine first factoring out the leading coefficient, a_n, and then breaking down the remaining polynomial into linear factors of the form $(x - z_1)$, $(x - z_2)$, and so forth. If there are fewer than n such factors, then when they are multiplied out we would get a polynomial of degree less than n. If there were more than n factors, the polynomial would have degree higher than n. So we need exactly n such factors to factor our polynomial into $y = a_n(x - z_1)(x - z_2) \ldots (x - z_n)$. By the factor theorem, each of those n factors corresponds to a zero. So there must be n zeros: $z_1, z_2, \ldots z_n$.

Be aware of a couple of things when using the fundamental theorem of algebra. Some polynomials have complex zeros. These complex zeros must be included in the total. Also, sometimes the same factor will be used more than once in a polynomial. When this happens, the corresponding zero must also be counted more than once. Such zeros are called double (or triple or higher) zeros and are said to have *multiplicity* 2 (or 3 or more).

COMPLEX ZEROS

First, let's look at two examples involving polynomials with complex zeros.

EXAMPLE 7.15 Find all the zeros of $y = -2x^3 - 3x^2 + 5$, graphed below.

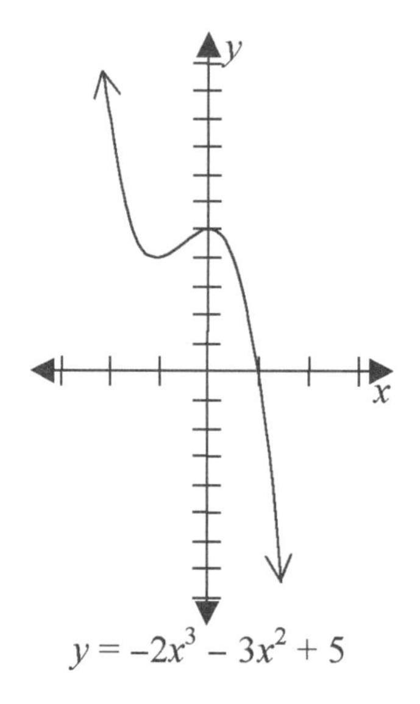

$y = -2x^3 - 3x^2 + 5$

SOLUTION

From the graph, we see there is only one real zero, $x = 1$. Use synthetic division to factor the polynomial.

$$\begin{array}{r|rrrr} 1 & -2 & -3 & 0 & 5 \\ & & -2 & -5 & -5 \\ \hline & -2 & -5 & -5 & 0 \end{array}$$

The polynomial factors into $y = (x - 1)\left(-2x^2 - 5x - 5\right)$. The quadratic formula can be used to find the remaining zeros, which turn out to be complex: $x = \frac{-5}{4} \pm i\frac{\sqrt{15}}{4}$.

The zeros of the function are 1 and $\frac{-5}{4} \pm i\frac{\sqrt{15}}{4}$. Note that this is a total of three zeros, in accordance with the fundamental theorem of algebra.

COMMON ERROR

If you wrote 1 | −2 −3 5 for the top row of the synthetic division, then you forgot the 0 coefficient for the *x*-term.

EXAMPLE 7.16

Find the zeros of $y = 2x^4 - x^3 - 17x - 20$.

SOLUTION

By the fundamental theorem, we expect to find four total zeros. The graph indicates real zeros at $x = -1$ and around $x = 2.5$.

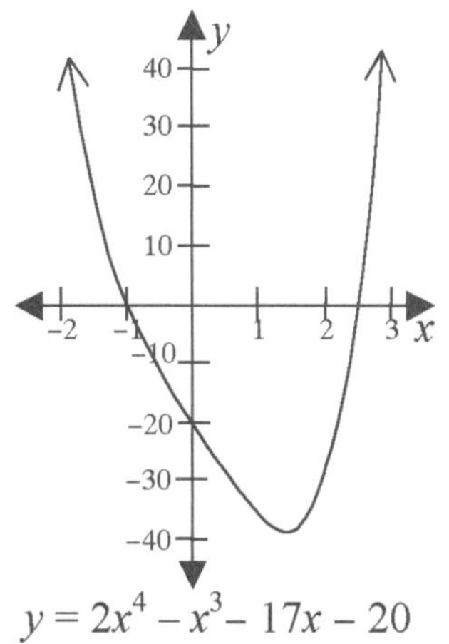

$y = 2x^4 - x^3 - 17x - 20$

Synthetic division gives us an easy way both to check that they really are zeros and to factor at the same time. Do not forget the 0 coefficient for the missing x^2-term. Factor out $x = -1$ first.

−1	2	−1	0	−17	−20
		−2	3	−3	20
	2	−3	3	−20	0

Extend the table and divide out $x = 2.5$.

−1	2	−1	0	−17	−20
		−2	3	−3	20
2.5	2	−3	3	−20	0
		5	5	20	
	2	2	8	0	

We have $y = (x + 1)(x - 2.5)\left(2x^2 + 2x + 8\right)$. Factoring out a greatest common factor of 2 from the trinomial gives $y = 2(x + 1)(x - 2.5)\left(x^2 + x + 4\right)$ or $y = (x + 1)(2x - 5)\left(x^2 + x + 4\right)$. The two linear factors contain the zeros we found graphically. The other two zeros are complex and can be found using the quadratic formula on the trinomial. They are $x = \frac{-1}{2} \pm \frac{i\sqrt{15}}{2}$.

The four zeros are $x = -1$, $x = 2.5$, and $x = \frac{-1}{2} \pm \frac{i\sqrt{15}}{2}$.

The previous two examples suggest another fact about the zeros of polynomials. If a polynomial function has all real coefficients, any complex zeros will occur in complex conjugate pairs. This in turn leads to one more useful theorem about polynomials. Any polynomial with real coefficients can be factored into a product of linear and/or quadratic factors, each having real coefficients. The polynomial in Example 7.16 was factored into one quadratic and two linear factors.

MULTIPLICITY OF ZEROS

Now let's look at some examples of polynomials with multiple roots.

EXAMPLE 7.17

Find all the zeros of $y = x^3 - 3x^2 + 4$.

SOLUTION

By the fundamental theorem, we expect three zeros. The graph shows one at $x = -1$ and what looks like another at $x = 2$.

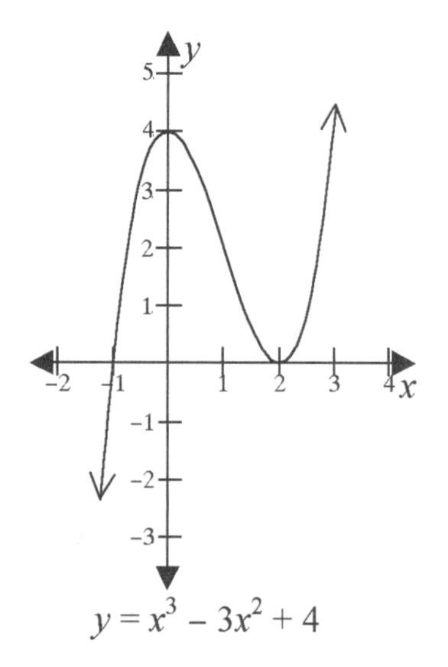

$y = x^3 - 3x^2 + 4$

Where is the third zero? It cannot be complex because complex zeros come in conjugate pairs, which would give too many total zeros. Do synthetic division for the two known zeros, $x = -1$ and $x = 2$. Remember that getting a 0 remainder in synthetic division confirms that these values are actual zeros and not just close approximations.

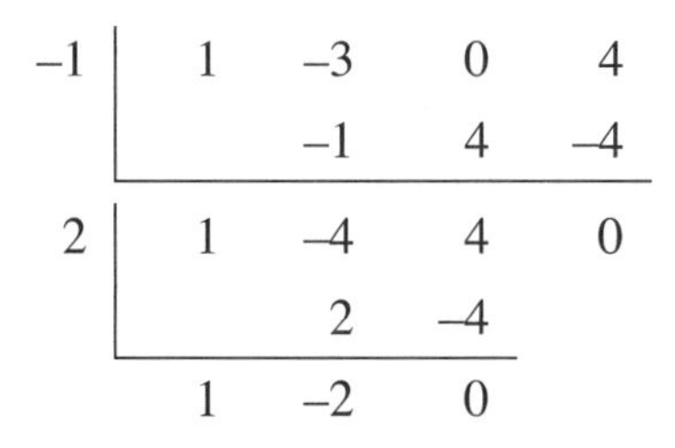

−1	1	−3	0	4
		−1	4	−4
2	1	−4	4	0
		2	−4	
	1	−2	0	

From this we see the polynomial factors into $y = (x + 1)(x - 2)(x - 2)$. By setting each factor equal to 0, we find the zeros are $x = -1$, $x = 2$, and $x = 2$. The third zero turns out to be the same as one of the other two. The factor $(x - 2)$, which occurs twice in the factored form of the polynomial, is said to have multiplicity 2. It's corresponding zero, $x = 2$, is called a double zero. The polynomial could be written as $y = (x + 1)(x - 2)^2$. In this form, the exponent on each factor gives that factor's multiplicity. The factor $(x + 1)$ has multiplicity 1, while $(x - 2)$ has multiplicity 2. In order for the total number of zeros to be three in accordance with the fundamental theorem, we need to count each zero according to its multiplicity; $x = -1$ counts once and $x = 2$ counts twice for a total of three zeros. We distinguish between total zeros and distinct zeros. This function has three total zeros, counting 2 twice, but only two distinct zeros.

Notice in the previous example that at the zero of multiplicity 1, the graph crossed the x-axis. At the zero of multiplicity 2, however, the graph was tangent to the x-axis. These are examples of a general rule about the behavior of the graph of a polynomial function near a zero. Look at the graphs below of $y = (x - 2)^n$ for $n = 1, 2, 3, 4$, and 5.

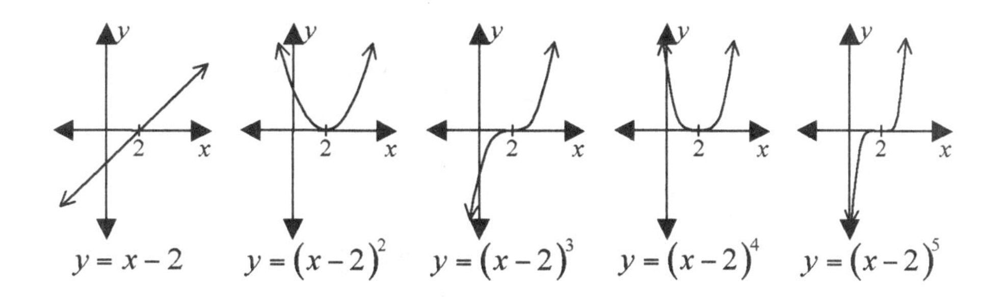

These graphs suggest the following rule:

- At zeros of odd multiplicity, the graph of a polynomial will cross the x-axis.
- At zeros of even multiplicity, the graph of a polynomial will be tangent to, but not cross, the x-axis.

EXAMPLE 7.18 Where will the graph of $y = x(x + 2)^3(x - 2)^2$ have zeros? At which zeros will it cross the x–axis? At which zeros will it be tangent to the x-axis?

SOLUTION

Because the function is already in factored form, finding the zeros is easy. The zeros will be at $x = -2$, $x = 0$, and $x = 2$. The zeros at $x = -2$ and $x = 0$ both have odd multiplicity (3 and 1, respectively), and so the graph will cross the x-axis at those points. The zero at $x = 2$ has even multiplicity 2 (remember, the multiplicity is the exponent, not the zero) and so should be tangent to the x-axis. The graph confirms this. In addition, note that at the zero $x = -2$, which has multiplicity 3, the graph seems to flatten out briefly, appearing to be close to horizontal, just as it crosses the x-axis. This is characteristic of zeros of multiplicity greater than 1 and becomes more pronounced as the multiplicity increases.

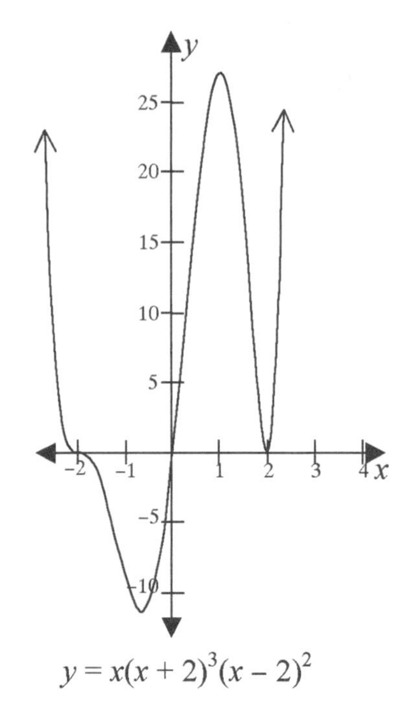

$y = x(x + 2)^3(x - 2)^2$

SKETCHING GRAPHS OF POLYNOMIAL FUNCTIONS

We can use information about the multiplicity of zeros, the y-intercept, and end behavior to sketch a rough graph of a polynomial function without using technology.

EXAMPLE 7.19 Sketch the graph of $y = \frac{1}{4}(x + 2)^2(x - 1)^3(x - 3)$ without using graphing technology.

SOLUTION

We want to consider three main things: the y-intercept, the zeros, and the end behavior (you don't have to do them in that order).

The y-intercept is $y = \frac{1}{4}(0 + 2)^2(0 - 1)^3(0 - 3) = 3$.

The zeros are at $x = -2$, $x = 1$, and $x = 3$. The zero at $x = -2$ has even multiplicity, so the graph will be tangent to the x-axis there. The other two zeros have odd multiplicity, so the graph will cross the x-axis. Because the zero at $x = 1$ has multiplicity greater than 1, the graph will flatten out as it crosses the axis there.

The degree of the function is $2 + 3 + 1 = 6$, which is even. So both sides will go in the same direction. The leading coefficient is positive, so both sides will rise to ∞.

Our graph so far looks like this, with points shown at the y-intercept and zeros and with arrows indicating end behavior.

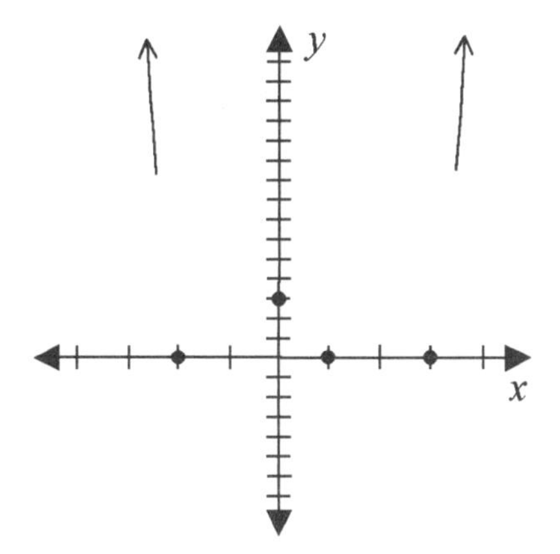

We could get a rough graph by connecting the dots with a smooth curve, remembering to make it tangent to the x-axis at $x = -2$ and cross at the other two zeros. We can refine the graph a little by evaluating a couple of extra points:

$$y(-1) = \frac{1}{4}(-1+2)^2(-1-1)^3(-1-3) = \frac{1}{4}(1)^2(-2)^3(-4) = 8 \text{ and}$$

$$y(2) = \frac{1}{4}(2+2)^2(2-1)^3(2-3) = \frac{1}{4}(4)^2(1)^3(-1) = -4.$$

Adding these points and connecting with a smooth curve gives the sketch below.

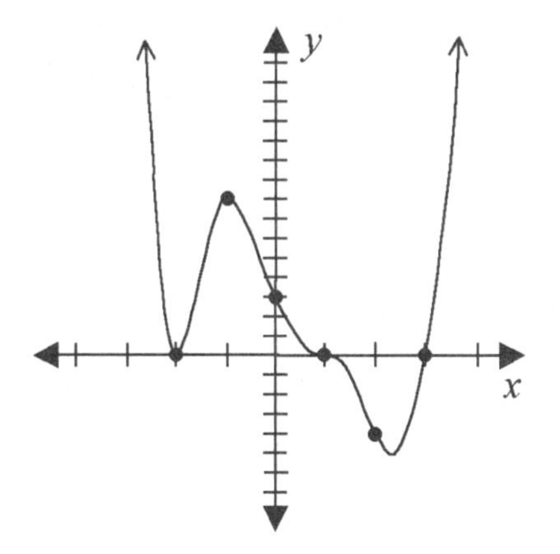

This graph should reasonably represent the shape of the function, but we shouldn't expect it to be perfect. In particular, without using either calculus or graphing technology (or both), we cannot know precisely where the relative maximum and minimum are located. In our sketch, we guessed that the relative maximum is close to the point (–1, 8) and that the relative minimum is a little to the right and below the point (2, –4). It turns out that the relative maximum is at approximately (–0.9, 8.1), which is close to our guess. The relative minimum is at approximately (2.6, –8.7), which is farther to the right and a fair amount lower than our guess.

Write a possible polynomial function for the graph below.

SOLUTION

The graph shows zeros at $x = -2$ and $x = 3$. The graph crosses the x-axis at -2, so the zero has odd multiplicity. The flattening of the graph at the zero $x = -2$ indicates a multiplicity greater than 1. The graph is tangent to the x-axis at 3, so that zero has an even multiplicity. On the principle of keeping things as simple as possible, we'll choose the lowest multiplicity for each zero. The function will look like $y = a(x + 2)^3(x - 3)^2$ for some nonzero value a. Note that this makes the degree of the function 5 (odd), which is consistent with the end behavior of the graph; the two sides go in opposite directions. Since the right side is going down, we know a will be negative. To find a, use the fact that the y-intercept is -9.

$$\begin{aligned} -9 &= a(0 + 2)^3(0 - 3)^2 \\ -9 &= 72a \\ -\frac{1}{8} &= a \end{aligned}$$

A polynomial function for the graph may be $y = -\frac{1}{8}(x + 2)^3(x - 3)^2$.

How confident are we of our answer? Is it possible that the function is of higher degree? One way to check is to evaluate the function at some convenient values of x. We find $y(-3) = 4.5$, $y(-1) = -2$, $y(1) = -13.5$, and $y(2) = -8$. All seem consistent with the graph. Experimenting with higher exponents on either or both factors clearly gives wrong y-values for some or all of those x-values.

SECTION EXERCISES

7–11. For the function $f(x) = x^3 - 3x^2 + 4x - 12$:

a. How many total zeros does it have?

b. Find all the zeros of the function.

For 12–14, find the real zeros using graphing technology, factor the polynomial function using synthetic division, and then find all its zeros.

7–12. $f(x) = 2x^3 - 11x^2 + 16x - 16$

7–13. $f(x) = x^4 - 4x^2 - 12x - 9$

7–14. $f(x) = 4x^5 + 18x^4 + 45x^2 - 25x$

7–15. A polynomial function has real zeros a (with multiplicity 1), b (with multiplicity 2), and c (with multiplicity 3). How many times does the function cross the x-axis?

7–16. Sketch a possible graph of $y = \frac{1}{6}(x + 3)^2(x + 1)(x - 2)^2$ without using graphing technology.

7–17. Write a possible polynomial function for the graph below.

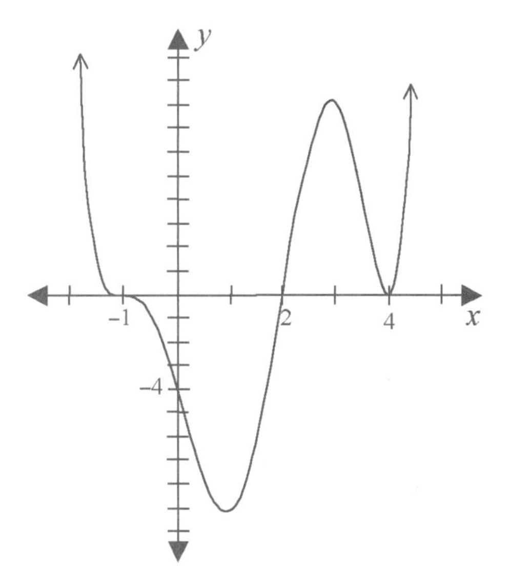

Solving Polynomial Equations and Inequalities

You should be able to solve all linear equations easily. Armed with the quadratic formula, you can solve all quadratic equations. Polynomial equations of higher degree are another story. Only a relative few can be solved strictly algebraically. The rest are either solved graphically with the help of technology or solved with a combination of both graphing and algebra.

EXAMPLE 7.21

Solve: $2x^3 + 3x^2 - 10x = 15$

SOLUTION

Set the equation equal to 0 and try factoring. Because it is a four-term polynomial, the only factoring technique we have is grouping. In this problem, grouping works.

$$2x^3 + 3x^2 - 10x - 15 = 0$$
$$x^2(2x + 3) - 5(2x + 3) = 0$$
$$\left(x^2 - 5\right)(2x + 3) = 0$$
$$x^2 - 5 = 0 \text{ or } 2x + 3 = 0$$
$$x = \pm\sqrt{5} \text{ or } x = -\frac{3}{2}$$

It was a cubic equation, and we have found all three zeros. So we are done.

The solution set is $\left\{-\sqrt{5}, -\frac{3}{2}, \sqrt{5}\right\}$.

EXAMPLE 7.22

Solve: $x^3 - 8x + 3 = 0$

SOLUTION

This is a trinomial. However, it is a cubic, so it will not factor using split the middle. Before we can factor this one, we need to find a zero. Graph $y = x^3 - 8x + 3$ using technology.

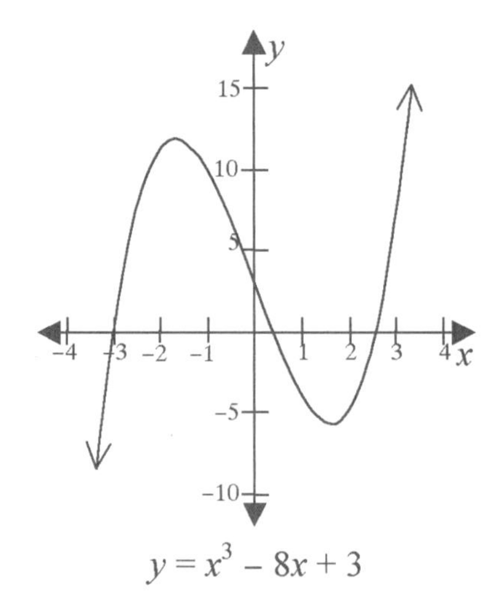

$y = x^3 - 8x + 3$

It appears there is an integer zero at $x = -3$. If good decimal approximations are all you need, a graphing calculator will find the other two zeros to be $x = 0.381966$ and $x = 2.618034$. If you want the exact values, use synthetic division to factor the polynomial and then use the quadratic formula.

–3	1	0	–8	3
		–3	9	–3
	1	–3	1	0

The equation factors into $(x + 3)(x^2 - 3x + 1) = 0$.

The zeros are $x = -3$ and $x = \dfrac{3 \pm \sqrt{5}}{2}$.

EXAMPLE 7.23

Solve $x^5 - 4x^4 + 5x^3 - 14x^2 + 44x - 40 = 0$ for all values of x, real and complex.

SOLUTION

Graph the function $y = x^5 - 4x^4 + 5x^3 - 14x^2 + 44x - 40$ using technology.

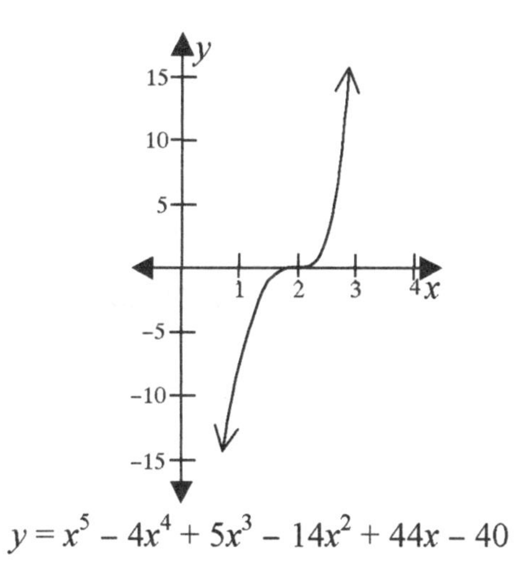

$y = x^5 - 4x^4 + 5x^3 - 14x^2 + 44x - 40$

We expect five total zeros. The graph seems to show only one real zero at $x = 2$. If the other four are all complex, we have a problem. However, the flattening out of the graph as it crosses the axis at $x = 2$ should lead you to suspect a multiple zero there of multiplicity at least three. Try synthetic division repeatedly.

2	1	−4	5	−14	44	−40
		2	−4	2	−24	40
2	1	−2	1	−12	20	0
		2	0	2	−20	
2	1	0	1	−10	0	
		2	4	10		
	1	2	5	0		

This equation can be written in factored form as $(x-2)^3\left(x^2+2x+5\right)=0$. The only real zero is at $x = 2$ (a triple zero). The two imaginary zeros can be found using the quadratic formula on the trinomial.

The zeros are $x = 2$ and $x = -1 \pm 2i$.

EXAMPLE 7.24

Solve $x^4 - 2x^3 + x^2 - 8x - 12 = 0$ for all values of x, real and complex.

SOLUTION

Graph $y = x^4 - 2x^3 + x^2 - 8x - 12$.

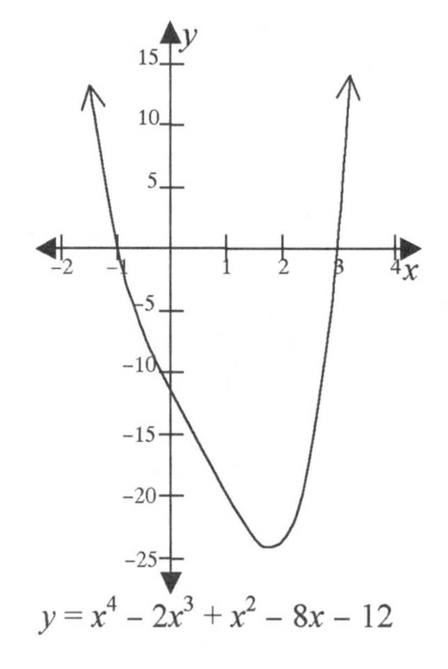

$y = x^4 - 2x^3 + x^2 - 8x - 12$

The graph shows two real zeros at $x = -1$ and $x = 3$. The other zeros must be complex. Use synthetic division to find the other factor.

−1	1	−2	1	−8	−12
		−1	3	−4	12
3	1	−3	4	−12	0
		3	0	12	
	1	0	4	0	

The equation factors into $(x + 1)(x - 3)\left(x^2 + 4\right) = 0$.

The real zeros are $x = -1$ and $x = 3$. The imaginary zeros are $x = \pm 2i$.

SOLVING POLYNOMIAL INEQUALITIES

Solving polynomial inequalities is a lot like solving polynomial equations but with an extra step of checking intervals. To solve an inequality, you must first find the zeros. The zeros divide the x-axis into intervals, each of which must be checked to see if it satisfies the inequality.

1. Find the zeros of the polynomial equation.
2. Graph the zeros on a number line.
3. Check each interval on the number line—numerically, algebraically, or graphically—to see if it satisfies the inequality.
4. Write the solution in appropriate inequality or interval notation.

You should know by now how to find the zeros. The following examples will focus mainly on methods for checking the intervals.

EXAMPLE 7.25

Solve $2x^3 - 3x^2 - 18x + 27 > 0$ algebraically.

SOLUTION

This polynomial can be factored by grouping followed by the difference of squares.

$$2x^3 - 3x^2 - 18x + 27 > 0$$
$$x^2(2x - 3) - 9(2x - 3) > 0$$
$$\left(x^2 - 9\right)(2x - 3) > 0$$
$$(x + 3)(x - 3)(2x - 3) > 0$$

The zeros of the polynomial are $x = -3$, $x = 3$, and $x = 1.5$. They divide the number line into four intervals. Since this problem is a strict inequality, use open circles on your number line. One way to check the intervals is to test a number in each interval in the inequality.

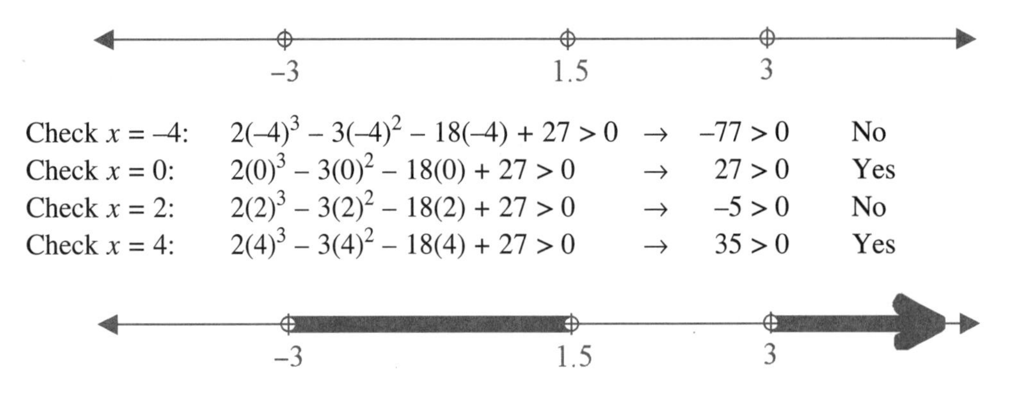

Check $x = -4$:	$2(-4)^3 - 3(-4)^2 - 18(-4) + 27 > 0$	$\rightarrow$	$-77 > 0$	No
Check $x = 0$:	$2(0)^3 - 3(0)^2 - 18(0) + 27 > 0$	$\rightarrow$	$27 > 0$	Yes
Check $x = 2$:	$2(2)^3 - 3(2)^2 - 18(2) + 27 > 0$	$\rightarrow$	$-5 > 0$	No
Check $x = 4$:	$2(4)^3 - 3(4)^2 - 18(4) + 27 > 0$	$\rightarrow$	$35 > 0$	Yes

The solution is $-3 < x < 1.5$ or $x > 3$. In interval notation, this is $(-3, 1.5) \cup (3, \infty)$.

Checking intervals by substituting values is straightforward but can be tedious, especially if you do not have a calculator handy. An alternative that does not require as much calculation is to make a sign chart. As long as the polynomial is being compared with 0, all we really care about is its sign, positive or negative, on each interval. We can find the sign of the polynomial by checking the signs of each of its factors. This is not as hard as it sounds.

SEE CALC TIP 7B

EXAMPLE 7.26 Solve $(x + 1)(x - 2)^2(x^2 - 10) > 0$ algebraically.

SOLUTION

This problem is conveniently already factored. The zeros are $x = -1$, $x = 2$, and $x = \pm\sqrt{10}$. Make the number line in the usual way, with open circles for the strict inequality (>). List each factor, with its exponent, down the left side, and write the whole polynomial at the bottom. Draw vertical dotted lines to separate the intervals and a horizontal one to separate the polynomial from its factors.

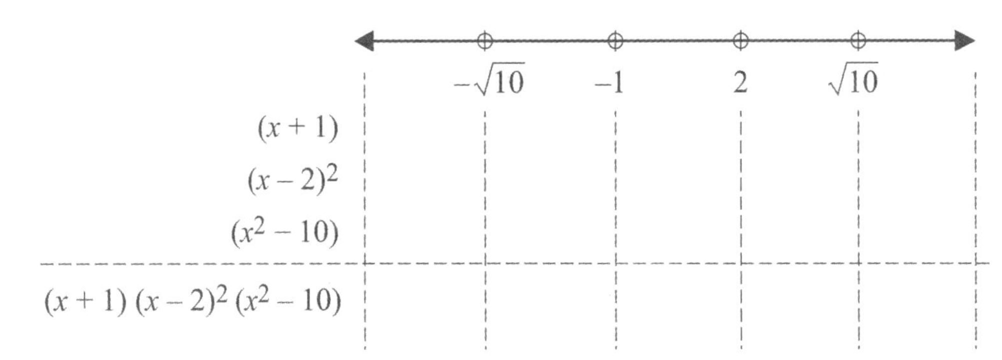

Now look at the factors one at a time and determine their signs on each interval. To determine the signs for the linear factor $x + 1$, mentally picture the line, $y = x + 1$.

The linear factor $x + 1$ will change sign exactly once, at its zero, -1. The line has a positive slope. So to the left of $x = -1$, $x + 1$ will be negative; to the right, $x + 1$ will be positive. Fill in the signs for each interval.

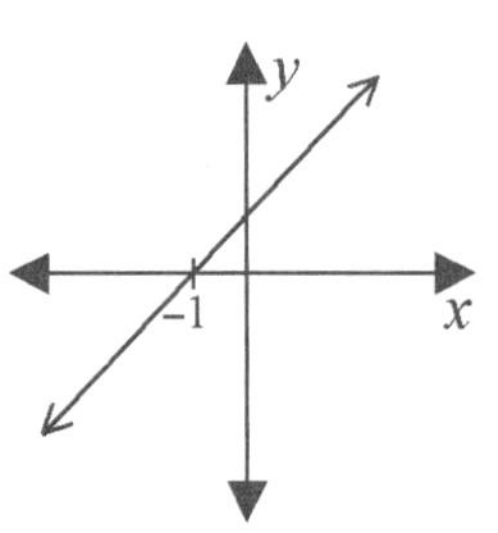

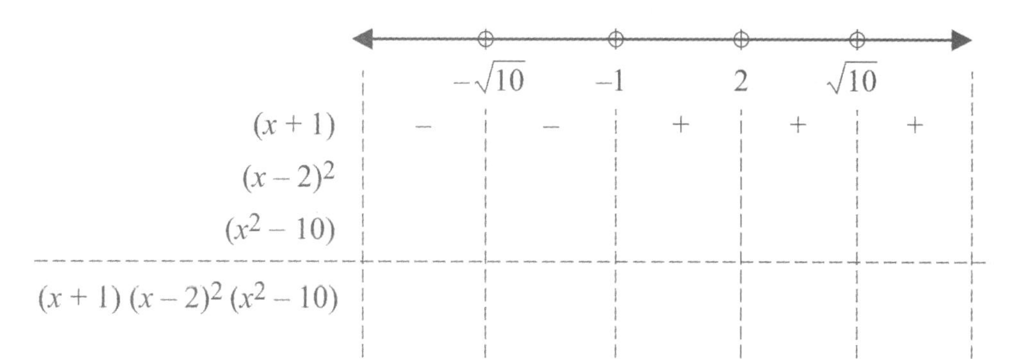

The factor $(x - 2)^2$ is easy. Because it is squared, it will be positive on all the intervals.

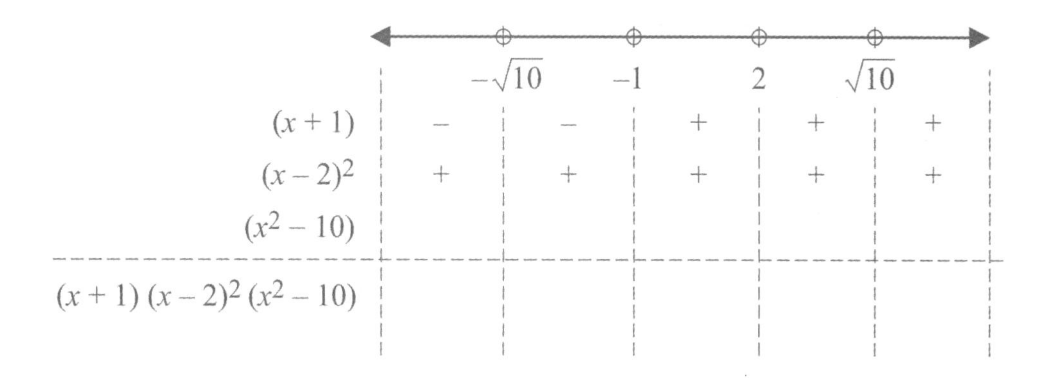

To determine the signs for the quadratic factor, mentally picture a parabola that opens up (since $a > 0$) and crosses the x-axis in two places (two distinct roots).

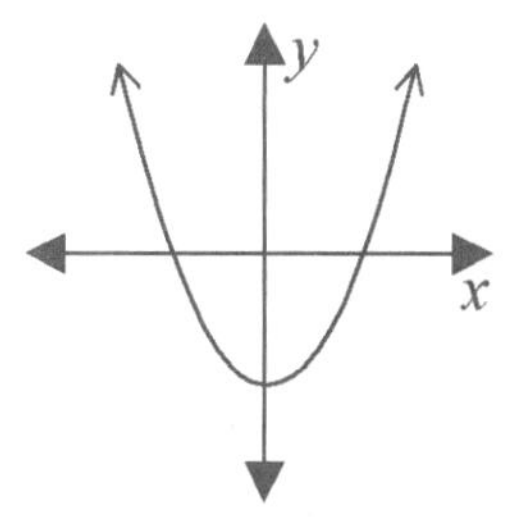

The parabola will be above the x-axis (positive) to the left and right of its zeros and below the x-axis (negative) between its zeros. This means the quadratic factor $x^2 - 10$ will change signs twice, once at each zero. Between the zeros, when $-\sqrt{10} < x < \sqrt{10}$, it will be negative. Outside that interval, it will be positive. Fill in the signs for each interval.

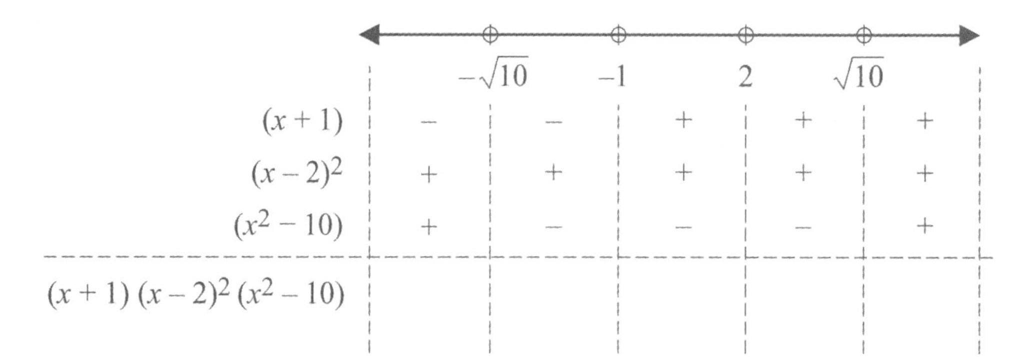

Now it is easy to find the sign of the whole polynomial. Remember that the sign of a product depends only on how many negative factors it contains. An odd number of negative factors makes a negative product; an even number of negative factors makes a positive product. For each interval, just count negative signs. The first interval, $x < -\sqrt{10}$, has just one negative sign; it will be negative. The next interval has two negative signs; it will be positive. Continue in the same way, filling in the bottom row of the sign chart.

	$x < -\sqrt{10}$	$-\sqrt{10}$ to -1	-1 to 2	2 to $\sqrt{10}$	$x > \sqrt{10}$
$(x+1)$	−	−	+	+	+
$(x-2)^2$	+	+	+	+	+
(x^2-10)	+	−	−	−	+
$(x+1)(x-2)^2(x^2-10)$	−	+	−	−	+

This problem asks where the polynomial is greater than 0. So we want all the positive intervals, excluding their endpoints. The solution is $\left(-\sqrt{10},\ -1\right) \cup \left(\sqrt{10},\ \infty\right)$.

Another way to make the sign chart is to use what you know about polynomials and where they change signs. This can be the quickest way to solve the problem but does require some careful thought.

EXAMPLE 7.27

Solve $-3x^2(x+2)(2x-5)^3(x-5)^4 \le 0$ algebraically.

SOLUTION

The zeros are $x = 0$, $x = -2$, $x = 2.5$, and $x = 5$. They each have a closed circle on the number line because of the inclusive inequality (≤).

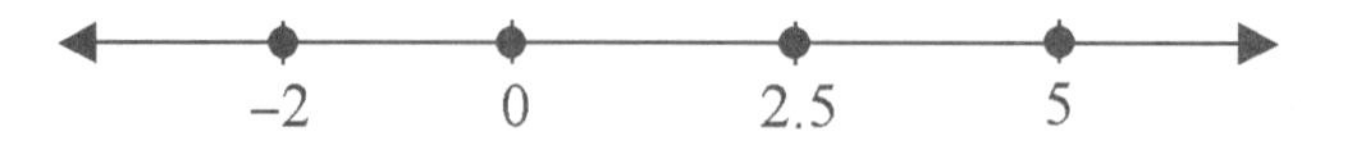

Now consider the polynomial function $y = -3x^2(x + 2)(2x - 5)^3(x - 5)^4$. By adding up the exponents, we find that it has degree 10, even. Both ends go in the same direction. Because of the –3 factor, the leading coefficient will be negative (it doesn't matter what the exact value is) so both ends go down. This means that for $x < -2$ and $x > 5$, the value of the polynomial will be negative.

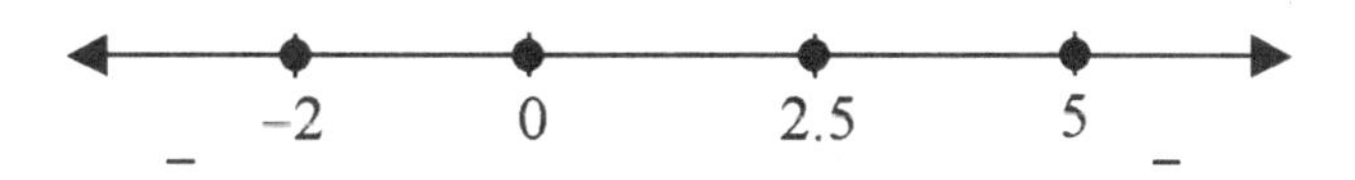

Each time the polynomial crosses the x-axis, it will change sign. Polynomials cross the axis at zeros with odd multiplicity. In this example, $x = -2$ and $x = 2.5$ each have odd multiplicity (odd exponents on the factors.) So the polynomial will change signs at $x = -2$ and $x = 2.5$ but not at $x = 0$ or $x = 5$. From this, we can complete the sign chart.

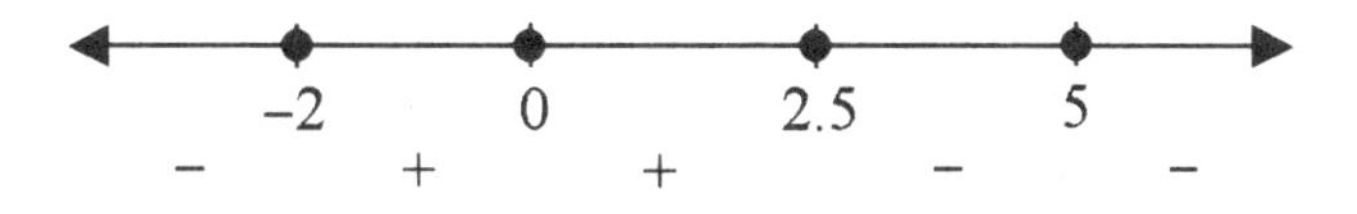

This problem asks where the polynomial is less than or equal to 0, so we want the intervals that are negative, including their endpoints. The solution can be written $(-\infty, -2] \cup [2.5, 5] \cup [5, \infty)$. Note that since the third interval starts where the second one ended and the common endpoint, $x = 5$, is included in the solution, the answer can be written more simply as $(-\infty, -2] \cup [2.5, \infty)$. (Had the problem been a strict inequality, "< 0" instead of "≤ 0," we could not do this. The endpoint $x = 5$ would not have been included, and the two intervals would have to be written separately.)

One final way, and perhaps the easiest way, to solve a polynomial inequality is graphically. The key is to remember that $p(x) > 0$ means we want intervals where $y = p(x)$ is above the x-axis while $p(x) < 0$ means we want intervals where $y = p(x)$ is below the x-axis.

Solve $x^4 - 3x^3 - x^2 + 3x \geq 0$ graphically.

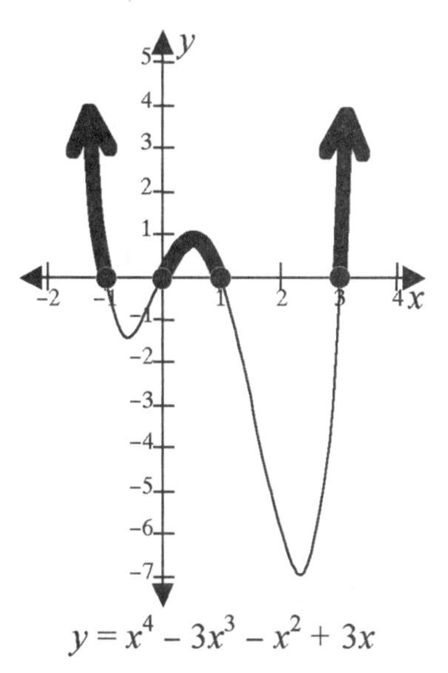

$y = x^4 - 3x^3 - x^2 + 3x$

SOLUTION

Graph $y = x^4 - 3x^3 - x^2 + 3x$ with graphing technology.

From the graph, we see the zeros are at $x = -1$, $x = 0$, $x = 1$, and $x = 3$. The problem asks for intervals where the value of the polynomial is greater than or equal to 0, in other words, where the graph of the polynomial is on or above the x-axis. These parts of the polynomial have been drawn bold. We see $x^4 - 3x^3 - x^2 + 3x \geq 0$ on the intervals $(-\infty, -1] \cup [0, 1] \cup [3, \infty)$.

SECTION EXERCISES

7–18. Solve: $5x^3 - 2x^2 - 35x = -14$

7–19. Solve: $x^3 - x^2 - 18x + 8 = 0$

7–20. Solve $x^5 + 9x^4 + 28x^3 + 36x^2 + 27x + 27 = 0$ for all values of x, real and complex.

7–21. Solve $x^4 - 2x^3 - 12x^2 - 13x - 10 = 0$ for all values of x, real and complex.

7–22. Solve for x algebraically. Express your answer in interval notation.
$x(x + 3)(x - 5)^2 < 0$

7–23. Solve for x algebraically using a sign chart. Express your answer in interval notation. $x^4(2x + 5)(x - 1)^2 (x - 4)^3 > 0$

7–24. Solve for all values of x that satisfy $x(x + 3)(x - 5)^2 \leq 0$ without using graphing technology. Use what you know about graphs of polynomial functions. Express your answer in interval notation.

7–25. Use graphing technology to solve $-2x^4 + 7x^3 + 45x^2 - 184x + 80 < 0$ graphically. Express your answer in interval notation.

CHAPTER EXERCISES

C7–1. For the polynomial function $y = -2x^5 - x^3 + 7x^2 - 5$:

a. What is the degree of the function?

b. What are the values of a_n and a_0? What is the value of a_4?

c. What is the largest number of real zeros the function could have? (Do not use your calculator.)

d. What is the largest number of turning points it could have? (Do not use your calculator.)

e. What is the y-intercept of the function? (Do not use your calculator.)

f. Describe the end behavior of the graph. (Do not use your calculator.)

C7–2. Find the y-intercept and degree of the function: $y = -3(x - 1)^3 (x + 2)^2 (x - 5)$

C7–3. If $f(x) = x^6 + x^4 + 1$ and $f(a) = 30$, find the value of $f(-a)$.

C7–4. Find the zeros: $f(x) = -2x^4 + 10x^3 + 2x^2 - 10x$

C7–5. Find all the zeros of the polynomial $y = 6x^4 + 13x^3 - 70x^2 + 71x - 20$. Write the function in factored form.

C7–6. Find a polynomial function that has zeros $x = -3$, $x = 4$, and $x = \frac{3}{4}$ and a y-intercept of –72.

C7–7. The polynomial $y = x^4 - 2x^3 - 28x^2 + 55x + 100$ has zeros at $x = -5$ and $x = 4$. Use synthetic division to factor the polynomial and find the other zeros.

C7–8. Consider the function $y = -3x^6 - 3x^5 + 5x^2 + a_1x + a_0$ where a_1 and a_0 are rational numbers.

a. How many total zeros does it have?
b. What are the possible combinations of numbers of real and complex zeros it could have?
c. At most, how many turning points does it have?
d. If $-1 + 3i$ is one zero, what is another zero?
e. Describe the end behavior of the graph.

C7–9. A polynomial function has the following zeros: $x = 1$ (multiplicity 3), $x = 3$ (multiplicity 4), and $x = 2 \pm 3i$.

a. What is the degree of the function?
b. How many times does the function cross the x-axis?

C7–10. A polynomial function has 5 total zeros.

a. At most, how many turning points does it have?
b. What is the largest number of imaginary zeros it could have?

C7–11. Write a polynomial function having a zero at $x = -\frac{2}{3}$ and a double zero at $x = 1$ and having a y-intercept of –8.

C7–12. Without using your calculator, sketch a representative graph of

$f(x) = 2x^3 - 7x^2 - 12x + 45$, given that $x = 3$ is a zero.

C7–13. A polynomial function with rational coefficients has 7 total zeros, counting possible complex zeros and multiplicities.

a. At most, how many turning points does it have?
b. What is the greatest number of times the function could cross the x-axis?
c. What is the least number of times the function could cross the x-axis?

C7–14. Given that 2 is a double zero of $y = 2x^4 - 11x^3 + 68x - 80$, factor the polynomial function completely and find all its zeros.

C7–15. Find all the zeros of the function $y = 2x^4 + x^3 - 5x^2 - 13x - 30$ and express the function in factored form using factors with only integral coefficients.

C7–16. Write a polynomial function for the graph below given that when $x = 1$, $y = 2$.

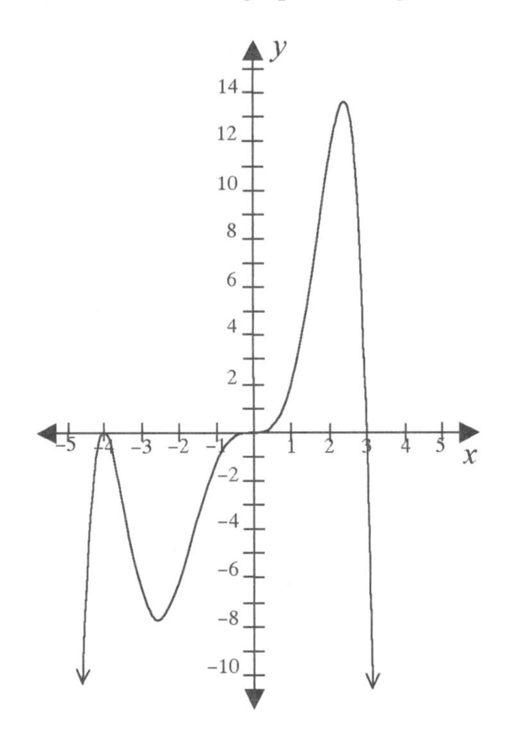

C7–17. Solve: $x^3 + 5x^2 - 9x = 45$

C7–18. Solve: $2x^4 + x^3 - 5x^2 - 2x + 2 = 0$

C7–19. Solve $3x^5 + x^4 - 44x^3 + 105x^2 - 115x + 50 = 0$ for all values of x, real and complex.

C7–20. Solve for x algebraically. Express your answer in interval notation.

$$x(x+1)^2(x-4)^3 \geq 0$$

C7–21. Solve for x algebraically using a sign chart. Express your answer in interval notation.

$$-2x(2x-3)(x+3)^2(x-1)^3 < 0$$

C7–22. Use graphing technology to solve $x^5 - 4x^4 - 54x^3 + 284x^2 - 395x + 168 > 0$ graphically. Express your answer in interval notation.

Radical Functions

WHAT YOU WILL LEARN

You have already learned about functions with powers like $y = x^2$ and $y = x^3$. The inverses of these functions are the square root and cube root functions, which are both examples of radical functions. To be able to make full use of powers, you must also be able to use radicals. In this chapter, you will learn to:

- Simplify radicals;
- Perform operations with radicals;
- Perform operations with rational exponents;
- Graph square root and cube root functions;
- Solve radical equations;
- Find inverses of radical functions.

SECTIONS IN THIS CHAPTER

- Radical Models
- Square Roots and Higher-Order Radicals
- Operations with Radicals
- Rational Exponents
- Graphing Square Root and Cube Root Functions
- Solving Radical Equations
- Inverses of Radical Functions

Radical Models

Most people know that the police sometimes use the length of skid marks left by a car to estimate how fast the car was moving at the start of the skid. Several factors go into these estimates, in particular the type of road surface and whether the road was wet or dry. However, the estimates all have one thing in common: the estimated speed is proportional to the square root of the length of the skid marks.

For example, on a dry asphalt road, the speed, s, in miles per hour might be estimated from the length of the skid marks, x, as shown in Figure 8.1.

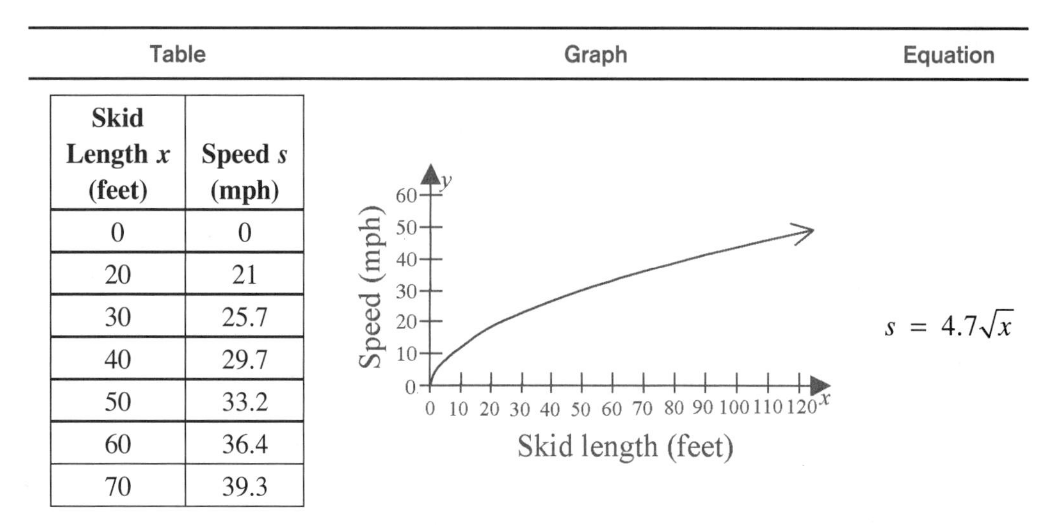

Skid Length x (feet)	Speed s (mph)
0	0
20	21
30	25.7
40	29.7
50	33.2
60	36.4
70	39.3

FIGURE 8.1

This is an example of a radical function. A physics-related example is the time an object takes to fall a certain distance, which is related to the square root of the distance. In geometry, the length of the side of a square of a certain area is equal to the square root of the area. Additionally, the radius of a sphere that contains a certain volume is proportional to the cube root of the volume.

Square Roots and Higher-Order Radicals

A number r is a *square root* of another number k if $r^2 = k$. For example, 4 is a square root of 16 because $4^2 = 16$. The number -4 is another square root of 16 because $(-4)^2 = 16$. You probably remember that all positive numbers have two square roots, one positive and one negative. The radical sign, $\sqrt{}$, is used to denote the positive square root, also called the *principal square root*. Thus $\sqrt{16} = 4$, not -4 or ± 4. Negative numbers have no real square roots because for any real number r, $r^2 \geq 0$. So $\sqrt{-16}$ is not real. (It is imaginary. See Chapter 6.)

The same idea can be extended to higher-order radicals. A number r is a *cube root* of another number k if $r^3 = k$. For example, 5 is a cube root of 125 because $5^3 = 125$. However, -5 is not a cube root of 125 because $(-5)^3 = -125 \neq 125$. Instead, -5 is a cube root of -125. This illustrates a big difference between cube roots and square roots. All real numbers have exactly one real cube root. Positive numbers have a positive cube root, and negative numbers have a negative cube root. The symbol for cube root is $\sqrt[3]{}$. Thus $\sqrt[3]{125} = 5$ and $\sqrt[3]{-125} = -5$.

In general, for any integer $n \geq 2$, we say that r is an nth *root* of a number k if $r^n = k$. The symbol for the principal nth root is $\sqrt[n]{}$. The n in the radical is called the *index*. (For a square root, the index is 2. However, it is rarely written, just as we seldom write a 1 for the exponent on x.) Roots with an even index have the same properties as square roots. Positive numbers will have two real roots; negative numbers will have no real roots. Thus both 3 and -3 are fourth roots of 81 because $3^4 = (-3)^4 = 81$. In contrast, -81 has no real fourth roots. Just as for the square root, the positive root is the principal root, $\sqrt[4]{81} = 3$, not -3 or ± 3. Roots with an odd index will behave like cube roots. All numbers will have one real root, with the same sign as the original number. For example, $\sqrt[5]{32} = 2$ because $2^5 = 32$ and $\sqrt[5]{-32} = -2$ because $(-2)^5 = -32$.

PROPERTIES OF RADICALS

Table 8.1 shows the properties of real numbers if the roots involved are real numbers.

TABLE 8.1
PROPERTIES OF RADICALS

Property*	Example
1. $\sqrt[n]{xy} = \sqrt[n]{x}\sqrt[n]{y}$	1. $\sqrt[3]{40} = \sqrt[3]{8}\sqrt[3]{5} = 2\sqrt[3]{5}$
2. $\sqrt[n]{\frac{x}{y}} = \frac{\sqrt[n]{x}}{\sqrt[n]{y}}$ $(y \neq 0)$	2. $\sqrt{\frac{25}{4}} = \frac{\sqrt{25}}{\sqrt{4}} = \frac{5}{2}$
3. $\sqrt[m]{\sqrt[n]{x}} = \sqrt[mn]{x}$	3. $\sqrt[3]{\sqrt{2}} = \sqrt[6]{2}$
4. $\sqrt[n]{x^m} = \left(\sqrt[n]{x}\right)^m$	4. $\sqrt[4]{16^5} = \left(\sqrt[4]{16}\right)^5 = (2)^5 = 32$
5. $\left(\sqrt[n]{x}\right)^n = \left(\sqrt[n]{x^n}\right) = x$ if $x \geq 0$	5. $\left(\sqrt[4]{7}\right)^4 = 7$ and $\sqrt[3]{(8)^3} = 8$

*Let x and y be real numbers. Let m and n be positive integers. The roots involved are all real numbers.

COMMON ERROR

Note that $\sqrt[n]{x+y} = \sqrt[n]{x} + \sqrt[n]{y}$ and $\sqrt[n]{x-y} = \sqrt[n]{x} - \sqrt[n]{y}$ are *not* on the list. This is because they *are not true*! Radicals do not distribute over addition or subtraction.

$$\sqrt[n]{x+y} \neq \sqrt[n]{x} + \sqrt[n]{y}$$
$$\sqrt[n]{x-y} \neq \sqrt[n]{x} - \sqrt[n]{y}$$

It is worth reemphasizing that these properties are guaranteed to hold only when all the roots involved are real numbers. For example, by the first property, $\sqrt{(4)(9)} = \sqrt{4}\sqrt{9}$ is true; both equal 6. However, $\sqrt{(-4)(-9)} = \sqrt{-4}\sqrt{-9}$ is not true. The quantity on the left equals 6, but the quantity on the right is either –6 or not defined depending whether you allow the use of imaginary numbers. Similarly, by the fourth property, $\sqrt{2^6} = \left(\sqrt{2}\right)^6$ is true (both sides equal 8) but $\sqrt{(-2)^6} = \left(\sqrt{-2}\right)^6$ is not true (8 is compared with –8 or is undefined). At the Algebra 2 level, it is not uncommon to avoid this issue by stating in radical problems that all variables represent nonnegative numbers as was done for the fifth property. However, if you study more math (which we encourage!), you may see radicands (the expressions inside radical symbols) containing variables that could be negative. In those cases, if the power is applied last, $\left(\sqrt[n]{x}\right)^n = x$ remains the same. If the power is first, though, you get: $\sqrt[n]{x^n} = \begin{cases} |x| & \text{if } n \text{ is even} \\ x & \text{if } n \text{ is odd} \end{cases}$. This definition applies to examples such as $\sqrt[4]{(-8)^4} = |-8| = 8$. So if it is possible that variables inside radicals represent negative numbers, you must be very careful when applying the radical properties.

SIMPLIFYING RADICALS

Although $\sqrt{25}$ is the same as 5, almost everyone agrees that 5 is the simpler expression. Even though $\sqrt{24}$ is not the same as any integer (or even any rational number), it can be simplified in the sense that it can be written with a smaller radicand: $\sqrt{24} = \sqrt{4 \cdot 6} = \sqrt{4}\sqrt{6} = 2\sqrt{6}$. We say a square root radical is in simplest form if:

1. There are no perfect square factors in the radicand;
2. There are no radicals in the denominator or denominators in the radical.

Techniques for dealing with radicals in a denominator are covered in a later section. Here we show how to break down the radicand so that some factors can be simplified out of the radical. The idea is to rewrite the radical as a product of prime numbers (you may remember prime factor trees from middle school). For square roots, after rewriting the radicand as a product of primes, the square root can be simplified as follows. If a factor appears twice inside the radical, then both factors are removed from inside the radical and one of them is written outside.

EXAMPLE 8.1

Simplify: $\sqrt{60}$

SOLUTION

The first step is to write 60 as a product of prime numbers. Here is a prime factor tree for 60 to remind you of what one looks like.

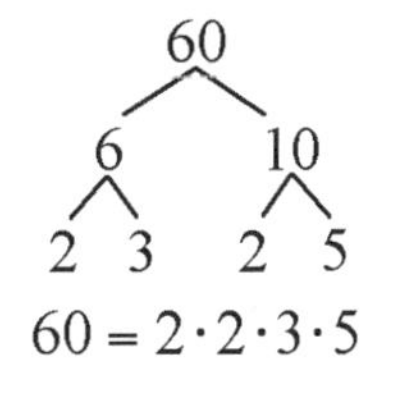

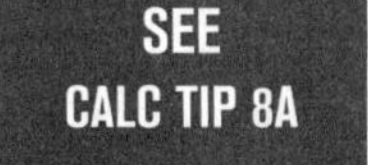

$\sqrt{60} = \sqrt{2 \cdot 2 \cdot 3 \cdot 5}$ — Rewrite 60 as a product of prime numbers.

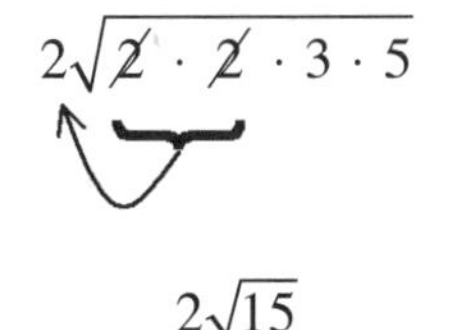

Any pairs of factors can be simplified. Here $\sqrt{2 \cdot 2} = \sqrt{4} = 2$, so the pair of 2s inside the radical becomes a single 2 outside.

$2\sqrt{15}$ — Simplify the result by multiplying values outside the radical and multiplying values remaining inside the radical.

This works with variables too.

EXAMPLE 8.2

Express $5\sqrt{72x^3}$, $x \geq 0$, in simplest radical form.

SOLUTION

$5\sqrt{72x^3} = 5\sqrt{2 \cdot 2 \cdot 2 \cdot 3 \cdot 3 \cdot x \cdot x \cdot x}$ — Rewrite 72 as a product of prime numbers, $72 = 2 \cdot 2 \cdot 2 \cdot 3 \cdot 3$. Write out the product for $x^3 = x \cdot x \cdot x$.

$5 \cdot 2 \cdot 3 \cdot x\sqrt{\not{2} \cdot \not{2} \cdot 2 \cdot \not{3} \cdot \not{3} \cdot \not{x} \cdot \not{x} \cdot x}$ — Any pairs of factors can be simplified. Here $\sqrt{2 \cdot 2} = \sqrt{4} = 2$, $\sqrt{3 \cdot 3} = \sqrt{9} = 3$, and $\sqrt{x \cdot x} = \sqrt{x^2} = x$ (because we know $x \geq 0$).

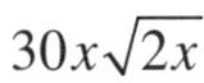

$30x\sqrt{2x}$ — Simplify the result by multiplying any values outside the radical and multiplying any values remaining inside the radical.

If a variable has a large exponent, you may want to use the following shortcut. To simplify the square root of a power, divide the exponent by 2. The whole number part of the answer becomes the exponent outside the radical. The remainder, if there is one, is the exponent that stays inside the radical. To simplify $\sqrt{x^8}$, note that $8 \div 2 = 4$ with no remainder. So $\sqrt{x^8} = x^4$ (with no radical left). For $\sqrt{x^7}$, $7 \div 2 = 3$ with a remainder of 1. So $\sqrt{x^7} = x^3\sqrt{x}$. (If you find writing out powers of variables is easier for you, then you should continue using that method.)

EXAMPLE 8.3

Express $\sqrt{54x^6y^9}$ in simplest radical form where x and y are both positive.

SOLUTION

$\sqrt{54x^6y^9} = \sqrt{2 \cdot 3 \cdot 3 \cdot 3 \cdot x^6y^9}$	Write 54 as a product of prime numbers, $54 = 2 \cdot 3 \cdot 3 \cdot 3$.
$3\sqrt{2 \cdot 3 \cdot \cancel{3} \cdot \cancel{3} \cdot x^6y^9}$	Simplify pairs. $3 = \sqrt{3 \cdot 3} = \sqrt{9}$.
$3x^3\sqrt{2 \cdot 3 \cdot y^9}$	Simplify the x^6; $6 \div 2 = 3$ with no remainder, so $\sqrt{x^6} = x^3$.
$3x^3y^4\sqrt{2 \cdot 3 \cdot y}$	Simplify the y^9. Since $9 \div 2 = 4$ with remainder 1, $\sqrt{y^9} = y^4\sqrt{y}$.
$3x^3y^4\sqrt{6y}$	Simplify the answer.

COMMON ERROR

Do not make the mistake of thinking that because 9 is a perfect square, $\sqrt{y^9} = y^3$. The square root of a power takes half of the inside exponent, with any remainder left inside. So $\sqrt{y^9} = y^4\sqrt{y}$.

Cube roots are simplified the same general way with the following difference. If a factor appears three times inside the radical, then all three factors are removed from inside the radical and one of them is written outside. The shortcut for powers is modified so the exponent is divided by 3. The whole-number answer is still the exponent outside the radical. The remainder, if any, is the exponent that stays inside the radical.

EXAMPLE 8.4 Express $\sqrt[3]{72x^{12}y^8}$ in simplest radical form.

SOLUTION

$\sqrt[3]{72x^{12}y^8} = \sqrt[3]{2 \cdot 2 \cdot 2 \cdot 3 \cdot 3 \cdot x^{12}y^8}$ — Rewrite 72 as a product of prime numbers.

$2\sqrt[3]{\not{2} \cdot \not{2} \cdot \not{2} \cdot 3 \cdot 3 \cdot x^{12}y^8}$ — Any triples of factors can be simplified. So, $\sqrt[3]{2 \cdot 2 \cdot 2} = \sqrt[3]{8} = 2$.

$2x^4\sqrt[3]{9y^8}$ — For the variables, $12 \div 3 = 4$ with no remainder. So $\sqrt[3]{x^{12}} = x^4$.

$2x^4y^2\sqrt[3]{9y^2}$ — Since $8 \div 3 = 2$ with remainder 2, $\sqrt[3]{y^8} = y^2\sqrt[3]{y^2}$.

These techniques can be used to simplify radicals of any order as long as you make appropriate changes based on the index.

EXAMPLE 8.5 Express $\sqrt[4]{80x^{10}y^{12}}$ in simplest radical form where both x and y are positive.

SOLUTION

$\sqrt[4]{80x^{10}y^{12}} = \sqrt[4]{2 \cdot 2 \cdot 2 \cdot 2 \cdot 5 \cdot x^{10}y^{12}}$ — Rewrite 80 as a product of prime numbers.

$2\sqrt[4]{\not{2} \cdot \not{2} \cdot \not{2} \cdot \not{2} \cdot 5 \cdot x^{10}y^{12}}$ — Any quadruples of factors can be simplified. So, $\sqrt[4]{2 \cdot 2 \cdot 2 \cdot 2} = \sqrt[4]{16} = 2$.

$2x^2\sqrt[4]{5 \cdot x^2y^{12}}$ — To simplify x^{10}, $10 \div 4 = 2$ with remainder 2. So $\sqrt[4]{x^{10}} = x^2\sqrt[4]{x^2}$.

$2x^2y^3\sqrt[4]{5x^2}$ — For y^{12}, $12 \div 4 = 3$ with no remainder. So $\sqrt[4]{y^{12}} = y^3$.

With practice, you may find you can do these in fewer steps.

SECTION EXERCISES

For 1–4, express in simplest radical form. Assume all variables represent positive numbers.

8–1. $\sqrt{20x^2}$

8–2. $\sqrt{81a^{10}b^{16}}$

8–3. $\sqrt[3]{54x^6y^8}$

8–4. $\sqrt[4]{48x^8y^9}$

Operations with Radicals

Radical expressions can be combined with the usual operations of addition, subtraction, multiplication, and division. Be careful though. The rules for adding and subtracting radicals are different from the rules for multiplying and dividing them.

ADDING AND SUBTRACTING RADICALS

Remember the rules for adding and subtracting monomials: we can combine only *like terms*. We add or subtract the coefficients but keep the variables the same. So $3x^2 + 5x^2 = 8x^2$ (not $8x^4$). However, $4x^2 + 2x^3$ cannot be simplified; neither can $6x + 4y$. Very similar rules apply to radicals. We can combine only like radicals; this means the same radicand with the same index.

To add or subtract radicals:

1. Simplify all radicals first.
2. Then combine like radicals. Add or subtract the coefficients, and keep the like radicands. *Do not add or subtract radicands!*

In the examples that follow, we omit the details of simplifying the radicals that were covered previously.

EXAMPLE 8.6

Express $\sqrt{75} + \sqrt{32} - 3\sqrt{12}$ in simplest radical form.

SOLUTION

Remember: Simplify all the radicals first. Then combine like terms.

$$\begin{aligned}\sqrt{75} + \sqrt{32} - 3\sqrt{12} &= 5\sqrt{3} + 4\sqrt{2} - 3(2)\sqrt{3} \\ &= 5\sqrt{3} + 4\sqrt{2} - 6\sqrt{3} \\ &= -\sqrt{3} + 4\sqrt{2}\end{aligned}$$

Since they were the only like radicals, the $\sqrt{3}$ terms were the only ones that could be combined.

EXAMPLE 8.7

Express $5\sqrt{28x^5} - 3x\sqrt{63x^3}$, $x \geq 0$, in simplest radical form.

SOLUTION

$$\begin{aligned}5\sqrt{28x^5} - 3x\sqrt{63x^3} &= 5\left(2x^2\right)\sqrt{7x} - 3x(3x)\sqrt{7x} \\ &= 10x^2\sqrt{7x} - 9x^2\sqrt{7x} \\ &= x^2\sqrt{7x}\end{aligned}$$

EXAMPLE 8.8

Simplify: $4\sqrt[3]{16x^4} - 5\sqrt[3]{8x^3} + \sqrt[3]{54x}$

SOLUTION

$$\begin{aligned}4\sqrt[3]{16x^4} - 5\sqrt[3]{8x^3} + \sqrt[3]{54x} &= 4(2x)\sqrt[3]{2x} - 5(2x) + 3\sqrt[3]{2x} \\ &= 8x\sqrt[3]{2x} - 10x + 3\sqrt[3]{2x} \\ &= (8x + 3)\sqrt[3]{2x} - 10x\end{aligned}$$

EXAMPLE 8.9

Find the sum of $3\sqrt{2x}$ and $5\sqrt[3]{2x}$.

SOLUTION

The sum of $3\sqrt{2x}$ and $5\sqrt[3]{2x}$ is $3\sqrt{2x} + 5\sqrt[3]{2x}$; it cannot be further simplified. Because one is a square root and the other is a cube root, we cannot combine them into a single term.

MULTIPLYING RADICALS

Recall that multiplying and dividing monomials follow different rules from adding and subtracting. In particular, we did not need like terms to multiply or divide. We multiplied or divided both the coefficients and the variable parts separately. For instance,

$\left(2x^3\right)(4x) = 8x^4$ and $\dfrac{6x^5}{2x^3} = 3x^2$. Similar rules apply to multiplying and dividing monomial radicals. In particular, we can multiply or divide radicals with different radicands, but they still must have the same index.

To multiply or divide radicals with the same index:

1. First multiply or divide the coefficients (including variables outside the radical), and multiply or divide the radicands. Do not multiply coefficients by radicands.
2. Simplify the answer.

Unlike adding and subtracting, you should simplify last when multiplying and dividing radicals. Simplifying first is rarely helpful and sometimes causes extra work.

EXAMPLE 8.10

Simplify: $\left(2x\sqrt{5x}\right)\left(3y\sqrt{10x^3}\right)$

SOLUTION

$\left(2x\sqrt{5x}\right)\left(3y\sqrt{10x^3}\right)$	
$6xy\sqrt{50x^4}$	Multiply the outsides; multiply the insides.
$6xy\left(5x^2\right)\sqrt{2}$	Simplify the radical.
$30x^3y\sqrt{2}$	Simplify outside the radical.

EXAMPLE 8.11

Simplify: $\left(5\sqrt[3]{6x^2y^5}\right)\left(3\sqrt[3]{12x^6y}\right)$

SOLUTION

$\left(5\sqrt[3]{6x^2y^5}\right)\left(3\sqrt[3]{12x^6y}\right)$	
$15\sqrt[3]{72x^8y^6}$	Multiply the outsides; multiply the insides.
$15\left(2x^2y^2\right)\sqrt[3]{9x^2}$	Simplify the radical.
$30x^2y^2\sqrt[3]{9x^2}$	Simplify outside the radical.

EXAMPLE 8.12 Which of the following is $\frac{\sqrt{72}}{\sqrt{12}}$ written in simplest radical form?

(1) $2\sqrt{3}$

(2) $3\sqrt{2}$

(3) $\frac{3\sqrt{2}}{\sqrt{3}}$

(4) $\sqrt{6}$

SOLUTION

Divide the radicands, $\frac{\sqrt{72}}{\sqrt{12}} = \sqrt{\frac{72}{12}} = \sqrt{6}$, and you are done. Note that if you simplify the numerator and denominator before you divide, you just make the problem more difficult: $\frac{\sqrt{72}}{\sqrt{12}} = \frac{6\sqrt{2}}{2\sqrt{3}} = \frac{3\sqrt{2}}{\sqrt{3}}$. This is not in simplest form because it has a radical in the denominator. (See "Rationalizing Denominators" in the next section.) The answer is choice (4).

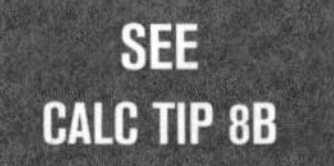

EXAMPLE 8.13 Simplify: $\frac{50\sqrt{75x^9y^8}}{5\sqrt{3xy}}$

SOLUTION

$$\frac{50\sqrt{75x^9y^8}}{5\sqrt{3xy}}$$

$10\sqrt{25x^8y^7}$	Reduce (divide) the outsides; reduce the insides.
$10\left(5x^4y^3\right)\sqrt{y}$	Simplify the radical.
$50x^4y^3\sqrt{y}$	Simplify outside the radical.

EXAMPLE 8.14 Simplify: $\dfrac{\sqrt{12ab^9}}{2\sqrt{3a^3b^7}}$ $a \geq 0,\ b \geq 0$

SOLUTION

$$\frac{\sqrt{12ab^9}}{2\sqrt{3a^3b^7}}$$

$\dfrac{\sqrt{4b^2}}{2\sqrt{a^2}}$ Reduce (divide) the outsides; reduce the insides.

$\dfrac{2b}{2a}$ Simplify the radicals.

$\dfrac{b}{a}$ Reduce.

EXAMPLE 8.15 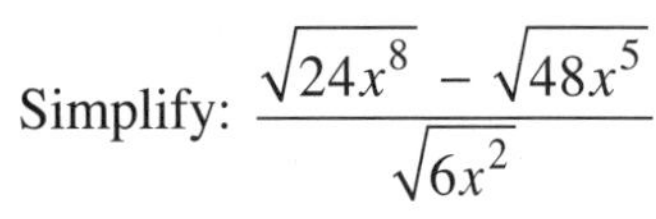

Simplify: $\dfrac{\sqrt{24x^8} - \sqrt{48x^5}}{\sqrt{6x^2}}$

SOLUTION

$$\frac{\sqrt{24x^8} - \sqrt{48x^5}}{\sqrt{6x^2}}$$

$\dfrac{\sqrt{24x^8}}{\sqrt{6x^2}} - \dfrac{\sqrt{48x^5}}{\sqrt{6x^2}}$ Break into two fractions with a common denominator.

$\sqrt{4x^6} - \sqrt{8x^3}$ Reduce (divide) the insides.

$2x^3 - 2x\sqrt{2x}$ Simplify the radicals.

RATIONALIZING DENOMINATORS

So far, the answers for all the division examples have not had radicals in the denominator. Obviously, this will not always happen. Consider $\dfrac{\sqrt{48}}{\sqrt{20}}$. We get $\dfrac{\sqrt{48}}{\sqrt{20}} = \sqrt{\dfrac{48}{20}} = \sqrt{\dfrac{12}{5}}$.

Do we leave it like that? Do we write $\dfrac{\sqrt{12}}{\sqrt{5}}$? Or maybe we should write $\sqrt{2.4}$? None of these is considered simplest radical form. In simplest form, radical expressions should have no radical in the denominator and no fraction or decimal in the radicand. So how do we deal with expressions like $\dfrac{\sqrt{12}}{\sqrt{5}}$? We need to *rationalize the denominator*. How we do that depends on whether we have a monomial or a binomial denominator.

Note that in this book, we only cover rationalizing square root denominators. It is possible to rationalize cube root and higher-order denominators but it is a little trickier and is not really used very often.

MONOMIAL DENOMINATORS

Rationalizing a monomial denominator relies on the fact that if a is a positive number, then $\sqrt{a} \cdot \sqrt{a} = \sqrt{a^2} = a$. This means that as long as a radical has a rational radicand, multiplying the radical by itself will give a rational number. To rationalize a monomial denominator that contains a radical, multiply both the numerator and denominator by the radical part of the denominator. Then simplify if necessary.

EXAMPLE 8.16 Rationalize the denominator and write in simplest form: $\dfrac{\sqrt{12}}{\sqrt{5}}$

SOLUTION

The denominator is $\sqrt{5}$. To rationalize it, multiply both the numerator and denominator by $\sqrt{5}$.

$$\frac{\sqrt{12}}{\sqrt{5}} = \frac{\sqrt{12}}{\sqrt{5}} \cdot \frac{\sqrt{5}}{\sqrt{5}} = \frac{\sqrt{60}}{5}$$

Now simplify.

$$\frac{\sqrt{60}}{5} = \frac{2\sqrt{15}}{5}$$

COMMON ERROR

Do not even think about reducing the 15 and the 5. The 15 is in a radicand; 5 is not. They cannot be combined.

EXAMPLE 8.17 Express $\dfrac{5x^2}{3\sqrt{10x}}$ in simplest form.

SOLUTION

We need to multiply by $\sqrt{10x}$. (Multiplying by $3\sqrt{10x}$ isn't wrong; it just makes a little extra work.)

$$\frac{5x^2}{3\sqrt{10x}} \cdot \frac{\sqrt{10x}}{\sqrt{10x}} = \frac{5x^2\sqrt{10x}}{3(10x)} = \frac{5x^2\sqrt{10x}}{30x} = \frac{x\sqrt{10x}}{6}$$

You will usually save yourself work if you do as much of the division as possible before rationalizing the denominator.

EXAMPLE 8.18

Express $\dfrac{3y\sqrt{120x^5y}}{4x\sqrt{18x^2y^3}}$ in simplest form.

SOLUTION

Instead of multiplying by $\sqrt{18x^2y^3}$, simplify first. Start by dividing common factors out of the radicands.

$$\frac{3y\sqrt{120x^5y}}{4x\sqrt{18x^2y^3}} = \frac{3y}{4x}\sqrt{\frac{120x^5y}{18x^2y^3}} = \frac{3y}{4x}\sqrt{\frac{20x^3}{3y^2}} = \frac{3y\sqrt{20x^3}}{4x\sqrt{3y^2}}$$

Then simplify the remaining radicals.

$$\frac{3y\sqrt{20x^3}}{4x\sqrt{3y^2}} = \frac{3y(2x)\sqrt{5x}}{4xy\sqrt{3}} = \frac{6xy\sqrt{5x}}{4xy\sqrt{3}} = \frac{3\sqrt{5x}}{2\sqrt{3}}$$

Now rationalize by multiplying both the numerator and the denominator by $\sqrt{3}$.

$$\frac{3\sqrt{5x}}{2\sqrt{3}} \cdot \frac{\sqrt{3}}{\sqrt{3}} = \frac{3\sqrt{15x}}{2(3)} = \frac{\sqrt{15x}}{2}$$

Simplify: $\dfrac{4x + \sqrt{2x}}{2\sqrt{6x}}$

SOLUTION

Rationalize by multiplying both the numerator and the denominator by $\sqrt{6x}$. Remember to distribute in the numerator.

$$\frac{\left(4x + \sqrt{2x}\right)}{2\sqrt{6x}} \cdot \frac{\sqrt{6x}}{\sqrt{6x}} = \frac{4x\sqrt{6x} + \sqrt{12x^2}}{2(6x)} = \frac{4x\sqrt{6x} + 2x\sqrt{3}}{12x}$$

Factor a GCF of $2x$ out of the numerator, and reduce the fraction.

$$\frac{4x\sqrt{6x} + 2x\sqrt{3}}{12x} = \frac{2x\left(2\sqrt{6x} + \sqrt{3}\right)}{12x} = \frac{\cancel{2x}\left(2\sqrt{6x} + \sqrt{3}\right)}{{}_{6}\cancel{12x}} = \frac{2\sqrt{6x} + \sqrt{3}}{6}$$

BINOMIAL DENOMINATORS

Binomial denominators need a slightly different method to rationalize them. If we try multiplying the numerator and denominator of $\dfrac{11}{5 - \sqrt{3}}$ by $\sqrt{3}$, we get

$\dfrac{11}{5 - \sqrt{3}} \cdot \dfrac{\sqrt{3}}{\sqrt{3}} = \dfrac{11\sqrt{3}}{5\sqrt{3} - 3}$. (Don't forget, we have to distribute in the denominator.)

We have not rationalized the denominator; we've just moved the radical to the other term. This is not helpful.

Recall from Chapter 2 that a pair of conjugates is two binomials that differ only in the sign of the second term, such as $a + b$ and $a - b$. Their product is $(a+b)(a-b) = a^2 - ab + ab - b^2 = a^2 - b^2$. This means that if either a or b (or both) start out as square roots of rational numbers, then the product of $(a+b)(a-b)$ will be rational. So to rationalize a binomial denominator, multiply both the numerator and the denominator by the conjugate of the denominator.

EXAMPLE 8.20 Rationalize the denominator: $\dfrac{11}{5-\sqrt{3}}$

SOLUTION

Multiply both the numerator and the denominator by $5+\sqrt{3}$, which is the conjugate of the denominator $5-\sqrt{3}$.

$$\begin{aligned}\frac{11}{5-\sqrt{3}}\cdot\frac{5+\sqrt{3}}{5+\sqrt{3}} &= \frac{11\left(5+\sqrt{3}\right)}{\left(5-\sqrt{3}\right)\left(5+\sqrt{3}\right)}\\ &= \frac{55+11\sqrt{3}}{25+5\sqrt{3}-5\sqrt{3}-3}\\ &= \frac{11\left(5+\sqrt{3}\right)}{22}\\ &= \frac{\cancel{11}\left(5+\sqrt{3}\right)}{{}_{2}\cancel{22}}\\ &= \frac{5+\sqrt{3}}{2}\end{aligned}$$

A couple of things are worth noting in this example. First, when multiplying out the denominator, the two middle terms were opposites and canceled each other. This will always happen when multiplying two conjugates; that is the whole idea. So we can take a shortcut in the denominator and just write

$$\left(5-\sqrt{3}\right)\left(5+\sqrt{3}\right) = 5^2 - \sqrt{3}^2 = 25 - 3 = 22.$$

Remember, use this shortcut only when multiplying two conjugates.

For a monomial numerator, it is often not worth distributing the numerator at first. In this example, after distributing the 11 through the conjugate, we just ended up factoring it back out to simplify our answer. This does not usually apply if you have a binomial numerator.

Rationalize the denominator: $\dfrac{6 + 2\sqrt{10}}{4 + \sqrt{10}}$

SOLUTION

Multiply both the numerator and the denominator by $4 - \sqrt{10}$, which is the conjugate of the denominator. In this case (with a binomial numerator), we must multiply out both the numerator and the denominator. We can use the shortcut for the denominator but not for the numerator.

$$\frac{6 + 2\sqrt{10}}{4 + \sqrt{10}} \cdot \frac{4 - \sqrt{10}}{4 - \sqrt{10}} = \frac{\left(6 + 2\sqrt{10}\right)\left(4 - \sqrt{10}\right)}{\left(4 + \sqrt{10}\right)\left(4 - \sqrt{10}\right)}$$

$$= \frac{24 - 6\sqrt{10} + 8\sqrt{10} - 2(10)}{16 - 10}$$

$$= \frac{4 + 2\sqrt{10}}{6}$$

By factoring the numerator, we can simplify this answer.

$$\frac{4 + 2\sqrt{10}}{6} = \frac{\cancel{2}\left(2 + \sqrt{10}\right)}{{}^{3}\cancel{6}} = \frac{2 + \sqrt{10}}{3}$$

COMMON ERROR

Do not even dream of reducing the $\frac{6}{4}$ or the $\frac{2\sqrt{10}}{\sqrt{10}}$ separately in the original problem. These are binomials; you cannot simplify just one part of a quotient of binomials.

SECTION EXERCISES

For 5–13, express in simplest form. Assume all variables represent positive numbers.

8–5. $2\sqrt{54x} - 3\sqrt{24x}$

8–6. $\left(\sqrt{8x}\right)\left(3\sqrt{2x^5}\right)$

8–7. $\left(2 + 3\sqrt{5y}\right)\left(4 - \sqrt{5y}\right)$

8–8. $\sqrt{2x}\left(\sqrt{18x^3} + 3\sqrt{6x}\right)$

8–9. $\dfrac{5y^2\sqrt{180x^5y}}{3\sqrt{10x^3y^5}}$

8–10. $2\sqrt[3]{250x^8} + x^2\sqrt[3]{54x^2}$

8–11. $\left(5\sqrt[3]{6x^2y^5}\right)\left(3\sqrt[3]{12x^6y}\right)$

8–12. $\dfrac{6x^2\sqrt[3]{48x^2y^5}}{4\sqrt[3]{6x^8y}}$

8–13. $3x\sqrt[4]{32x} - 5\sqrt[4]{162x^5}$

For 14–19, express in simplest form with rational denominators. Assume all variables represent positive numbers.

8–14. $\sqrt{\dfrac{2}{15}}$

8–15. $\dfrac{4\sqrt{7}}{\sqrt{8}}$

8–16. $\dfrac{3\sqrt{2x^5y^7}}{\sqrt{8x^3y}}$

8–17. $\dfrac{\sqrt{2}}{\sqrt{6} - \sqrt{2}}$

8–18. $\dfrac{10}{3\sqrt{5} - 5}$

8–19. $\dfrac{1 - 5\sqrt{7}}{3 - \sqrt{7}}$

Rational Exponents

In many problems involving radicals, it is convenient to express the radicals as exponents. To do this, we need to introduce the idea of *rational exponents*, exponents that are fractions.

We know that $a^1 = 1 \cdot a$, $a^2 = 1 \cdot a \cdot a$, and so forth. The exponent tells how many factors of a are in the product. The 1s are optional but help make sense of $a^0 = 1$, with no factors of a in the product. However, what does $a^{\frac{1}{2}}$ mean?

Whatever it means, we want it to obey the usual rules of exponents, one of which is $\left(a^m\right)^n = a^{mn}$. By using that rule, we get $\left(a^{\frac{1}{2}}\right)^2 = a^{\frac{2}{2}} = a^1 = a$. If $\left(a^{\frac{1}{2}}\right)^2 = a$, then $a^{\frac{1}{2}}$ must be either $\sqrt{a}$ or $-\sqrt{a}$. We choose the positive one and define $a^{\frac{1}{2}} = \sqrt{a}$. In a very similar way, we get $a^{\frac{1}{3}} = \sqrt[3]{a}$. In general, for positive integer n:

$$a^{\frac{1}{n}} = \sqrt[n]{a}$$

We can now go one step further using the same exponent property (in reverse) and find, for positive integers m and n:

$$\begin{aligned} a^{\frac{m}{n}} &= \left(a^m\right)^{\frac{1}{n}} = \sqrt[n]{a^m} \\ &= \left(a^{\frac{1}{n}}\right)^m = \left(\sqrt[n]{a}\right)^m \end{aligned}$$

DEFINITION OF RATIONAL EXPONENTS

If m and n are positive integers with $\frac{m}{n}$ in reduced form (lowest terms) and if all the roots involved are real:

$$a^{\frac{m}{n}} = \sqrt[n]{a^m} = \left(\sqrt[n]{a}\right)^m$$

The key to dealing with rational exponents is to remember "power over root." The numerator of the exponent is the power, and the denominator is the root.

Note the caveat that all the roots involved are real. As long as $a \geq 0$, this is not a problem. If $a < 0$ and the denominator, n, of the exponent is even, however, we get imaginary roots and this rule should not be applied. In short, we will allow negative bases only when we have an odd denominator in the exponent.

EXAMPLE 8.22

Evaluate: $32^{\frac{3}{5}}$

SOLUTION

Power over root. So the exponent $\frac{3}{5}$ means third power and fifth root.

$$32^{\frac{3}{5}} = \left(\sqrt[5]{32}\right)^3 = 2^3 = 8$$

Note that according to the rule, the order in which you evaluate the power and the root does not matter. However, evaluating the root first will avoid radicals of unnecessarily large numbers. Here is the same example with the power evaluated before the root.

$$32^{\frac{3}{5}} = \sqrt[5]{32^3} = \sqrt[5]{32{,}768} = 8$$

EXAMPLE 8.23

Evaluate: $(-8)^{\frac{2}{3}}$

SOLUTION

Power over root; 2 is the power and 3 is the root. By choosing to do the root first to keep the numbers smaller, we get:

$$(-8)^{\frac{2}{3}} = \left(\sqrt[3]{-8}\right)^2 = (-2)^2 = 4$$

If you choose to do it in the opposite order, remember that you are squaring the base -8, not just 8:

$$(-8)^{\frac{2}{3}} = \sqrt[3]{(-8)^2} = \sqrt[3]{64} = 4$$

COMMON ERROR

If you forget that the entire base, -8, is squared, you will get

$$(-8)^{\frac{2}{3}} = \sqrt[3]{-8^2} = \sqrt[3]{-64} = -4,$$

which is wrong.

Note that the somewhat similar looking problem $(-9)^{\frac{3}{2}}$ involves a negative base and an even denominator in the exponent. This leads to imaginary numbers. The exponent rules in this chapter are guaranteed to work only when all roots are real numbers. Remember in Algebra 2, we allow negative bases only when we have an odd denominator in the exponent.

Write $6x^{\frac{2}{3}}$ in radical form.

SOLUTION

Use power over root to write $6x^{\frac{2}{3}} = 6\sqrt[3]{x^2}$.

COMMON ERROR

The exponent is acting on only the x in Example 8.24. $6x^{\frac{2}{3}} = \sqrt[3]{6x^2}$ is incorrect, and so is $6x^{\frac{2}{3}} = \sqrt[3]{(6x)^2}$.

EXAMPLE 8.25

Write $\left(\sqrt{9x^2 + 1}\right)^3$ in exponent form.

SOLUTION

Use power over root to change the square root to an exponent with a denominator of 2.

$$\left(\sqrt{9x^2 + 1}\right)^3 = \left(9x^2 + 1\right)^{\frac{3}{2}}$$

COMMON ERROR

$$\left(\sqrt{9x^2 + 1}\right)^3 \neq (3x + 1)^3$$

Square roots (or any other kind of radical) cannot be distributed over addition.

EXAMPLE 8.26

Simplify: $\left(8x^{12}\right)^{\frac{2}{3}}$

SOLUTION

Distribute the power, and then simplify.

$$\left(8x^{12}\right)^{\frac{2}{3}} = 8^{\frac{2}{3}}\left(x^{12\left(\frac{2}{3}\right)}\right) = \left(\sqrt[3]{8}\right)^2 x^8 = (2)^2 x^8 = 4x^8$$

EXAMPLE 8.27

Simplify: $\dfrac{8x^{\frac{1}{2}}y}{4x^{\frac{1}{4}}\sqrt{y^3}}$

SOLUTION

The problem will be easiest if you convert the radical to exponent form and then use the rules of exponents. Remember that a square root is an exponent of $\frac{1}{2}$ (power over root), so $\sqrt{y^3} = \left(y^3\right)^{\frac{1}{2}} = y^{\frac{3}{2}}$.

$$\frac{8x^{\frac{1}{2}}y}{4x^{\frac{1}{4}}\sqrt{y^3}} = \frac{8x^{\frac{1}{2}}y}{4x^{\frac{1}{4}}y^{\frac{3}{2}}} = \frac{8x^{\frac{1}{2}-\left(\frac{1}{4}\right)}}{4y^{\frac{3}{2}-1}} = \frac{2x^{\frac{1}{4}}}{y^{\frac{1}{2}}}$$

The answer can also be expressed in radical form (power over root): $\dfrac{2\sqrt[4]{x}}{\sqrt{y}}$. If you need to rationalize the answer, $\dfrac{2\sqrt[4]{x}}{\sqrt{y}} \cdot \dfrac{\sqrt{y}}{\sqrt{y}} = \dfrac{2\sqrt{y}\sqrt[4]{x}}{y}$.

SECTION EXERCISES

For 20–21, evaluate the following (without using a calculator).

8–20. $100^{\frac{3}{2}}$

8–21. $(-125)^{\frac{2}{3}}$

For 22–23, write the following in radical form.

8–22. $8x^{\frac{2}{3}}$

8–23. $\left(x^3 - 8\right)^{\frac{1}{3}}$

For 24–25, write the following in exponent form.

8–24. $-2\sqrt[3]{x^2} + 9\sqrt[4]{y^5}$

8–25. $12\sqrt[3]{x^2 - 5}$

For 26–27, express in simplest radical form. Assume all variables represent positive numbers.

8–26. $\left(9x^6y^2\right)^{\frac{3}{2}}$

8–27. $\dfrac{4\sqrt{xy^3}}{12x^{\frac{3}{2}}y^{\frac{1}{2}}}$

Graphing Square Root and Cube Root Functions

The square root and cube root functions and transformations of them are considered basic functions that Algebra 2 students should recognize and be able to graph.

BASIC GRAPHS OF SQUARE ROOT AND CUBE ROOT FUNCTIONS

The graphs of simple square root and cube root functions are shown on the next page in Figure 8.2.

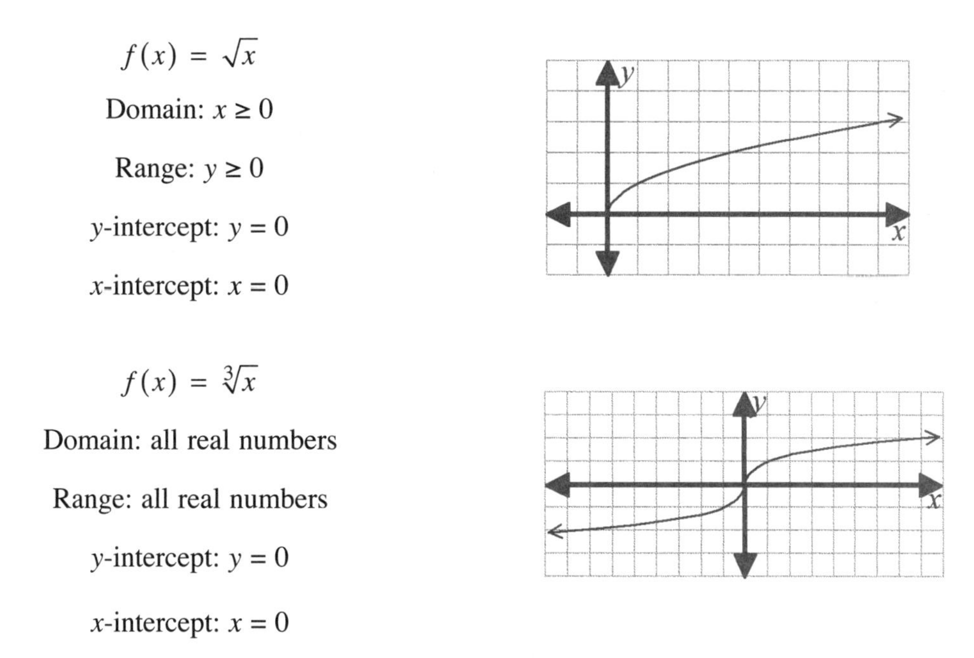

FIGURE 8.2

The graph of the cube root function has an interesting feature at the origin, a *point of inflection*. On the left of the origin, the graph is concave up. On the right of the origin, the graph is concave down. A point of inflection marks the location where a graph changes concavity. Many other functions have points of inflection, sometimes more than one. However, points of inflection are not always as obvious on a graph as is the one in $f(x) = \sqrt[3]{x}$. Finding them is usually a calculus problem.

TRANSFORMATIONS OF SQUARE ROOT AND CUBE ROOT GRAPHS

All the transformations covered in Chapter 3 can be applied to radical functions. The following examples illustrate the concepts.

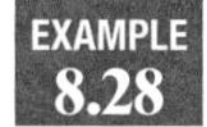

Compare the graph of the function $y = \sqrt[3]{x}$ after a reflection over the y-axis to the graph after a reflection over the x-axis.

SOLUTION

The two graphs are shown below.

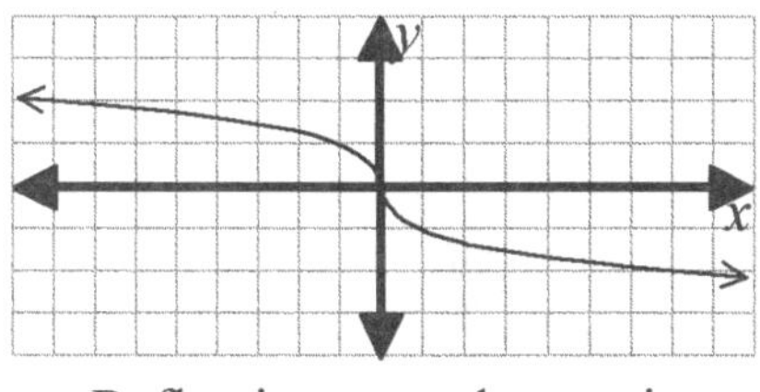

Reflection over the y-axis.

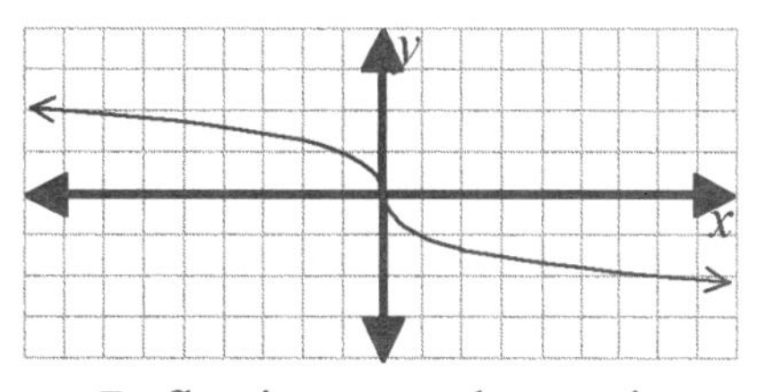

Reflection over the x-axis.

They appear the same. Why? To reflect $y = f(x)$ over the y-axis, change it to $y = f(-x)$. To reflect it over the x-axis, change the function to $y = -f(x)$. For this problem, that means $y = \sqrt[3]{x}$ becomes $y = \sqrt[3]{-x}$ after a reflection over the y-axis and becomes $y = -\sqrt[3]{x}$ after a reflection over the y-axis. Notice that $y = \sqrt[3]{-x} = \sqrt[3]{-1}\sqrt[3]{x} = -\sqrt[3]{x}$, so these two functions are indeed the same. (Note that we cannot do the same thing with the square root function.) The cube root function has origin symmetry and is an odd function. For any function with origin symmetry, a reflection over the x-axis will be the same as a reflection over the y-axis.

EXAMPLE 8.29

Write the equation for the graph of $y = \sqrt{x}$ after a reflection over the origin.

SOLUTION

To reflect $y = f(x)$ over the origin, change it to $y = -f(-x)$. For this problem, that means $y = \sqrt{x}$ becomes $y = -\sqrt{-x}$.

We should note a couple of things. First, we cannot cancel those two negative signs. One is inside the radical, and the other is not. We can't combine them. Also, some students are bothered by the negative sign inside the radical, thinking it makes all the y-values imaginary. That is not so. Remember that x can be either positive or negative and that $-x$ means "the opposite of x," not "x is a negative number." The domain of the transformed function can be found by applying our usual rule for finding domain with a radical: the radicand of an even index root must be nonnegative. So we know $-x \geq 0$, from which we get $x \leq 0$ (don't forget to flip the inequality symbol). The graph is shown at right.

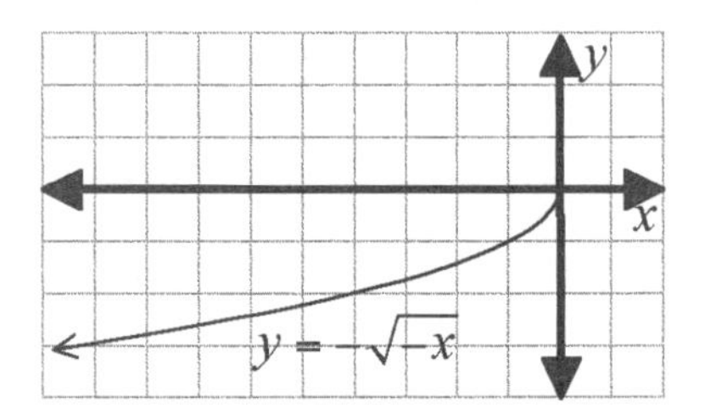

EXAMPLE 8.30

Describe how to transform $y = \sqrt{x}$ to get the graph of $y = \sqrt{x+5} - 3$.

SOLUTION

The graph of $y = f(x + h)$ is the translation of the graph of $y = f(x)$ by h units left. The graph of $y = f(x) - k$ is a translation of the graph of $y = f(x)$ by k units down. When taken together, these mean that the graph of $y = \sqrt{x+5} - 3$ is the graph of $y = \sqrt{x}$ moved to the left 5 and down 3.

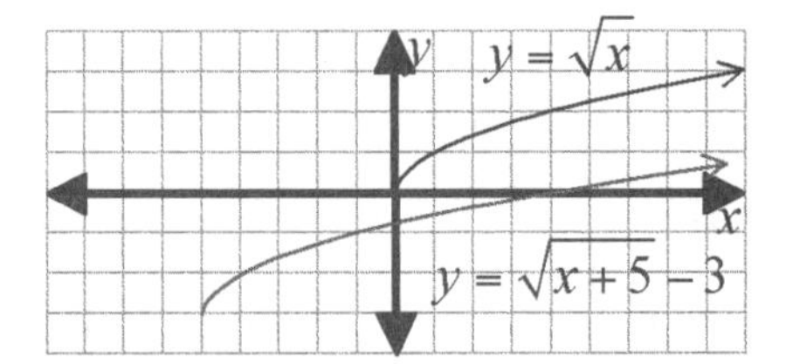

EXAMPLE 8.31 The graph of $f(x) = \sqrt[3]{x}$ has been moved 4 units to the right and 1 unit up. Write the equation of this new graph.

SOLUTION

In the equation $y = \sqrt[3]{x - h} + k$, h represents the number of units translated to the right. So $h = 4$. In the equation, k represents the number of units translated up. So $k = 1$. The new equation is $y = \sqrt[3]{x - 4} + 1$.

Square root and cube root functions can usually be rewritten so that horizontal dilations are written as vertical dilations. Many students find vertical dilations easier to graph.

EXAMPLE 8.32 Sketch the graph of $f(x) = \sqrt{4x}$.

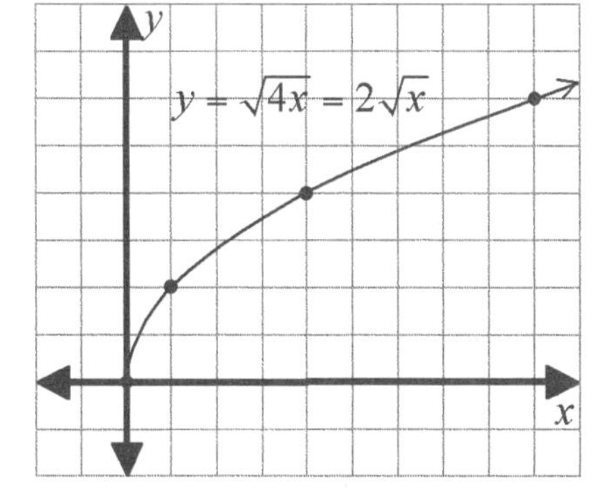

SOLUTION

The function $f(x) = \sqrt{4x}$, as written, is a horizontal dilation of $y = \sqrt{x}$ by a factor of $\frac{1}{4}$. However, we can rewrite it: $f(x) = \sqrt{4x} = \sqrt{4}\sqrt{x} = 2\sqrt{x}$. This is the same as a vertical dilation of $y = \sqrt{x}$ by a factor of 2. Each y-coordinate on the graph of $y = \sqrt{x}$ is multiplied by 2.

EXAMPLE 8.33 Sketch the graph of $y = 3\sqrt[3]{x - 2}$.

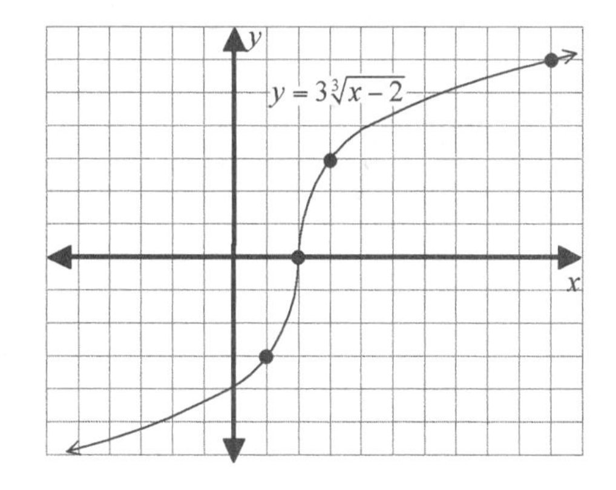

SOLUTION

Compared with the graph of $y = \sqrt[3]{x}$, the graph of $y = 3\sqrt[3]{x - 2}$ will be vertically dilated by a factor of 3 and translated 2 units to the right. Because these transformations work in different directions, one vertical and one horizontal, it doesn't matter which order you do them in. The x-coordinates are increased by 2, and the y-coordinates are multiplied by 3.

SECTION EXERCISES

For 28–29, describe the transformations on $y = \sqrt{x}$ to obtain the following functions.

8–28. $y = \sqrt{x + 4} - 1$

8–29. $y = \sqrt{-x}$

8–30. Describe the transformations on $y = \sqrt[3]{x}$ to obtain the following function.

$$y = -\frac{1}{2}\sqrt[3]{x}$$

For 31–32, sketch graphs of the following functions.

8–31. $y = \sqrt{x - 2} + 3$

8–32. $y = \sqrt[3]{x + 4} - 1$

8–33. Write the equation for the function after the described transformations.
The graph of $y = \sqrt{x}$ is reflected over the x-axis and moved left 6 and up 5.

Solving Radical Equations

A radical equation is an equation where the variable appears in a radicand. Thus $\sqrt{2x - 5} = 3$ is a radical equation. However, $x\sqrt{2} + \sqrt{8} = \sqrt{18}$, although it contains several radicals, is not a radical equation.

To solve a radical equation:

1. Isolate the radical term on one side of the equation.
   (It is not necessary to divide off numbers multiplying the radical. However, you must move anything being added or subtracted from the radical term to the other side of the equation.)
2. Raise both sides to the appropriate power (square to get rid of square roots, cube to get rid of cube roots, and so on).
3. Solve the new equation for the variable.
4. Check the answer(s) in the original equation.

EXAMPLE 8.34 Solve: $\sqrt{2x-5} = 3$

SOLUTION

$\sqrt{2x-5} = 3$	The radical term is already by itself on one side of the equation.
$\left(\sqrt{2x-5}\right)^2 = 3^2$	Square both sides to eliminate the squarc root.
$2x - 5 = 9$	Remember that when you square a radical, you end up with just the radicand.
$2x = 14$	Solve for x.
$x = 7$	

Check the answer in the original equation (in this case, you may be able to do it mentally).

$$\sqrt{2(7)-5} = 3$$
$$\sqrt{14-5} = 3$$
$$\sqrt{9} = 3$$
$$3 = 3$$

The solution is $x = 7$.

When solving algebraically, the check is very important. Squaring both sides of an equation can sometimes lead to extraneous roots that seem perfectly reasonable until you do the check and find out they do not work.

EXAMPLE 8.35 Solve: $x + 2\sqrt{7-2x} = 1$

SOLUTION

When isolating the radical term, it is not necessary to divide off the 2. (It is not wrong, but it can lead to fractions that most people prefer to avoid.) Be careful when squaring both sides. Do not forget to square the 2, and remember to distribute when squaring the binomial.

$$x + 2\sqrt{7-2x} = 1$$

$2\sqrt{7-2x} = 1 - x$ — Isolate the radical term.

$\left(2\sqrt{7-2x}\right)^2 = (1-x)^2$ — Square both sides.

$$4(7-2x) = (1-x)(1-x)$$
$$28 - 8x = 1 - 2x + x^2$$

$0 = x^2 + 6x - 27$ Solve the resulting equation.

$0 = (x + 9)(x - 3)$

$x = -9$ or $x = 3$

Now check both answers in the original equation.

Check $x = -9$:

$$-9 + 2\sqrt{7 - 2(-9)} = 1$$
$$-9 + 2\sqrt{25} = 1$$
$$-9 + 2(5) = 1$$
$$1 = 1$$

Check $x = 3$:

$$3 + 2\sqrt{7 - 2(3)} = 1$$
$$3 + 2\sqrt{1} = 1$$
$$3 + 2(1) = 1$$
$$5 = 1$$

Only $x = -9$ checks; $x = 3$ does not. The solution is $x = -9$.

The issue of extraneous roots can be avoided by solving the equation graphically. To solve graphically:

1. Graph each side of the equation separately on the same set of axes.
2. Find the x-value(s) of the point(s) of intersection.

EXAMPLE 8.36

Solve $x + 2\sqrt{7 - 2x} = 1$ graphically.

SOLUTION

SEE CALC TIP 8C

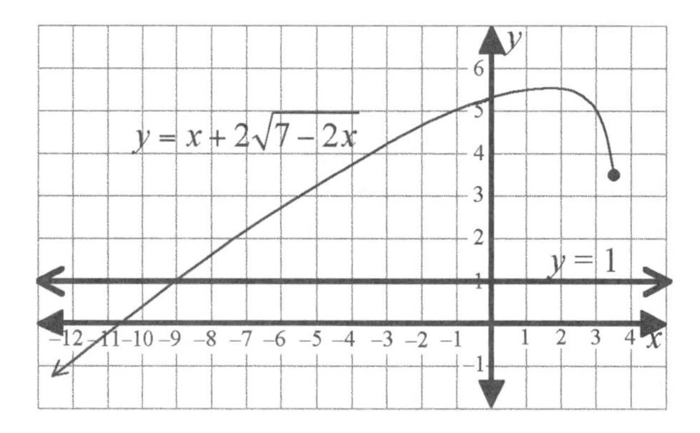

The graph clearly shows that only $x = -9$ is a solution.

To solve equations with rational exponents, rewrite the equation in radical form and then solve.

EXAMPLE 8.37 Solve for x: $2x^{\frac{3}{2}} + 5 = 21$

SOLUTION

Use power over root to rewrite the exponential term in radical form.

$$2x^{\frac{3}{2}} + 5 = 21$$
$$2\sqrt{x^3} + 5 = 21$$
$$2\sqrt{x^3} = 16$$
$$\sqrt{x^3} = 8$$
$$\left(\sqrt{x^3}\right)^2 = (8)^2$$
$$x^3 = 64$$
$$\sqrt[3]{x^3} = \sqrt[3]{64}$$
$$x = 4$$

As for any radical equation, you should check your answer in the original equation.

Check $x = 4$:

$$2(4)^{\frac{3}{2}} + 5 = 21$$
$$2(8) + 5 = 21$$
$$16 + 5 = 21$$
$$21 = 21$$

$x = 4$ checks and is the only solution.

EXAMPLE 8.38 Solve for x: $(x - 2)^{\frac{4}{3}} = 16$

SOLUTION

$$(x - 2)^{\frac{4}{3}} = 16$$
$$\sqrt[3]{(x - 2)^4} = 16$$

Rewrite the rational exponent in radical form using power over root.

$$\left(\sqrt[3]{(x - 2)^4}\right)^3 = 16^3$$

Cube both sides to remove the cube root.

$$(x - 2)^4 = 4{,}096$$

$\sqrt[4]{(x-2)^4} = \pm\sqrt[4]{4{,}096}$ Take the fourth root of each side to undo the fourth power.

$x - 2 = \pm 8$ Remember that an even root of a positive number will have two real answers, one positive and one negative. So don't forget the ± symbol.

$x = 2 \pm 8$ Solve for x by adding 2.

$x = 10$ or $x = -6$ Evaluate the ± expressions separately.

Both these answers check.

There is a shortcut for solving simple rational exponent equations that many people like. By the properties of exponents, for a rational exponent $\frac{m}{n}$, $\left(x^{\frac{m}{n}}\right)^{\frac{n}{m}} = x$. This means that to solve a rational exponent equation of the form $x^{\frac{m}{n}} = k$, we can raise both sides to the reciprocal of the exponent: $x^{\frac{m}{n}} = k \rightarrow \left(x^{\frac{m}{n}}\right)^{\frac{n}{m}} = k^{\frac{n}{m}}$ or $x = k^{\frac{n}{m}}$.

If we want the complete solution to the problem, we need to pay attention to the exponent. If the numerator of the original exponent is even, then after taking the reciprocal, the denominator will be even. The denominator of a rational exponent corresponds to the root. Even roots (of positive numbers) have two real answers, one positive and one negative. So if the numerator of the original exponent is even, you need to include ± in your answer when solving. Let's redo the last two examples with the shortcut.

Solve for x: $2x^{\frac{3}{2}} + 5 = 21$

SOLUTION

$2x^{\frac{3}{2}} + 5 = 21$

$x^{\frac{3}{2}} = 8$ First isolate the variable with the exponent.

$\left(x^{\frac{3}{2}}\right)^{\frac{2}{3}} = 8^{\frac{2}{3}}$ Since we are taking an odd root, we do not need a ± sign.

$x = 4$ As we already know, it checks.

Solve for x: $(x - 2)^{\frac{4}{3}} = 16$

SOLUTION

$(x - 2)^{\frac{4}{3}} = 16$

$\left((x - 2)^{\frac{4}{3}}\right)^{\frac{3}{4}} = \pm 16^{\frac{3}{4}}$ Since we are taking an even root, we need a ± sign to get both answers.

$x - 2 = \pm 8$

$x = 2 \pm 8$

$x = 10$ or $x = -6$ Both check.

SECTION EXERCISES

For 34–39, solve each of the following equations algebraically.

8–34. $\sqrt{3 - x} = 4$

8–35. $\sqrt{x^2 - 9} = x + 1$

8–36. $x + 2\sqrt{x - 1} = 9$

8–37. $\sqrt{16 - 5x} - x = 10$

8–38. $4x^{\frac{1}{2}} = 36$

8–39. $-3x^{\frac{2}{3}} + 16 = -11$

8–40. One way to produce artificial gravity on a space station is to rotate it. To produce Earth-normal gravity on a space station with radius r meters, the station must rotate N times per minute where $N = \dfrac{210}{\pi\sqrt{5r}}$. If rotating a space station once a minute produces Earth-normal gravity, what is the radius of the station, to the nearest tenth of a meter?

Inverses of Radical Functions

Both the square root and cube root functions are one-to-one functions, so their inverses are also functions. You may want to review inverse functions in Chapter 3. When finding the inverse of a square root function, pay special attention to the domain of the inverse.

EXAMPLE 8.41

Find $f^{-1}(x)$ for $f(x) = \sqrt{x}$.

SOLUTION

$y = \sqrt{x}$	Write the function with $y = f(x)$.
$x = \sqrt{y}$	Switch x and y.
$(x)^2 = \left(\sqrt{y}\right)^2$	Solve for y.
$y = x^2$	

So far, this seems straightforward. However, if you compare the graphs of $y = \sqrt{x}$ and $y = x^2$, you will notice that the graphs are not reflections of each other over the line $y = x$. We want only part of the parabola for the inverse. Since the right side of the parabola gives $y = \sqrt{x}$ when reflected over $y = x$, the inverse function must be $y = x^2, x \geq 0$.

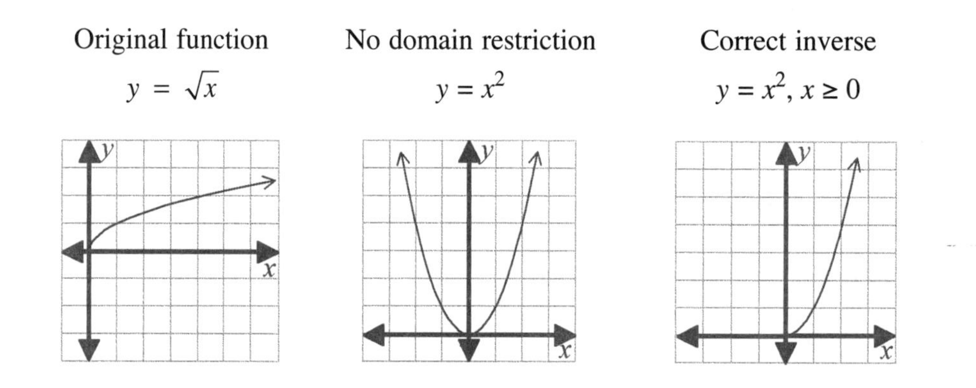

Could we have figured that out without graphing? Consider the range of the original function $y = \sqrt{x}$. The range of the square root function is $y \geq 0$. When we find the inverse by switching x and y, the domain and range switch. The domain of the inverse becomes $x \geq 0$.

For $f(x) = \sqrt{x}$, $f^{-1}(x) = x^2, x \geq 0$.

EXAMPLE 8.42 Find the inverse: $f(x) = 5\sqrt[3]{2x - 3}$

SOLUTION

$y = 5\sqrt[3]{2x - 3}$ Write the function with $y = f(x)$.

$x = 5\sqrt[3]{2y - 3}$ Switch x and y.

$\frac{x}{5} = \sqrt[3]{2x - 3}$ Solve for y.

$$\left(\frac{x}{5}\right)^3 = \left(\sqrt[3]{2x - 3}\right)^3$$

$$\frac{x^3}{125} = 2y - 3$$

$$\frac{x^3}{125} + 3 = 2y$$

$$\frac{\frac{x^3}{125} + 3}{2} = y$$

$$\frac{x^3}{250} + \frac{3}{2} = y$$

$f^{-1}(x) = \frac{x^3}{250} + \frac{3}{2}$ Write in inverse notation.

Since the range of the original cube root function is all real numbers, the domain of the inverse is also all real numbers. This inverse does not need a domain restriction.

EXAMPLE 8.43 Find the inverse: $f(x) = \frac{1}{2}\sqrt{x + 1} + 6$

SOLUTION

$y = \frac{1}{2}\sqrt{x + 1} + 6$ Write the function with $y = f(x)$.

$x = \frac{1}{2}\sqrt{y + 1} + 6$ Switch x and y.

$$x - 6 = \frac{1}{2}\sqrt{y + 1} \quad \text{Solve for } y.$$
$$2(x - 6) = \sqrt{y + 1}$$
$$(2(x - 6))^2 = \left(\sqrt{y + 1}\right)^2$$
$$4(x - 6)^2 = y + 1$$
$$4(x - 6)^2 - 1 = y$$

SEE CALC TIP 8D

$$f^{-1}(x) = 4(x - 6)^2 - 1 \quad \text{Write in inverse notation.}$$

Note the original function was a square root function translated up 6 units (the + 6). The range of a simple square root function in is $y \geq 0$. A translation up to 6 makes the range $y \geq 6$. (The vertical dilation and horizontal translation do not change the range.) Since the range of the original function is $y \geq 6$, the domain of the inverse function must be $x \geq 6$.

For $f(x) = \frac{1}{2}\sqrt{x + 1} + 6$, the inverse function is $f^{-1}(x) = 4(x - 6)^2 - 1, x \geq 6$.

SECTION EXERCISES

For 41–43, find the inverse of the following functions. State any domain restrictions.

8–41. $f(x) = \frac{1}{2}\sqrt{x + 5}$

8–42. $f(x) = 5\sqrt{x} - 1$

8–43. $f(x) = \sqrt[3]{x + 4} - 1$

CHAPTER EXERCISES

For 1–2, express in simplest radical form. Assume all variables represent positive numbers.

C8–1. $\sqrt{108x^2y^6}$

C8–2. $\sqrt[3]{48x^8y^9}$

For 3–8, express in simplest form. Assume all variables represent positive numbers.

C8–3. $\left(5 - 2\sqrt{3x}\right)^2$

C8–4. $\left(-5\sqrt{3x^5y}\right)\left(\sqrt{6x^3y^5}\right)$

C8–5. $3x^2\sqrt{96x^2y} - 5\sqrt{54x^6y}$

C8–6. $\dfrac{6x\sqrt{240x^3y^6}}{4\sqrt{12x^5y^3}}$

C8–7. $5\sqrt[3]{16x^7} - 2x\sqrt[3]{54x^4}$

C8–8. $\dfrac{\sqrt[3]{24x^4} + 2x\sqrt[3]{192x}}{\sqrt[3]{4x}}$

For 9–12, express in simplest form with rational denominators. Assume all variables represent positive numbers.

C8–9. $\dfrac{3}{2}\sqrt{\dfrac{5}{18}}$

C8–10. $\dfrac{\sqrt{75x^9y^5}}{\sqrt{3xy}}$

C8–11. $\dfrac{-7}{3 + \sqrt{10}}$

C8–12. $\dfrac{3 + \sqrt{12}}{3 - \sqrt{3}}$

For 13–14, write the following in radical form.

C8–13. $(25xy)^{\frac{1}{2}}$

C8–14. $5\left(x^2 - 4\right)^{\frac{3}{2}}$

For 15–16, write the following in exponent form.

C8–15. $3\sqrt{1 - 2x}$

C8–16. $5\sqrt[4]{16x^3}$

C8–17. Express in simplest radical form. Assume all variables represent positive numbers:

$$\frac{8r\sqrt{s}}{4r^{\frac{1}{3}}s^{\frac{3}{2}}}$$

C8–18. Describe the transformations on $y = \sqrt{x}$ to obtain the given function: $y = -3\sqrt{x}$.

C8–19. Describe the transformations on $y = \sqrt[3]{x}$ to obtain the given function:

$y = \sqrt[3]{x-5} + 2$.

For 20–21, sketch graphs of the following functions.

C8–20. $y = \sqrt{x+1} - 4$

C8–21. $y = \sqrt[3]{x-2} + 3$

For 22, write the equation for the function after the described transformations.

C8–22. The graph of $y = \sqrt[3]{x}$ after a vertical dilation by a factor of 0.25 followed by a move right 10 and down 2.

For 23–25, solve each of the following equations algebraically.

C8–23. $5 + \sqrt{x-1} = 2$

C8–24. $\sqrt{x+5} - x = 3$

C8–25. $\dfrac{x^{\frac{3}{2}} - 1}{7} = 9$

C8–26. The period of a pendulum is the time the pendulum takes to swing back and forth once. The period, T, in seconds, of a pendulum of length, L, in feet, is $T = 2\pi\sqrt{\frac{L}{32}}$. What is the length of a pendulum, to the nearest tenth of a foot, if its period is 5 seconds?

For 27–28, find the inverse of the following functions. State any domain restrictions.

C8–27. $f(x) = \sqrt{x-2} + 3$

C8–28. $f(x) = 2\sqrt[3]{x-3} + 4$

9 Rational Functions

WHAT YOU WILL LEARN

Rational expressions are used to write functions that represent ratios such as miles per gallon or miles per hour. The graphs of rational functions introduce a new graphing concept, asymptotes. The algebra of rational functions requires you pay careful attention to detail. In this chapter, you will learn to:

- Perform operations on rational expressions;
- Use negative exponents;
- Graph rational functions;
- Solve rational equations and inequalities using a variety of methods;
- Find inverses of rational functions.

SECTIONS IN THIS CHAPTER

- Rational Models
- Operations with Rational Expressions
- Negative Exponents
- Graphing Rational Functions
- Solving Rational Equations
- Solving Rational Inequalities

Rational Models

Suppose Steady Steve can run at a speed of x feet per second. Steve's friend Fast Freddy can run 2 feet per second faster than Steve. How fast can each run 300 feet and what will be the difference in their times?

Distance divided by speed gives time. Steve's time will be $t_S = \frac{300}{x}$. Freddy's time will be $t_F = \frac{300}{x+2}$. The difference between their times will be $t = \frac{300}{x} - \frac{300}{x+2}$. These are all examples of rational functions. The third one we might like to express in a simpler form as a single rational expression: $t = \frac{600}{x(x+2)}$.

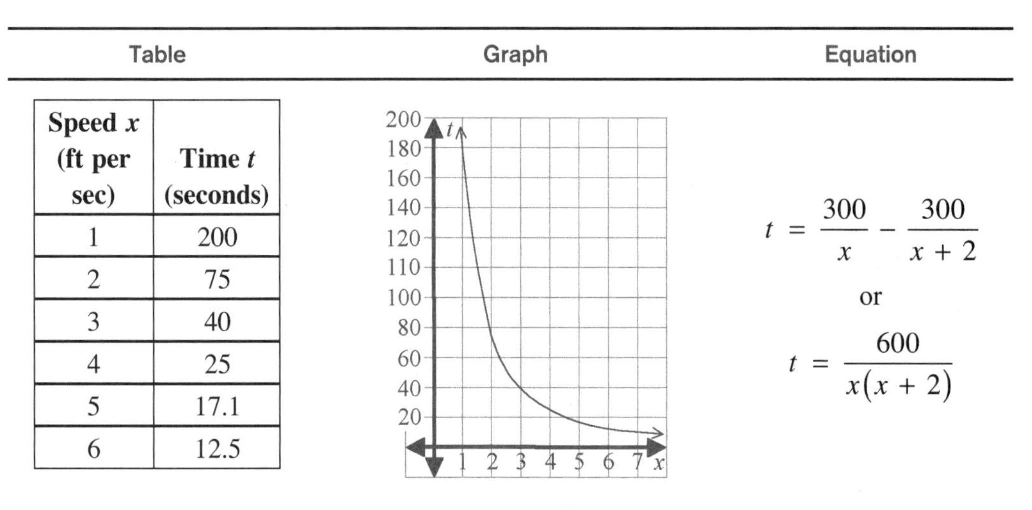

Table

Speed x (ft per sec)	Time t (seconds)
1	200
2	75
3	40
4	25
5	17.1
6	12.5

Graph

Equation

$$t = \frac{300}{x} - \frac{300}{x+2}$$

or

$$t = \frac{600}{x(x+2)}$$

FIGURE 9.1

Operations with Rational Expressions

A *rational expression*, also called an *algebraic fraction*, is a ratio of two polynomials such as $\frac{2}{x}$ or $\frac{x+6}{x^2-3x}$. Like ordinary numerical fractions, rational expressions can be added, subtracted, multiplied, divided, and sometimes simplified. First, though, we need to make sure our expression has meaning.

UNDEFINED EXPRESSIONS

Because division by zero has no meaning, a rational expression is *undefined* if its denominator equals zero. When working with rational expressions, you want to be aware of which x-values make the expressions undefined.

EXAMPLE 9.1 For what value(s) of x is the expression $\frac{2x-6}{x^2-9}$ undefined?

SOLUTION

A fraction is undefined when the denominator equals zero.

$$x^2 - 9 = 0$$
$$x = \pm 3$$

The way the question was worded, $x = \pm 3$ is the correct answer. What it really means is the numbers 3 and –3 will make the expression undefined, so they should NOT be substituted in for x. The domain for the function $y = \frac{2x-6}{x^2-9}$ is all real numbers except $x = \pm 3$. You may often see that written explicitly as $y = \frac{2x-6}{x^2-9}, x \neq \pm 3$.

In the example above, $x = 3$ makes both the denominator and numerator of the expression equal zero. This is a special case called the *indeterminate form.* You can learn more about this if you take calculus. For now, do not assume $\frac{0}{0} = 1$. The denominator is zero, so the fraction is not defined.

SIMPLIFYING RATIONAL EXPRESSIONS

Just as fractions can be simplified to lowest terms, for example $\frac{8}{12} = \frac{2}{3}$, rational expressions may sometimes be reduced to simpler equivalent expressions. Simplifying rational expressions with monomial denominators was covered in Chapter 3. This section will show how to simplify expressions with polynomial denominators.

To simplify a rational expression:

1. Factor both the numerator and denominator completely.
2. Simplify by dividing out (canceling) like factors in the numerator and the denominator.

It is absolutely essential that you factor first. Remember never to cancel terms of a polynomial; always cancel matching factors.

EXAMPLE 9.2 Express $\frac{x^2-6x+8}{x^2-16}$ in simplest form.

SOLUTION

$$\frac{x^2-6x+8}{x^2-16} = \frac{(x-2)(x-4)}{(x-4)(x+4)}$$ Factor first.

$$\frac{(x-2)\cancel{(x-4)}}{\cancel{(x-4)}(x+4)}$$ Cancel matching factors.

$$\frac{(x-2)}{(x+4)}$$

COMMON ERROR

The most common mistake students make is to cancel terms, $\frac{\cancel{x^2}-6x+8}{\cancel{x^2}-16}$, or reduce the $\frac{8}{16}$, as in $\frac{x^2-6x+\cancel{8}}{x^2-\cancel{16}_2}$. Doing this is a guaranteed way to get the problem wrong and drive your math teacher insane.

A subtle detail to note from the previous example is the domain. The original expression is not defined for $x = \pm 4$. The fact that the simplified expression appears to be defined when $x = 4$ does not change the original domain. The expressions are equivalent only for $x \neq \pm 4$.

If a factor in a numerator is the exact opposite of a factor in a denominator, –1 may be factored out of one of them (usually the one in the numerator) so that the expression can be simplified.

EXAMPLE 9.3

Simplify: $\frac{8-4x}{x^2-4}$

SOLUTION

$$\frac{8-4x}{x^2-4} = \frac{4(2-x)}{(x-2)(x+2)} = \frac{4(-1)\cancel{(x-2)}}{\cancel{(x-2)}(x+2)} = \frac{-4}{x+2}$$

Reducing opposite factors such as $x-2$ and $2-x$ to –1 is done so frequently that factoring out –1 is not usually written as a separate step. The factors are just reduced to –1.

$$\frac{4\overset{-1}{\cancel{(2-x)}}}{\cancel{(x-2)}(x+2)} = \frac{4(-1)}{(x+2)} = \frac{-4}{x+2}$$

SEE CALC TIP 9A

MULTIPLYING RATIONAL EXPRESSIONS

To multiply simple fractions, first cancel common factors in the numerator and denominator, and then multiply both numerators and denominators straight across.

For example: $\frac{14}{15}\cdot\frac{5}{8} = \frac{\cancel{14}^7}{\cancel{15}_3}\cdot\frac{\cancel{5}^1}{\cancel{8}_4} = \frac{7}{12}$. The procedure for multiplying rational expressions is the same except you must remember to factor first.

To multiply rational expressions:

1. Factor all polynomial numerators and denominators.
2. Simplify by canceling like factors in any numerator and denominator.
3. Multiply the remaining factors in numerators and denominators straight across. Usually the result can be left in factored form. Multiplying out the numerator or denominator is not generally necessary.

EXAMPLE 9.4 Express $\frac{2x^2 + 10x}{2x^2 + 7x - 10} \cdot \frac{4x^2 - 9}{6x}$ as a single fraction in simplest form.

SOLUTION

$$\frac{2x^2 + 10x}{2x^2 + 7x - 10} \cdot \frac{4x^2 - 9}{6x}$$

$$\frac{2x(x + 5)}{(2x - 3)(x + 5)} \cdot \frac{(2x - 3)(2x + 3)}{6x} \quad \text{Factor first.}$$

$$\frac{\cancel{2x}\,\cancel{(x + 5)}}{\cancel{(2x - 3)}\,\cancel{(x + 5)}} \cdot \frac{\cancel{(2x - 3)}(2x + 3)}{\cancel{6x}_{\,3}} \quad \text{Cancel like factors.}$$

$$\frac{2x + 3}{3}$$

The original fractions are undefined when $x = -5$, $x = 0$, or $x = 1.5$. The simplified expression is equivalent to the original only when $x \neq -5$, $x \neq 0$, and $x \neq 1.5$.

COMMON ERROR

Do not cancel the 3s in the final answer!

DIVIDING RATIONAL EXPRESSIONS

If you can multiply rational expressions, you can divide them. Dividing one rational expression by another is the same as multiplying the first by the reciprocal of the second. Some students like to remember "copy-dot-flip." Copy the first expression, change division to multiplication (dot), and flip the second expression. Now it is a multiplication problem.

EXAMPLE 9.5

Express the quotient $\frac{3x^2 + x - 2}{6x^2 - 4x} \div \frac{2x^2 - 2}{8x^2}$ as a single fraction in simplest form.

SOLUTION

$$\frac{3x^2 + x - 2}{6x^2 - 4x} \div \frac{2x^2 - 2}{8x^2}$$

$$\frac{3x^2 + x - 2}{6x^2 - 4x} \cdot \frac{8x^2}{2x^2 - 2}$$ Copy-dot-flip.

$$\frac{(3x - 2)(x + 1)}{2x(3x - 2)} \cdot \frac{8x^2}{2(x + 1)(x - 1)}$$ Factor.

$$\frac{\cancel{(3x - 2)}\cancel{(x + 1)}}{\cancel{2x}\cancel{(3x - 2)}} \cdot \frac{\cancel{8x^2}\,2x}{\cancel{2}\cancel{(x + 1)}(x - 1)}$$ Reduce.

$$\frac{2x}{x - 1}$$

The original problem has the domain of all real numbers except $x = 0$, $x = \frac{2}{3}$, and $x = \pm 1$. The $x = \pm 1$ terms are included because, in addition to each fraction not being undefined, the divisor (the second expression) must not equal zero. You cannot divide an expression by zero.

ADDING AND SUBTRACTING RATIONAL EXPRESSIONS

The rules for adding and subtracting rational expressions are quite different than those for multiplying or dividing. The three biggest differences are the following:

1. To add or subtract rational expressions, you must have common denominators.
2. Only the numerators are added or subtracted, never the denominators.
3. Simplification is done last. (Never "cross cancel" in an addition or subtraction problem.)

The key step for adding and subtracting is rewriting the fractions with common denominators. Remember how you rewrite ordinary fractions: you multiply both the numerator and denominator by the same quantity. Never rewrite by adding to the numerator and denominator.

To add or subtract rational expressions:

1. Get common denominators.
   - Factor all polynomial denominators. (Do not factor numerators at this time.)
   - Rewrite the fractions by multiplying numerators and denominators by appropriate factors until all the denominators are the same.
2. Multiply out the numerators, and then add or subtract. (Review the rules for adding and subtracting polynomials if you need to.) Keep the common denominator; do not multiply it out.
3. Factor the numerator and simplify.

EXAMPLE 9.6

Express $\frac{x+5}{x^2-9} - \frac{4}{x^2-3x}$ as a single fraction in simplest form.

SOLUTION

$$\frac{x+5}{x^2-9} - \frac{4}{x^2-3x}$$

$$\frac{x+5}{(x-3)(x+3)} - \frac{4}{x(x-3)}$$ Factor denominators.

$$\frac{x(x+5)}{x(x-3)(x+3)} - \frac{4(x+3)}{x(x-3)(x+3)}$$ Rewrite fractions with a common denominator. Note that both denominators already contain $(x-3)$. To make the denominators the same, the first fraction needs a factor of x and the second needs a factor of $(x+3)$.

$$\frac{x^2+5x-4x-12}{x(x-3)(x+3)}$$ Multiply and subtract the numerators. Remember to distribute the minus sign. Keep the denominator.

$$\frac{x^2+x-12}{x(x-3)(x+3)}$$ Simplify the numerator.

$$\frac{(x+4)(x-3)}{x(x-3)(x+3)}$$ Factor the numerator.

$$\frac{(x+4)\cancel{(x-3)}}{x\cancel{(x-3)}(x+3)}$$ Reduce matching factors.

$$\frac{x+4}{x(x+3)}$$ Write the answer in factored form.

EXAMPLE 9.7 Express $\frac{2x-2}{x+5} - \frac{4}{x}$ as a single fraction in simplest form.

SOLUTION

COMMON ERROR

Don't even think about adding 5 to the numerator and denominator of the second fraction!

$$\frac{2x-2}{(x+5)} - \frac{4}{(x)}$$

Writing the simple denominators in parentheses may help you see them as separate factors.

$$\frac{x(2x-2)}{x(x+5)} - \frac{4(x+5)}{x(x+5)}$$

The left fraction needs to be multiplied by $\frac{x}{x}$. The right fraction needs to be multiplied by $\frac{(x+5)}{(x+5)}$.

$$\frac{2x^2 - 2x - 4x - 20}{x(x+5)}$$

Multiply and write as a single numerator. Remember to distribute the minus sign. Keep the denominator.

$$\frac{2x^2 - 6x - 20}{x(x+5)}$$

Simplify the numerator.

$$\frac{2(x-5)(x+2)}{x(x+5)}$$

Factor the numerator. No reducing is possible.

Note that $(x - 5)$ and $(x + 5)$ are not the same factor. They are not opposite factors either; the opposite of $x - 5$ is $5 - x$. This answer cannot be simplified further.

COMPLEX FRACTIONS

A *complex fraction* is a fraction with fractions in the numerator and/or denominator. There are two equivalent ways to simplify complex fractions into simple fractions. The first method is faster for some problems, while the second method can be easier for other types of problems. Two examples are shown below; each is done two different ways. Method 1 appears easier for the first example. Method 2 appears easier for the second example. Choose the method that makes the most sense to you. Even if it takes more steps, you will be more accurate if you understand the steps.

Method 1: Multiply the numerator and denominator by the lowest common denominator (LCD) of all of the fractions. Then factor and reduce.

EXAMPLE 9.8A Simplify $\dfrac{\frac{2}{x^2}-\frac{1}{2}}{2-\frac{4}{x}}$

SOLUTION

$$\frac{\frac{2}{x^2}-\frac{1}{2}}{2-\frac{4}{x}}$$

The denominators are 2, x, and x^2. The LCD is $2x^2$.

$$\frac{\left(2x^2\right)\frac{2}{x^2}-\left(2x^2\right)\frac{1}{2}}{\left(2x^2\right)2-\left(2x^2\right)\frac{4}{x}}$$

Multiply all terms by $2x^2$.

$$\frac{\left({}^{2}\cancel{2x^2}\right)\frac{2}{\cancel{x^2}}-\left({}^{x^2}\cancel{2x^2}\right)\frac{1}{\cancel{2}}}{\left(2x^2\right)2-\left({}^{2x}\cancel{2x^2}\right)\frac{4}{\cancel{x}}}$$

$$\frac{(2)2-\left(x^2\right)1}{\left(2x^2\right)2-(2x)4}$$

Simplify each term. Then multiply and simplify the numerator and denominator.

$$\frac{4-x^2}{4x^2-8x}$$

$$\frac{(2-x)(2+x)}{4x(x-2)}$$

Factor the numerator and denominator.

$$\frac{{}^{-1}\cancel{(2-x)}(2+x)}{4x\cancel{(x-2)}}$$

$$\frac{-1(2+x)}{4x}$$

Then reduce.

$$\frac{-2-x}{4x}$$

Write the simplified answer.

Method 2: Using common denominators, rewrite both the numerator and the denominator as single fractions. Change the complex fraction to multiplication by the reciprocal of the denominator (copy-dot-flip). Factor and reduce.

EXAMPLE 9.8B Simplify: $\dfrac{\dfrac{2}{x^2} - \dfrac{1}{2}}{2 - \dfrac{4}{x}}$

SOLUTION

$$\frac{\dfrac{2}{x^2} - \dfrac{1}{2}}{2 - \dfrac{4}{x}}$$

The LCD for the numerator is $2x^2$. The LCD for the denominator is x.

$$\frac{\left(\dfrac{2}{2}\right)\dfrac{2}{x^2} - \left(\dfrac{x^2}{x^2}\right)\dfrac{1}{2}}{\left(\dfrac{x}{x}\right)\dfrac{2}{1} - \dfrac{4}{x}}$$

$$\frac{\dfrac{4}{2x^2} - \dfrac{x^2}{2x^2}}{\dfrac{2x}{x} - \dfrac{4}{x}}$$

Rewrite the numerator as a single fraction. Rewrite the denominator as a single fraction.

$$\frac{\dfrac{4 - x^2}{2x^2}}{\dfrac{2x - 4}{x}}$$

$$\frac{4 - x^2}{2x^2} \cdot \frac{x}{2x - 4}$$

Copy-dot-flip.

$$\frac{(2 - x)(2 + x)}{2x^2} \cdot \frac{x}{2(x - 2)}$$

Factor the numerator and denominator.

$$\frac{{}^{-1}\cancel{(2 - x)}(2 + x)}{{}_{2x}\cancel{2x^2}} \cdot \frac{{}^{1}\cancel{x}}{2\cancel{(x - 2)}}$$

$$\frac{-1(2 + x)}{4x}$$

Then reduce.

$$\frac{-2 - x}{4x}$$

Write the simplified answer.

The two methods will always give the same answer. In the preview example, many students feel the first method is a little easier. That will often be the case but not always, as the next example illustrates.

Simplify $\dfrac{\dfrac{3}{x}+\dfrac{1}{x+2}}{\dfrac{7}{x-9}-\dfrac{1}{x}}$ by multiplying both the numerator and denominator by the LCD of all terms (Method 1).

SOLUTION

$$\frac{\frac{3}{x}+\frac{1}{x+2}}{\frac{7}{x-9}-\frac{1}{x}}$$

The denominators are $x + 2$, $x - 9$, and x. The LCD is $x(x+2)(x-9)$.

$$\frac{x(x+2)(x-9)\frac{3}{x}+x(x+2)(x-9)\frac{1}{x+2}}{x(x+2)(x-9)\frac{7}{x-9}-x(x+2)(x-9)\frac{1}{x}}$$

Multiply all terms by $x(x+2)(x-9)$.

$$\frac{\cancel{x}(x+2)(x-9)\frac{3}{\cancel{x}}+x\cancel{(x+2)}(x-9)\frac{1}{\cancel{x+2}}}{x(x+2)\cancel{(x-9)}\frac{7}{\cancel{x-9}}-\cancel{x}(x+2)(x-9)\frac{1}{\cancel{x}}}$$

$$\frac{(x+2)(x-9)3+x(x-9)1}{x(x+2)7-(x+2)(x-9)1}$$

$$\frac{3x^2-21x-54+x^2-9x}{7x^2+14x-x^2+7x+18}$$

$$\frac{4x^2-30x-54}{6x^2+21x+18}$$

Reduce each term, multiply, and simplify the numerator and denominator.

$$\frac{2\left(2x^2-15x-27\right)}{3\left(2x^2+7x+6\right)}$$

Factor the numerator and denominator.

$$\frac{2(x-9)(2x+3)}{3(2x+3)(x+2)}$$

$$\frac{2(x-9)\cancel{(2x+3)}}{3\cancel{(2x+3)}(x+2)}$$

Then reduce.

$$\frac{2(x-9)}{3(x+2)}$$

Write the simplified answer.

Now let's do the same problem with the second method.

EXAMPLE 9.9B Simplify $\dfrac{\dfrac{3}{x}+\dfrac{1}{x+2}}{\dfrac{7}{x-9}-\dfrac{1}{x}}$ by writing the numerator and denominator each as a single fraction and then use copy-dot-flip (Method 2).

SOLUTION

$$\frac{\dfrac{3}{x}+\dfrac{1}{x+2}}{\dfrac{7}{x-9}-\dfrac{1}{x}}$$

The LCD for the numerator is $x(x+2)$. The LCD for the denominator is $x(x-9)$. You do not need to use the same LCD for both the numerator and denominator in this method.

$$\frac{\dfrac{(x+2)}{(x+2)}\dfrac{3}{x}+\dfrac{1}{(x+2)}\dfrac{x}{x}}{\dfrac{7}{(x-9)}\dfrac{x}{x}-\dfrac{1}{x}\dfrac{(x-9)}{(x-9)}}$$

$$\frac{\dfrac{3x+6+x}{x(x+2)}}{\dfrac{7x-x+9}{x(x-9)}}$$

Rewrite the numerator as a single fraction. Rewrite the denominator as a single fraction.

$$\frac{\dfrac{4x+6}{x(x+2)}}{\dfrac{6x+9}{x(x-9)}}$$

$$\frac{4x+6}{x(x+2)}\cdot\frac{x(x-9)}{6x+9}$$

Copy-dot-flip.

$$\frac{2(2x+3)}{x(x+2)}\cdot\frac{x(x-9)}{3(2x+3)}$$

Factor the numerator and denominator.

$$\frac{2\cancel{(2x+3)}}{\cancel{x}(x+2)}\cdot\frac{\cancel{x}(x-9)}{3\cancel{(2x+3)}}$$

Then reduce.

$$\frac{2(x-9)}{3(x+2)}$$

Write the simplified answer.

In this example, the second method was easier. It didn't require factoring any hard trinomials. You can solve all problems knowing only one method (either one), but you should know both methods.

SECTION EXERCISES

For 1–2, determine for what value(s) of the variable, if any, the expression is undefined.

9–1. $\dfrac{4x - 2}{x^2 - x - 12}$

9–2. $\dfrac{x + 2}{x^2 + 2x}$

9–3. Reduce to lowest terms: $\dfrac{9 - y^2}{4y - 12}$

For 4–8, perform the following calculations. Express all answers in lowest terms.

9–4. $\dfrac{x^2 + 5x + 4}{7x^2 + 7x} \cdot \dfrac{28x^2}{x^2 - 16}$

9–5. $\dfrac{18 - 6a}{30a^3} \div \dfrac{a^2 - 9}{5a^2 + 15a}$

9–6. $\dfrac{x^2 - 2x - 8}{4 - x^2} \div \dfrac{x^2 - 4x}{2x - 4}$

9–7. $\dfrac{3}{2x - 2} + \dfrac{5}{6x - 6}$

9–8. $\dfrac{3a}{2a - 4} - \dfrac{6}{a^2 - 2a}$

For 9–11, express each complex rational expression in simplest form.

9–9. $\dfrac{\dfrac{2}{xy} + \dfrac{2}{y^2}}{\dfrac{4}{xy}}$

9–10. $\dfrac{1 + \dfrac{7}{x - 2}}{1 + \dfrac{3}{x + 2}}$

9–11. $\dfrac{\dfrac{a}{b} - \dfrac{b}{a}}{\dfrac{1}{a} + \dfrac{1}{b}}$

9–12. The legs of a right triangle measure $\frac{8}{x}$ and $\frac{9x}{2x-2}$, while the hypotenuse measures $\frac{x+40}{2x}$.

a. Express the perimeter of the triangle as a single fraction in terms of x.
b. Express the area of the triangle in simplest form in terms of x.

Negative Exponents

You may already know about negative exponents from Algebra 1, but here is a quick review. Negative exponents indicate *reciprocals* of the base. Negative exponents follow the usual exponent rules, including $\frac{x^m}{x^n} = x^{m-n}$. For example, $\frac{x^3}{x^5} = x^{3-5} = x^{-2}$.

We also know that $\frac{x^3}{x^5} = \frac{1}{x^2}$. The logical conclusion is that x^{-2} must be the same as $\frac{1}{x^2}$.

In general, we have:

$$x^{-n} = \left(\frac{1}{x}\right)^n = \frac{1}{x^n}$$

Positive exponents represent multiplication, and negative exponents represent division (reciprocals).

EXAMPLE 9.10 Simplify: $\left(\frac{3x^4}{5}\right)^{-2}$

SOLUTION

By the negative exponent rule, find the reciprocal of the base and then distribute the power.

$$\left(\frac{3x^4}{5}\right)^{-2} = \left(\frac{5}{3x^4}\right)^2 = \frac{5^2}{(3)^2\left(x^4\right)^2} = \frac{25}{9x^8}$$

EXAMPLE 9.11 Simplify the expression: $\frac{3x^{-2}}{5y^{-4}}$

SOLUTION

The safe way to do any problem with negative exponents is to rewrite the expression with reciprocals, which gives a complex fraction. Then simplify.

$$\frac{3x^{-2}}{5y^{-4}} = \frac{\frac{3}{x^2}}{\frac{5}{y^4}} = \frac{3}{x^2} \cdot \frac{y^4}{5} = \frac{3y^4}{5x^2}$$

COMMON ERROR

Do not write $\frac{\frac{1}{3x^2}}{\frac{1}{5y^4}}$ in the first step. The exponents are on only the variables in this example, not the coefficients.

There is a shortcut. If the numerator and denominator involve only multiplication—no addition or subtraction—then a negative exponent in the numerator becomes a positive exponent in the denominator while a negative exponent in the denominator becomes a positive exponent in the numerator. With that shortcut, the previous example can be done in your head in one step.

$$\frac{3x^{-2}}{5y^{-4}} = \frac{3y^4}{5x^2}$$

Be careful, this shortcut does not work if the numerator or denominator involves addition or subtraction. If you apply the shortcut to $\frac{3 + x^{-2}}{5 - y^{-4}}$, you may get $\frac{3 + y^4}{5 - x^2}$, which is wrong. Any negative exponents with addition or subtraction must be rewritten as reciprocals before simplifying.

$$\frac{3 + x^{-2}}{5 - y^{-4}} = \frac{3 + \frac{1}{x^2}}{5 - \frac{1}{y^4}} = \frac{y^4\left(3x^2 + 1\right)}{x^2\left(5y^4 - 1\right)}$$

EXAMPLE 9.12

If $f(x) = 2x^{-2} + (6x)^{-1}$, evaluate $f\left(\frac{1}{2}\right)$.

SOLUTION

$$f\left(\frac{1}{2}\right) = 2\left(\frac{1}{2}\right)^{-2} + \left(6 \cdot \frac{1}{2}\right)^{-1} = 2(2)^2 + 3^{-1} = 8 + \frac{1}{3} = \frac{25}{3}$$

Pay careful attention to the base for each exponent. In the first term, $2x^{-2}$, the exponent is on only the value of x. In the second term, $(6x)^{-1}$, the exponent is on the entire base within the parentheses.

COMMON ERROR

Negative exponents can also result in complex fractions. Be careful that you use exponent rules accurately.

$$x^{-1}y^{-1} = \frac{1}{xy}$$

$$x^{-1} + y^{-1} = \frac{1}{x} + \frac{1}{y} = \frac{y+x}{xy}$$

$$x^{-1} + y^{-1} \neq \frac{1}{x+y}$$

Simplify: $\dfrac{x^{-1} - y^{-1}}{(xy)^{-1}}$

SOLUTION

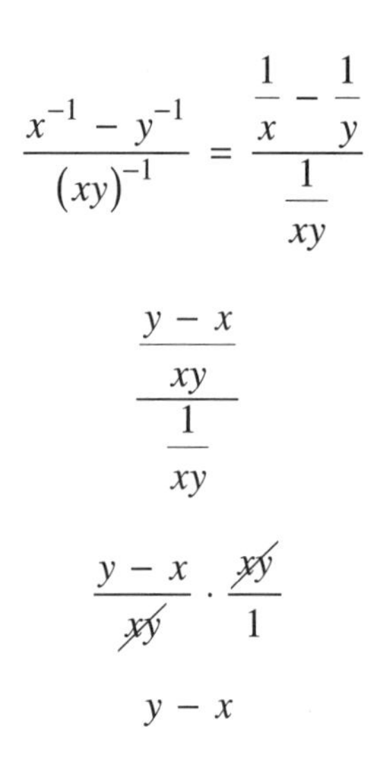

$$\frac{x^{-1} - y^{-1}}{(xy)^{-1}} = \frac{\frac{1}{x} - \frac{1}{y}}{\frac{1}{xy}}$$

Rewrite negative exponents in fraction form. Remember the exponent applies to only the base it touches.

$$\frac{\frac{y-x}{xy}}{\frac{1}{xy}}$$

Rewrite the numerator with a common denominator.

$$\frac{y-x}{\cancel{xy}} \cdot \frac{\cancel{xy}}{1}$$

Rewrite the complex fraction as multiplication by the reciprocal of the denominator. Then reduce.

$$y - x$$

SECTION EXERCISES

For 13–16, simplify. Express all answers without *using negative exponents.*

9–13. $\dfrac{5x^{-2}y}{20xy^{-4}}$

9–14. $18x^4\left(9x^6\right)^{-\frac{3}{2}}$

9–15. $\dfrac{6x^{\frac{1}{2}}y^{\frac{2}{3}}z^{-\frac{1}{4}}}{2x^{\frac{3}{2}}y^{-\frac{4}{3}}z^{-\frac{3}{4}}}$

9–16. $\left(\dfrac{8x^{-3}}{27}\right)^{-\frac{2}{3}}$

9–17. If $f(x) = 4x^{-2} + (2x)^{-1}$, evaluate $f(4)$.

9–18. If $f(x) = 3x^2 + 2x + x^0 - x^{-1}$, evaluate $f\left(\dfrac{1}{3}\right)$.

For 19–20, simplify. Express all answers in lowest terms.

9–19. $\dfrac{1 + x^{-1}}{1 - x^{-2}}$

9–20. $\dfrac{x^{-2} - 2x^{-3} + x^{-4}}{x^{-1} - x^{-2}}$

Graphing Rational Functions

A *rational function* is one that has the form "y = a rational expression." Since a rational expression is a ratio of two polynomials, we can define a rational function a little more precisely as a ratio of two polynomial functions, $r(x) = \dfrac{p(x)}{q(x)}$, where $p(x)$ and $q(x)$ are both polynomial functions and $q(x) \neq 0$. Rational functions can also be expressed in the form $y = \dfrac{a_n x^n + a_{n-1}x^{n-1} + \cdots a_1 x + a_0}{b_m x^m + b_{m-1}x^{m-1} + \cdots b_1 x + b_0}$, where all coefficients are real numbers.

For example, $y = \dfrac{x^2 - x - 12}{x^3 - 2x^2}$ is a rational function. Note that the degrees of the numerator and denominator do not have to be the same. In this section, we will assume that the degree of the denominator is at least 1 since a rational function with a constant denominator is no different than a polynomial function.

DOMAIN OF A RATIONAL FUNCTION

A rational function is a ratio of two polynomial functions, and the domain of a polynomial function is all real numbers. Because division by zero is undefined, the domain of a rational function must exclude x-values that would make the denominator zero. Therefore, the domain of a rational function is the set of all real numbers except the zeros of the denominator.

Find the domain of the function: $y = \dfrac{x^2 - x - 12}{x^3 - 2x^2}$

SOLUTION

The domain is all real numbers except where the denominator equals zero.

$$x^3 - 2x^2 \neq 0$$
$$x^2(x - 2) \neq 0$$
$$x \neq 0 \text{ and } x \neq 2$$

The domain is all real numbers except $x = 0$ and $x = 2$. This can be written in interval notation as $(-\infty,0) \cup (0,2) \cup (2,\infty)$.

ZEROS AND *y*-INTERCEPT OF A RATIONAL FUNCTION

The only way a fraction can equal zero is if its numerator is 0. Therefore, zeros of a rational function can occur only at zeros of the numerator. However, zeros can occur only at values of x in the domain of the function. So zeros of the denominator are excluded from being zeros of the function. Thus, zeros of a rational function occur at those values of x where the numerator equals 0 and the denominator does not.

Finding the domain of a rational function involves finding the zeros of its denominator. Finding the zeros of the function involves finding zeros of both the numerator and denominator. For these reasons, it is very helpful to express rational functions with both numerators and denominators in factored form.

The y-intercept of a rational function is found the same way as you find the y-intercept of any other function: substitute $x = 0$ and evaluate. When the polynomials in the numerator and denominator are written in standard form, the y-intercept is $\dfrac{a_0}{b_0}$. It is often convenient to express rational functions in other forms, though, such as factored form. In those cases, the easiest way to find the y-intercept is by setting $x = 0$.

EXAMPLE 9.15

Find the y-intercept and the zeros: $y = \dfrac{x^2 + x - 6}{x^2 - 4}$

SOLUTION

To find the y-intercept, substitute $x = 0$.

$$y = \frac{0^2 + 0 - 6}{0^2 - 4} = \frac{-6}{-4} = \frac{3}{2}$$

The y-intercept is $\dfrac{3}{2}$ or 1.5.

To find the zeros, factor both the numerator and the denominator.

$$y = \frac{x^2 + x - 6}{x^2 - 4} = \frac{(x + 3)(x - 2)}{(x + 2)(x - 2)}$$

The zeros of the numerator are $x = -3$ and $x = 2$. However, $x = 2$ is also a zero of the denominator, so it is not in the domain of the function. It cannot be a zero of the function. The only zero is $x = -3$.

The behavior of the graph of a rational function at a zero is the same as the behavior of the graph of a polynomial function at a zero. If the zero has odd multiplicity (an odd exponent on its factor), the graph will cross the x-axis at the zero. If the zero has even multiplicity (an even exponent on its factor), the graph will be tangent to the x-axis at the zero. In Example 9.15, the zero at $x = -3$ has multiplicity one (odd) so the graph will cross the x-axis at $x = -3$.

EXAMPLE 9.16

Find the zeros of $f(x) = \dfrac{(x-3)^2(x+5)}{x^2+4}$, and determine if the graph will cross the x-axis or be tangent to the x-axis at each zero.

SOLUTION

To find the zeros, find the x-values where the numerator equals 0 and the denominator does not equal 0.

$$(x-3)^2(x+5) = 0$$
$$x-3 = 0 \text{ or } x+5 = 0$$
$$x = 3 \text{ or } x = -5$$

For this function, the denominator will never equal 0. Zeros of the function occur at $x = 3$ and $x = -5$.

Look at the multiplicity of each zero to determine if the graph will cross the x-axis (odd multiplicity) or be tangent to it (even multiplicity). The zero at $x = -5$ has multiplicity 1, so the graph will cross the x-axis at $x = -5$. The zero at $x = 3$ has multiplicity 2, so the graph will be tangent to the x-axis at $x = 3$. The graph is shown below.

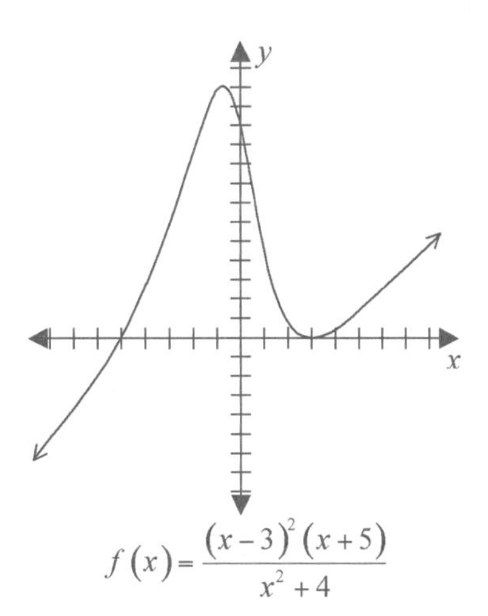

VERTICAL ASYMPTOTES AND REMOVABLE DISCONTINUITIES

Let's consider the function $y = \frac{6}{(x-2)(x+1)^2}$. Since it is expressed in factored form, you should be able to easily tell that its domain is all real numbers except $x = -1$ and $x = 2$. At either of these x-values, y will be undefined. We are interested in how the y-values of the graph behave at x-values near these two points. A graph of the function is shown in Figure 9.2.

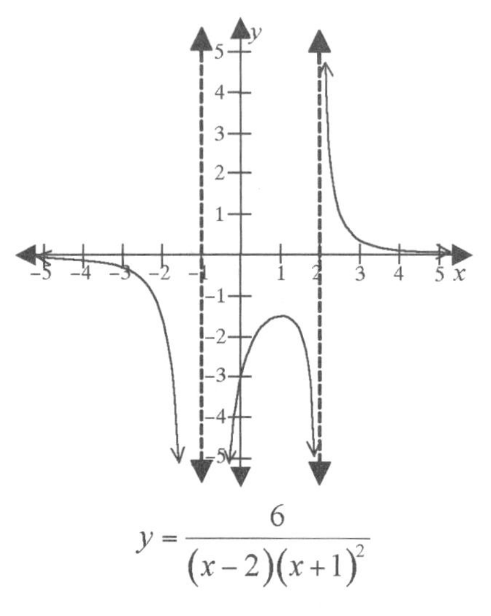

$$y = \frac{6}{(x-2)(x+1)^2}$$

FIGURE 9.2

Look at $x = 2$. For values of x just to the left of 2, the graph falls quickly toward $-\infty$. For x-values just to the right of 2, the graph rises quickly toward ∞. Of course, there is no point on the graph exactly at $x = 2$. Near $x = -1$, the graph behaves similarly except that both sides go in the same direction, falling toward $-\infty$. Again, there is no point on the graph exactly at $x = -1$.

Why does the graph behave like this? Evaluate the function $y = \frac{6}{(x-2)(x+1)^2}$ for x-values very close to 2.

TABLE 9.1

VALUES FOR $y = \frac{6}{(x-2)(x+1)^2}$

x	1.9	1.99	1.999	1.9999	. . .	2.0001	2.001	2.01	2.1
y	–7.1	–67.1	–667.1	–6667.1	. . .	6,666.2	666.2	66.2	6.2

As x gets closer to 2 from the left, the y-values go to $-\infty$. As x gets closer to 2 from the right, the y-values go to $+\infty$.

Now evaluate the function $y = \frac{6}{(x-2)(x+1)^2}$ for x-values very close to –1.

TABLE 9.2

VALUES FOR $y = \dfrac{6}{(x-2)(x+1)^2}$

x	−1.1	−1.01	−1.001	. . .	−0.999	−0.99	−0.9
y	−193.5	−19,933.6	−1,999,333.6	. . .	−2,000,666.9	−20,066.9	−206.9

As x gets closer to −1 from the left, the y-values go to $-\infty$. As x gets closer to −1 from the right, the y-values go to $-\infty$.

These are two examples of *vertical asymptotes* of a graph. A graph has a vertical asymptote at a point $x = a$ if as x approaches a, written $x \to a$, y either increases without bound, $y \to \infty$, or decreases without bound, $y \to -\infty$. Vertical asymptotes are usually indicated graphically by vertical dashed lines (see Figure 9.2). These lines are not actually part of the graph of the function; they simply help clarify the behavior of the graph nearby. (Note: Some older graphing calculators will show vertical or near-vertical lines at the asymptotes of rational functions. You need to be aware that those lines are not really part of the graph but are a limitation of older calculators. Newer graphing calculators do not show these lines.)

Vertical asymptotes are classified as either odd or even. At an odd vertical asymptote, the y-values go in opposite directions on each side of the asymptote, to $+\infty$ on one side and $-\infty$ on the other. At an even vertical asymptote, the y-values go in the same direction on both sides, either both sides to $+\infty$ or both sides to $-\infty$. The function $y = \dfrac{6}{(x-2)(x+1)^2}$ has an odd vertical asymptote at $x = 2$ and an even vertical asymptote at $x = -1$.

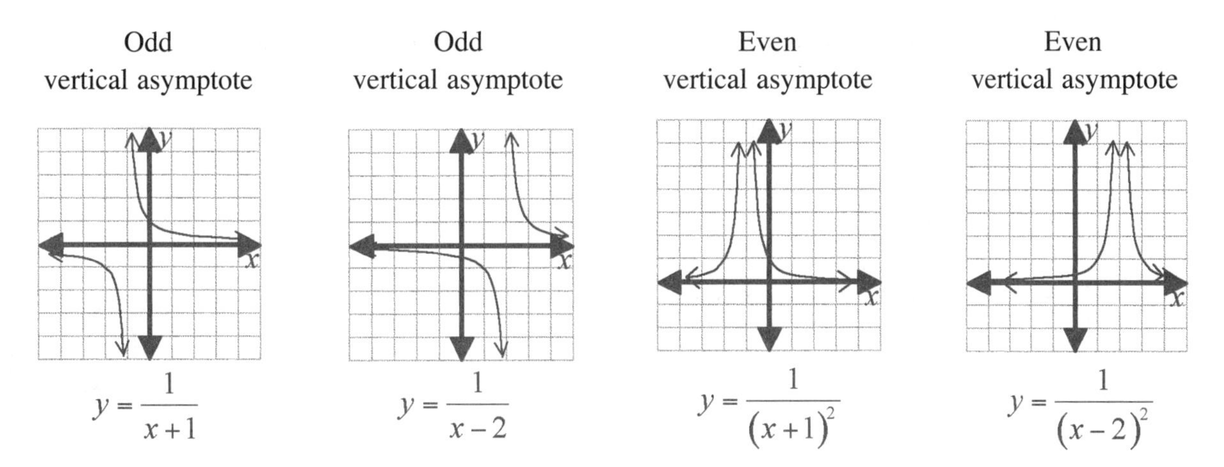

FIGURE 9.3

At this point, you might suspect that a rational function will always have a vertical asymptote wherever the denominator has a zero. You would be almost right.

Let's look at a second example. The rational function $y = \dfrac{x^2 + x - 2}{x^2 - 4} = \dfrac{(x+2)(x-1)}{(x+2)(x-2)}$ has a domain of all real numbers except $x = \pm 2$. Its graph is shown in Figure 9.4.

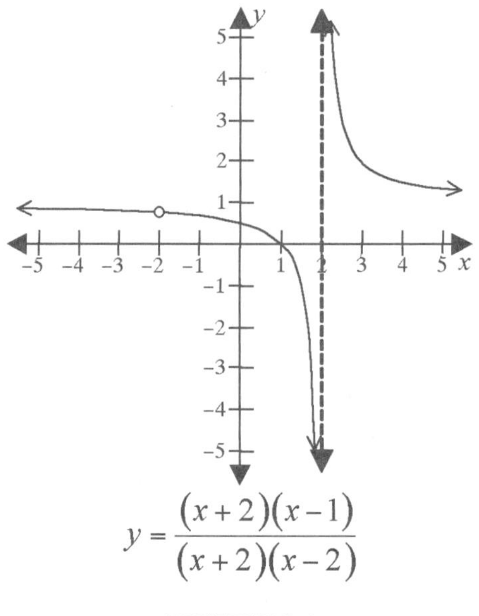

$$y = \frac{(x+2)(x-1)}{(x+2)(x-2)}$$

FIGURE 9.4

As expected, the graph does have a vertical asymptote at $x = 2$. Something different happens at $x = -2$, though. Instead of a vertical asymptote, the graph shows a tiny hole (actually called a *removable discontinuity*) at $x = -2$, represented by an open circle on the graph. (This feature can be very difficult to see on a graphing calculator.) When x is very close to -2, the y-values, instead of going to ∞ or $-\infty$, stay very close to 0.75. Why does the graph behave so differently at $x = 2$ and at $x = -2$? At $x = 2$, the denominator has a zero but the numerator does not. This leads to the vertical asymptote. At $x = -2$, though, both denominator and numerator have zeros. This is the indeterminate fraction, $\frac{0}{0}$.

By looking at the factored form of the function, $y = \frac{(x + 2)(x - 1)}{(x + 2)(x - 2)}$, we can simplify the rational expression to $y = \frac{x - 1}{x - 2}$, $x \neq -2$. For x close to -2, we have y close to $\frac{-2 - 1}{-2 - 2} = \frac{-3}{-4} = \frac{3}{4}$. However, right at $x = -2$, the original function is not defined ($x = -2$ is not in its domain), so the graph has a removable discontinuity at $(-2, 0.75)$.

SEE CALC TIP 9B

SUMMARY OF DENOMINATOR ZEROS FOR RATIONAL FUNCTIONS

1. If $x = a$ is a zero of the denominator but not a zero of the numerator, the function will have a vertical asymptote at $x = a$. It will be an odd vertical asymptote if the multiplicity of the zero at a is odd. It will be an even vertical asymptote if the multiplicity is even.

2. If $x = a$ is a zero of both the numerator and the denominator, one of three things will happen.

   - If the multiplicity of the zero in the denominator is greater than the multiplicity of the zero in the numerator, the function will have a vertical asymptote.
   - If the multiplicity of the zero in the denominator is equal to the multiplicity of the zero in the numerator (as is the case in the previous example), the function will have a removable discontinuity (a hole in the graph) not on the x-axis.
   - If the multiplicity of the zero in the denominator is less than the multiplicity of the zero in the numerator, the function will have a removable discontinuity on the x-axis (do not confuse this with a zero; it is not one).

Removable discontinuities might not be included in your Algebra 2 course. In Algebra 2, rational functions will usually not have any common factors in the numerator and denominator and so will not have any removable discontinuities. Still, you should know that if the numerator and denominator of a rational function are both zero, there may be a removable discontinuity instead of an asymptote.

EXAMPLE 9.17

Find the zeros and the vertical asymptotes of the function: $f(x) = \dfrac{x^2 - x - 12}{x^3 - 4x}$

SOLUTION

Rewrite the function in factored form.

$$f(x) = \frac{x^2 - x - 12}{x^3 - 4x} = \frac{(x + 3)(x - 4)}{x(x + 2)(x - 2)}$$

Note that the numerator and denominator have no factors in common, so the function has no removable discontinuities. Set the numerator equal to 0 to find the zeros.

$$(x + 3)(x - 4) = 0$$
$$x + 3 = 0 \text{ or } x - 4 = 0$$
$$x = -3 \text{ or } x = 4$$

Set the denominator equal to 0 to find the vertical asymptotes.

SEE CALC TIP 9C

$$x(x - 2)(x + 2) = 0$$
$$x = 0 \text{ or } x - 2 = 0 \text{ or } x + 2 = 0$$
$$x = 0 \text{ or } x = 2 \text{ or } x = -2$$

The zeros are $x = -3$ and $x = 4$. The vertical asymptotes are $x = -2$, $x = 0$, and $x = 2$.

COMMON ERROR

Some students write vertical asymptotes as –2, 0, and 2, omitting the $x =$. Vertical asymptotes are vertical lines, and equations of vertical lines must include $x =$.

Write a rational function that has a zero at $x = 3$, an odd vertical asymptote at $x = -1$, and an even vertical asymptote at $x = 5$.

SOLUTION

A zero at $x = 3$ means $(x - 3)$ is a factor in the numerator.

An odd vertical asymptote at $x = -1$ means that $(x + 1)$ or that $(x + 1)^3$ or that another odd power of $(x + 1)$ is a factor in the denominator. Keep it simple and use $(x + 1)$ as a factor in the denominator.

An even vertical asymptote at $x = 5$ means that $(x - 5)^2$ or that $(x - 5)^4$ or that an even power of $(x - 5)$ is a factor in the denominator. Again, keep it simple. Use $(x - 5)^2$ as the factor.

Putting it together gives $f(x) = \dfrac{x - 3}{(x + 1)(x - 5)^2}$; this is a rational function that meets these conditions.

END BEHAVIOR, HORIZONTAL AND SLANT ASYMPTOTES

By *end behavior*, we mean what happens to the function values (the y-values) as $|x| \to \infty$. (Note that $|x| \to \infty$ is a short way of saying either $x \to \infty$ or $x \to -\infty$.) In polynomial functions, the ends either rise to ∞ or fall to $-\infty$. The ends of rational functions may go to ∞ or $-\infty$, but there is a third possibility; the two ends may approach some finite number. In this case, we say the graph has a *horizontal asymptote*.

Consider the function $y = \dfrac{2x}{x^2 - 1}$. We already know how to find its y-intercept of 0, its zero of 0, and its vertical asymptotes of $x = -1$ and $x = 1$. What can we learn about its end behavior?

The graph of the function is shown below.

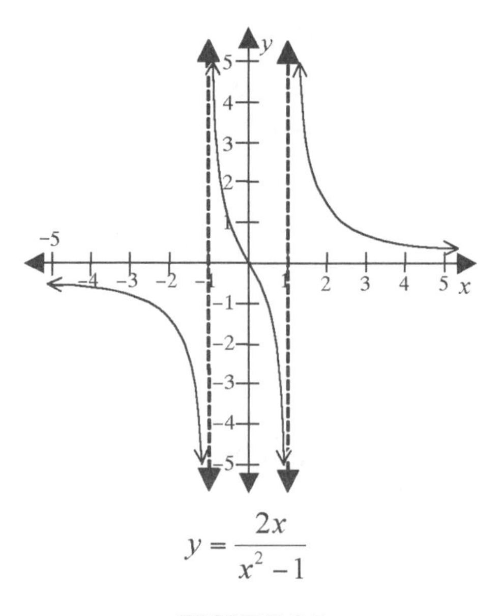

$$y = \frac{2x}{x^2 - 1}$$

FIGURE 9.5

The graph suggests that when x gets very large, either in the positive or in the negative direction, the y-values get closer to 0. One way to see this is to remember that the end behavior of a polynomial is determined by its highest exponent term. For very large values of x, all the lower-degree terms are insignificant compared with the highest-degree term and can be ignored. In this example, for very large values of x (either positive or negative), the numerator will behave like $2x$ and the denominator will behave like x^2. This means that for very large x, the rational function will behave like $y \approx \frac{2x}{x^2} = \frac{2}{x}$. As x goes to either ∞ or $-\infty$, $\frac{2}{x} \to 0$. When x is a large positive number, $\frac{2}{x}$ will be positive (but small). So as $x \to +\infty$, the graph of the function will stay above the x-axis. When x is a large negative number, $\frac{2}{x}$ will be negative (but small). So as $x \to -\infty$, the graph of the function will stay below the x-axis. So the right side of the graph approaches $y = 0$ (the x-axis) from above while the left side approaches $y = 0$ from below, exactly as shown in Figure 9.5.

The line $y = 0$ is called a horizontal asymptote of the graph. A graph has a horizontal asymptote at $y = b$ if as x either increases without bound ($x \to \infty$) or decreases without bound ($x \to -\infty$), the y-values approach extremely close to b, $y \to b$. In our example, as $x \to \infty$ or $x \to -\infty$, $y \to 0$. So $y = 0$, the x-axis, is the equation of the horizontal asymptote.

The reasoning above should suggest the following generalization. If the degree of the denominator of a rational function is greater than the degree of the numerator, the function will have the x-axis as a horizontal asymptote.

EXAMPLE 9.19 Describe the end behavior of the rational function: $y = \frac{8 - 2x^2}{x^4 - 1}$

SOLUTION

Since the degree of the denominator is greater than the degree of the numerator, we know the function will have the x-axis as its horizontal asymptote. By looking at the ratio of the highest-degree terms in the numerator and denominator, we see that for large x, $y \approx \frac{-2x^2}{x^4} = \frac{-2}{x^2}$. From this we see that for large values of x, either positive or negative, y will be negative. Both ends of the function approach the x-axis from below. This is shown graphically below.

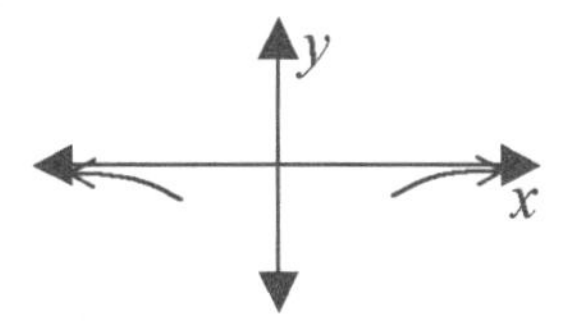

We can always use the ratio of the highest-degree terms in the numerator and denominator to get a general description of the end behavior of a rational function.

EXAMPLE 9.20 Describe the end behavior of the rational function: $y = \dfrac{2x^2 - 8}{x^2 - 2x + 1}$

SOLUTION

By looking at the ratio of the highest-degree terms in the numerator and denominator, we see that for very large values of x, $y \approx \dfrac{2x^2}{x^2} = 2$. As x goes to ∞ or $-\infty$, both sides of the graph will approach $y = 2$. The graph has a horizontal asymptote at $y = 2$.

SEE CALC TIP 9D

In general, any rational function having the degree of its numerator the same as the degree of its denominator, $y = \dfrac{a_n x^n + \cdots}{b_n x^n + \cdots}$, will have a horizontal asymptote given by the ratio of its leading coefficients, $y = \dfrac{a_n}{b_n}$.

In Example 9.20, comparing the highest-degree terms tells us that the function has a horizontal asymptote of $y = 2$. In many cases, particularly in Algebra 2, this may be all we really need to know about the end behavior. However, there may be times when you want to know more, such as are the ends of the graph above or below the asymptote (Figure 9.6)? Except for the special case when the asymptote is $y = 0$, simply comparing highest-degree terms will not tell us that. If that is important to know, you can use long division.

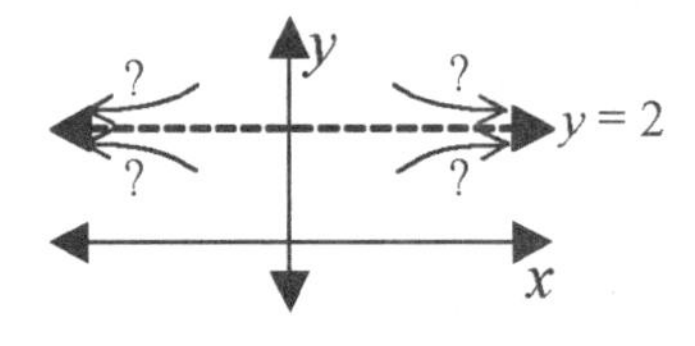

FIGURE 9.6

If a rational function, $f(x)$, has a numerator of degree greater than or equal to the degree of the denominator, long division can be used to express the function as the sum of a polynomial and a *proper rational remainder*: $f(x) = p(x) + r(x)$, where $p(x)$ is a polynomial and $r(x)$ is a rational expression having a denominator of higher degree than its numerator. The polynomial describes the end behavior of the function. The remainder can be used to determine if the ends are above or below the polynomial. For the function in Example 9.20, long division (Chapter 2) gives:

$$y = \frac{2x^2 - 8}{x^2 - 2x + 1} = 2 + \frac{4x - 10}{x^2 - 2x + 1}$$

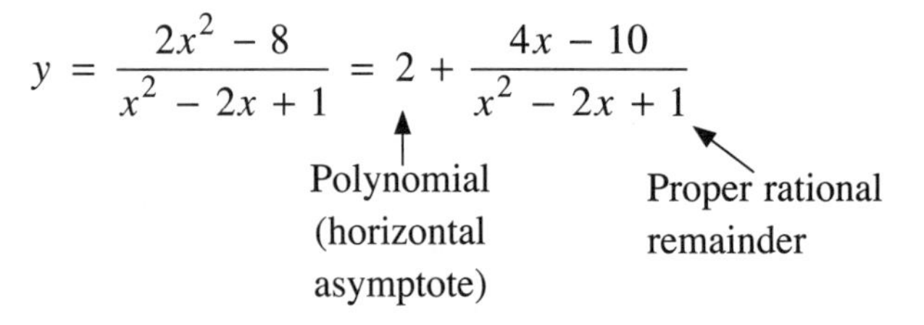

The ends of the function will act like the (constant) polynomial $y = 2$; this is the horizontal asymptote we had already found. By comparing the highest-degree terms in the remainder, we find that for very large values of x, $r(x) \approx \frac{4x}{x^2} = \frac{4}{x}$. When x is a large positive number, this remainder will be small but positive, meaning the graph of the original function will be slightly above its asymptote. When x is a large negative number, the remainder is small but negative, meaning the graph of the function will be slightly below its asymptote. The ends behave as in the graph in Figure 9.7.

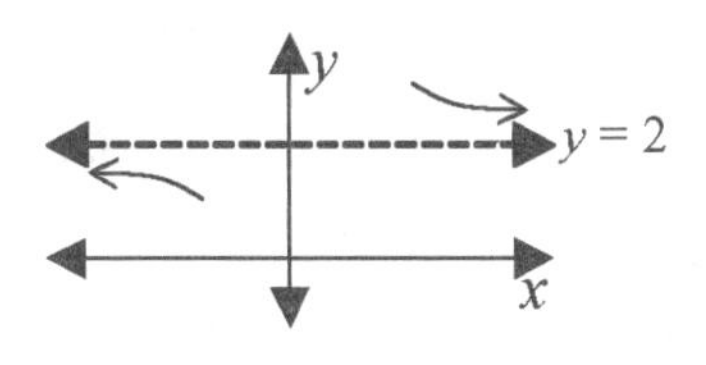

FIGURE 9.7

The complete graph is shown in Figure 9.8.

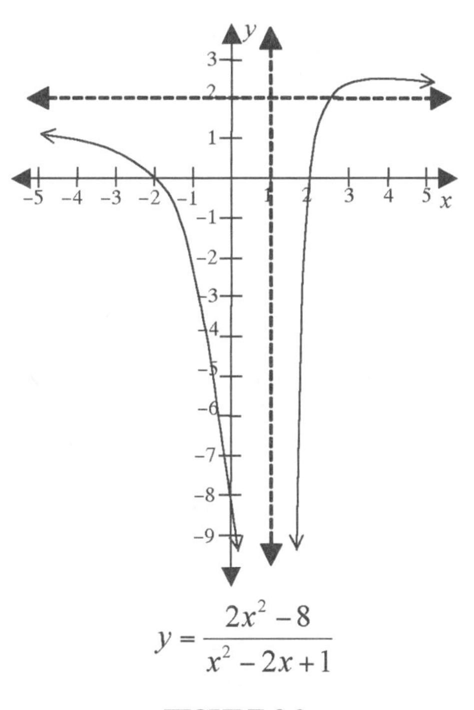

$$y = \frac{2x^2 - 8}{x^2 - 2x + 1}$$

FIGURE 9.8

Students often mistakenly believe that the graph of a rational function can never cross its horizontal asymptote. As the graph above shows, this is not true. The horizontal asymptote only describes what happens to the function for very large values of x. For smaller values of x, near the zeros and vertical asymptotes, the horizontal asymptote plays no role at all.

If a rational function has a horizontal asymptote, both sides of the function will approach the same y-value. However, not all rational functions have horizontal asymptotes. Some of them have end behavior going to ∞ or $-\infty$.

EXAMPLE 9.21 Describe the end behavior of the rational function: $y = \dfrac{x^2 - 4}{2x - 2}$

SOLUTION

To get a general idea of the end behavior, we can look at the ratio of the highest-degree term in the numerator and denominator.

$$\text{For } |x| \to \infty, \; y \approx \frac{x^2}{2x} = \frac{x}{2}.$$

In this case, the right side of the graph rises to ∞ and the left side falls to $-\infty$. Specifically, both sides approach a diagonal line with a slope of $\frac{1}{2}$. This function is said to have a *slant* (or *oblique*) *asymptote*.

If we need more information than just this, we can do the long division.

$$y = \frac{x^2 - 4}{2x - 2} = \frac{1}{2}x + \frac{1}{2} + \frac{-3}{2x - 2}$$

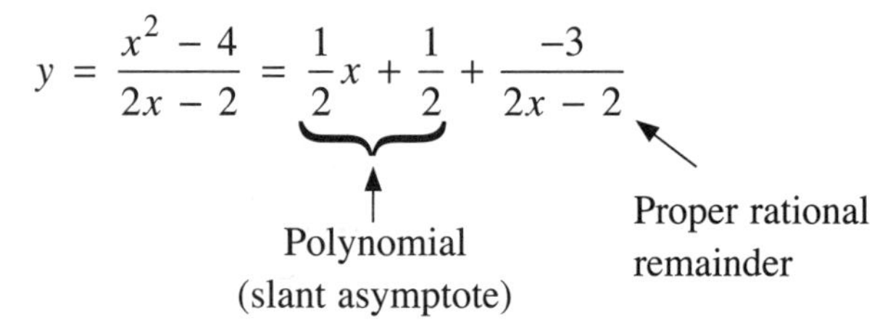

From this we find that the actual equation of the slant asymptote is $y = \frac{1}{2}x + \frac{1}{2}$. The asymptote does not pass through the origin but has a y-intercept of $\frac{1}{2}$. By looking at the highest-degree terms in the remainder, $r(x) \approx \frac{-3}{2x}$, we see that when x is a large positive number, the remainder will be negative, meaning the graph will be slightly below the asymptote. When x is a large negative number, the remainder will be positive and the graph will be slightly above the asymptote.

The graph of the function is shown below.

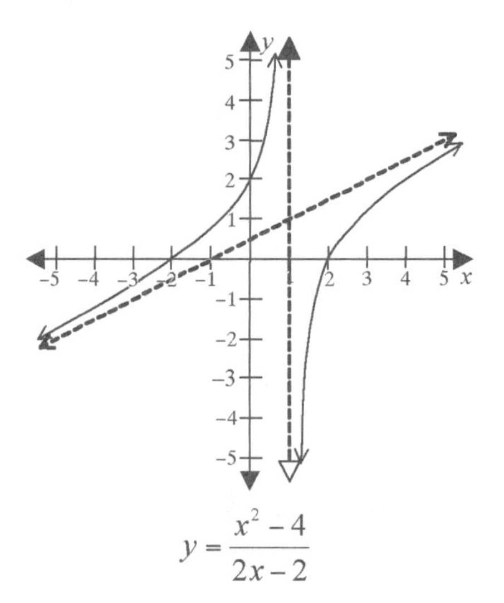

$$y = \frac{x^2 - 4}{2x - 2}$$

A rational function will have a slant asymptote whenever the degree of the numerator is exactly one more than the degree of the denominator. Finding the ratio of the highest-degree terms in the numerator and denominator will tell you the slope of the slant asymptote, which in many problems may be all you really need to know. If you need the exact equation of the asymptote (including its y-intercept) or want to know if the ends of the function are above or below the asymptote, do long division and rewrite the function as the sum of a polynomial and a proper rational function.

If the degree of the numerator exceeds the degree of the denominator by 2 or more, the function will have *nonlinear* end behavior. For example, by comparing the highest-degree terms of $y = \frac{x^3}{x - 3}$, we find that for very large values of x, $y \approx x^2$. The ends of this function will behave like a parabola; both sides will rise to ∞. To find the actual equation of the parabola and tell whether the ends of the function are above or below the parabola, we could do long division (try it if you are curious).

SUMMARY OF END BEHAVIOR FOR RATIONAL FUNCTIONS

The end behavior of a rational function will fall into one of four categories depending on how the degree of the numerator compares with the degree of the denominator.

1. The degree of the numerator is less than the degree of the denominator. The rational function will have a horizontal asymptote at $y = 0$, the x-axis. The sign of the ratio of the highest-degree terms in the numerator and denominator tells whether the ends of the graph lie above or below the asymptote.
2. The degree of the numerator is the same as the degree of the denominator. The rational function will have a horizontal asymptote at $y = b$ for some value of $b \neq 0$ (in other words, not the x-axis). The ratio of the highest-degree terms in the numerator and denominator gives the value of b. If you want to know whether the ends of the graph lie above or below the asymptote, you need to do long division.
3. The degree of the numerator is one more than the degree of the denominator. The rational function will have a slant asymptote of the form $y = mx + b$ where $m \neq 0$. The ratio of the highest-degree terms in the numerator and denominator gives the value of m. If you need to know the value of b or if you need to know if the ends of the graph lie above or below the asymptote, do long division.
4. The degree of the numerator exceeds the degree of the denominator by two or more. The rational function will have nonlinear end behavior. The ratio of the highest-degree terms in the numerator and denominator will tell you the general form of the behavior: quadratic, cubic, or other. To find the equation of the end behavior or to determine if the ends of the graph lie above or below this equation, do long division.

SKETCHING RATIONAL FUNCTIONS BY HAND

By far the easiest way to graph a rational function is with technology. However, we now understand enough about rational functions that we could sketch a pretty reasonable graph by hand if required.

To get a rough sketch of a rational function, first factor the function completely. Let's assume for simplicity that the function has no common factors in the numerator and denominator so that it has no removable discontinuities (these are rare in Algebra 2 anyway). Then find the following. (You don't have to find them in this order.)

- **The *y*-intercept:** Put in $x = 0$ and evaluate.
- **The zeros:** Find the x-values where the numerator equals 0. Note whether each zero has even or odd multiplicity.
- **The vertical asymptotes:** Find the x-values where the denominator equals 0. Note whether each will be an even or an odd asymptote.
- **The end behavior:** Find the ratio of the highest-degree term in the numerator and denominator. Consider how the ratio behaves for very large values of x.

Sketch in the vertical asymptotes and the end behavior. Then plot the y-intercept and zeros. Sometimes it may be helpful to evaluate the function at a few more x-values to help complete the graph.

If you want a more accurate sketch, particularly with respect to the end behavior, you can do long division of the function. This will give the exact equation of the asymptote (if it's other than a horizontal asymptote) and also tell whether the ends of the graph are above or below the asymptote.

EXAMPLE 9.22

Sketch the graph: $y = \dfrac{2x}{x^2 - 4x + 4} = \dfrac{2x}{(x-2)^2}$

SOLUTION

Note that the function was given in both standard and factored form. If it had been given in only standard form, the first thing to do would be to factor it. Now find some details about the graph.

- The y-intercept: When $x = 0$, $y = 0$. So the y-intercept is 0.
- Zeros: The numerator has a zero at $x = 0$ only. The zero has multiplicity 1.
- Vertical asymptotes: From the factored form, we see that the denominator has a zero of even multiplicity at $x = 2$.
- End behavior: From the standard form, when x is very large, $y \approx \dfrac{2x}{x^2} = \dfrac{2}{x}$. The function will have a horizontal asymptote at $y = 0$ (the x-axis). For large positive values of x, $\dfrac{2}{x}$ will be positive and the function will be above the x-axis. For large negative values of x, $\dfrac{2}{x}$ will be negative and the function will be below the x-axis.

The vertical asymptote, the y-intercept/zero, and the end behavior are sketched below.

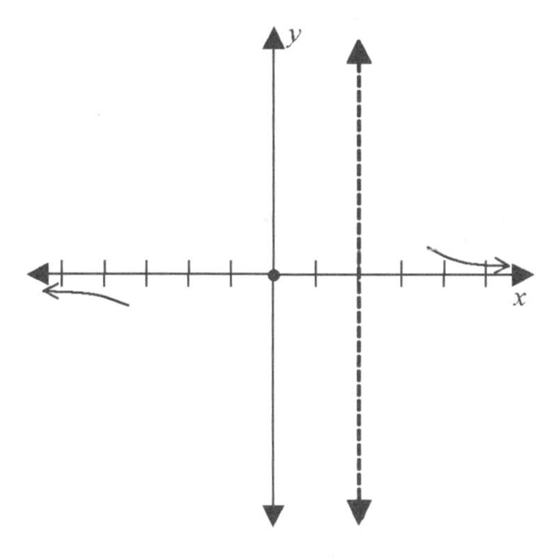

The only thing left to do is connect the pieces we have so far. This is easy if you keep in mind the following:

1. The vertical asymptote is even, so both sides will go in the same direction, either both up toward ∞ or both down toward $-\infty$.
2. The zero at $x = 0$ has odd multiplicity, so the graph must cross the x-axis there.
3. There cannot be any other zeros.

Work from the left. The function must cross the x-axis at the origin.

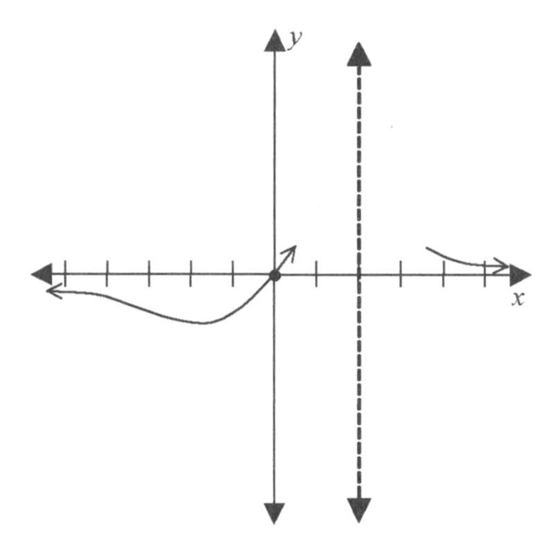

Since there are no more zeros, the function must go to ∞ on the left side of the vertical asymptote. Because the asymptote is even (and also because there are no more zeros), the function must also go to ∞ on the right side of the asymptote.

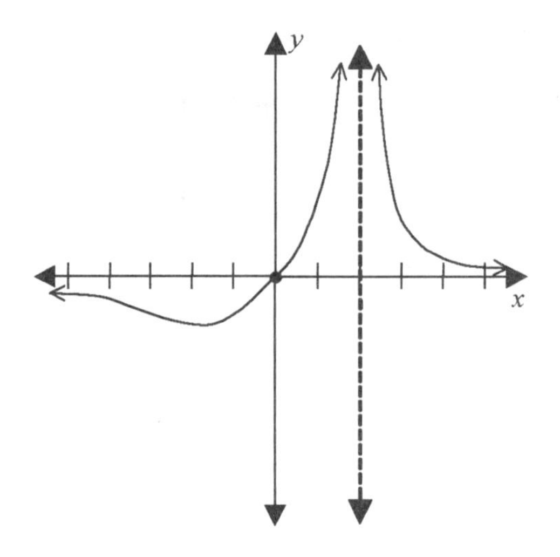

Note that there is no scale on the y-axis. This is deliberate. Our sketch shows the important features of the graph, its intercepts, asymptotes, and end behavior. However, it should not be taken as a perfect representation of the function. Some of the y-values have been exaggerated to show the features. The exact location and value of the relative minimum to the left of the zero is a calculus problem, not Algebra 2. If you need a really accurate graph, you should expect to use technology. A graph of the function made with a computer graphing program is shown below to see how it compares with our sketch.

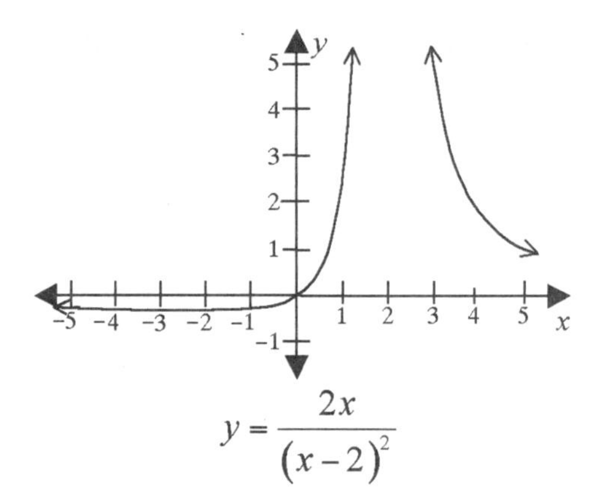

$$y = \frac{2x}{(x-2)^2}$$

EXAMPLE 9.23 Sketch the graph: $y = \dfrac{2x^2 - 2x - 4}{x^2 - 9} = \dfrac{2(x+1)(x-2)}{(x+3)(x-3)}$

SOLUTION

The function is given in both standard and factored form. Find some details about the graph:

- The y-intercept: From the standard form, when $x = 0$, $y = \dfrac{-4}{-9} = \dfrac{4}{9}$. The y-intercept is $\dfrac{4}{9}$.

- Zeros: From the factored form, we see the numerator has zeros at $x = -1$ and $x = 2$. Both have odd multiplicity.
- Vertical asymptotes: From the factored form, we see that the denominator has zeros at $x = -3$ and $x = 3$. Both have odd multiplicity.
- End behavior: From the standard form, when x is very large, $y \approx \frac{2x^2}{x^2} = 2$. The function will have a horizontal asymptote at $y = 2$.

The y-intercept, zeros, and asymptotes are sketched below.

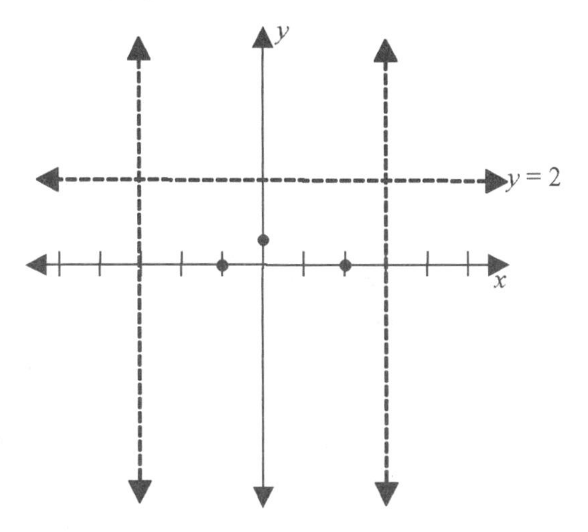

The function must cross the x-axis at each zero (they are both odd multiplicity) and must pass through the y-intercept. Since there are no other zeros, this leaves the middle part of the graph no choice but to go down both vertical asymptotes toward $-\infty$.

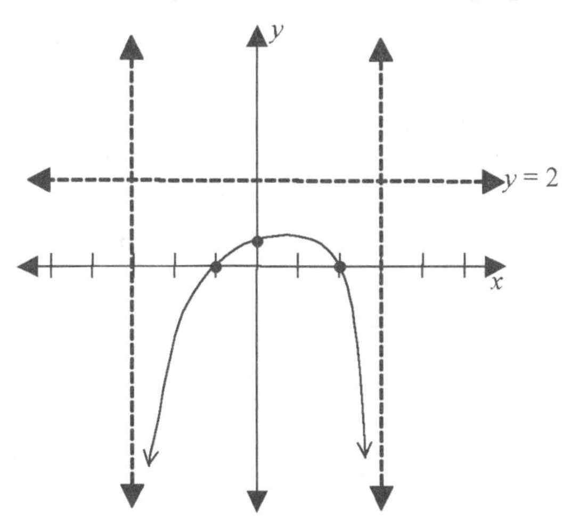

Since both vertical asymptotes are odd, the graph will go in opposite directions on the two sides of each asymptote. At $x = -3$, the function is going down on the right side of the asymptote so it will go up on the left side. At $x = 3$, the function is going down on the left side so it will go up on the right.

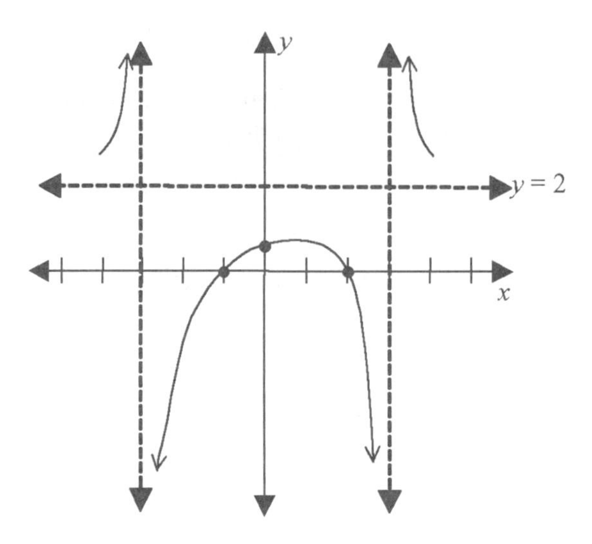

The only feature left is the end behavior, and we know the function has $y = 2$ as a horizontal asymptote. Many students assume that both sides of this function will approach the horizontal asymptote from above. This is not necessarily true, but it may give a good enough sketch for many purposes. If precision about the end behavior is important, we need to do long division:

$$y = \frac{2x^2 - 2x - 4}{x^2 - 9} = 2 + \frac{-2x + 14}{x^2 - 9}$$

By keeping just the highest-degree terms in the numerator and denominator of the remainder, we find that for large x, $r(x) \approx \frac{-2x}{x^2} = \frac{-2}{x}$. From this we see that when x is a large positive number, the remainder is negative. This means the right end of the graph will cross below the asymptote and then come back up toward it. When x is a large negative number, the remainder is positive. So the left end will stay above the horizontal asymptote. Note that this analysis will not tell us where the function crosses its asymptote on the right. If you are curious, the remainder must equal 0 at the point where the function intersects the line $y = 2$. Set the numerator of the remainder equal to 0 and solve for x. At $x = 7$, the graph crosses the horizontal asymptote.

Our final sketch is shown below.

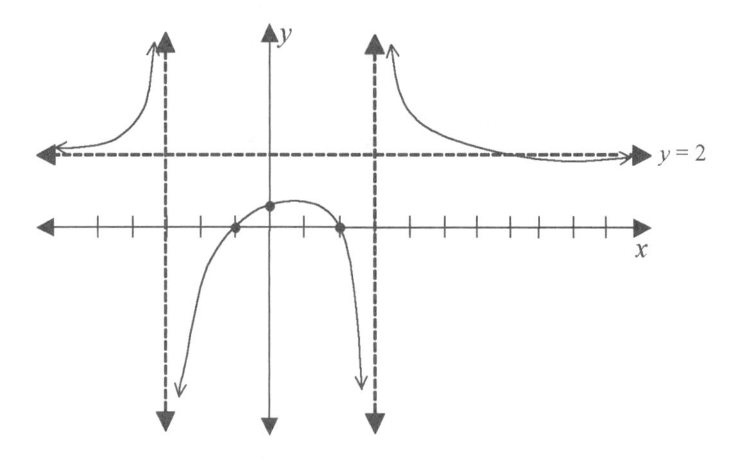

Again, a computer-drawn graph of the function is shown for comparison. Though not perfect, our sketch looks pretty good. The computer graph shows that although the function does cross its horizontal asymptote at $x = 7$, the graph never gets very far below the asymptote. In fact, the graph gets only about 0.075 below the asymptote before going back toward it.

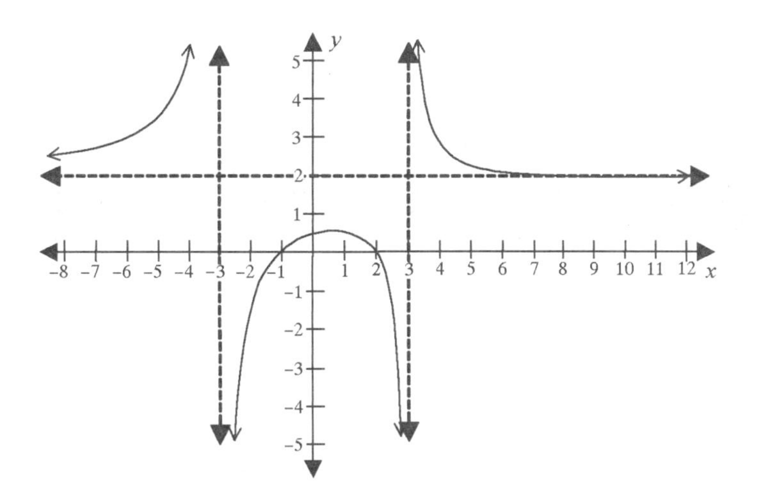

SECTION EXERCISES

9–21. Construct a single rational function, $f(x)$, that fits all of the following criteria:

- $f(x)$ has an odd vertical asymptote at $x = 2$
- $f(x)$ has an even vertical asymptote at $x = -3$
- $f(x)$ has a root at $x = 3$
- $f(x)$ has a y-intercept of $+4$

9–22. Consider the function $f(x) = \dfrac{(x-a)(x-c)(x-d)(x-e)^2}{(x-b)(x-c)(x-d)^2(x-e)}$ where a, b, c, d, and e are all distinct real numbers. Tell whether f should have a root, a vertical asymptote, or neither at each of the points a, b, c, d, and e.

9–23. Let g be the function $g(x) = \dfrac{2x^3 - 3x^2 - 4x + 5}{x^2 - 4}$. Express g as a sum of a polynomial and a proper rational function.

For 24–25, determine the y*-intercept, all real roots, all vertical asymptotes (specify even or odd), and end behavior. Then sketch a graph of each function without using your graphing calculator. (Check your graph with your calculator* after *you have drawn it.)*

9–24. $f(x) = \dfrac{1 - x}{x^2 - 4}$

9–25. $f(x) = \dfrac{25(x + 2)}{(x - 2)(x + 5)^2}$

9–26. The total cost, C, of producing x widgets at the ACME Widget Company is $C = 0.02x^2 + 10x + 50$. (Use graphing technology.)

a. Write a formula for the average cost per widget $\overline{C}$ as a function of x.

b. What is the domain of the function in part a?

c. Sketch the graph of the average cost function.

d. How many widgets should be produced to minimize the average cost?

e. Would producing the number of widgets found in part d maximize the company's profits?

Solving Rational Equations

A *rational equation* is any equation that contains at least one rational expression. So technically, $x + \frac{1}{3} = \frac{1}{2}$ is a rational equation. In this section, we will skip easy problems like that and consider equations that have at least one term where the variable is in a denominator.

The basic strategy for solving a rational equation algebraically is to multiply the equation by the least common denominator of all the terms in the equation. This will lead to an equation with no fractions, usually either linear or quadratic, that can be solved with familiar techniques. There are two slightly different procedures for carrying out this strategy; both lead to the same nonfractional equation. The procedure outlined below requires a few more steps than the other but may help beginners make fewer mistakes. The slightly shorter procedure is discussed at the end of the first example.

To solve a rational equation algebraically:

1. Identify the least common denominator for all the terms in the equation. Use the same procedure as for finding the LCD in an addition or subtraction problem.
2. Rewrite the equation so that all the terms have the LCD as their denominator.
3. Equate the numerators. (This is equivalent to multiplying the whole equation by its LCD.)
4. Solve the resulting equation.
5. Check the candidate solution(s) in the *original* equation to make sure they are defined. Any undefined results are extraneous solutions that must be rejected.

EXAMPLE 9.24

Solve for x: $\frac{x+4}{2x} = \frac{1}{3x} + 3$

SOLUTION

Rewrite the constant as a fraction by writing it as $\frac{3}{1}$. This will help you keep the numerators and denominators lined up and remind you that all the terms need a common denominator.

$$\frac{x+4}{2x} = \frac{1}{3x} + \frac{3}{1}$$

The LCD for all three terms is $6x$. The numerator and denominator of the first term has to be multiplied by 3, the second term by 2, and the third term by $6x$. This gives:

$$\frac{3(x+4)}{3(2x)} = \frac{2(1)}{2(3x)} + \frac{6x(3)}{6x}$$

Now all three denominators are the same. Multiplying the whole equation by $6x$ eliminates all the denominators and leaves

$$\begin{aligned} 3(x+4) &= 2(1) + 6x(3) \\ 3x + 12 &= 2 + 18x \\ 10 &= 15x \\ x &= \frac{2}{3} \end{aligned}$$

Check this answer in the original equation.

$$\frac{\frac{2}{3}+4}{2\left(\frac{2}{3}\right)} = \frac{1}{3\left(\frac{2}{3}\right)} + 3$$
$$3.5 = 3.5$$

Do not skip the check. If you are very confident of your algebra, you can save time by checking only to see if your candidate solution makes any of the denominators of the original equation equal to 0, which would make part of the equation undefined. In this equation, only $x = 0$ would give an undefined term, so $x = \frac{2}{3}$ is fine. Just remember, this shortcut will not check your algebra for you. It will only eliminate extraneous solutions.

Many people prefer a slightly quicker method for solving the equation. Find the LCD as above. Then instead of actually rewriting all the terms to have the same denominator, multiply the original equation by the LCD and carefully simplify the fractions. If you use this method, remember to multiply all the terms of the equation by the LCD. In this problem, we would have:

$$6x\left(\frac{x+4}{2x}\right) = 6x\left(\frac{1}{3x}\right) + 6x(3)$$
$${}^{3}\cancel{6x}\left(\frac{x+4}{\cancel{2x}}\right) = {}^{2}\cancel{6x}\left(\frac{1}{\cancel{3x}}\right) + 6x(3)$$
$$3(x+4) = 2(1) + 6x(3)$$

As you can see, both methods lead to the same equation. Which method you use is a matter of personal preference. Use the method that works best for you.

Solve for x: $\dfrac{7}{2x+4} - \dfrac{x+4}{x^2+2x} = \dfrac{1}{4x}$

SOLUTION

Remember that to find common denominators when you have polynomial denominators, factor first!

$$\frac{7}{2(x+2)} - \frac{x+4}{x(x+2)} = \frac{1}{4x}$$

The LCD is $4x(x + 2)$. To get common denominators, the numerator and denominator of the first fraction need to be multiplied by $2x$, the second fraction by 4, and the third fraction by $(x + 2)$.

$$\frac{(2x)7}{(2x)2(x+2)} - \frac{4(x+4)}{4x(x+2)} = \frac{1(x+2)}{4x(x+2)}$$

Since all three denominators are the same, we can eliminate them and solve the numerator equation.

$$(2x)7 - 4(x+4) = 1(x+2)$$
$$14x - 4x - 16 = x + 2$$
$$10x - 16 = x + 2$$
$$9x = 18$$
$$x = 2$$

This answer checks in the original equation.

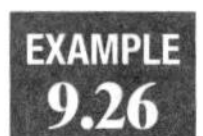

Solve: $\dfrac{x-1}{x-5} - \dfrac{1}{x} = \dfrac{20}{x^2 - 5x}$

SOLUTION

Factor denominators first!

$$\frac{x-1}{x-5} - \frac{1}{x} = \frac{20}{x(x-5)}$$

The LCD is $x(x - 5)$.

$$\frac{x(x-1)}{x(x-5)} - \frac{1(x-5)}{x(x-5)} = \frac{20}{x(x-5)}$$
$$x(x-1) - 1(x-5) = 20$$
$$x^2 - x - x + 5 = 20$$
$$x^2 - 2x - 15 = 0$$
$$(x-5)(x+3) = 0$$

$$x = 5 \text{ or } x = -3$$

The answer $x = 5$ does not check. It makes the original equation undefined and so it cannot be a solution. (The fact that both sides of the equation become undefined does NOT mean it checks.) The value $x = -3$ does check in the original equation and is the only solution.

FINDING INVERSES OF RATIONAL FUNCTIONS

We find the inverse of a rational function the same way we find the inverse of any other function (see Chapter 3): switch x and y, and then solve for y. We can use all of our usual algebraic techniques to solve for y. An additional technique that is sometimes helpful is to take reciprocals of both sides of an equation. This is valid as long as it does not make a denominator equal 0.

EXAMPLE 9.27 Find the inverse: $f(x) = \dfrac{1}{x-2}$

SOLUTION

$y = \dfrac{1}{x-2}$ Rewrite the function using y instead of $f(x)$

$x = \dfrac{1}{y-2}$ Switch x and y.

$\dfrac{1}{x} = y - 2$ Take the reciprocal of each side.

$\dfrac{1}{x} + 2 = y$ Solve for y.

The inverse function is $f^{-1}(x) = \dfrac{1}{x} + 2$.

Note that taking reciprocals of both sides was a convenient shortcut but not actually necessary for solving the problem. After switching x and y, we could have cross multiplied and continued from there. It would have taken an extra step or two, but we would have gotten the same (or at least an equivalent) answer.

EXAMPLE 9.28 Find the inverse: $y = 2 + \dfrac{5}{x-1}$

SOLUTION

$y = 2 + \dfrac{5}{x-1}$

$x = 2 + \dfrac{5}{y-1}$ Switch x and y.

$x - 2 = \dfrac{5}{y-1}$ Subtract 2 to get a single fraction on each side.

$$\frac{1}{x-2} = \frac{y-1}{5}$$ Take the reciprocal of each side.

$$\frac{5}{x-2} = y - 1$$

$$\frac{5}{x-2} + 1 = y$$ Solve for y.

COMMON ERROR

When taking the reciprocal of each side of an equation, you must take the reciprocal of the entire side. You cannot take reciprocals of each term individually. The reciprocal of $x - 2$ in the example above is $\frac{1}{x-2}$, not $\frac{1}{x} - \frac{1}{2}$.

Sometimes finding inverses of rational functions will require factoring out a monomial to help solve for y.

EXAMPLE 9.29 Find the inverse: $y = \frac{2x}{x-3}$

SOLUTION

$$y = \frac{2x}{x-3}$$

$$x = \frac{2y}{y-3}$$ Switch x and y.

$$x(y-3) = 2y$$ Cross multiply.

$$xy - 3x = 2y$$ Distribute.

$$xy - 2y = 3x$$ Move all terms with y to one side and all terms with no y to the other.

$$y(x-2) = 3x$$ Factor out the common y.

$$y = \frac{3x}{x-2}$$ Solve for y.

SECTION EXERCISES

For 27–31, solve the following equations for x.

9–27. $\dfrac{10}{x-2} - \dfrac{9}{x} = 2$

9–28. $\dfrac{x+4}{3x-6} + \dfrac{10}{x^2-2x} = \dfrac{1}{3}$

9–29. $\dfrac{x}{x-2} - \dfrac{7}{x+2} = \dfrac{8}{x^2-4}$

9–30. $\dfrac{3x+2}{x^2-9} - \dfrac{2}{x-3} = \dfrac{x}{x+3}$

9–31. $\dfrac{x}{a} + b = \dfrac{x}{c}$

9–32. Find the inverse of $f(x) = \dfrac{1}{2x+1}$.

9–33. Find the inverse of $f(x) = \dfrac{5x}{x+4}$.

Solving Rational Inequalities

In earlier chapters, we learned how to solve absolute value, quadratic, and polynomial inequalities. The strategy for solving rational inequalities is the same but with just one small extra step.

To solve a rational inequality algebraically:

1. Find the zeros of the associated equation.
2. Find the values of the variable that will make the equation undefined.
3. Plot both the zeros and the undefined points on a number line. Undefined points will always be open circles. The zeros will be open or closed circles depending on the type of inequality in the original inequality.
4. Check each interval to see if it satisfies the original inequality.
   As with polynomial inequalities, there are several ways you can check the intervals: numerically, graphically, with a sign chart, or by applying your knowledge of rational functions.
5. Write the answer in appropriate notation.

COMMON ERROR

For nonlinear inequalities, do not just use the original inequality sign without checking the intervals.

EXAMPLE 9.30 Solve for x: $\dfrac{x}{x-2} \leq x$

SOLUTION

First find the zeros of the associated equation. Since the equation is in the form of a proportion, the easiest way is to cross multiply.

$$\frac{x}{x-2} = x$$
$$x = x(x-2)$$
$$x = x^2 - 2x$$
$$0 = x^2 - 3x$$
$$0 = x(x-3)$$
$$x = 0 \text{ or } x = 3$$

Identify x-values that will make any term in the equation undefined. These are any x-values that will make the denominator equal zero. In this problem:

$$x - 2 = 0$$
$$x = 2$$

Put the zeros and the undefined points on a number line. The undefined point, $x = 2$, gets an open circle. Undefined points will never be included as part of the solution set. In this problem, the two zeros, $x = 0$ and $x = 3$ are included because of the $\geq$ symbol, so they get closed circles. Then check a value from each interval in the original inequality.

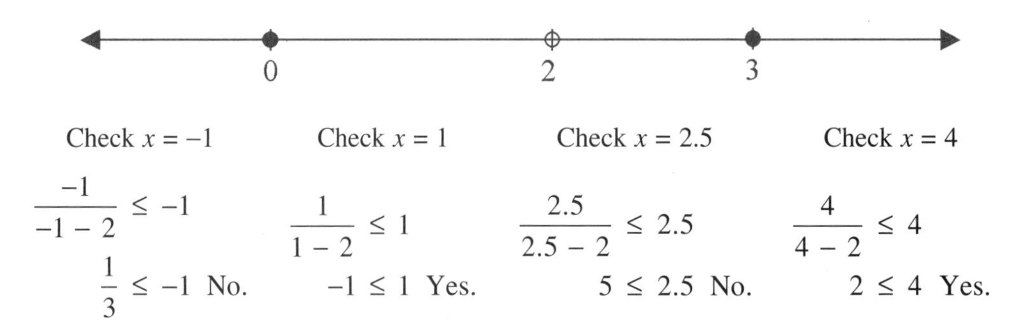

Sketch the solution on your number line, and then write the answer.

The solution set is $0 \le x < 2$ or $x \ge 3$.

It is not a good idea to check one interval and then just assume that the signs will alternate. Rational functions do not necessarily change sign at every zero or asymptote. The next example illustrates this.

EXAMPLE 9.31

Solve for x: $\dfrac{x^2 - 2x - 8}{(x - 3)^2} > 0$

SOLUTION

Because this problem involves a single rational expression compared with 0, it is a perfect candidate for using a sign chart. First factor the numerator and denominator.

$$\frac{(x + 2)(x - 4)}{(x - 3)(x - 3)} > 0$$

Zeros will be where the numerator equals 0: $x = -2$ and $x = 4$. Undefined points will be where the denominator equals 0: $x = 3$. Because of the strict inequality, all three get open circles on the number line. Create a sign chart below the number line. Each row represents where a single factor is positive or negative. Sign charts are explained in Chapter 7.

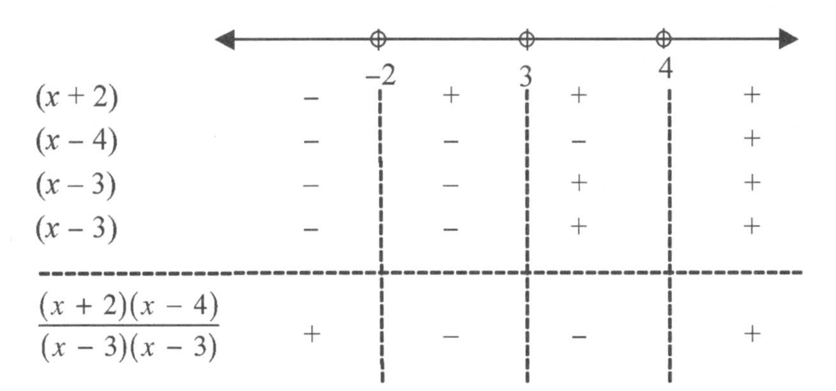

This problem asks for greater than 0, so we want the positive intervals.

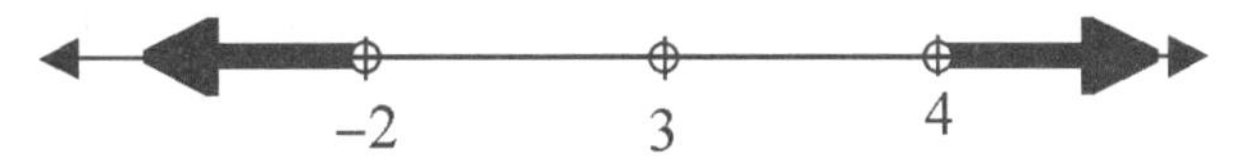

The solution set is $x < -2$ or $x > 4$.

EXAMPLE 9.32 Solve for x: $\frac{x}{x-2} \le \frac{8}{x}$

SOLUTION

This example can be solved exactly like the first example. Instead, though, let's see how to solve the inequality by thinking about the behavior of the graph.

First rearrange the equation and rewrite with common denominators so that it is in the form of a single fraction compared with 0. Factor the numerator.

$$\frac{x}{x-2} \le \frac{8}{x}$$

$$\frac{x}{x-2} - \frac{8}{x} \le 0$$

$$\frac{x(x)}{x(x-2)} - \frac{8(x-2)}{x(x-2)} \le 0$$

$$\frac{x^2 - 8x + 16}{x(x-2)} \le 0$$

$$\frac{(x-4)(x-4)}{x(x-2)} \le 0$$

$$\frac{(x-4)^2}{x(x-2)} \le 0$$

Think about what you know about the graph of $y = \frac{(x-4)^2}{x(x-2)}$. The solution to $\frac{(x-4)^2}{x(x-2)} \le 0$ will be all the intervals where the graph is at or below the x-axis. The function will have a zero (of multiplicity 2) at $x = 4$. This point will have a closed circle on the number line because of the $\le$. The function is undefined for $x = 0$ and $x = 2$. Draw open circles at those points on the number line.

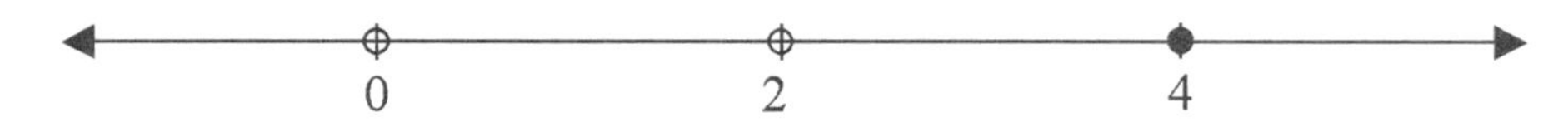

By looking at the highest-order terms in the numerator and denominator, we see the function $y = \frac{(x-4)^2}{x(x-2)}$ has a horizontal asymptote of $y = 1$. This means the function will be close to $y = 1$ on both the right and left ends and therefore positive for $x < 0$ and $x > 4$.

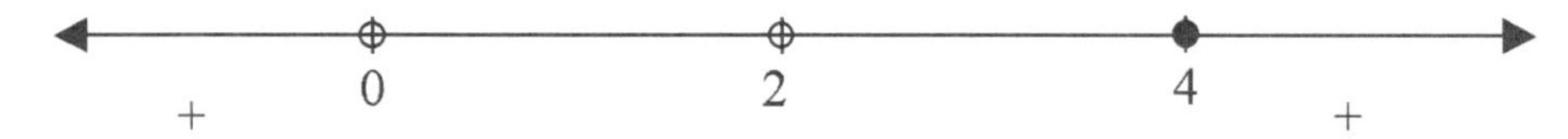

The only places a rational function can change sign are at odd zeros or odd vertical asymptotes. This function will change sign at both asymptotes, $x = 0$ and $x = 2$, but not at the zero, $x = 4$. So our final sign chart looks like this.

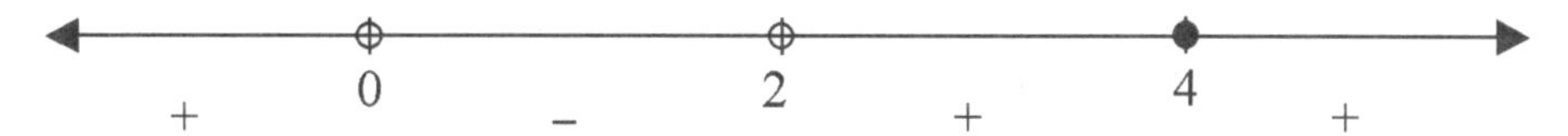

We want all x-values where the function is less than or equal to 0. The solution set is $0 < x < 2$ or $x = 4$. This solution set is a little unusual in that instead of just intervals, it contains an isolated point, $x = 4$. In interval notation, the solutions would be written as $(0, 2) \cup \{4\}$. If the problem had been $<$ instead of $\leq$, the 4 would not be included.

EXAMPLE 9.33

Solve $\dfrac{(x-4)(x+5)}{(x-1)(x+1)(x+3)} \geq 0$ graphically using the graph shown below.

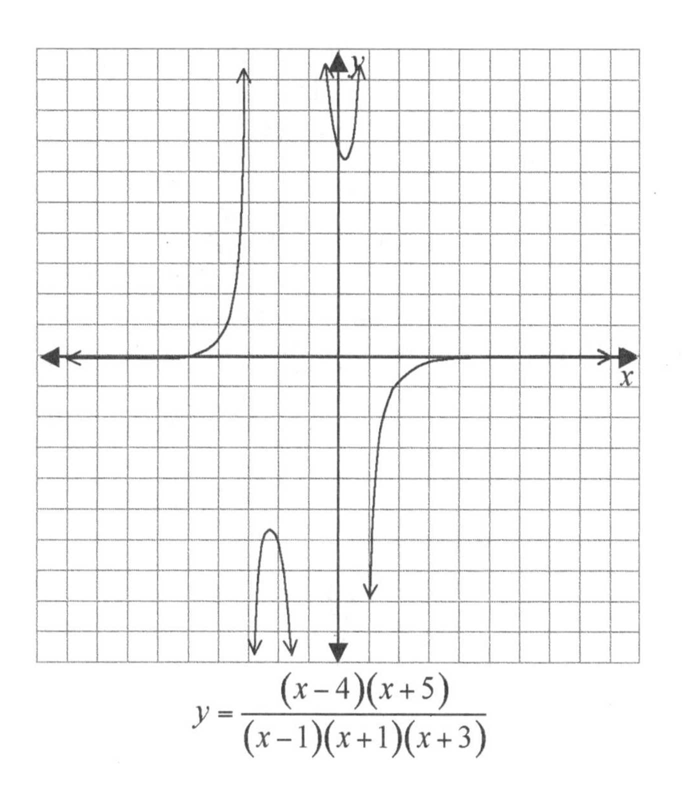

$$y = \frac{(x-4)(x+5)}{(x-1)(x+1)(x+3)}$$

SOLUTION

Since the inequality asks for $y \geq 0$, we want all the parts of the graph that are at or above the x-axis. The part on the left that is at or above the x-axis is drawn in bold on the following graph.

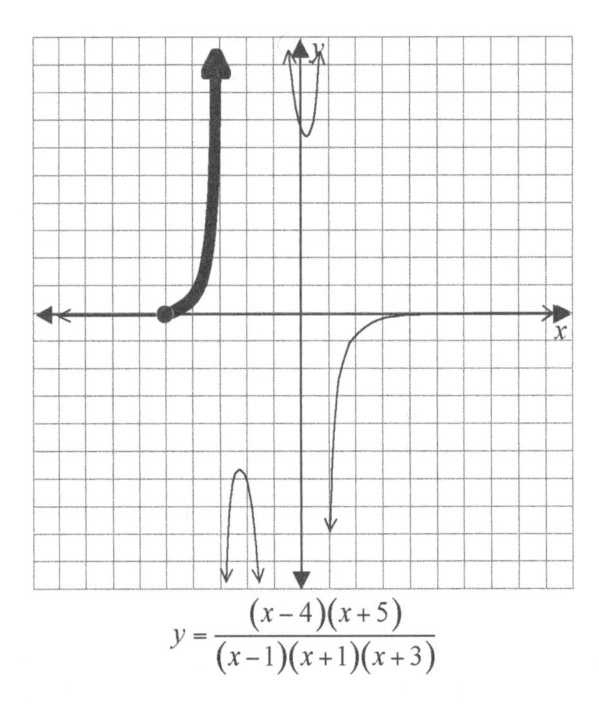

On the left, the graph between –5 and –3 is above the x-axis. Look at the function to see that $x = -5$ is a zero and $x = -3$ is a vertical asymptote. These values are the boundaries for the interval. Since the original inequality symbol is ≥, any zeros are included but vertical asymptotes are not. The interval [–5, –3) matches this part of the graph.

The part in the middle of the graph that is at or above the x-axis is drawn in bold below.

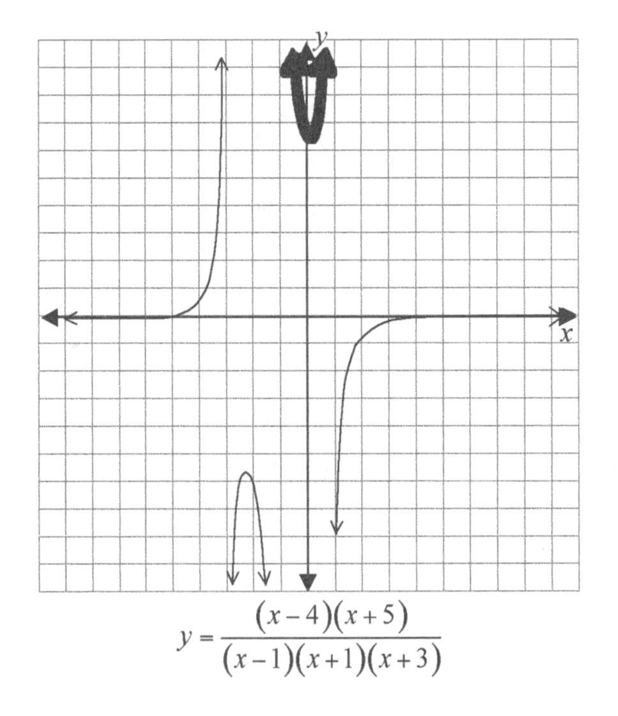

This part of the graph is between two vertical asymptotes, $x = -1$ and $x = 1$. The exact values of these vertical asymptotes can be most easily seen in the equation rather than on the graph. The interval $(-1, 1)$ matches this part of the graph.

The part on the right side of the graph that is at or above the x-axis is drawn in bold below.

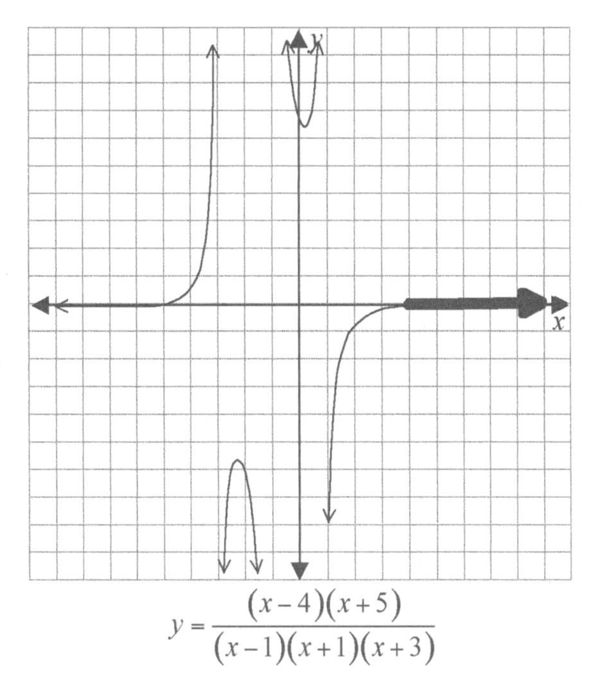

$$y = \frac{(x-4)(x+5)}{(x-1)(x+1)(x+3)}$$

This part of the graph is to the right of the zero, $x = 4$. The interval $[4,\infty)$ matches this part of the graph.

The solution set is the union of the three intervals $[-5,-3) \cup (-1,1) \cup [4,\infty)$.

SECTION EXERCISES

For 34–36, solve for x *and express the solution in interval notation.*

9–34. $\frac{2}{3} + \frac{x+7}{x} > 4$

9–35. $\frac{2(x+6)(1-x)}{3(x+5)} \leq 0$

9–36. $\frac{x+9}{x^3 - 4x^2} < 0$

9–37. Solve for x graphically, $\frac{x-4}{x-5} - \frac{2x}{x+4} \geq 0$. $f(x) = \frac{x-4}{x-5} - \frac{2x}{x+4}$ is graphed below. The roots are 2 and 8.

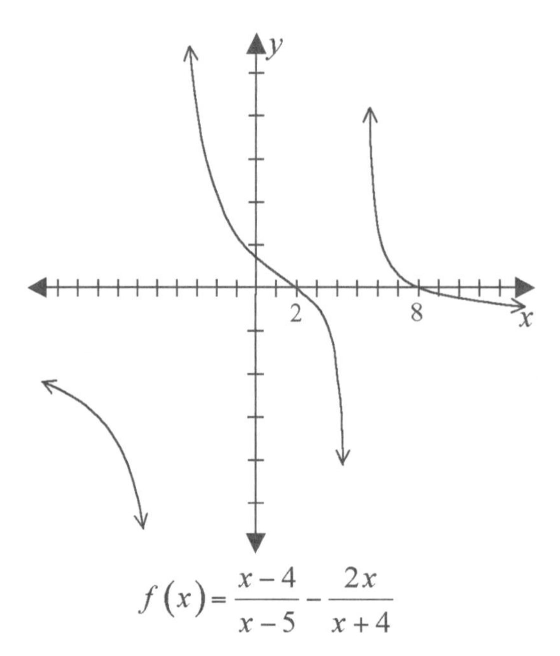

CHAPTER EXERCISES

C9–1. Determine for what value(s) of the variable, if any, the expression is undefined:

$$\frac{x^2 - 9}{x^3 - 4x}$$

C9–2. Reduce to simplest form: $\frac{1 - a^2}{a^2 - 4a + 3}$

For 3–5, perform the following calculations and express all answers in simplest form.

C9–3. $\frac{3x}{2x^2 + x - 3} \cdot \frac{6 + 4x}{6 - 3x}$

C9–4. $\frac{x^2 - 1}{x} \div (x + 1)$

C9–5. $\dfrac{2x - 1}{x^2 - x - 12} - \dfrac{1}{x + 3}$

For 6–7, express each complex rational expression in simplest form.

C9–6. $\dfrac{\dfrac{2}{x} - 1}{1 - \dfrac{4}{x^2}}$

C9–7. $\dfrac{x^{-2} - y^{-2}}{x^{-1}y^{-1}}$

For 8–9, simplify. Express all answers without using negative exponents.

C9–8. $\left(\dfrac{3}{5x^2}\right)^{-2}$

C9–9. $\dfrac{5x^{-2}y}{20xy^{-4}}$

C9–10. If $f(x) = 4x^2 - 5x + x^0 - x^{-1}$, evaluate $f\left(-\dfrac{1}{2}\right)$.

For 11–12, determine the y*-intercept (if any), all real roots, all vertical asymptotes (specify even or odd), and end behavior. Then sketch a graph of each function without using your graphing calculator. (Check your graph with your calculator after you have drawn it.)*

C9–11. $f(x) = \dfrac{x^2 - x - 2}{x^3 - 8x^2 + 16x} = \dfrac{(x + 1)(x - 2)}{x(x - 4)^2}$

C9–12. $f(x) = \dfrac{5x - x^2}{x^3 - 8}$

C9–13. Construct a single rational function, $f(x)$, that fits all of the following criteria:

- $f(x)$ has an odd vertical asymptote at $x = -3$
- $f(x)$ has an even vertical asymptote at $x = 5$
- $f(x)$ has a root at $x = 1$
- $f(x)$ has a y-intercept of $+1$

C9–14. Let f be the function $f(x) = \dfrac{2x^2 - 3x - 2}{x^2 + 4x + 4}$. Express f as a sum of a polynomial and a proper rational function.

For 15–17, solve for x.

C9–15. $\dfrac{2}{x} + \dfrac{5 - x}{x^2 - 9} = \dfrac{1}{x^2 - 3x}$

C9–16. $\dfrac{1}{x + 4} + \dfrac{45}{x^2 - x - 20} = \dfrac{x}{x - 5}$

C9–17. $\dfrac{a}{x} + \dfrac{1}{b} = c$

C9–18. Find the inverse: $f(x) = \dfrac{2}{x + 3}$

C9–19. Find the inverse: $f(x) = \dfrac{x - 1}{x + 4}$

C9–20. The Department of Environmental Conservation has introduced 50 deer into newly acquired state land. They predict the population of the herd will grow according to the function $P(t) = \dfrac{10(5 + 3t)}{1 + 0.04t}$, where t is years with $t \geq 0$. (Use graphing technology.)

a. Evaluate $P(0)$. Does this make sense?
b. Find the predicted deer population after 5 years.
c. According to the model, how long does it take the population to reach 400?
d. What is the limiting size of the herd?

For 21–22, solve for x, *and express the solution in interval notation.*

C9–21. $\dfrac{(x - 4)(x + 1)}{(x - 3)^2} \leq 0$

C9–22. $\dfrac{x^2 + 3x - 10}{x^3 - 9x} \geq 0$

Exponential Functions

WHAT YOU WILL LEARN

You have undoubtedly heard of exponential growth, perhaps in the context of world population. All other things being equal, the more people there are in the world, the faster the population will grow. Exponential functions model growth of anything where the rate of growth is proportional to the amount present. If you want to understand how compounding interest makes your money grow, you need to understand exponential growth. In this chapter, you will learn to:

- Evaluate exponential functions;
- Graph exponential functions and transformations;
- Solve exponential equations.

SECTIONS IN THIS CHAPTER

- Exponential Models
- Exponential Functions
- Graphing Exponential Functions
- Transformations of Exponential Functions
- Solving Exponential Equations with Common Bases

Exponential Models

Math is often used to model the growth of quantities. For example, suppose a bucket with 100 milliliters of water in it is left out in a steady rain where it gains an extra 100 milliliters each hour. The amount of water in the bucket after t hours is modeled like this:

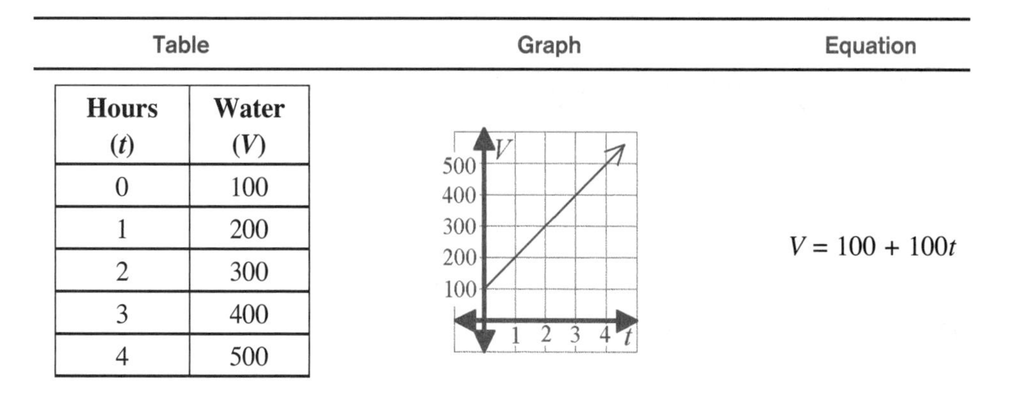

Hours (t)	Water (V)
0	100
1	200
2	300
3	400
4	500

$$V = 100 + 100t$$

FIGURE 10.1

The rate of growth of the amount of water is constant. No matter how much water is in the bucket, another 100 milliliters gets added each hour. This is an example of linear growth. Linear models were covered in Chapter 1.

Now suppose there are 100 bacteria in the water in the bucket and in one hour the number of bacteria doubles. There are now twice as many bacteria in the bucket. During the second hour, the population doubles again so that there are four times the original number. If this pattern continues, the number of bacteria in the bucket can be modeled as follows:

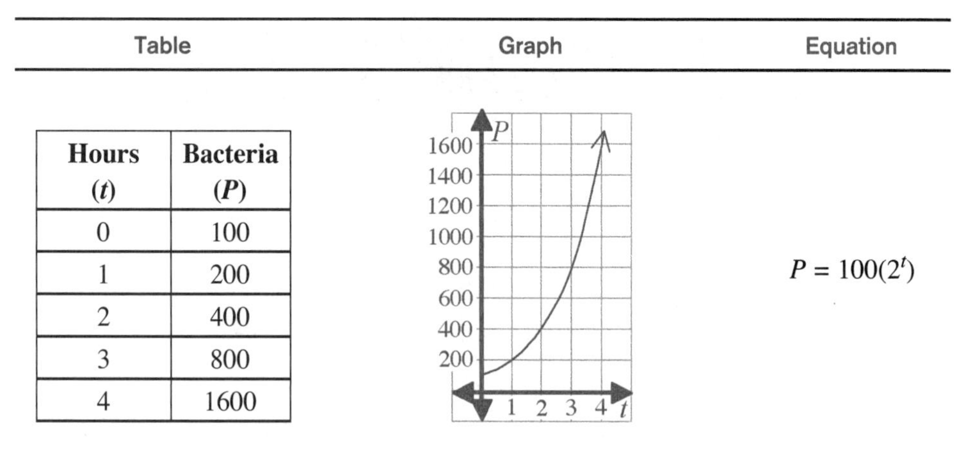

Hours (t)	Bacteria (P)
0	100
1	200
2	400
3	800
4	1600

$$P = 100(2^t)$$

FIGURE 10.2

The rate of growth of the number of bacteria is not constant. It is instead proportional to the number of bacteria present. This is an example of exponential growth. Before studying exponential functions in depth, you should review the rules of exponents summarized below. (These rules were covered in Chapters 2, 8, and 9 if you need a more thorough review with examples.)

EXPONENT RULES SUMMARY

For positive bases a and b, the following rules of exponents hold for all real numbers m and n:

1. $b^m b^n = b^{m+n}$
2. $\dfrac{b^m}{b^n} = b^{m-n}$
3. $(b^m)^n = b^{mn}$
4. $b^0 = 1$
5. $b^{-n} = \left(\dfrac{1}{b}\right)^n = \dfrac{1}{b^n}$
6. $b^{m/n} = \left(\sqrt[n]{b}\right)^m$
7. $(ab)^n = a^n b^n$
8. $\left(\dfrac{a}{b}\right)^n = \dfrac{a^n}{b^n}$

Exponential Functions

A simple exponential function has the form $y = b^x$, where b is a positive number called the base. Usually $b = 1$ is excluded since $y = 1^x$ is the same as $y = 1$, which is a horizontal line. So an exponential function can be written as $f(x) = b^x, b > 0, b \neq 1$.

The difference between an exponential function and polynomial functions is that in an exponential function, the base is a constant and the independent variable is the exponent. In a polynomial function, the base is the variable and the exponent is a constant. The function $y = 2^x$ is an exponential function. The function $y = x^2$ is a polynomial function.

EXAMPLE 10.1

Which of the following is an exponential function?

(1) $y = 3x$ (2) $y = x^3$ (3) $y = 3^x$ (4) $y = \sqrt[3]{x}$

SOLUTION

Only choice (3) has the variable as the exponent.

The exponential function is often modified to $y = a \cdot b^x$, where a is the initial value (the value of y when $x = 0$) and b is the multiplying factor. In an exponential function, as the value of x increases by 1, the value of y is multiplied by b. By contrast, in a linear function $y = mx + b$, when the value of x increases by 1, the value of y is increased by the slope, m (increased by addition, not multiplication).

For each table below, determine if it represents an exponential function of x. If so, write its equation.

a.

x	y
0	1
1	4
2	16
3	64

b.

x	y
0	1
1	5
2	9
3	13

c.

x	y
0	4
1	2
2	1
3	0.5

SOLUTION

a. Each time x increases by 1, y increases by a factor of 4 (multiply by 4). This is an exponential function with base 4: $y = 4^x$.

b. Each time x increases by 1, y increases by 4 (add 4). This is a linear function, $y = 4x + 1$, and is not exponential.

c. Each time x increases by 1, y decreases by a factor of $\frac{1}{2}$ (multiply by $\frac{1}{2}$). This is an exponential function with base $\frac{1}{2}$. When $x = 0$, $y = 4$, so the initial value is 4. The equation is $y = 4\left(\frac{1}{2}\right)^x$.

EXAMPLE 10.3

Some rabbits are set free on a remote island and begin to reproduce. The rabbit population on the island is modeled by $P = 25(1.2)^t$, where t is time in years since the rabbits were released.

a. How many rabbits were released on the island?

b. What is the average rate of change of the rabbit population over the first five years?

c. How is the rabbit population on the island changing?

SOLUTION

a. At time $t = 0$, there were $P = 25(1.2)^0 = 25(1) = 25$ rabbits on the island.

b. The average rate of change of the population is $\frac{\text{change in population}}{\text{change in time}}$.

In this example, the average rate of change over the first five years is $\frac{P(5) - P(0)}{5 - 0} = \frac{62.208 - 25}{5} = \frac{37.208}{5} = 7.4416$ rabbits per year. During the first five years, the population increased, on average, about 7.4 rabbits per year.

c. Each year, the rabbit population increases by a factor of 1.2.

There is another way to think about the rate of growth of the rabbit population in Example 10.3. The population started at 25. After one year, it had grown to $25(1.2)^1 = 30$ rabbits, a 20% increase. The second year, it grew from 30 to $25(1.2)^2 = 36$ rabbits, another 20% increase. In general, the population is increasing by 20% each year. This can be seen by rewriting the base 1.2 as $1 + 0.20$. The 1 represents the proportion, 100%, of the population present at the beginning of each year. The 0.20 represents the amount of change, 20%, during each year. Any quantity that grows by a fixed percent each time period can be modeled by an exponential function of the form $y = a(1 + r)^t$, where a is the initial amount present at time $t = 0$ and the r is the growth rate, expressed in decimal form.

EXAMPLE 10.4 You invest $1,000 in an account that earns 5% interest per year. Write an equation for the balance as a function of time in years.

SOLUTION

Your initial investment (the principal) was $1,000. Your money increases at a rate of 5%, so $r = 0.05$. Your balance after t years is $y = 1{,}000(1 + 0.05)^t$ or $y = 1{,}000(1.05)^t$.

Many people are interested in understanding how their money grows with interest. In Example 10.4, the interest was paid once each year. A more general model would allow for the possibility that interest is compounded more than once a year. In this case, the formula is $A = P\left(1 + \frac{r}{n}\right)^{nt}$, where

A = the amount of money in the account after t years
P = the principal; the initial deposit in the account
r = the annual interest rate, expressed as a decimal
n = the number of times per year interest is compounded
t = the number of years since the principal was deposited

EXAMPLE 10.5 Suppose you put $1,000 into an account that pays 2.75% interest compounded monthly. How much money will be in the account after 8 years?

SOLUTION

Use the formula $A = P\left(1 + \frac{r}{n}\right)^{nt}$, and identify values for the variables.

A = the money in the account after 8 years
P = $1,000 (the principal)
r = 0.0275 (the interest rate, 2.75%, converted to a decimal)
n = 12 (monthly compounding means twelve times a year)
t = 8 years

Substituting the values into the formula gives $A = \$1{,}000\left(1 + \frac{0.0275}{12}\right)^{12(8)} = \$1{,}245.76$.

After 8 years, the account will be worth $1245.76 to the nearest cent.

In addition to modeling how money grows, exponential models are used to model how the value of an item decreases with time. Cars (and many other things) *depreciate* over time. This means they lose value. The car you paid $15,000 for four years ago is not worth $15,000 today. One commonly used formula for depreciation is $V = C(1 - r)^t$, where V

is what the car is worth after t years given that you paid C for it originally and r is the annual depreciation rate expressed as a decimal.

Write a model for the value of your car as a function of time if it cost $24,000 new and depreciates 20% each year.

SOLUTION

The initial value of your car is $24,000. Each year it is losing 20% of its value, so $r = 0.20$. The value after t years is modeled by $V = 24{,}000(1 - 0.20)^t$ or $V = 24{,}000(0.8)^t$.

The previous examples illustrate an important idea of exponential functions. If the base b is greater than 1, the function will be increasing. If the base is between 0 and 1, the function will be decreasing. This will be seen again graphically in the next section.

SECTION EXERCISES

10–1. Identify each of the following functions as linear, exponential, or neither.

a.

Hours (x)	0	1	2	3	4
Population (y)	5	10	15	20	25

b.

Hours (x)	0	1	2	3	4
Population (y)	3	6	12	24	48

c.

Hours (x)	0	1	2	3	4
Population (y)	0	5	20	45	80

10-2. Are the functions $y = 4(2^x)$ and $y = 8^x$ the same? Justify your answer.

10–3. In the year 2000, world population was estimated at 6.1 billion with an estimated growth rate of 1.3% per year. If this growth rate were to continue, the world population, P, in billions, could be estimated by the formula $P = 6.1(1.013)^t$, where t is the number of years since 2000.

a. Estimate the world population in 2020. (Note that it is believed that the growth rate is actually slowing.)

b. Determine the average rate of change of the population from 2000 to 2010, rounded to the nearest thousandth.

10–4. Newton's law of cooling (yes, that's the same Newton who studied gravity) says that if you take an object with initial temperature T_0 and put it into an environment where the surrounding temperature is T_s, then after a time t minutes its temperature, T, in degrees Fahrenheit, will be $T = T_s + (T_0 - T_s)0.82^t$. Suppose Newton took a cup of 170°F coffee outside and sat under an apple tree where the temperature was about 40°F. What will be the temperature of the coffee 5 minutes after Newton takes it outside?

Graphing Exponential Functions

The graphs of two exponential functions are shown in Figure 10.3. From the graphs and the rules of exponents, we can learn several things about exponential functions.

x	2^x	$\left(\frac{1}{2}\right)^x$
−3	$\frac{1}{8}$	8
−2	$\frac{1}{4}$	4
−1	$\frac{1}{2}$	2
0	1	1
1	2	$\frac{1}{2}$
2	4	$\frac{1}{4}$
3	8	$\frac{1}{8}$

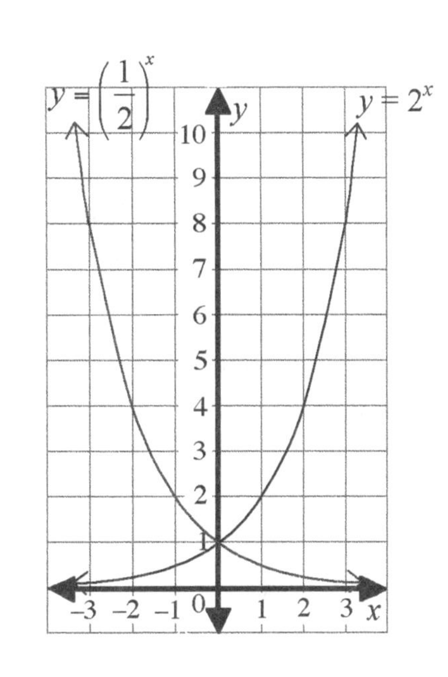

FIGURE 10.3

DOMAIN AND RANGE

An exponent can be any value of x: positive, zero, or negative. This means the domain for an exponential function is all real numbers $(-\infty, \infty)$. However, a positive base raised to any real number will never give 0 or a negative result. The range for an exponential function is limited to positive real numbers $(0, \infty)$.

y-INTERCEPT

From the graphs, it appears that all simple exponential functions pass through the point (0, 1). This is because any positive number raised to the 0 power is 1. Simple exponential functions have a *y*-intercept of 1.

ZEROS

Because b^x is positive for all real numbers x, a simple exponential function will never have a zero.

INCREASING OR DECREASING

When the base is greater than 1, exponential functions are increasing. When the base is between 0 and 1, exponential functions are decreasing.

END BEHAVIOR AND ASYMPTOTES

The end behavior of a simple exponential function depends on the base b. For $b > 1$, as x increases without bound, $x \to \infty$, y also increases without bound, $y \to \infty$. As x decreases without bound, $x \to -\infty$, y always decreases but never reaches 0, $y \to 0$. The left side of the graph has a horizontal asymptote at the x-axis, $y = 0$.

For $0 < b < 1$, the end behavior is reversed. As $x \to \infty$, y decreases toward 0 but never reaches 0. As $x \to -\infty$, y increases without bound. In this case, the right side of the graph has a horizontal asymptote at the x-axis, $y = 0$.

INVERSES

Because simple exponential functions are either always increasing or always decreasing, they are all one-to-one. This means all simple exponential functions will have inverse functions. These inverses, called logarithms, are the subject of the next chapter.

These properties of simple exponential functions are summarized on the following page.

THE NATURAL BASE e

Any positive number except 1 can be used as the base of an exponential function. Because we use a base 10 number system, 10 is a common base for exponential functions. Scientific notation is an example. However, another base, a number named e, appears so often in mathematics that it is called the natural base for exponential functions. Like π, the number e is irrational. The value of e is approximately 2.71828. While this certainly doesn't seem like a number you would "naturally" use for a base, it turns out that the natural exponential function $y = e^x$ has properties that help simplify many problems in math and science.

PROPERTIES OF EXPONENTIAL FUNCTIONS

$f(x) = b^x$, $b > 0$, $b \neq 1$

1. Domain: All real numbers $(-\infty, \infty)$
2. Range: All positive real numbers $(0, \infty)$
3. y-intercept: 1
4. Zeros: None
5. The function increases for $b > 1$
   As $x \to \infty$, $y \to \infty$
   As $x \to -\infty$, $y \to 0$
6. The function decreases for $0 < b < 1$
   As $x \to \infty$, $y \to 0$
   As $x \to -\infty$, $y \to \infty$
7. Horizontal asymptote: x-axis ($y = 0$)
8. The function is one-to-one; it has an inverse

EXAMPLE 10.7

Sketch the graph of $y = e^x$ over the interval $-2 \leq x \leq 2$.

SOLUTION

Use your calculator to make a table of values accurate to the nearest tenth.

x	e^x
–2	0.1
–1	0.4
0	1
1	2.7
2	7.4

Graph these points, and connect them with a smooth curve.

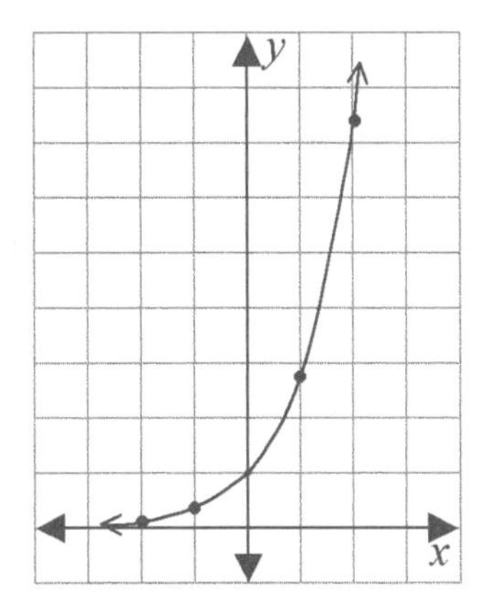

Transformations of Exponential Functions

Like all functions, simple exponential functions can be transformed by reflections, translations, and dilations. These transformations are covered in detail in Chapter 3. The following examples show how they apply to exponential functions.

a. Sketch the reflection of the graph of $y = e^x$ over the x-axis.
b. What is the equation of the graph from part a?
c. Sketch the reflection of the graph of $y = e^x$ over the y-axis.
d. What is the equation of the graph from part c?

SOLUTION

Refer back to Example 10.7 for a table of values for $y = e^x$ on the interval $[-2, 2]$. It has been graphed below for comparison.

a. To reflect over the x-axis, change the sign of the y-value of each of the points graphed for $y = e^x$. The reflected graph is labeled *a*.
b. The reflection of $y = f(x)$ over the x-axis is $y = -f(x)$. In this case, we get $y = -e^x$.
c. To reflect over the y-axis, change the sign of the x-value of each of the points graphed for $y = e^x$. The reflected graph is labeled *c*.
d. The reflection of $y = f(x)$ over the y-axis is $y = f(-x)$. In this case, we get $y = e^{-x}$.

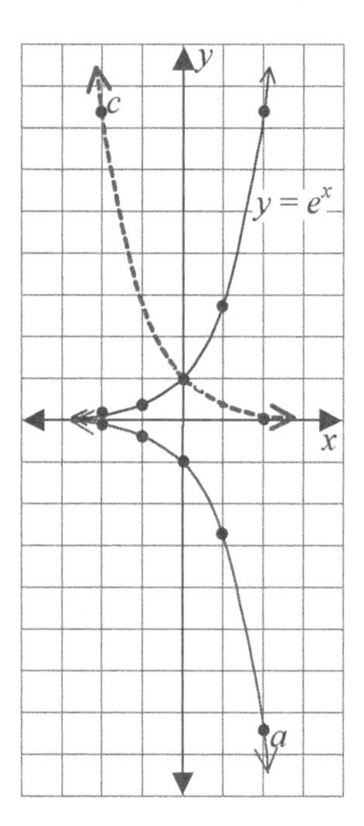

EXAMPLE 10.9

a. Sketch the graph of $y = e^x$ after a shift 2 units to the right.
b. What is the equation of the graph from part a?
c. Sketch the graph of $y = e^x$ after a shift 3 units down.
d. What is the equation of the graph from part c?

SOLUTION

a. Refer to the table of values made in Example 10.7. (The graph is shown below.) To shift the graph two units to the right, add 2 to each x-value. The translated graph is labeled a.
b. The translation of $y = f(x)$ h units to the right is $y = f(x - h)$. In this case, $h = 2$ and we get $y = e^{x-2}$.
c. To shift the graph 3 units down, subtract 3 from each y-value in the original table. The reflected graph is labeled c.
d. The translation of $y = f(x)$ k units down is $y = f(x) - k$. In this case, $k = 3$ and we get $y = e^x - 3$.

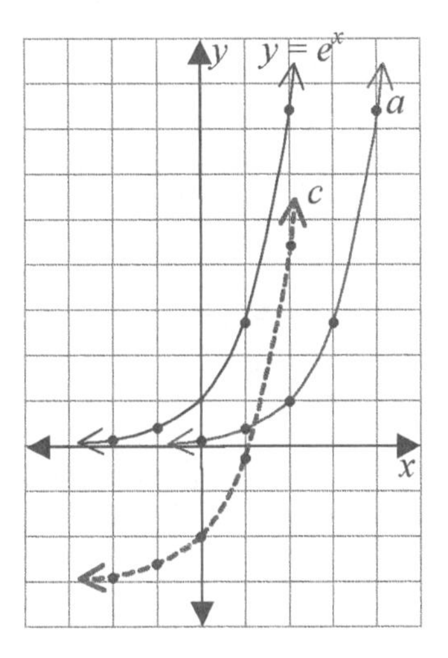

EXAMPLE 10.10

a. Sketch the graph of $y = e^x$ after a vertical dilation by a factor of 2.
b. What is the equation of the graph from part a?
c. Sketch the graph of $y = e^x$ after a horizontal dilation by a factor of 2.
d. What is the equation of the graph from part c?

SOLUTION

a. Refer again to the table of values made in Example 10.7. (The graph is shown below.) To stretch the graph vertically by a factor of 2, multiply each y-value by 2. The new graph is labeled a.
b. The dilation of $y = f(x)$ vertically by a factor of a is $y = af(x)$. In this case, $a = 2$ and we get $y = 2e^x$.

c. To dilate the graph horizontally by a factor of 2, multiply each x-value in the original table by 2. The new graph is labeled c.

d. The dilation of $y = f(x)$ horizontally by a factor of b is $y = f\left(\frac{x}{b}\right)$. In this case, $b = 2$ and we get $y = e^{\frac{x}{2}}$.

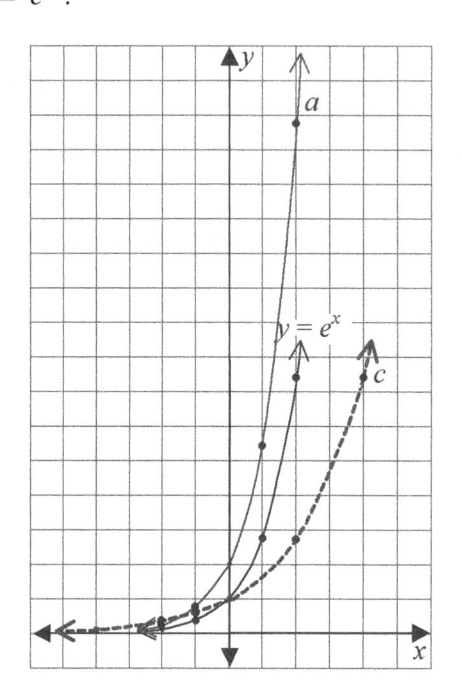

EXAMPLE 10.11 Describe the transformations on $y = 2^x$ represented in these functions. Use the given graph of $y = 2^x$ to sketch graphs of these functions.

a. $y = 2^x - 4$
b. $y = 2^{x-4}$
c. $y = 3(2^x)$
d. $y = 2^{\frac{x}{3}}$

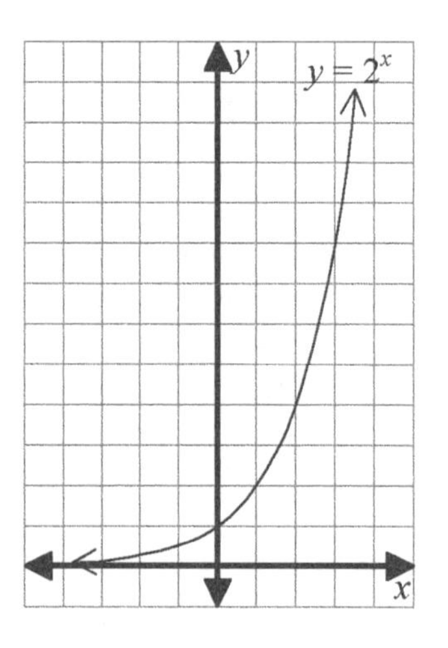

SOLUTION

a. This is a vertical translation down 4 units.
b. This is a horizontal translation right 4 units.
c. This is a vertical dilation by a factor of 3.
d. This is a horizontal dilation by a factor of 3.

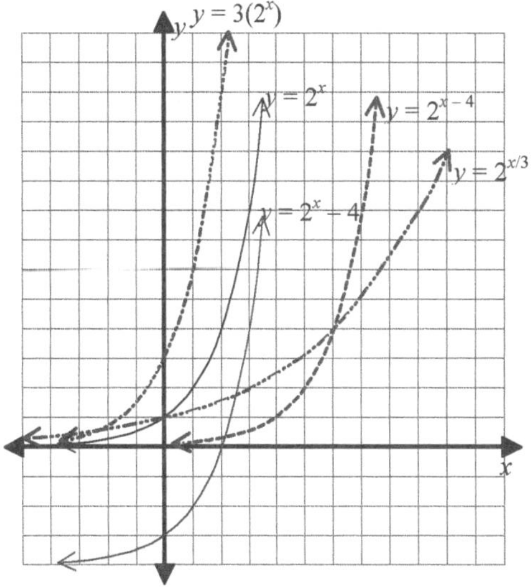

EXAMPLE 10.12 Find the horizontal asymptote and range of $y = e^{-x} + 3$.

SOLUTION

The graph of $y = e^{-x} + 3$ is the graph of $y = e^x$ after a reflection over the y-axis and a shift up 3 units.
The graph of $y = e^x$ has a horizontal asymptote at $y = 0$ and range of $(0, \infty)$.
Reflecting over the y-axis does not change the horizontal asymptote or range.
The shift up 3 units changes the horizontal asymptote to $y = 3$ and the range to $(3, \infty)$.

SEE CALC TIP 10A

SECTION EXERCISES

10–5. The function f is given by $f(x) = a^x$ $(a > 0;\ a \neq 1)$.

a. What is the domain of f?
b. What is the range of f?
c. What is the y-intercept of f?
d. What are the zero(s) of f?
e. For what values of a will the function be increasing?
f. For what values of a will the function be decreasing?
g. How will the graph of $y = a^{-x} = \left(\frac{1}{a}\right)^x$ compare with the graph of $y = a^x$?
h. How will the graph of $y = a^x - 2$ compare with the graph of $y = a^x$?
i. How will the graph of $y = a^{(x-2)}$ compare with the graph of $y = a^x$?
j. How will the graph of $y = 2a^x$ compare with the graph of $y = a^x$?

10–6. What is the range of the function $y = 2^x - 3$? What is the equation of its horizontal asymptote?

10–7. **a.** Sketch the graph of $y = 3^x$ over the interval $-2 \leq x \leq 2$.
b. From your graph, estimate the value of $3^{1.5}$
c. Sketch the reflection of the graph from part *a* over the line $y = x$.

10–8. **a.** Sketch the graph of $y = \left(\frac{9}{4}\right)^x$ on the interval $-3 \leq x \leq 3$.
b. From your graph, estimate the value of $\left(\frac{9}{4}\right)^{\frac{2}{3}}$.
c. From your graph, estimate the solution to the equation $\left(\frac{9}{4}\right)^x = 6$.

Solving Exponential Equations with Common Bases

An *exponential equation* is an equation with the variable in an exponent. The equations $9^x = 27$, $2^{x+2} = 2^{4x-7}$, and $5^{2x} = 25^{3x-4}$ are all exponential equations. The equations $x^6 = 64$ and $(3x - 2)^{3/4} = 8$, although they contain exponents, are not exponential equations. (These equations are covered in Chapters 7 and 8.)

There are two main ways to solve exponential equations algebraically. In this chapter, we will discuss using common bases. The method does not work for all exponential equations. The second method uses logarithms, which is the subject of the next chapter.

USING COMMON BASES

If you had to solve $3^x = 3^5$, your common sense would say $x = 5$ is the solution since $3^5 = 3^5$.

COMMON BASES

$b^x = b^y$ if and only if $x = y$

EXAMPLE 10.13

Solve: $2^{x+2} = 2^{4x-7}$

SOLUTION

Since the bases are the same, the exponents must be equal:

$$2^{x+2} = 2^{4x-7}$$
$$x + 2 = 4x - 7$$
$$x = 3$$

Check:
$$2^{3+2} = 2^{4(3)-7}$$
$$2^5 = 2^5$$
$$32 = 32$$

With help from the rules of exponents, this method for solving exponential equations can be used for any equation where both sides can be expressed in terms of the same base.

SOLVING EXPONENTIAL EQUATIONS WITH COMMON BASES

1. Rewrite the equation (if necessary) so both sides have the same base.
2. Set the exponents equal to each other.
3. Solve.

Solve: $3^{2x} = 9^{3x-4}$

SOLUTION

Since $9 = 3^2$, the equation can be rewritten using only base 3:

$3^{2x} = 9^{3x-4}$	
$3^{2x} = (3^2)^{3x-4}$	Substitute 3^2 for 9. Use parentheses.
$3^{2x} = 3^{2(3x-4)}$	Use the power property for exponents, and multiply the exponents on the right side.
$3^{2x} = 3^{6x-8}$	Distribute 2 through the binomial.
$2x = 6x - 8$	The bases are equal, so set the exponents equal to each other.
$8 = 4x$	Solve for x.
$x = 2$	

Check:

$$3^{2(2)} = 9^{3(2)-4}$$
$$3^4 = 9^2$$
$$81 = 81$$

EXAMPLE 10.15

Solve: $\left(\frac{1}{4}\right)^{x+2} = 8^{\frac{2x}{3}}$

SOLUTION

Both $\frac{1}{4}$ and 8 can be expressed as powers of 2: $\frac{1}{4} = 2^{-2}$ and $8 = 2^3$.

$$\left(\frac{1}{4}\right)^{x+2} = 8^{\frac{2x}{3}}$$

$\left(2^{-2}\right)^{x+2} = \left(2^3\right)^{\frac{2x}{3}}$ Substitute $\frac{1}{4} = 2^{-2}$ on the left side and $8 = 2^3$ on the right side.

$2^{-2(x+2)} = 2^{3\left(\frac{2x}{3}\right)}$ Use the power property, and multiply the exponents.

$2^{-2x-4} = 2^{2x}$	Simplify the exponent expressions.
$-2x - 4 = 2x$	The bases are equal, so set the exponents equal to each other.
$-4 = 4x$	Solve for x.
$x = -1$	

SEE CALC TIP 10B

Check: $\left(\frac{1}{4}\right)^{-1+2} = 8^{\frac{2(-1)}{3}}$

$$\left(\frac{1}{4}\right)^{1} = 8^{-\frac{2}{3}}$$

$$\frac{1}{4} = \frac{1}{\left(\sqrt[3]{8}\right)^{2}}$$

$$\frac{1}{4} = \frac{1}{4}$$

EXAMPLE 10.16 Solve: $10^x = 5$

SOLUTION

It is not possible to express both 5 and 10 as integer powers of the same base. So the method of common bases is not going to be as convenient for this problem as it was for the previous two examples. That doesn't mean the method doesn't work. It turns out that $5 \approx 10^{0.69897}$.

$$10^x = 5$$
$$10^x \approx 10^{0.69897}$$
$$x \approx 0.69897$$

The obvious question is how did we get that 0.69897? The answer is we got it by using logarithms, which is the subject of the next chapter.

SECTION EXERCISES

In problems 9–14, solve for x.

10–9. $2^{x+1} = 8$

10–10. $4^{2x} = 2^{3x+2}$

10–11. $3^{x-4} = 1$

10–12. $16^{x+2} = 8^{-x}$

10–13. $27^{x+3} = 9^{x^2}$

10–14. $4^{3x-4} = \left(\frac{1}{8}\right)^{x-1}$

Solve for x *and* y.

10–15. $\begin{cases} 2^y = 8^x \\ 3^y = 3^{x+4} \end{cases}$

CHAPTER EXERCISES

C10–1. Identify the following function as linear, exponential, or neither.

Hours (x)	0	1	2	3	4
Population (y)	1	2/3	4/9	8/27	16/81

C10–2. Suppose you bought a compact car five years ago for $16,000 and it depreciated 15% per year. What is it worth today?

C10–3. The interest formula for interest compounded continuously is $A = Pe^{rt}$, where A is the value of the investment, P is the principal, r is the annual rate expressed as as a decimal, and t is the number of years.

a. If you deposit $1,000 in an account that gives 5% annual interest compounded continuously, how much will you have after four years?

b. What is the average rate of change of the value of the account over the first four years? How does this compare to the 5% annual interest rate?

C10–4. **a.** Sketch the graph of $y = \left(\frac{2}{3}\right)^x$ over the interval $-3 \le x \le 3$.

b. Sketch the reflection of the graph from part a over the x-axis.

c. What is the equation of the graph from part b?

d. Sketch the reflection of the graph from part a over the y-axis.

e. What is the equation of the graph from part d?

C10–5. The graph of the function $y = 2^x$ is shown below.

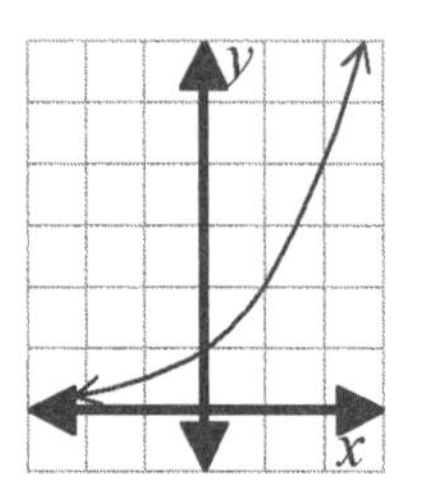

Each of the following graphs shows a transformation of $y = 2^x$. Write an appropriate equation for each.

a.

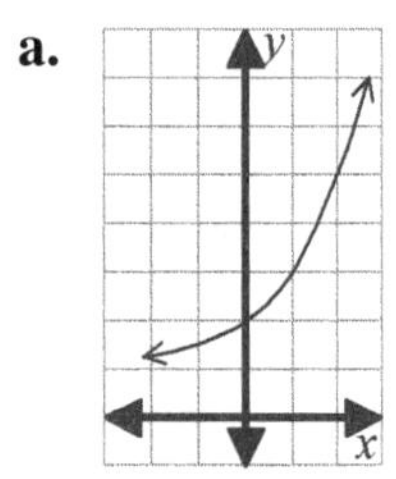

b.

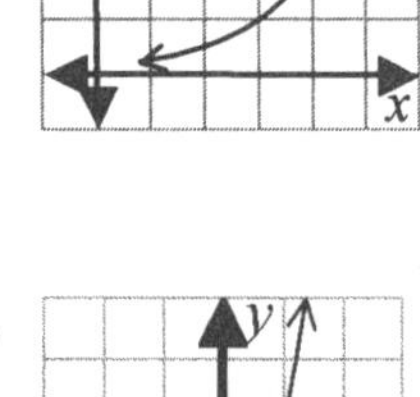

c.

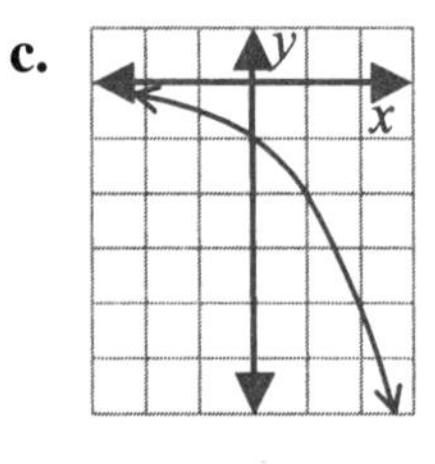

d.

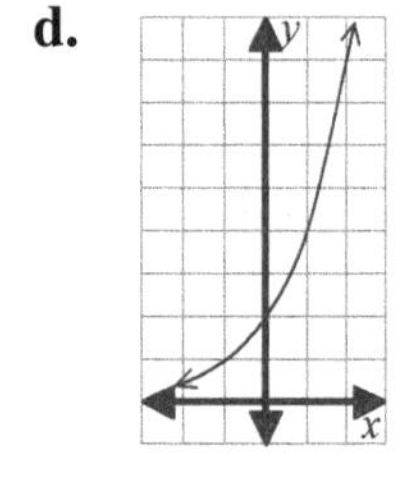

e.

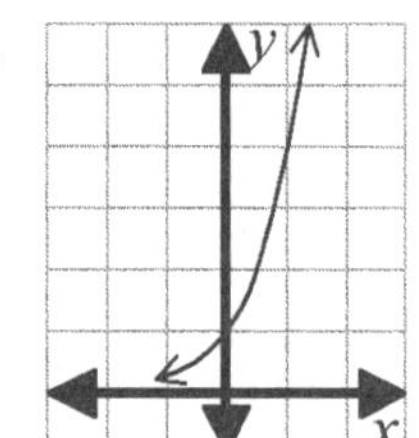

In problems 6–9, solve for x.

C10–6. $3^x = \frac{1}{9}$

C10–7. $2^{x(x-5)} = 64$

C10–8. $81^{x+2} = 27^{5x+4}$

C10–9. $\left(\frac{1}{2}\right)^{2x} = 16^{x+3}$

Logarithmic Functions

WHAT YOU WILL LEARN

Logarithms are one of the most challenging topics in Algebra 2. Just the name seems strange and unfamiliar. While you read this chapter, remember that logarithms, usually called just logs, are exponents. The rules for logs are directly related to the rules for exponents. In this chapter, you will learn to:

- Find the inverse of exponential functions;
- Apply logarithm rules to simplify expressions;
- Graph logarithmic functions;
- Solve exponential and logarithmic equations;
- Find logarithms using any base.

SECTIONS IN THIS CHAPTER

- Logarithmic Models
- Inverses of Exponential Functions
- Logarithm Rules
- Graphing Logarithmic Functions
- Solving Exponential and Logarithmic Equations
- Change of Base Formula

Logarithmic Models

The inverse of an exponential function is called a logarithmic function. As suggested at the end of the previous chapter, logarithms are helpful for solving exponential equations. However, logarithmic functions are also useful for measuring certain quantities and for modeling some real-life phenomena. You may know that sound levels are measured in decibels. Sound level in decibels is a logarithmic function of the intensity of the sound in watts per square meter. Earthquake magnitude is another familiar example of a logarithmic measurement scale. Who has not seen a TV mystery where authorities estimate the time of death based on the temperature of the victim's body? These estimates are based on logarithmic models of cooling times.

The table in Figure 11.1 shows the range of sound pressures you might normally expect to encounter. The lowest one in the table, 2×10^{-5} pascals, corresponds roughly to the quietest sound a typical person can hear. As you move down the table, the numbers represent, in order, a soft whisper, normal conversational speaking, a busy road, loud power tools, the front row at a rock concert, and finally a jet engine at 100 meters. Note the pressures have a range of seven orders of magnitude; the jet engine is about 10 million times as loud as the softest sound that can be heard. To avoid dealing with such a wide range of values, engineers use logarithms to convert the sound pressures to decibels. With decibels, everyday sound levels can be expressed on a scale of 0 to about 140.

The graph in Figure 11.1 is typical of the graphs of logarithmic functions. One of the characteristic features is that except near the vertical asymptote, y increases very slowly. To get an idea of just how slowly, note that the graph covers only the first two rows of data in the table, the threshold of hearing and a soft whisper. On the same scale, the conversational voice would be graphed about 13 feet to the right of the vertical axis and 1.5 inches above the horizontal axis. The jet engine would be about 25 miles to the right of the vertical axis but still only about 3.5 inches above the horizontal axis.

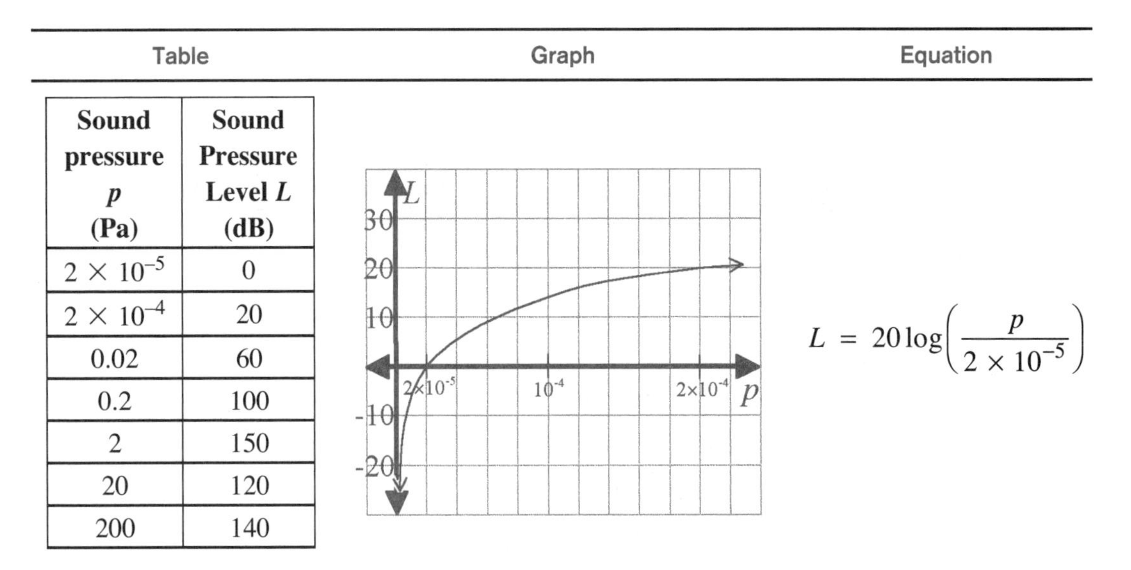

Table	Graph	Equation

Sound pressure p (Pa)	Sound Pressure Level L (dB)
2×10^{-5}	0
2×10^{-4}	20
0.02	60
0.2	100
2	150
20	120
200	140

$$L = 20\log\left(\frac{p}{2 \times 10^{-5}}\right)$$

FIGURE 11.1

Inverses of Exponential Functions

Any exponential function, $y = b^x$, where $b > 0$ and $b \neq 1$, is a one-to-one function and so must have an inverse function. What is this function? Recall that to find the inverse of a function, we switch x and y and then solve for y. Let's review some really basic examples.

TABLE 11.1
FINDING INVERSES

Function	Inverse	Description
$y = x + 3$	$x = y + 3 \rightarrow y = x - 3$	The inverse of addition is subtraction.
$y = 3x$	$x = 3y \rightarrow y = \dfrac{x}{3}$	The inverse of multiplication is division.
$y = x^3$	$x = y^3 \rightarrow y = \sqrt[3]{x}$	The inverse of cubing is taking the cube root.
$y = 3^x$	$x = 3^y \rightarrow y = ?$	The inverse of raising to an exponent is what?

We need a new function to describe the inverse of an exponential function. It is called a *logarithmic function*. In Table 11.1, we want a logarithmic function with base 3 for the inverse of $y = 3^x$. This function is written $y = \log_3 x$ and is spoken "y is the logarithm base 3 of x," which is often shortened to "log base 3 of x." We can have a logarithmic function for any positive base $b \neq 1$.

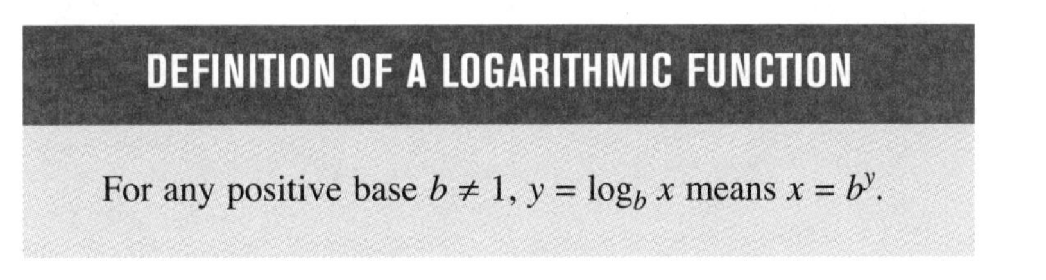

DEFINITION OF A LOGARITHMIC FUNCTION

For any positive base $b \neq 1$, $y = \log_b x$ means $x = b^y$.

This definition says that a logarithm is an exponent. In particular, the logarithm base b of x, $\log_b x$, is the exponent that goes on base b to give x, as shown in Figure 11.2.

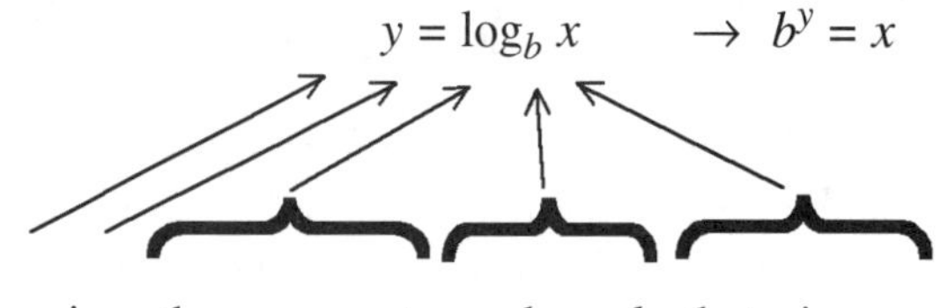

y is the exponent on base b that gives x

FIGURE 11.2

Since $y = \log_b x$ is equivalent to $x = b^y$ and $x = b^y$ is the inverse of $y = b^x$, we see that the inverse of an exponential function is a logarithmic function with the same base.

$$\text{If } f(x) = b^x \text{, then } f^{-1}(x) = \log_b x.$$

An exponential function has domain all real x and range $y > 0$. When finding an inverse function, domain and range are switched. So a logarithmic function has domain $x > 0$ and range all real y.

EXAMPLE 11.1

Rewrite each of the following in equivalent logarithmic form.

a. $2^5 = 32$
b. $10^x = N$
c. $A = e^t$

SOLUTION

a. Think "5 is the exponent on base 2 that makes 32."

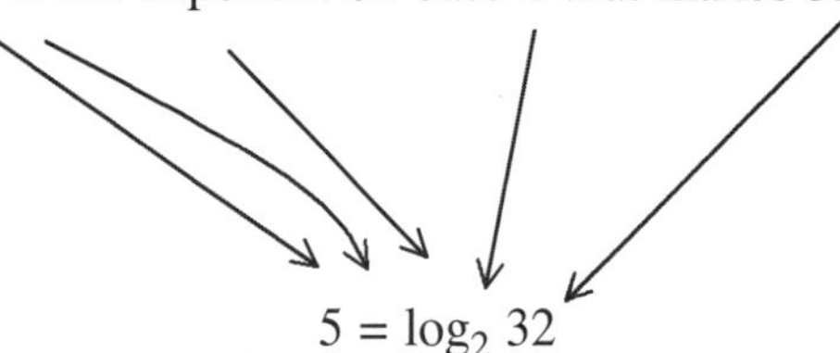

b. Think "x is the exponent on base 10 that makes N": $x = \log_{10} N$. Because we use a base 10 number system, base 10 is one of the most commonly used bases for logarithms. Base 10 logarithms are called *common logarithms* and are often written without the subscript 10; $\log N$ is understood to mean $\log_{10} N$. The answer to this question is usually written $x = \log N$.

c. Think "t is the exponent on base e that makes A": $t = \log_e A$. Recall that because of some special properties it has, e is called the natural base for the exponential function. Similarly, logarithms with base e are called *natural logarithms* and are usually abbreviated ln; $\ln A$ means $\log_e A$. So the answer to this question is usually written $t = \ln A$.

EXAMPLE 11.2

Rewrite each of the following in equivalent exponential form.

a. $t = \log_a z$

b. $\log 1{,}000 = 3$

c. $\ln k = 0.7$

SOLUTION

a. In words, it says "t is the exponent that goes on base a to make z; a is the base, t is the exponent, and z is the answer." So $a^t = z$.

b. Remember, log without an explicitly written base means base 10. So 10 is the base, 3 is the exponent, and 1,000 is the answer: $10^3 = 1{,}000$.

c. The symbol ln means base e; 0.7 is the exponent and k is the result: $e^{0.7} = k$.

EXAMPLE 11.3

Write the *inverse* of each function.

a. $f(x) = 2^x$

b. $f(x) = \log_8 x$

SOLUTION

a. The inverse of an exponential function is a logarithmic function with the same base. So the inverse of $f(x) = 2^x$ is $f^{-1}(x) = \log_2 x$.

b. The inverse of a logarithmic function is an exponential function with the same base. So the inverse of $f(x) = \log_8 x$ is $f^{-1}(x) = 8^x$.

THE DEFINING PROPERTY OF INVERSES

If f^{-1} is the inverse of f:

- $f^{-1}(f(x)) = x$ for all x in the domain of f
- $f\left(f^{-1}(x)\right) = x$ for all x in the domain of f^{-1} (the range of f)

For exponential and logarithmic functions, this means:

$$\log_b b^x = x \text{ for all real } x$$

$$b^{\log_b x} = x \text{ for all } x > 0$$

SECTION EXERCISES

For 1–2, rewrite each of the following as an equivalent log equation.

11–1. $y = 4^x$

11–2. $N = b^k$

For 3–4, rewrite each of the following as an equivalent exponential equation.

11–3. $y = \log_2 x$

11–4. $\log_a x = k$

For 5–6, write the inverse of each of the following functions.

11–5. $f(x) = 6^x$

11–6. $f(x) = \ln x$

For 7–8, simplify each of the following.

11–7. $2^{\log_2 5}$

11–8. $\ln e^{-2x}$

Logarithm Rules

Since logarithmic functions are inverses of exponential functions, it should come as no surprise that each property of exponents has a corresponding property of logarithms. For example, since $b^0 = 1$, we have $\log_b 1 = 0$. This and other important properties of logarithms are listed in Table 11.2. Anyone who wants to understand and use logarithms needs to know these properties literally forward and backward.

TABLE 11.2
PROPERTIES OF EXPONENTS AND LOGARITHMS

Properties of Exponents	Properties of Logarithms	Name of Property
$b^0 = 1$	$\log_b 1 = 0$	Log of 1
$b^1 = b$	$\log_b b = 1$	Log of the base
$b^{\log_b x} = x$	$\log_b b^x = x$	Inverse property
$b^x b^y = b^{x+y}$	$\log_b (xy) = \log_b x + \log_b y$	Product property
$\dfrac{b^x}{b^y} = b^{x-y}$	$\log_b \left(\dfrac{x}{y}\right) = \log_b x - \log_b y$	Quotient property
$\left(b^x\right)^n = b^{nx}$	$\log_b x^n = n\log_b x$	Power property

When working with expressions and equations involving logarithms, it is often helpful to use the properties of logarithms to rewrite expressions in more convenient forms. Depending on the problem, we might want to take a logarithm of a complicated expression and expand it as much as possible into logarithms of simpler terms. Instead, we might want to do the reverse and combine several logarithmic expressions into a single logarithm.

EXAMPLE 11.4 Assuming x, y, and z are all positive, expand $\log_b\left(\frac{xy^3}{\sqrt{z}}\right)$ as much as possible.

SOLUTION

When expanding logarithms, it is usually best to apply the quotient property first, followed by the product property and then the power property. When applying the power property, remember that radicals can be written in exponent form.

$\log_b\left(\frac{xy^3}{\sqrt{z}}\right) = \log_b\left(xy^3\right) - \log_b\sqrt{z}$	Apply the quotient property.
$= \log_b x + \log_b y^3 - \log_b\sqrt{z}$	Apply the product property.
$= \log_b x + \log_b y^3 - \log_b z^{\frac{1}{2}}$	Rewrite the radical in exponent form.
$= \log_b x + 3\log_b y - \frac{1}{2}\log_b z$	Apply the power property.

EXAMPLE 11.5 Expand $\log 100x^5$ as much as possible.

SOLUTION

$\log 100x^5 = \log 100 + \log x^5$	Apply the product property.
$= \log 10^2 + \log x^5$	Because this is a base 10 log, rewrite 100 as 10^2.
$= 2\log 10 + 5\log x$	Apply the power property.
$= 2 + 5\log x$	Substitute $\log 10 = 1$.

EXAMPLE 11.6 If $x > 0$, expand $\ln\left(e^{-kx}\sqrt{x^2+1}\right)$ as much as possible.

SOLUTION

$$\ln\left(e^{-kx}\sqrt{x^2+1}\right) = \ln e^{-kx} + \ln\sqrt{x^2+1} \quad \text{Apply the product property.}$$

$$= \ln e^{-kx} + \ln\left(x^2+1\right)^{\frac{1}{2}} \quad \text{Rewrite the radical in exponent form.}$$

$$= -kx\ln e + \frac{1}{2}\ln\left(x^2+1\right) \quad \text{Apply the power property.}$$

$$= -kx + \frac{1}{2}\ln\left(x^2+1\right) \quad \text{Substitute } \ln e = 1.$$

Note that this is as far as we can go. There is no property that allows us to expand the logarithm of a sum.

COMMON ERROR

$$\ln\left(x^2+1\right) \neq \ln x^2 + \ln 1$$

Logs cannot be distributed over addition or subtraction.

EXAMPLE 11.7 Write as a single logarithm: $\frac{1}{3}\log x + 2\log(x-1) - \log(x+2)$

SOLUTION

In general, when combining several logarithmic expressions into one, apply the power property first and then the product and quotient properties.

$$\frac{1}{3}\log x + 2\log(x-1) - \log(x+2)$$

$$\log x^{\frac{1}{3}} + \log(x-1)^2 - \log(x+2) \quad \text{Apply the power property.}$$

$$\log\left(x^{\frac{1}{3}}(x-1)^2\right) - \log(x+2) \quad \text{Apply the product property.}$$

$$\log\left(\frac{x^{\frac{1}{3}}(x-1)^2}{(x+2)}\right) \quad \text{Apply the quotient property.}$$

EXAMPLE 11.8 Write as a single logarithm: $3\ln x - \frac{1}{2}(\ln(x-1) + \ln(x+1))$

SOLUTION

Simplify inside the parentheses first. Then do the rest of the problem.

$$3\ln x - \frac{1}{2}(\ln(x-1) + \ln(x+1))$$

$$3\ln x - \frac{1}{2}\ln((x-1)(x+1))$$ Apply the product property.

$$3\ln x - \frac{1}{2}\ln\left(x^2 - 1\right)$$ Multiply.

$$\ln x^3 - \ln\left(x^2 - 1\right)^{\frac{1}{2}}$$ Apply the power property.

$$\ln x^3 - \ln\sqrt{x^2 - 1}$$ Rewrite the exponent in radical form.

$$\ln\left(\frac{x^3}{\sqrt{x^2-1}}\right)$$ Apply the quotient rule.

COMMON ERROR

$$\sqrt{x^2 - 1} \neq \sqrt{x^2} - \sqrt{1}$$

Radicals do not distribute over addition or subtraction.

EXAMPLE 11.9 Write as a single logarithmic function: $\frac{1}{2}\log x + 3$

SOLUTION

The 3 looks tricky because it is not obviously a logarithm. However, we can rewrite 3 as $\log 10^3$, using the same base as the log. After that, the rest of the problem should be easy.

$$\frac{1}{2}\log x + 3$$

$$\frac{1}{2}\log x + \log 10^3$$ Substitute $3 = \log 10^3$.

$$\log x^{\frac{1}{2}} + \log 10^3$$ Apply the power property.

$$\log\left(10^3 x^{\frac{1}{2}}\right)$$ Apply the product property.

$$\log\left(1{,}000\sqrt{x}\right)$$ Rewrite the exponent in radical form, and substitute $10^3 = 1{,}000$

EXAMPLE 11.10 Write as a single log: $\dfrac{\log x}{\log y}$

SOLUTION

This is a trick question, but it is an important one. The quotient property of logarithms applies to the logarithm of a quotient, $\log\left(\dfrac{x}{y}\right)$. There is no property dealing with the quotient of two logarithms. This problem cannot be further simplified.

COMMON ERROR

Don't rewrite $\dfrac{\log x}{\log y}$ as $\log\left(\dfrac{x}{y}\right)$ or as $\log x - \log y$. Don't even think about canceling the logs to get $\dfrac{x}{y}$.

SECTION EXERCISES

For 9–12, use the properties of logarithms to expand each of the following as much as possible.

11–9. $\log\left(\dfrac{x^3}{yz^2}\right)$

11–10. $\log(xy)^3$

11–11. $\log(1{,}000x^3)$

11–12. $\ln\left(\dfrac{x^2}{\sqrt{x^2 - 1}}\right)$

For 13–16, express each of the following as the log of a single quantity.

11–13. $2\ln x - \ln 3$

11–14. $\dfrac{1}{2}\log x + 2\log(x - 1)$

11–15. $3\log x + \dfrac{1}{2}\log y - 1$

11–16. $2(\log a + \log b) - \log c$

11–17. If you know only the values of $\log_b 2$ and $\log_b 3$, which of the following cannot be evaluated?

(1) $\log_b 6$ (2) $\log_b 7$ (3) $\log_b 8$ (4) $\log_b 9$

11–18. If $\log_b x = 0.8$ and $\log_b y = 0.6$, evaluate the following:

a. $\log_b x^2$

b. $\log_b \dfrac{1}{x\sqrt[3]{y}}$

Graphing Logarithmic Functions

Since a logarithmic function is the inverse of an exponential function, the graph of a logarithmic function will be the reflection of the graph of an exponential function over the line $y = x$. Figure 11.3 shows the function $f(x) = 2^x$ and its inverse $f^{-1}(x) = \log_2 x$. All logarithmic functions with base greater than 1 will look similar to the graph of $y = \log_2 x$. For bases between 0 and 1, the graph of $y = \log_b x$ will appear to be reflected over the x-axis compared with the graph of $y = \log_2 x$. Logarithms with bases between 0 and 1 are rarely used.

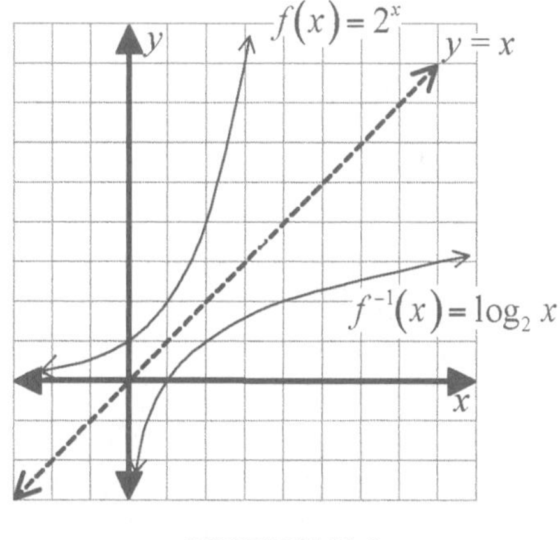

FIGURE 11.3

Each of the basic features of the exponential function leads to a corresponding feature of the logarithmic function. These are summarized in Table 11.3.

TABLE 11.3

FEATURES OF EXPONENTIAL AND LOGARITHMIC FUNCTIONS

	Exponential Function $y = b^x$, $b > 1$	Logarithmic Function $y = \log_b x$, $b > 1$
Domain	All real x	$x > 0$
Range	$y > 0$	All real y
y-intercept	1	None
Zeros	None	1
Horizontal Asymptote	$y = 0$ (x-axis)	None
Vertical Asymptote	None	$x = 0$ (y-axis)

Notice how any feature of x or y, such as domain and range, is switched between the exponential and logarithmic functions.

Simple logarithmic functions can be transformed by reflections, translations, and dilations. For a detailed discussion of transformations of functions, see Chapter 3. The following examples show how transformations apply to logarithmic functions.

EXAMPLE 11.11 The graph of $y = \log_2 x$ is shown along with four simple transformations labeled a–d. Identify each transformation, and write its equation.

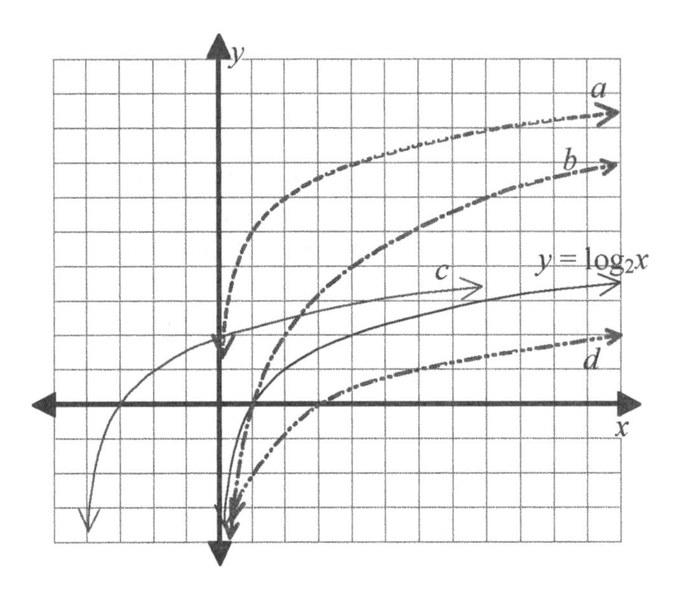

SOLUTION

a. This appears to be a vertical translation up 5 units. Note that $y = \log_2 x$ includes the points (1, 0), (2, 1), (4, 2), and (8, 3) while graph a passes through the points (1, 5), (2, 6), (4, 7), and (8, 8). To translate a graph vertically k units, we add k to the function. The equation of graph a is $y = \log_2 x + 5$.

b. Note that the zero at 1 is unchanged. This is a vertical dilation. The original graph passes through (2, 1); graph b passes through (2, 2). The dilation factor is 2. To dilate the graph of a function vertically by a, we multiply the function by a. The equation of graph b is $y = 2\log_2 x$.

c. The zero has moved from 1 to –3, left 4 units; this appears to be a horizontal translation. As confirmation, note that the point (2, 1) on the original graph has moved to (–2, 1) on graph c. To translate the graph of $y = f(x)$ to the left h, we change the equation to $y = f(x + h)$. The equation of graph c is $y = \log_2 (x + 4)$.

d. The zero has moved from 1 to 3, which is right 2 units. However, the point (2, 1) did not move to (4, 1); this is not another horizontal translation. Instead, consider a horizontal dilation. The zero changed by a factor of 3. Under the same dilation, the point (2, 1) should have an image at (6, 1). Graph d does pass through (6, 1), confirming that this is a dilation. To dilate the graph of a function horizontally by a factor of b, we change the equation to $y = f\left(\frac{x}{b}\right)$. The equation of graph d is $y = \log_2\left(\frac{x}{3}\right)$.

EXAMPLE 11.12

Sketch the graphs.

a. $f(x) = \log_3 x$

b. $f(x) = \log_3 (x + 1)$

c. $f(x) = \log_3 x - 2$

d. $f(x) = -\log_3 x$

SOLUTION

a. One way to graph a simple logarithmic function is to make a table for its inverse and then switch all the x-values and y-values. The inverse of $y = \log_3 x$ is $y = 3^x$.

x	$y = 3^x$
−2	0.111
−1	0.333
0	1
1	3
2	9

→

x	$y = \log_3 x$
0.111	−2
0.333	−1
1	0
3	1
9	2

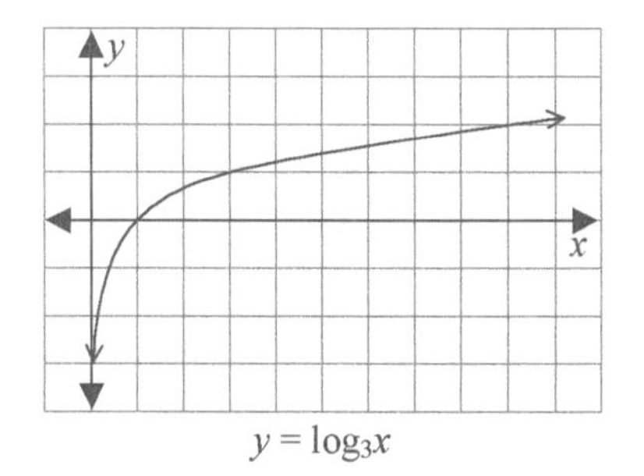

$y = \log_3 x$

b. $f(x) = \log_3 (x + 1)$ is a translation of $f(x) = \log_3 x$ one unit to the left. Simply subtract 1 from each of the x-values in the second table in part a.

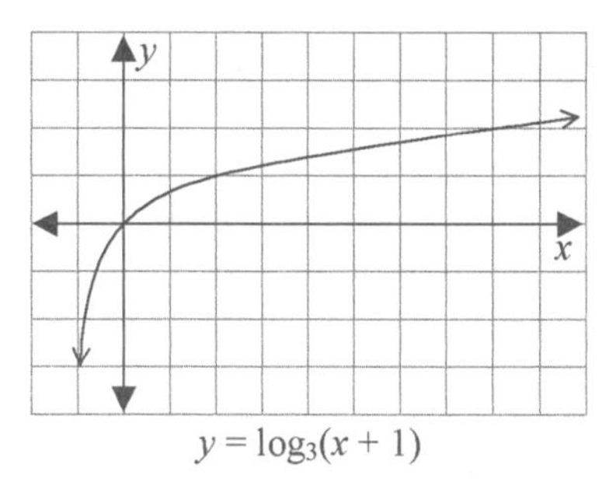

$y = \log_3(x + 1)$

c. $f(x) = \log_3 x - 2$ is a translation of $f(x) = \log_3 x$ down by 2 units. Subtract 2 from each of the y-values in the second table in part a.

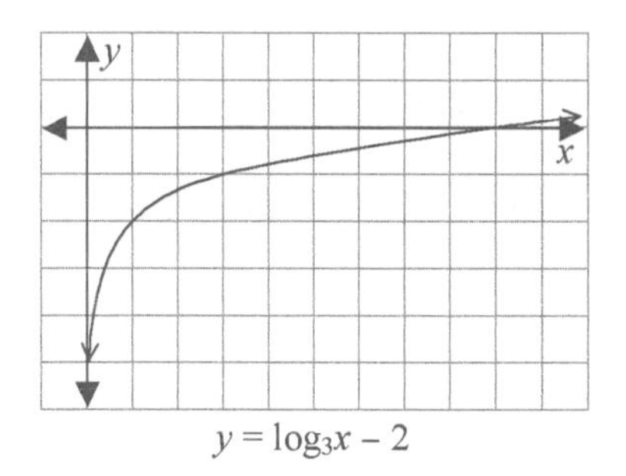

$y = \log_3 x - 2$

d. $f(x) = -\log_3 x$ is a reflection of $f(x) = \log_3 x$ over the x-axis. Negate each y-value in the second table in part a.

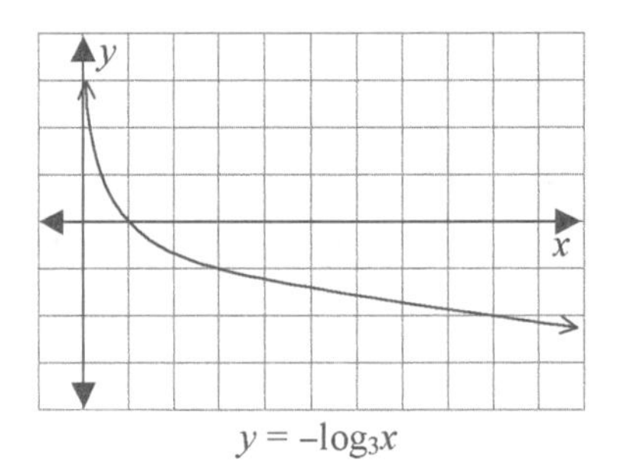

$y = -\log_3 x$

EXAMPLE 11.13

For $f(x) = \ln(x - 3) + 1$, identify the following:

a. The domain

b. The range

c. The vertical asymptote

SOLUTION

The graph of $f(x) = \ln(x - 3) + 1$ is the graph of $f(x) = \ln x$ moved right 3 units and up 1 unit.

SEE CALC TIP 11A

a. Domain: The domain of $f(x) = \ln x$ is $x > 0$. After moving the graph to the right 3 units, the domain for $f(x) = \ln(x - 3) + 1$ is $x > 3$.

b. Range: The range of $f(x) = \ln x$ is all real numbers. Moving right and up will not change the range. The range for $f(x) = \ln(x - 3) + 1$ is all real numbers.

c. Vertical asymptote: $f(x) = \ln x$ has a vertical asymptote at $x = 0$. After moving the graph to the right 3 units, the vertical asymptote for $f(x) = \ln(x - 3) + 1$ is $x = 3$.

You can find the domain and vertical asymptote algebraically without a graph of the function. In Chapter 3, we defined the natural domain of a function to be all real values of x for which the function was both defined and real. This led to two algebraic rules for finding the domain of a function.

1. The denominator cannot equal 0.
2. The radicand (of even index root) must be nonnegative.

With the introduction of logarithmic functions, we need to extend these rules. For any real base, the logarithm of 0 is undefined and the logarithm of a negative number is imaginary. Therefore, we must add the following rule.

3. The argument of a logarithm must be positive.

EXAMPLE 11.14

For the function, $y = \log(2x - 3)$, find the following:

a. The domain

b. The equation of the vertical asymptote of the function

SOLUTION

a. In this example, the first two domain rules pose no problem; there is no denominator or radical. The third rule says the argument, $2x - 3$, must be positive. Write this as an inequality, and solve for x.

$$2x - 3 > 0$$
$$2x > 3$$
$$x > 1.5$$

The domain for this function is $\{x \mid x > 1.5\}$ or, in interval notation, $(1.5, \infty)$.

b. To find the vertical asymptote for any logarithmic function, set the argument equal to 0 and solve for x.

$$2x - 3 = 0$$
$$2x = 3$$
$$x = 1.5$$

For $y = \log(2x - 3)$, the vertical asymptote is $x = 1.5$. The vertical asymptote for this example is at the boundary of the domain. This will be true whenever the argument is a linear function, $ax + b$, which includes almost all the problems you are likely to see in Algebra 2.

For the function, $y = \log_2\left(\frac{x}{4} + 8\right) - 4$, find the following:

a. The domain

b. The equation of the vertical asymptote of the function

c. The y-intercept

SOLUTION

a. Again, the first two domain rules pose no problem. The denominator is a constant, and there is no radical. The third rule says the argument, $\frac{x}{4} + 8$, must be positive. Write this as an inequality, and solve for x.

$$\frac{x}{4} + 8 > 0$$
$$\frac{x}{4} > -8$$
$$x > -32$$

The domain for this function is $\{x \mid x > -32\}$ or, in interval notation, $(-32, \infty)$.

b. To find the vertical asymptote for any logarithmic function, set the argument equal to 0 and solve for x.

$$\frac{x}{4} + 8 = 0$$

$$\frac{x}{4} = -8$$

$$x = -32$$

For $y = \log_2\left(\frac{x}{4} + 8\right) - 4$, the vertical asymptote is $x = -32$.

c. Remember that to find the y-intercept for any function, substitute $x = 0$ and evaluate.

$$\begin{aligned} y &= \log_2\left(\frac{0}{4} + 8\right) - 4 \\ &= \log_2 8 - 4 \\ &= 3 - 4 \\ &= -1 \end{aligned}$$

The y-intercept is -1.

SECTION EXERCISES

11–19. The graph of $y = \log_3 x$ is shown below. The following graphs all show transformations (reflections and translations, no dilations) of the graph of $y = \log_3 x$. Label each graph with its correct equation.

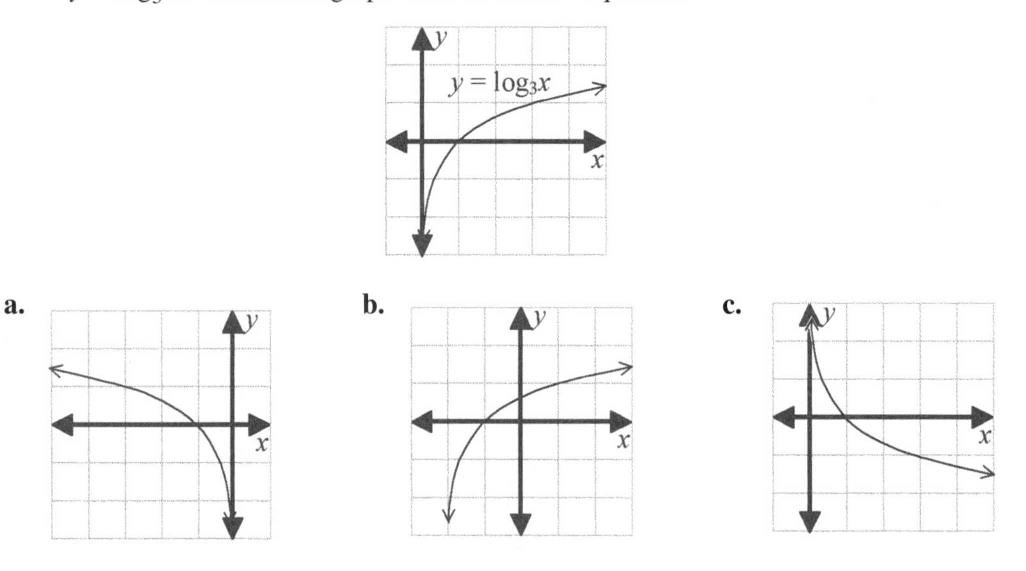

For 20–21, identify the domain, range, asymptotes, and intercepts. Sketch the following. Try to do these without *using your calculator.*

11–20. $f(x) = \log_4 (x - 2)$

11–21. $f(x) = \ln(8 - x)$

11–22. The graphs of the functions $y = \log_2 x$ and $y = \log_3 x$ intersect at what point?

11–23. If $\log x < 0$, what do we know about x?

11–24. Which point is not on the graph of $y = \log_4 x$?

(1) (1, 0)

(2) (2, 0.5)

(3) (0.5, –0.5)

(4) (–4, –1)

Solving Exponential and Logarithmic Equations

In the previous chapter, we were able to solve some exponential equations by rewriting the two sides using the same base. This works for only certain cases. Now that we have learned about logarithms, we can solve a greater variety of exponential equations algebraically. In addition, we can solve logarithmic equations, which are equations with the variable in the argument of a logarithm.

EXPONENTIAL EQUATIONS

To solve an exponential equation algebraically, do the following.

1. Isolate the base and the exponent.
2. Take the logarithm of each side using an appropriate base (see the examples).
3. Solve for the variable.

EXAMPLE 11.16

Solve for x: $10^x = 35$.

SOLUTION

The base and exponent are already isolated, so we just need to take the logarithm of each side. We can use any base logarithm we want. Since the problem involves base 10, it makes sense (and saves a little work) to use the common logarithm.

$$10^x = 35$$
$$\log 10^x = \log 35$$

By the properties of logarithms, $\log 10^x = x$, so we have:

$$x = \log 35 \approx 1.544$$

EXAMPLE 11.17

Solve for x to the nearest thousandth: $8e^{0.3x} = 300$

SOLUTION

First isolate $e^{0.3x}$. In this problem, just divide each side by 8.

$$e^{0.3x} = \frac{300}{8} = 37.5$$

Now take the logarithm of both sides. Again, any base will do. However, the problem involves base e, so it will be easiest to use the natural logarithm.

$$\ln e^{0.3x} = \ln 37.5$$

By the properties of logarithms, $\ln e^{0.3x} = 0.3x$, so we have the following.

$$0.3x = \ln 37.5$$

$$x = \frac{\ln 37.5}{0.3} \approx 12.081$$

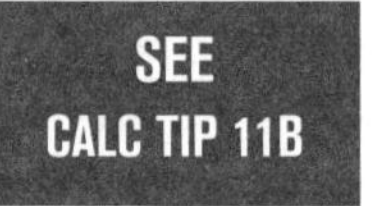

Suppose you could invest your money at a constant 5% annual interest rate. To find out how long it would take for your initial investment to double, solve $1.05^t = 2$. Round your answer to the nearest thousandth.

SOLUTION

The base and exponent are already isolated, so we are ready to take logarithms. The question is what base to use. If your calculator will take logarithms of any base, use base 1.05 and this problem will take just one step.

$$\log_{1.05} 1.05^t = \log_{1.05} 2$$

$$t = \log_{1.05} 2 \approx 14.207 \text{ years.}$$

Some calculators will do logs only in base 10 or in base e. This will add one step to your solution. Choose a base. It does not matter which one; we will use base 10. Take logs of both sides of the equation.

$$\log 1.05^t = \log 2$$

By the power property, $\log 1.05^t = t \log 1.05$. So we need to divide by $\log 1.05$ to finish solving for t.

$$t \log 1.05 = \log 2$$

$$t = \frac{\log 2}{\log 1.05} \approx 14.207 \text{ years.}$$

EXAMPLE 11.19 Suppose a new species of fish is introduced into a lake and its population doubles every three years. The fish population can be modeled by $P = P_0(2)^{\frac{t}{3}}$, where P is the population, t is time in years, and P_0 is the number of fish originally released into the lake. If 1,000 fish were originally released, how long will it take the fish population to reach 5,000?

SOLUTION

Substitute the known information, $P_0 = 1{,}000$ and $P = 5{,}000$, into the given formula.

$$5000 = 1000(2)^{\frac{t}{3}}$$

Isolate the base and exponent by dividing each side by 1,000.

$$5 = 2^{\frac{t}{3}}$$

Now take the logarithm of each side. If your calculator will take logarithms of any base, use base 2:

$$\log_2 5 = \log_2 2^{\frac{t}{3}}$$

$$\log_2 5 = \frac{t}{3}$$

$$3\log_2 5 = t$$

$$t \approx 6.966 \text{ years}$$

If your calculator does not do logarithms of any base, use your choice of 10 or e. We will use base e just for variety. Note that this gives us an extra factor of ln 2 that we need to divide by when solving for t.

$$5 = 2^{\frac{t}{3}}$$

$$\ln 5 = \ln 2^{\frac{t}{3}}$$

$$\ln 5 = \frac{t}{3}\ln 2$$

$$\frac{3\ln 5}{\ln 2} = t$$

$$t \approx 6.966 \text{ years}$$

EXAMPLE 11.20

Solve for x: $2^{2x+5} = 6^{x-3}$

SOLUTION

This problem is a little unusual in that it involves two different bases. Fortunately, the two bases are each isolated on their own side of the equation. So take the logarithm of each side using the same base. (We cannot take a logarithm base 2 on the left and a logarithm base 6 on the right.) We will use base 10.

$\log 2^{2x+5} = \log 6^{x-3}$	Take the log of each side.
$(2x+5)\log 2 = (x-3)\log 6$	Apply the power property.
$2x\log 2 + 5\log 2 = x\log 6 - 3\log 6$	Distribute.
$2x\log 2 - x\log 6 = -3\log 6 - 5\log 2$	Move all terms with x to left side, and all terms without x to the right.
$x(2\log 2 - \log 6) = -3\log 6 - 5\log 2$	Factor out x on the left.
$x = \dfrac{-3\log 6 - 5\log 2}{2\log 2 - \log 6}$	Solve for x by dividing.
$x \approx 21.805$	Evaluate on a calculator.

LOGARITHMIC EQUATIONS

To solve a logarithmic equation algebraically, do the following.

1. Use the properties of logarithms if necessary to rewrite the equation so it has only one logarithmic term.
2. Isolate the logarithm.
3. Rewrite the equation in exponential form.
4. Solve for the variable.
5. Check the solution in the original equation to make sure the equation is defined. Remember, logarithms are defined only for positive arguments.

EXAMPLE 11.21

Solve for x: $\log_4 x = 3$

SOLUTION

Remember the definition of logarithms: $y = \log_b x$ means $x = b^y$ or, in words, "y is the exponent that goes on base b to make x." In this problem, we have "3 is the exponent that goes on base 4 to make x":

$$4^3 = x$$
$$x = 64$$

There is an alternate way to solve the problem that some students prefer. Write both sides of the original equation as exponents on the base of the logarithm. In this problem, the base is 4. So write both sides of the equation, $\log_4 x = 3$, as exponents on base 4:

$$4^{\log_4 x} = 4^3$$

Because exponential functions and logarithmic functions of the same base are inverses, $4^{\log_4 x} = x$. So our equation becomes:

$$x = 4^3 = 64$$

As you can see, this alternate solution method leads to exactly the same result. However, this method makes it clearer that we are undoing a logarithm with its inverse, exponentiation, and following the familiar rule of equation solving "do the same thing to both sides."

EXAMPLE 11.22

Solve for x: $3\ln(2x + 1) = 5$

SEE CALC TIP 11C

SOLUTION

Isolate the logarithm by dividing by 3.

$$\ln(2x + 1) = \frac{5}{3}$$

Rewrite in exponential form (or write both sides as exponents on base e and simplify).

$$2x + 1 = e^{\frac{5}{3}}$$

Solve for x.

$$x = \frac{e^{\frac{5}{3}} - 1}{2} \approx 2.147$$

This solution checks in the original equation.

EXAMPLE 11.23

Find the exact value of the x-intercept: $f(x) = \ln(x - 3) + 1$

SOLUTION

Find the exact value of the x-intercept by setting $f(x) = 0$ and solving for x.

$$0 = \ln(x - 3) + 1$$

$$-1 = \ln(x - 3)$$

$$e^{-1} = x - 3$$

$$\frac{1}{e} = x - 3$$

$$3 + \frac{1}{e} = x$$

EXAMPLE 11.24

Solve for x: $2\log_4 x - 1 = \log_4 (24 - x)$

SOLUTION

This problem has two logarithmic terms. We need to use algebra to get both terms on the same side of the equation and then use the properties of logarithms to express them as a single logarithm. This can be done in more than one way.

Get both logarithms on the left and the constant on the right.

$$2\log_4 x - 1 = \log_4 (24 - x)$$
$$2\log_4 x - \log_4 (24 - x) = 1$$

Use the properties of logarithms to write the left side as a single logarithm.

$$\log_4 x^2 - \log_4 (24 - x) = 1 \quad \text{Apply the power property}$$

$$\log_4\left(\frac{x^2}{24 - x}\right) = 1 \quad \text{Apply the quotient property}$$

Now rewrite in exponential form (or write both sides as exponents on base 4).

$$\frac{x^2}{24 - x} = 4^1 = 4$$

Solve for x.

$$x^2 = 4(24 - x)$$
$$x^2 = 96 - 4x$$
$$x^2 + 4x - 96 = 0$$
$$(x + 12)(x - 8) = 0$$
$$x = -12 \text{ or } x = 8$$

SEE CALC TIP 11D

Looking back to the original equation, $2\log_4 x - 1 = \log_4 (24 - x)$, we see that $x = -12$ cannot be a solution because $\log_4 (-12)$ is not real. The only solution is $x = 8$.

COMMON ERROR

Students who do not check all solutions to log equations risk including extraneous solutions in their answer. Make sure candidate solutions do not lead to logarithms of negative numbers when substituted into the original equation.

SECTION EXERCISES

For 25–33, solve for x *or* t *algebraically. Round your answers to the nearest thousandth.*

11–25. $5^x = 12$

11–26. $3^{-x} = 16$

11–27. $4{,}000e^{-0.025t} = 20$

11–28. $100\left(\frac{1}{2}\right)^{\frac{t}{5}} = 20$

11–29. $\log_x 25 = 4$

11–30. $\log x = -1.35$

11–31. $2\ln(100 - 3x) - 8 = 0.5$

11–32. $\log(10 - 3x) - 2\log x = 0$

11–33. $\log_2 (6x - 1) - 3 = 2\log_2 x$

11–34. If Annabelle invests \$200 at 3% interest for t years, she will have the amount $A = 200(1.03)^t$ dollars. How many whole years must Annabelle wait for the amount to reach \$500?

11–35. The interest formula for interest compounded continuously is $A = Pe^{rt}$, where A is the value of the investment, P is the principal, r is the annual interest rate expressed as a decimal, and t is the number of years. If you deposit \$1,000 in an account that gives 5% annual interest compounded continuously, how many years, to the nearest tenth, will it take your money to grow to \$2,000?

11–36. Find the domain of the function $y = \log(3x + 9)$ algebraically.

Change of Base Formula

Some calculators will find logarithms of only base 10 or base e. These calculators can still be used to find logarithms of other bases using the change of base formula.

Suppose you need to evaluate $\log_b N$ but your calculator finds only logarithms of base a. How can you do it?

$\log_b N = x$	Give $\log_b N$ a convenient name, such as x.
$b^x = N$	Rewrite in exponential form.
$\log_a b^x = \log_a N$	Take the logarithm of base a on each side.
$x\log_a b = \log_a N$	Apply the power property.
$x = \dfrac{\log_a N}{\log_a b}$	Solve for x.

This is called the change of base formula: $\log_b N = \dfrac{\log_a N}{\log_a b}$. If your calculator will find logarithms of only base 10 or base e and you need to take a logarithm with another base, b, you can use either of the following:

$$\log_b N = \frac{\log N}{\log b}$$

or

$$\log_b N = \frac{\ln N}{\ln b}$$

EXAMPLE 11.25 Evaluate $\log_2 50$ using base 10.

SEE CALC TIP 11E

SOLUTION

By the first formula above, we have $\log_2 50 = \dfrac{\log 50}{\log 2} \approx 5.644$

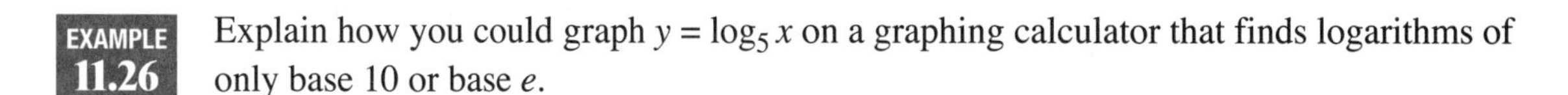

EXAMPLE 11.26 Explain how you could graph $y = \log_5 x$ on a graphing calculator that finds logarithms of only base 10 or base e.

SOLUTION

Choose either base 10 or base e, it does not matter which, and use the change of base formula. Graph either $y = \dfrac{\log x}{\log 5}$ or $y = \dfrac{\ln x}{\ln 5}$.

SECTION EXERCISES

11–37. Rewrite the following in base e.

a. $\log_2 64$

b. $\log 0.6$

11–38. Rewrite $\log_6 72$ in base 10.

CHAPTER EXERCISES

C11–1. Rewrite the following as an equivalent log equation.
$5^4 = 625$

C11–2. Rewrite the following as an equivalent exponential equation.
$n = \log_x y$

For 3–4, write the inverse of each of the following functions.

C11–3. $f(x) = \log_8 x$

C11–4. $f(x) = e^x$

For 5–6, simplify each of the following.

C11–5. $\log 10^{(2x-1)}$

C11–6. $e^{\ln f(x)}$

For 7–8, use the properties of logarithms to expand each of the following as much as possible.

C11–7. $\log\frac{\sqrt{xy}}{z}$

C11–8. $\log\frac{\sqrt{x}}{100}$

For 9–11, express each of the following as the log of a single quantity.

C11–9. $3\log a - \frac{1}{2}\log b$

C11–10. $3\log x - \frac{1}{2}(\log y + \log z)$

C11–11. $1 + 2\log x$

C11–12. Which is the expression $\log 4x$ is equivalent to?

(1) $4\log x$
(2) $4 + \log x$
(3) $(\log 4)(\log x)$
(4) $\log 4 + \log x$

C11–13. If $\log_b x = 0.8$ and $\log_b y = 0.6$, evaluate $\log_b xy$.

C11–14. The graph of $y = \log_3 x$ is shown below. The following graphs all show transformations (reflections and translations, no dilations) of the graph of $y = \log_3 x$. Label each graph with its correct equation.

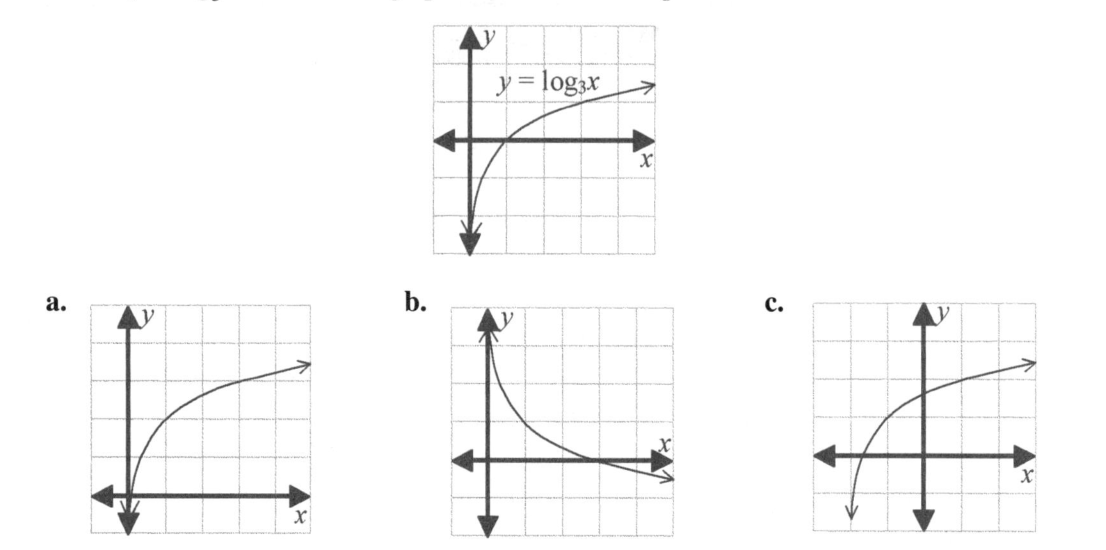

C11–15. For $y = \log_2 (x + 5) + 1$, identify the domain, range, asymptotes, and intercepts and sketch. Try to do these without using your calculator.

C11–16. Find the domain of the function $y = \ln(8 - 4x)$ algebraically.

C11–17. For $y = \log_9\left(\frac{x + 6}{2}\right)$, find the following algebraically.

a. The domain

b. The y-intercept

c. The zero(s)

d. The asymptote(s)

For 18–22, solve for x *or* t *algebraically. Round your answers to the nearest thousandth.*

C11–18. $4^{x+1} = 23$

C11–19. $3{,}000(1 - e^{-2t}) = 1{,}200$

C11–20. $\log_3 x = 2.5$

C11–21. $\log_2 (x - 4) + \log_2 x = 5$

C11–22. $\log_4\left(x^2 + 3x\right) - \log_4(x + 5) = 1$

C11–23. A scientist has 10 grams of a radioactive substance with a half-life of 8 days. The formula $A = 10(0.5)^{t/8}$ tells how much will be left after t days. How many days, to the nearest tenth, will it take until only 1 gram is left?

C11–24. A colony of 25 bacteria grows exponentially so that its population after t hours is given by $P = 25(2.72)^{0.26t}$. How many hours, to the nearest hundredth, will the colony take to grow to 25,000 bacteria?

C11–25. Rewrite the following in base e.

a. $\log_3 12$

b. $\log_{16}\left(\frac{1}{32}\right)$

Trigonometric Functions

WHAT YOU WILL LEARN

You have likely already learned about using trigonometry to solve problems with right triangles. We want to modify the definitions of the trigonometric functions so they can be applied to any real number instead of just acute angles. Having done that, we will have powerful tools for modeling many periodic, or cyclical, phenomena (discussed in the next chapter). In this chapter, you will learn to:

- Measure angles in radians;
- Find trigonometric functions of any angle;
- Find exact trigonometric values for special angles;
- Apply the Pythagorean identity for trigonometric functions;
- Apply the sum and difference formulas.

SECTIONS IN THIS CHAPTER

- Radian Measure and Arc Length
- Angles of Any Size and Coterminal Angles
- Unit Circle and Trigonometric Function Values
- Trigonometric Function Values for Special Angles
- Reciprocal Trigonometric Functions
- Trigonometric Identities
- The Sum and Difference Formulas

Radian Measure and Arc Length

The word *trigonometry* means "measurement of triangles," and that is how trigonometry was first developed. Its early applications included surveying, navigation, and astronomy. More recently, the use of trigonometry has expanded to describe all types of physical phenomena of a periodic nature, ranging from swinging pendulums to orbiting planets and from sound waves to electricity and magnetism. To see how trigonometry can be used in all these ways, we must somehow extend our ideas of angles beyond 0°–180° and extend the definitions of the trigonometric functions beyond just the angles in a right triangle. In addition, we will find it helpful to have a new way to measure angles.

RADIAN MEASURE

Measuring angles in degrees is fine for many geometry applications. However, another way of measuring angles, *radian measure*, is more convenient in calculus and is therefore widely used in other applications. One *radian* is defined to be the measure of the central angle of a circle that intercepts an arc with the same length as the radius of the circle, as shown in the diagram.

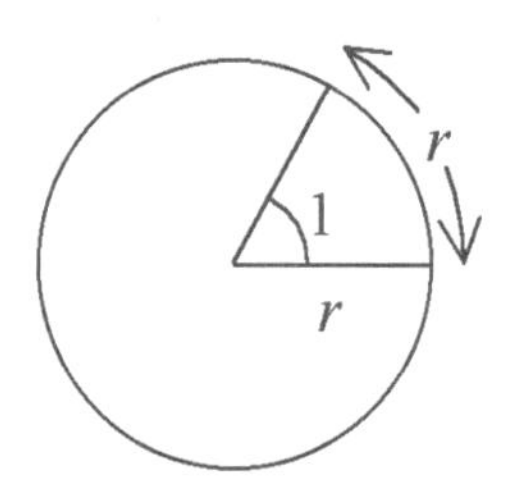

FIGURE 12.1

This can be generalized to define the radian measure of any central angle of a circle, θ (the Greek letter *theta)*, as the ratio of the length of the intercepted arc, s, to the length of the radius, r, where s and r must both be measured in the same units of length.

$$\theta = \frac{s}{r}$$

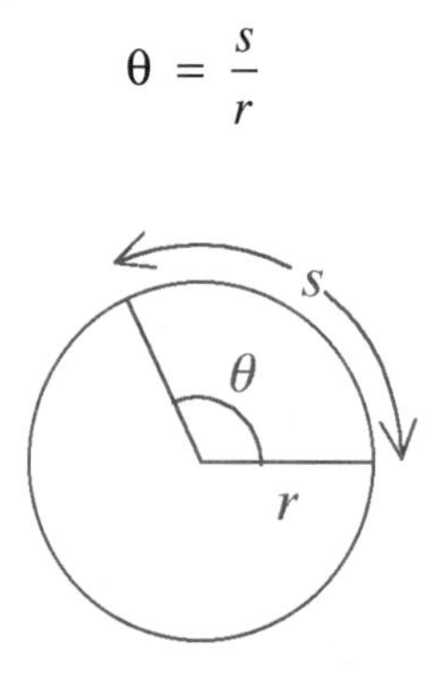

FIGURE 12.2

The symbol for radian is rad. However, radians are different from other units of measurement such as centimeters, grams, or seconds in that radians are dimensionless. For example, suppose in a circle of radius 4 cm, a central angle intercepts an arc of length 6 cm. The radian measure of the central angle is $\theta = \dfrac{6\text{ cm}}{4\text{ cm}} = 1.5$. Note that the units, centimeters, cancel out so that the final answer is just a number without units. The symbol rad is sometimes used to clarify that the number represents an angle measure. However, the symbol is often omitted, especially if other physical dimensions are in the problem.

Now consider one full circle. The arc, s, of a full circle is its circumference, $C = 2\pi r$. Therefore the radian measure of one full circle is $\theta = \dfrac{C}{r} = \dfrac{2\pi r}{r} = 2\pi$. One full circle is also 360°. This means we can compare radians and degrees.

RADIAN AND DEGREE COMPARISON

$360° = 2\pi$ radians
$180° = \pi$ radians

We can convert back and forth from radian measure to degree measure.

1. To change from degree measure to radian measure, multiply the number of degrees by $\dfrac{\pi}{180°}$.
2. To change from radian measure to degree measure, multiply the number of radians by $\dfrac{180°}{\pi}$.

EXAMPLE 12.1 Convert 75° to radians.

SOLUTION

To convert degrees to radians, multiply the degrees by $\dfrac{\pi}{180°}$. If you have trouble remembering which goes on top, π or 180°, think about the units. In this problem, we want degrees to cancel out so we will be left with radians. This means we need 180° in the denominator and π radians in the numerator. Writing the 75° as a fraction over 1 can help.

$$75° = \frac{75°}{1} \cdot \frac{\pi}{180°} = \frac{75\pi}{180} = \frac{5\pi}{12} \text{ radians}$$

Radian measures are often left in terms of π but may also be expressed as decimals; $\dfrac{5\pi}{12} \approx 1.309$.

EXAMPLE 12.2 How many degrees are in $\frac{7\pi}{6}$ radians?

SOLUTION

In this problem, we want the radians to cancel out. So π goes in the denominator and 180° in the numerator.

$$\frac{7\pi}{6} = \frac{7\pi}{6} \cdot \frac{180^\circ}{\pi} = 210^\circ$$

EXAMPLE 12.3 Convert 2 radians to degrees. Round your answer to the nearest thousandth.

SOLUTION

Don't get the mistaken idea that all radian measures involve the symbol π or that the word radian somehow means π. This means 2 radians is just 2, not 2π (which is a little more than 6). To convert to degrees, multiply by $\frac{180^\circ}{\pi}$ even though π does not cancel out.

$$2 = \frac{2}{1} \cdot \frac{180^\circ}{\pi} = \left(\frac{360}{\pi}\right)^\circ \approx 114.592^\circ$$

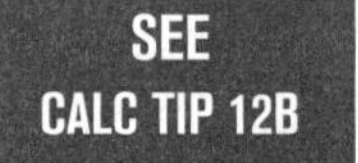

ARC LENGTH

The formula $\theta = \frac{s}{r}$, which was used to define the radian measure of a central angle of a circle, can also be written $s = r\theta$ and be used to find the length of an arc in a circle. Just remember when using this formula that the central angle θ must be measured in radians. If the angle is given in degrees, you must convert to radians before using the formula.

EXAMPLE 12.4 In a circle of radius 12 centimeters, what is the length of the arc intercepted by a central angle of 1.75 radians?

SOLUTION

A diagram is optional but may be helpful.
We know that $\theta = 1.75$ rad and $r = 12$ cm:

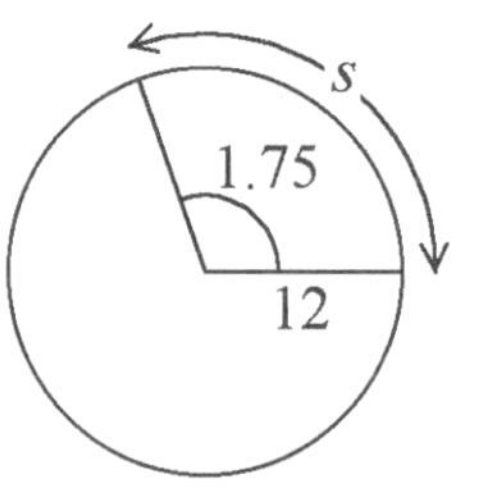

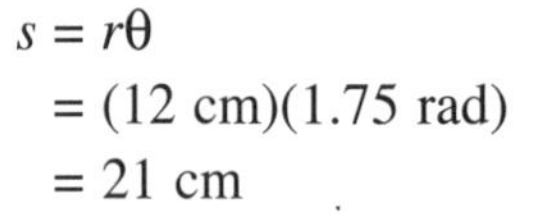

$$\begin{aligned} s &= r\theta \\ &= (12 \text{ cm})(1.75 \text{ rad}) \\ &= 21 \text{ cm} \end{aligned}$$

Recall that radians are dimensionless, so the units of the radius and the arc are the same.

A pendulum with a length of 24 inches swings through an angle of 40°. How far does the bob travel during one swing of the pendulum?

SOLUTION

Again, a diagram is optional but could be helpful.
In this problem, the angle is given in degrees. Before using $s = r\theta$, we must convert the angle to radians.

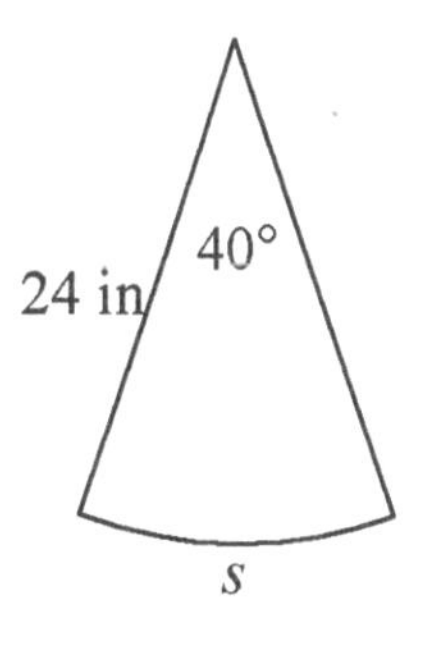

$$\theta = 40° = \frac{40°}{1} \cdot \frac{\pi}{180°} = \frac{2\pi}{9}$$

$$s = r\theta$$

$$= (24 \text{ in.})\left(\frac{2\pi}{9}\right) = \frac{16\pi}{3} \text{ in.} \approx 16.755 \text{ in.}$$

EXAMPLE 12.6

A hoist consists of a cable wrapping around a motor-driven drum 6 inches in diameter. How far must the drum turn to lift a load 36 inches?

SOLUTION

Here a diagram is very helpful.

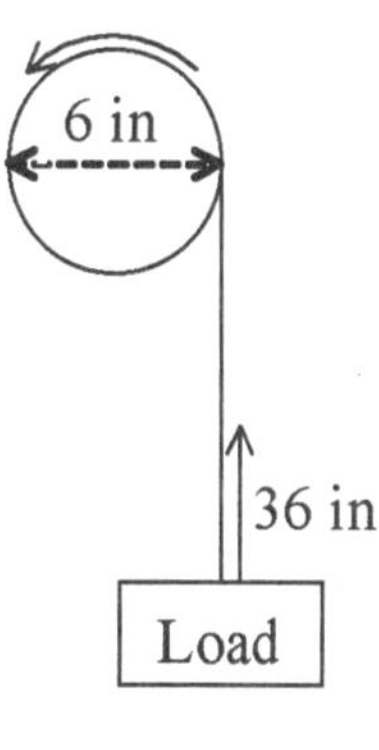

To raise the load 36 inches, 36 inches of cable must wrap around the drum. This means $s = 36$ inches. Since the diameter is 6 inches, we get $r = 3$ inches.

$$s = r\theta$$

$$36 \text{ in} = (3 \text{ in})\theta$$

$$\theta = 12$$

Thus the drum must turn through 12 radians. That can be converted to degrees if you prefer.

$$12 = \frac{12}{1} \cdot \frac{180°}{\pi} \approx 687.5°$$

The drum must turn 687.5°, almost two complete revolutions, to raise the load 36 inches.

SECTION EXERCISES

For 1–2, convert the following angles into radians.

12–1. 50°

12–2. 200°

For 3–4, convert the following angles into degrees. If necessary, round to the nearest tenth.

12–3. $\frac{9}{10}\pi$ rad

12–4. 3 rad

For 5–6, find the measure of each unknown. Include appropriate units and, if appropriate, round to the nearest hundredth.

12–5. $r = 5$ inches, $\theta = 0.6$ radians, $s = ?$

12–6. $\theta = 106°$, $s = 37$ cm, $r = ?$

12–7. Rufus and Goofus are sharing a 16″ diameter pizza that has been sliced through the center into wedge-shaped pieces.

a. Goofus's piece makes an angle of 1.2 radians at its vertex. How long is the outer crust (the round edge) on his piece?

b. Rufus measured along the round edge of his piece and found it was 6.5 inches long. What is the measure of the angle his piece makes at the vertex, in radians?

Angles of Any Size and Coterminal Angles

In geometry, an *angle* is made up of two rays with a common endpoint called the *vertex* of the angle. Most angles in geometry have measures between 0° and 180°. In a 180° angle, the two rays form a straight line. If your geometry course included circle geometry, then you are familiar with angles of up to 360° (one full circle).

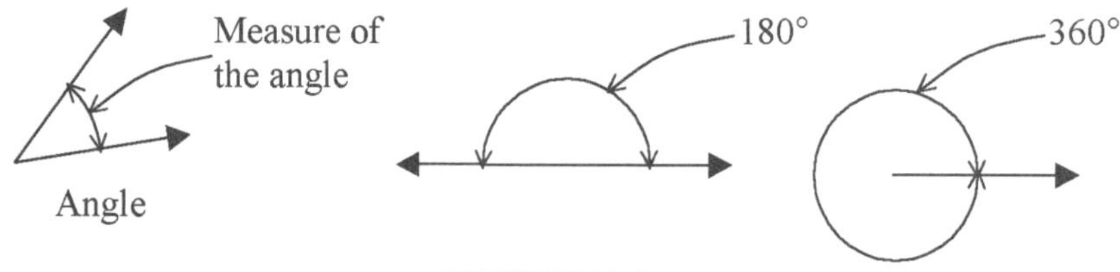

FIGURE 12.3

As illustrated in Example 12.6, if an angle is considered a rotation, then it can make sense to talk about angles greater than 360°. By considering the direction of the rotation, we can also have angles with negative measure.

Start with a single ray, called the *initial side* of the angle, and let that ray rotate about its vertex. Its final position is called the *terminal side* of the angle. The measure of the angle, which can be in either degrees or radians, tells how far the ray rotated. In mathematics, positive angles always represent counterclockwise rotations; negative angles represent clockwise rotations.

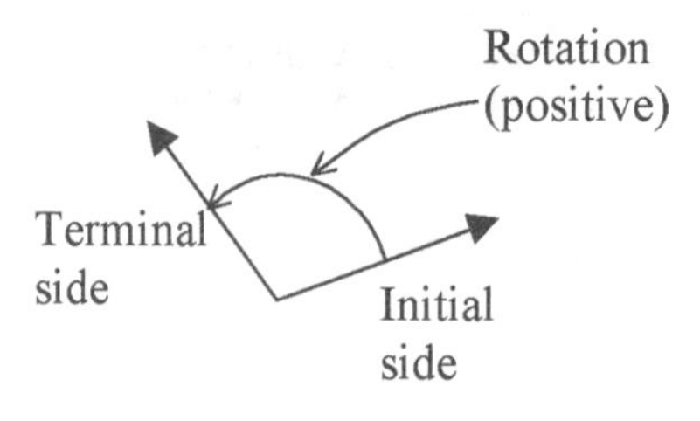

FIGURE 12.4

It is often convenient in trigonometry to represent an angle on coordinate axes. An angle is said to be in *standard position* if its vertex is at the origin and its initial side is along the positive x-axis. Two examples of angles in standard position are shown in Figure 12.5.

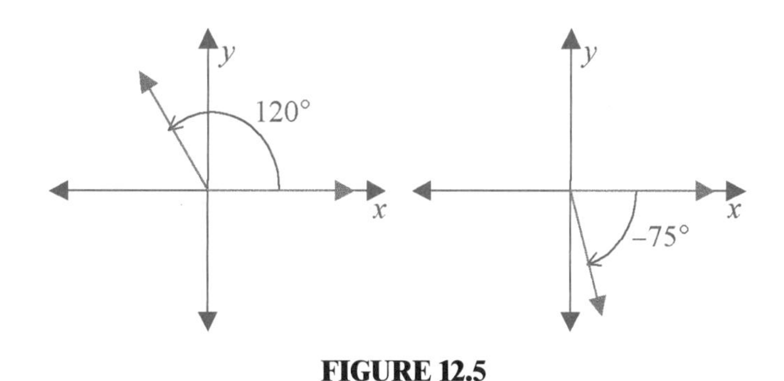

FIGURE 12.5

Interpreting an angle as a rotation allows us to think about angles of any size, both positive and negative. However, this has an interesting consequence. It is possible for two (or many) angles to have different measures but have the same initial and terminal sides. Such angles are called *coterminal angles*. Three coterminal angles are represented in the Figure 12.6. However, it should be clear that there are infinite coterminal angles having the same initial and terminal sides.

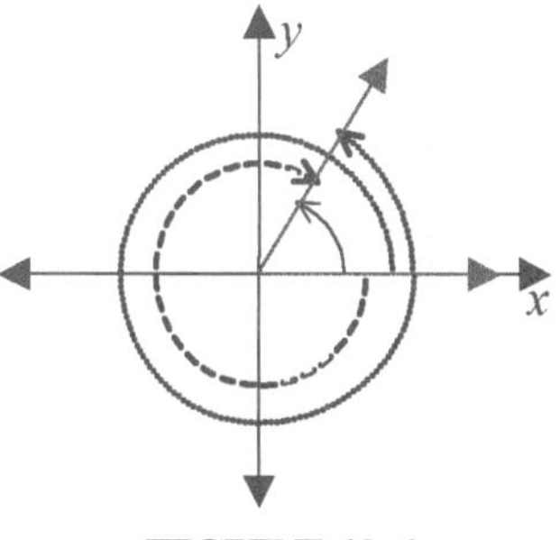

FIGURE 12.6

EXAMPLE 12.7

Find angles that are coterminal with $\theta = 50°$ in standard position.

SOLUTION

A 50° angle is sketched in standard position in the diagram to the right.

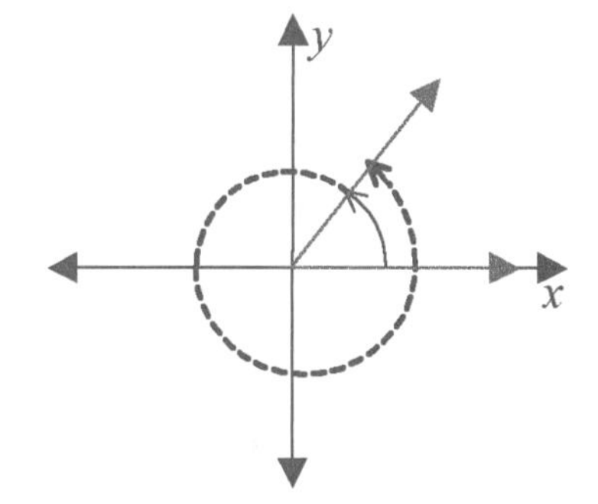

To find a coterminal angle, we can continue counterclockwise one complete rotation from the terminal side as shown in the diagram. In effect, this adds 360° to the angle. So an angle of $50° + 360° = 410°$ is coterminal with the 50° angle. We can continue adding complete rotations as many times as we want to find more coterminal angles: 770°, 1,130°, 1,490°, and so on. We can also start at the terminal side and go one or more rotations *clockwise*, subtracting 360° each time, to find still more coterminal angles: $50° - 360° = -310°$, then −670°, then −1,030°, and so on.

All these coterminal angles differ from the original angle by an integral multiple of 360°. A convenient way to write the set of all the angles is $(50 + 360n)°$ where n can be any integer.

Find angles that are coterminal with $\theta = \frac{4\pi}{3}$ in standard position.

SOLUTION

The idea here is exactly the same as in the previous example. To find coterminal angles, we may add or subtract any number of complete rotations to the original angle. The difference here is that the given angle is measured in radians. In radians, one complete rotation is 2π. Therefore, we find coterminal angles by adding any integer multiple of 2π to the original angle: $\frac{4\pi}{3} + 2\pi n$, where n is an integer. For $n = 1$, this gives an angle of $\frac{4\pi}{3} + 2\pi = \frac{10\pi}{3}$. For $n = -1$, we get $\frac{4\pi}{3} - 2\pi = -\frac{2\pi}{3}$.

EXAMPLE 12.9

In what quadrant does an angle of 1,230° terminate when in standard position?

SOLUTION

Remember, Quadrant I is where both x and y are positive and the other quadrants are numbered in counterclockwise order.

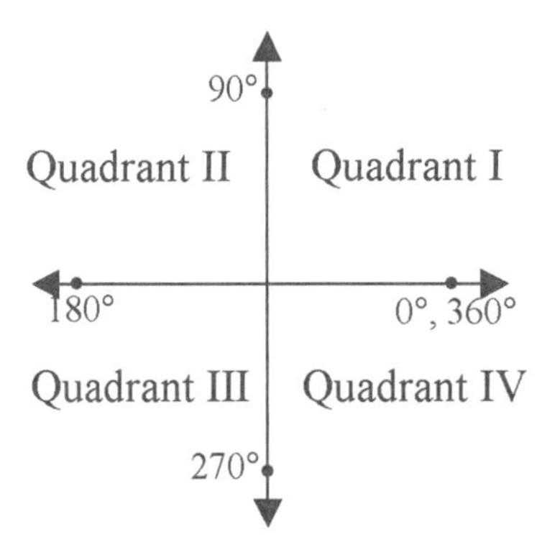

Divide the given angle by 360° (one full rotation): $\frac{1230°}{360°} = 3.41\overline{6}$ or $3\frac{5}{12}$. This means that the angle makes three complete rotations, bringing it back to the positive x-axis, and then an extra $\frac{5}{12}$ of a rotation; $\frac{5}{12}$ of a rotation is $\frac{5}{12}(360°) = 150°$. Since $90° < 150° < 180°$, the angle terminates in Quadrant II.

If you are good at thinking in terms of fractions of a rotation, you do not need to convert $\frac{5}{12}$ of a rotation to degrees. Think of it as a fraction of a complete circle. Since $\frac{5}{12}$ is more than $\frac{1}{4}$ and less than $\frac{1}{2}$, the angle will terminate in Quadrant II.

EXAMPLE 12.10 Sketch an angle of $\frac{10}{9}\pi$ in standard position. Then name two other angles, one positive and one negative, that are coterminal with it.

SOLUTION

We know that $\pi < \frac{10}{9}\pi < \frac{3}{2}\pi$, so the angle is in Quadrant III. For many purposes in trigonometry, just sketching your angle in the correct quadrant is good enough. If you want a more accurate sketch, note that $\frac{10}{9}\pi - \pi = \frac{1}{9}\pi$ which is 20°. So the angle is 20° into Quadrant III.

To find coterminal angles, we must add or subtract one or more complete rotations to the given angle. In radian measure, that means we can find coterminal angles by adding any integral multiple of 2π to the angle.

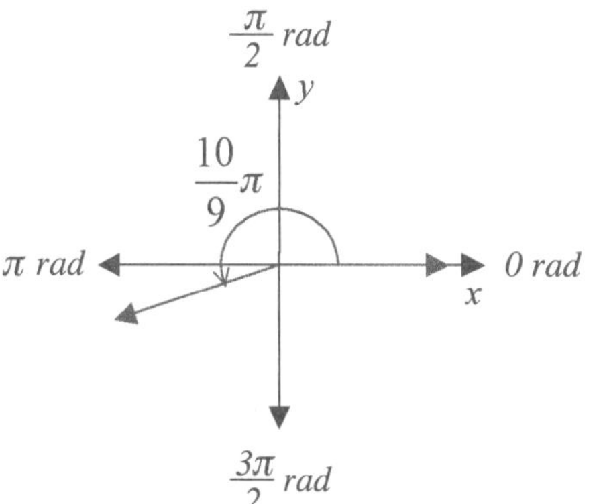

$\frac{10}{9}\pi + 2\pi = \frac{28}{9}\pi$ is one positive coterminal angle

$\frac{10}{9}\pi - 2\pi = -\frac{8}{9}\pi$ is one negative coterminal angle

SECTION EXERCISES

12–8. Find the measure of the smallest positive value of an angle coterminal with an angle of –165°.

12–9. Find the measures of all angles that are coterminal with an angle of $\frac{3}{4}\pi$ rad.

12–10. Portnoy is riding the Ferris wheel shown to the right. The wheel rotates in the positive direction.

a. If Portnoy is at position 1, how many degrees must the wheel turn before he first reaches the lowest point of his ride?

b. If Portnoy is at position 12, where will he be after the wheel turns 2,010°?

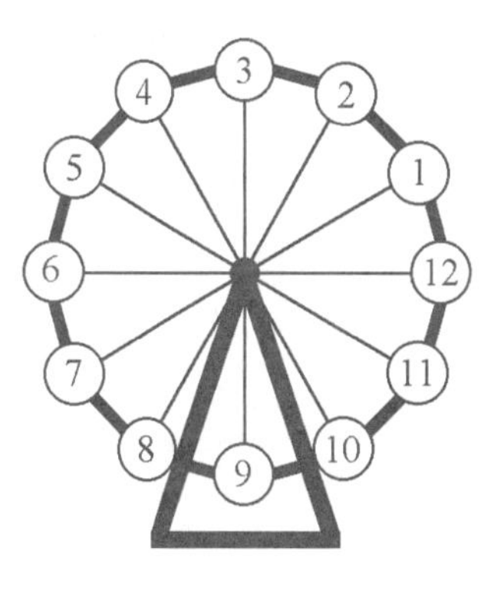

Unit Circle and Trigonometric Function Values

Almost everyone's first introduction to trigonometric functions involves defining the functions as ratios of the lengths of the sides of a right triangle. Figure 12.7 shows how to define trigonometric functions for an acute angle θ in a right triangle.

The sine of θ: $\sin\theta = \dfrac{\text{Opposite}}{\text{Hypotenuse}}$

The cosine of θ: $\cos\theta = \dfrac{\text{Adjacent}}{\text{Hypotenuse}}$

The tangent of θ: $\tan\theta = \dfrac{\text{Opposite}}{\text{Adjacent}}$

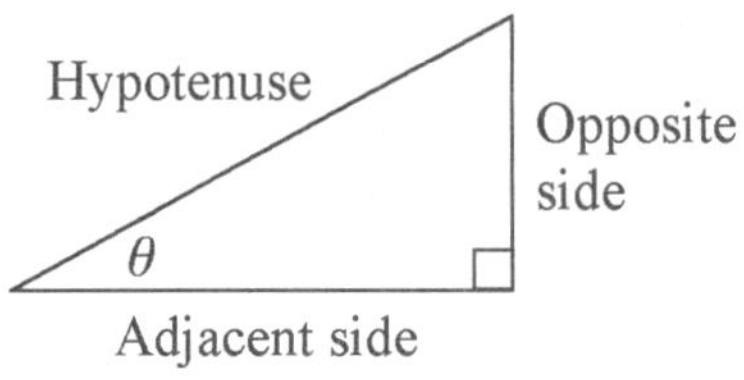

FIGURE 12.7

Remember that in the definitions, θ represents the measure of an angle and "opposite," "adjacent," and "hypotenuse" represent lengths of sides of the right triangle.

EXAMPLE 12.11 Use the triangle shown in the diagram to find the values of sin 60°, cos 60°, and tan 60°.

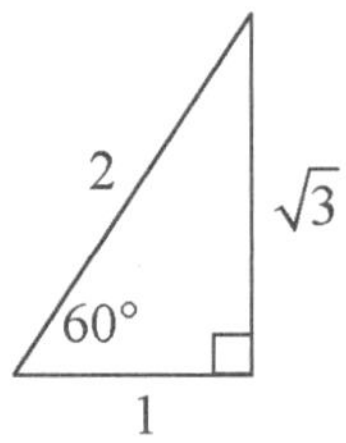

SOLUTION

From the definitions, we have the following:

$$\sin 60^\circ = \frac{\sqrt{3}}{2} \approx 0.8660 \qquad \cos 60^\circ = \frac{1}{2} = 0.5 \qquad \tan 60^\circ = \frac{\sqrt{3}}{1} \approx 1.7321$$

A couple of things are worth noting from Example 12.11 above. First, the number $\frac{\sqrt{3}}{2}$ is the exact value of sin 60°; 0.8660 is a decimal approximation of this irrational number. It is difficult if not impossible to find exact values for all but a handful of special angles. The decimal approximations provided by a calculator are adequate for most applications.

Second, since $60^\circ = \frac{\pi}{3}$ rad, we also have $\sin\frac{\pi}{3} = \frac{\sqrt{3}}{2}$, $\cos\frac{\pi}{3} = \frac{1}{2}$, and $\tan\frac{\pi}{3} = \sqrt{3}$.

Trigonometric functions may be done in either degrees or radians. Just make sure when using your calculator that it is set in the right mode for your problem.

The problem with the right triangle definitions is that they apply only to angles strictly between 0° and 90° (0 and $\frac{\pi}{2}$ radians). We would like a definition that can work for any angle. To do this, we move from the right triangle to the *unit circle*, a circle of radius 1 centered at the origin.

Suppose θ is an acute angle in standard position. Let its terminal side intersect the unit circle at the point $P(x, y)$ as shown on the left in Figure 12.8. Drop a perpendicular segment down from point P to the x-axis to create right triangle OPR as shown on the right in Figure 12.8. This triangle has a hypotenuse of length 1 and two sides of lengths x and y.

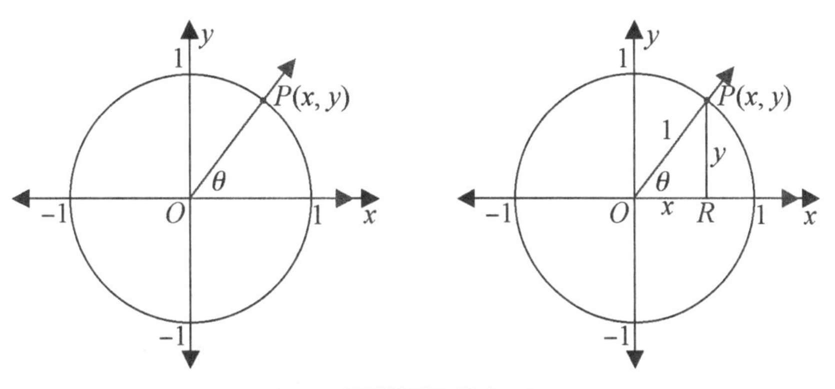

FIGURE 12.8

From the right triangle trigonometric definitions, we now have the following:

$$\sin\theta = \frac{y}{1} = y \qquad \cos\theta = \frac{x}{1} = x \qquad \tan\theta = \frac{y}{x}$$

So far we have assumed that θ is an acute angle. Now we can extend the trigonometric functions to any angle by using the above formulas to define the trigonometric functions.

UNIT CIRCLE DEFINITIONS OF THE TRIGONOMETRIC FUNCTIONS

Let θ be any angle in standard position and let the terminal side of θ intersect the unit circle at the point $P(x, y)$. Figure 12.9 shows the trigonometric functions for any angle in standard position.

$$\sin\theta = y$$

$$\cos\theta = x$$

$$\tan\theta = \frac{y}{x},\ x \neq 0$$

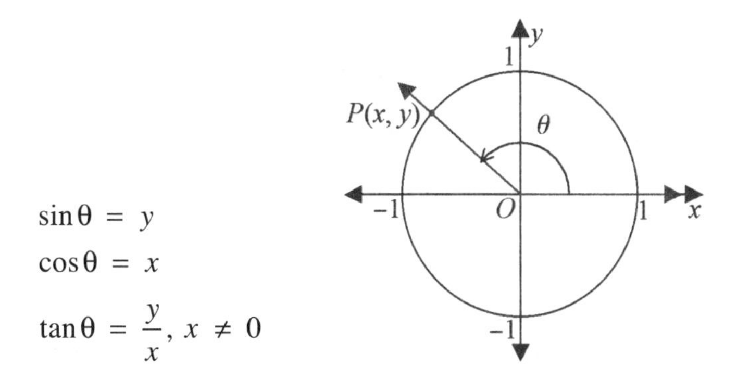

FIGURE 12.9

EXAMPLE 12.12

The terminal side of a 210° angle in standard position intersects the unit circle at the point $\left(-\frac{\sqrt{3}}{2}, -\frac{1}{2}\right)$. Find sin 210°, cos 210°, and tan 210°.

SOLUTION

A diagram is optional but is shown for clarity. From the unit circle definitions:

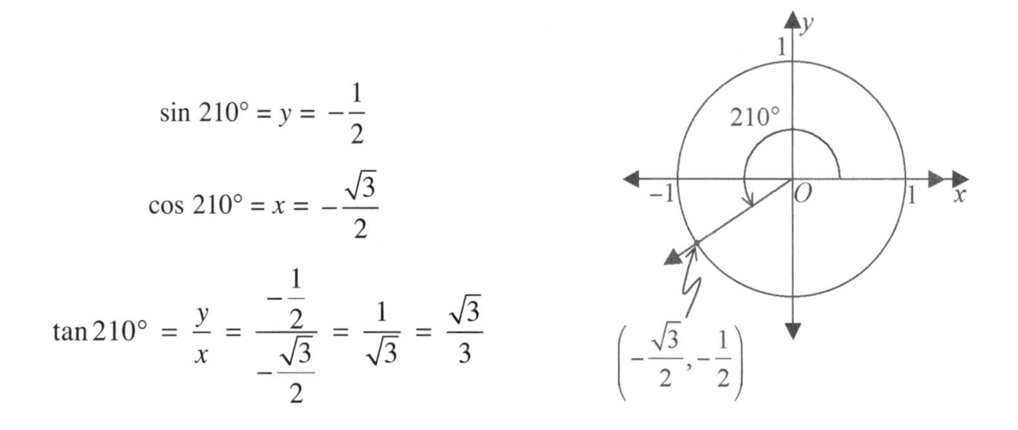

$$\sin 210° = y = -\frac{1}{2}$$

$$\cos 210° = x = -\frac{\sqrt{3}}{2}$$

$$\tan 210° = \frac{y}{x} = \frac{-\frac{1}{2}}{-\frac{\sqrt{3}}{2}} = \frac{1}{\sqrt{3}} = \frac{\sqrt{3}}{3}$$

For some problems, it is more convenient to use a point on the terminal side of the angle that is not on the unit circle. It is easy to adjust the definitions for this case. If θ is in standard position, $P(x, y)$ is any point on the terminal side of θ, and $r = \sqrt{x^2 + y^2}$ is the distance from P to the origin:

$$\sin\theta = \frac{y}{r}$$

$$\cos\theta = \frac{x}{r}$$

$$\tan\theta = \frac{y}{x},\ x \neq 0$$

FIGURE 12.10

When using these definitions, x and y may be positive or negative depending on which quadrant θ terminates in, but r should always be positive.

The terminal side of an angle, θ, in standard position passes through the point (3,–5). Find sin θ, cos θ, and tan θ.

SOLUTION

A diagram is optional but is shown for clarity. To use the general definitions, you need to know $x = 3$, $y = -5$, and r. Find r from the values given for x and y.

$$r = \sqrt{x^2 + y^2}$$
$$r = \sqrt{(3)^2 + (-5)^2}$$
$$r = \sqrt{34}$$

Then substitute the values into the general definition for sine, cosine, and tangent.

$$\sin\theta = \frac{y}{r} = \frac{-5}{\sqrt{34}} = \frac{-5\sqrt{34}}{34}$$
$$\cos\theta = \frac{x}{r} = \frac{3}{\sqrt{34}} = \frac{3\sqrt{34}}{34}$$
$$\tan\theta = \frac{y}{x} = \frac{-5}{3} = -\frac{5}{3}$$

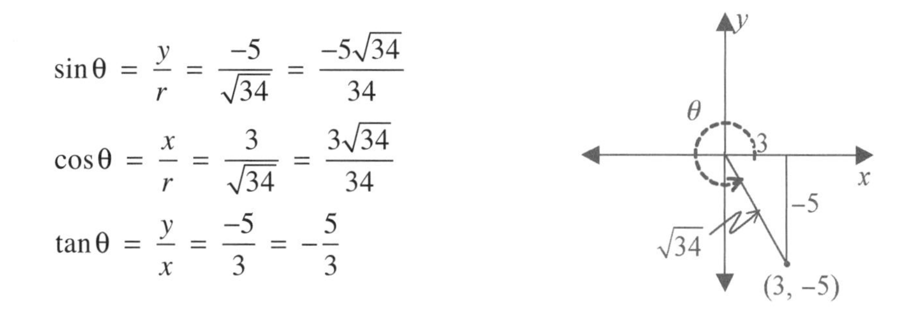

The ratios in the Example 12.13 were rationalized. Although it is customary to rationalize trigonometric values, it is not always required.

COMMON ERROR

Students familiar with the 3–4–5 Pythagorean triple sometimes jump to the conclusion that if a triangle has sides 3 and 5, the missing side is necessarily of length 4. This is true only if 5 is the length of the hypotenuse, not a leg.

SIGNS OF TRIGONOMETRIC FUNCTIONS

You may have noticed that some trigonometric function values are negative. It is easy to determine the signs of trigonometric function values. All we need to know is the unit circle definitions and in which quadrant the angle terminates.

Since $\cos\theta = x$ and x is positive in Quadrants I and IV, $\cos\theta$ is positive for any angle θ that terminates in Quadrants I or IV and is negative for angles that terminate in Quadrants II or III. In the same way, since $\sin\theta = y$ and y is positive in Quadrants I and II, $\sin\theta$ is positive for angles terminating in Quadrants I or II and negative for angles terminating in Quadrants III or IV. Finally, $\tan\theta = \frac{y}{x}$ is positive in quadrants where x and y have the same sign. These are Quadrants I and III. In Quadrants II and IV, $\tan\theta$ is negative. It is easiest to remember only in which quadrants the trigonometric function values are positive. These are summarized in Figure 12.11.

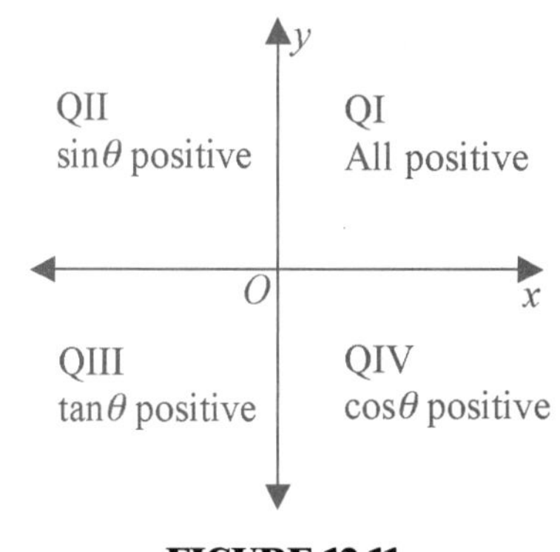

FIGURE 12.11

There are several mnemonics to help remember this. A common one is "**a**ll **s**tudents **t**ake **c**alculus."

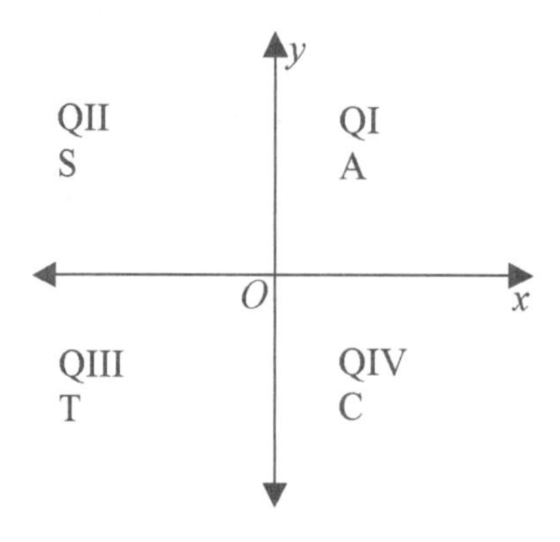

FIGURE 12.12

Use the first letter of each word to help you remember.

Quadrant I:	**A**	$\rightarrow$ All trigonometric function values are positive.
Quadrant II:	**S**	$\rightarrow$ Sine is positive (cosine and tangent are negative).
Quadrant III:	**T**	$\rightarrow$ Tangent is positive (sine and cosine are negative).
Quadrant IV:	**C**	$\rightarrow$ Cosine is positive (sine and tangent are negative).

EXAMPLE 12.14

If $\sin \theta < 0$, in what quadrant(s) could θ terminate?

SOLUTION

If $\sin \theta < 0$, sine is negative. Since $\sin \theta$ is positive in Quadrants I and II, it must be negative in Quadrants III and IV.

If $\cos \theta > 0$ and $\tan \theta < 0$, in what quadrant(s) can θ terminate?

SOLUTION

If $\cos \theta > 0$ and $\tan \theta < 0$, cosine is positive and tangent is negative. Cosine is positive in Quadrants I and IV. Tangent is negative in Quadrants II and IV. The only quadrant in which both cosine is positive and tangent is negative is Quadrant IV.

EXAMPLE 12.16

Find $\sin\theta$ and $\cos\theta$ if $\tan\theta = \dfrac{15}{8}$ and $\cos\theta < 0$.

SOLUTION

We need to know in what quadrant θ terminates. We know tangent is positive and cosine is negative. Tangent is positive in Quadrants I and III. Cosine is negative in Quadrants II and III. Therefore, θ is in Quadrant III.

Sketch a triangle with a leg on the x-axis in Quadrant III. Since $\tan\theta = \dfrac{y}{x}$, we have $\dfrac{y}{x} = \dfrac{15}{8}$. In Quadrant III, both y and x are negative, so let $x = -8$ and $y = -15$. To find sine and cosine, you must know r, the length of the hypotenuse. Use the Pythagorean theorem to find the hypotenuse.

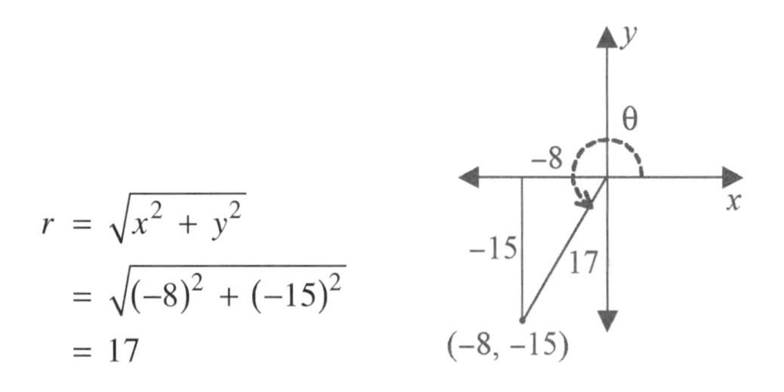

$$\begin{aligned} r &= \sqrt{x^2 + y^2} \\ &= \sqrt{(-8)^2 + (-15)^2} \\ &= 17 \end{aligned}$$

From the definitions of sine and cosine, we have:

$$\sin\theta = \frac{y}{r} = \frac{\text{Opposite}}{\text{Hypotenuse}} = \frac{-15}{17} = -\frac{15}{17}$$

$$\cos\theta = \frac{x}{r} = \frac{\text{Adjacent}}{\text{Hypotenuse}} = \frac{-8}{17} = -\frac{8}{17}$$

REFERENCE ANGLES

Our general definitions of sine, cosine, and tangent for any point (x, y) do not require a right triangle. However, as we saw in Example 12.16, drawing a diagram is very helpful. For any angle θ in standard position whose terminal side does not lie along one of the axes, we can draw a perpendicular from a point on the terminal side to the x-axis (not the y-axis!) and create a right triangle called the *reference triangle*. The acute angle between the terminal side of θ and the x-axis is called the *reference angle*.

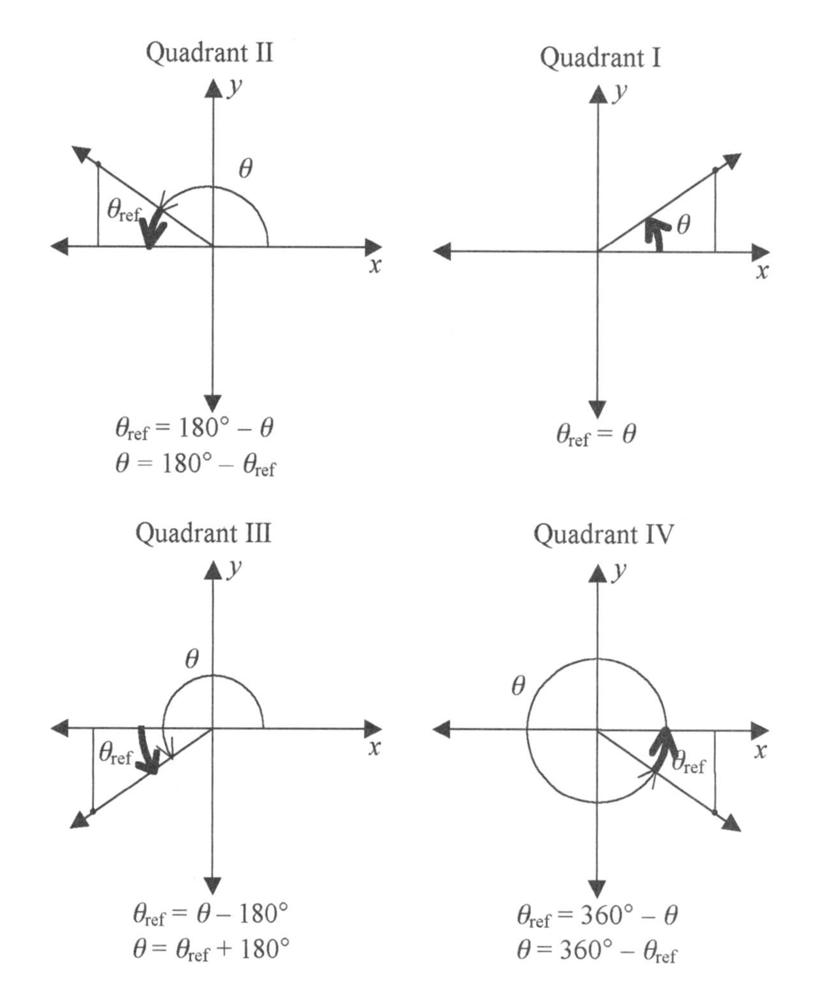

FIGURE 12.13

EXAMPLE 12.17

Find the reference angle for each of the following angles.

a. 120°
b. 340°
c. 255°

SOLUTION

It is not necessary to memorize all the relationships between θ and θ_{ref} shown in Figure 12.13. Instead, sketching a diagram for each problem will help you to easily find the desired reference angle.

a. The angle 120° is in Quadrant II. The reference angle is the additional amount needed to make 180°: $\theta_{ref} = 180° - 120° = 60°$.

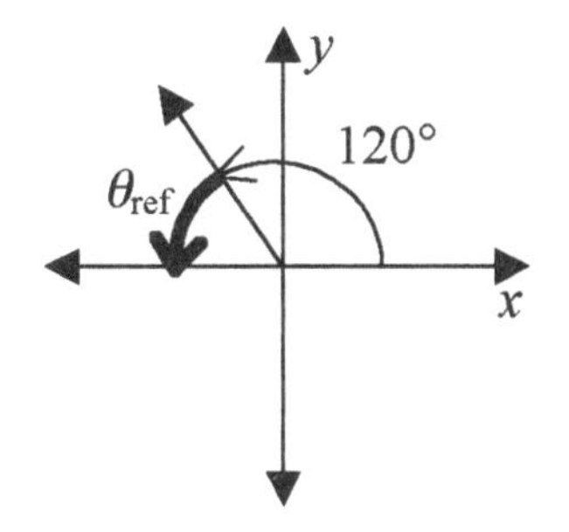

b. The angle 340° is in Quadrant IV. The reference angle is the additional amount needed to make 360°: $\theta_{ref} = 360° - 340° = 20°$.

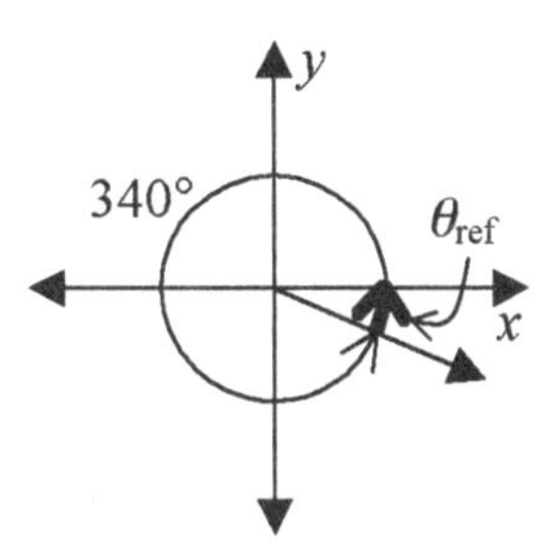

c. The angle 255° is in Quadrant III. The reference angle is the excess beyond 180°: $\theta_{ref} = 255° - 180° = 75°$.

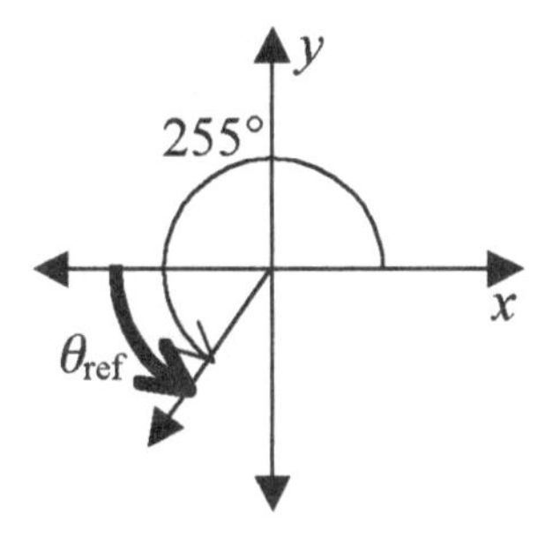

COMMON ERROR

When drawing the diagrams to determine the reference angle, some students draw a right angle with the *y*-axis instead of the *x*-axis. Do not do this! Reference angles are always derived from comparisons with 180° and 360°, never with 90° or 270°.

The same diagrams and formulas are used to determine reference angles in radians. The difference is that 180° is replaced with π and 360° is replaced with 2π.

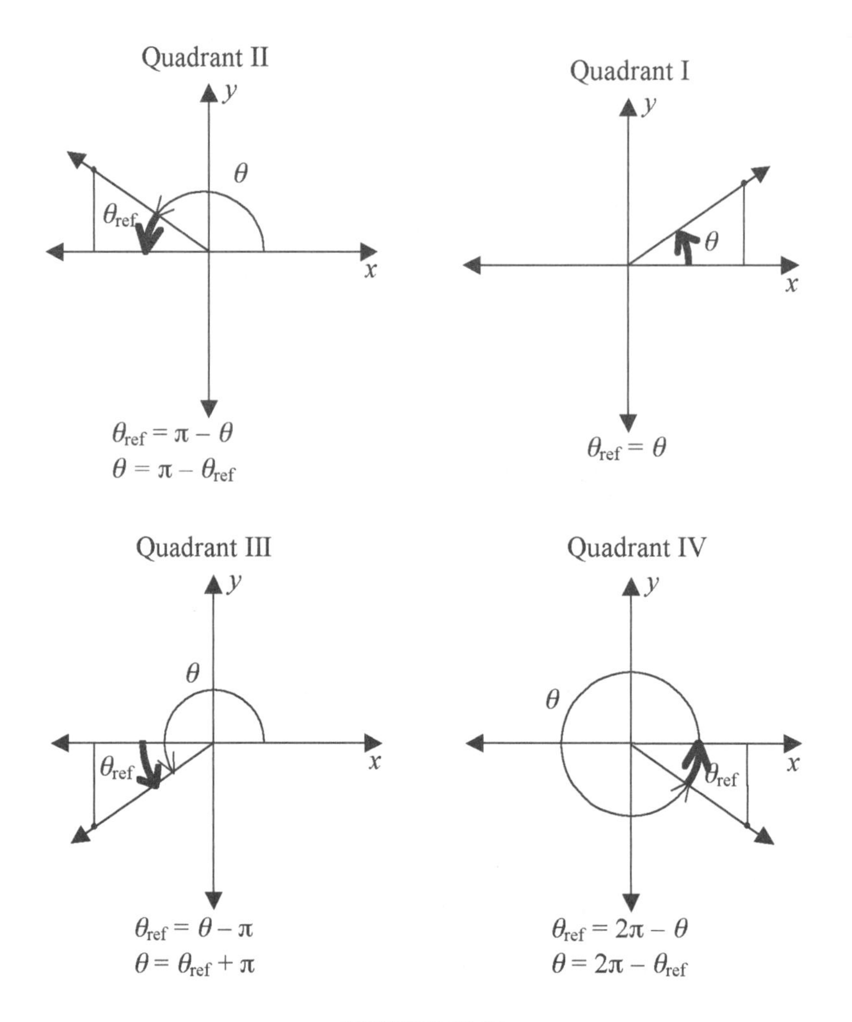

FIGURE 12.14

EXAMPLE 12.18

Find the reference angle for each of the following angles.

a. $\frac{5}{6}\pi$ **b.** $\frac{5}{4}\pi$ **c.** $\frac{5}{3}\pi$

SOLUTION

Sketch a diagram for each problem and see how to find the desired reference angle by comparing it to π or 2π.

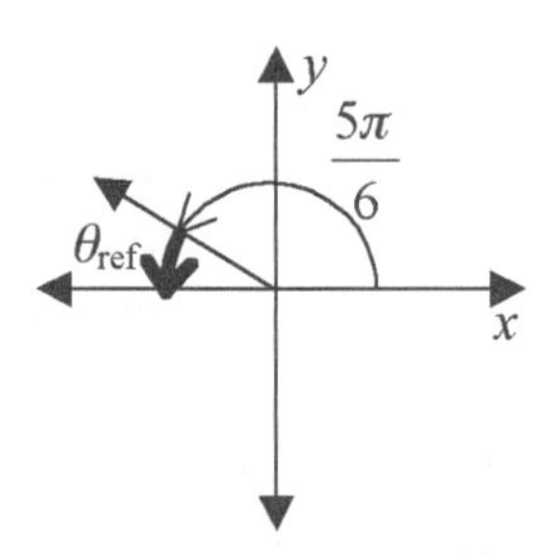

a. The angle $\frac{5}{6}\pi$ is in Quadrant II. The reference angle is the additional amount needed to make π:

$$\theta_{ref} = \pi - \frac{5}{6}\pi = \frac{6\pi}{6} - \frac{5\pi}{6} = \frac{\pi}{6}.$$

Rewriting π as a fraction with a common denominator with the angle will make the subtraction easier.

b. The angle $\frac{5}{4}\pi$ is in Quadrant III. The reference angle is the excess beyond π:

$$\theta_{ref} = \frac{5}{4}\pi - \pi = \frac{5\pi}{4} - \frac{4\pi}{4} = \frac{\pi}{4}.$$

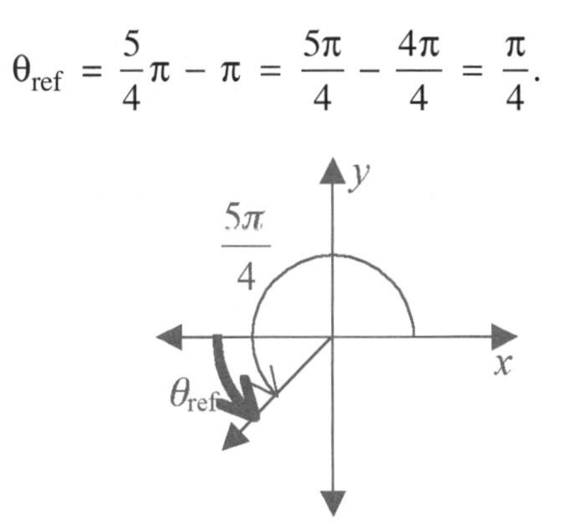

c. The angle $\frac{5}{3}\pi$ is in Quadrant IV. The reference angle is the additional amount needed to make 2π:

$$\theta_{ref} = 2\pi - \frac{5}{3}\pi = \frac{6\pi}{3} - \frac{5\pi}{3} = \frac{\pi}{3}.$$

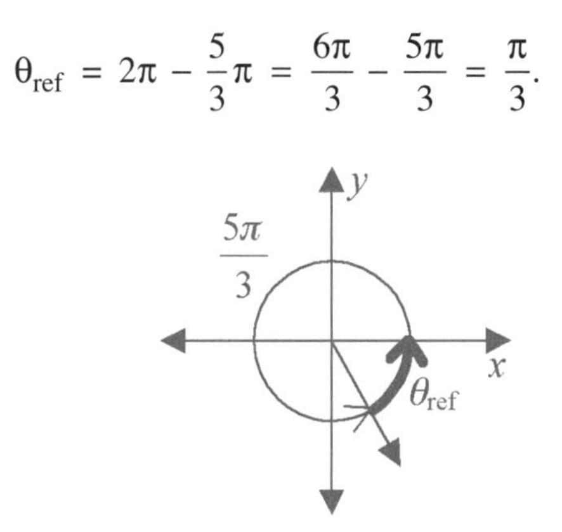

Reference angles allow us to relate the trigonometric function values of any angle θ to the values of an acute angle θ_{ref}. The trigonometric functions values for θ will be the same as the trigonometric function values for θ_{ref} except possibly for the sign. The sign will be determined by the quadrant where θ is. In Example 12.17, we found the reference angle for 120° to be 60°. In Quadrant II, sine is positive; cosine and tangent are negative. So sin 120° = sin 60°, cos 120° = –cos 60°, and tan 120° = –tan 60°.

EXAMPLE 12.19 Express sin 320°, cos 320°, and tan 320° as functions of an acute angle.

SOLUTION

First find the reference angle. 320° is in Quadrant IV, so the reference angle is 360° – 320° = 40°. In Quadrant IV, cosine is positive and sine and tangent are both negative:

sin 320° = –sin 40°

cos 320° = cos 40°

tan 320° = –tan 40°

SECTION EXERCISES

12–11. The terminal side of an angle θ in standard position intersects the unit circle at the point $(-0.341, k)$ where k is a positive number.

a. Find the value of k to the nearest thousandth.

b. Find the values of $\cos \theta$, $\sin \theta$, and $\tan \theta$ to the nearest thousandth.

12–12. If θ is in standard position and $\cos \theta = -\frac{\sqrt{3}}{2}$, in what quadrant(s) could θ terminate?

12–13. The angle θ is in standard position. In what quadrant does θ terminate if $\sin \theta < 0$ and $\tan \theta > 0$?

12–14. Find the measure of the reference angle for each of the following angles:

a. $255°$

b. $320°$

12–15. An angle θ in standard position has a reference angle measuring $35°$. Find the measure of θ if θ terminates:

a. In Quadrant II

b. In Quadrant IV

12–16. Find the measure of the reference angle for each of the following angles:

a. $\frac{4}{5}\pi$

b. $\frac{11}{9}\pi$

Trigonometric Function Values for Special Angles

Your calculator will evaluate trigonometric functions very quickly and accurately enough for almost any application. Nevertheless, you should know the exact trigonometric function values of certain special angles without having to rely on your calculator. Think of them as the trigonometry equivalent of multiplication tables. Your calculator will multiply for you, but you still ought to be able to do 2×3 in your head.

QUADRANTAL ANGLES

Angles having their terminal sides on a coordinate axis are called *quadrantal angles.* These angles include $0°$, $90°$, $180°$, $270°$, and $360°$. There are others, of course, such as $450°$ and $-90°$, but these five are the important ones. It is not even strictly necessary to memorize them. You just need to understand that you can easily find trigonometric function values of these angles using the unit circle. All you need are the coordinates of the points where the unit circle intersects the coordinate axes and the unit circle definitions of the trigonometric function values.

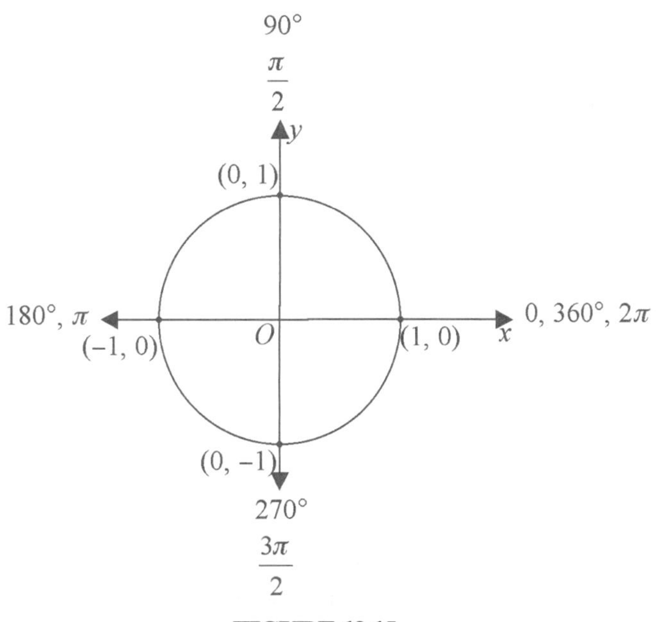

FIGURE 12.15

TABLE 12.1
TRIGONOMETRIC VALUES OF QUADRANTAL ANGLES

	0° (0)	**90° $\left(\frac{\pi}{2}\right)$**	**180° (π)**	**270° $\left(\frac{3\pi}{2}\right)$**	**360° (2π)**
$\sin \theta = y$	0	1	0	−1	0
$\cos \theta = x$	1	0	−1	0	1
$\tan \theta = \frac{y}{x}$	0	Undefined	0	Undefined	0

30°, 45°, AND 60° ANGLES

The other special angles worth knowing come from two special triangles, the isosceles right triangle and the 30°–60°–90° triangle. By using these right triangles, we can determine the exact value of sine, cosine, and tangent of these angles. You should either memorize these exact values or, even better, know how to find them quickly using the appropriate triangles.

An isosceles right triangle can be used to find the trigonometric function values for 45°. Just remember that the two legs in an isosceles triangle are congruent; it is convenient to let them both have length 1. Then you can either remember that the length of the hypotenuse will be $\sqrt{2}$ or you can easily work it out with the Pythagorean theorem.

The triangle is shown in Figure 12.16 with the exact trigonometric function values for 45° (which is the same as $\frac{\pi}{4}$ radians).

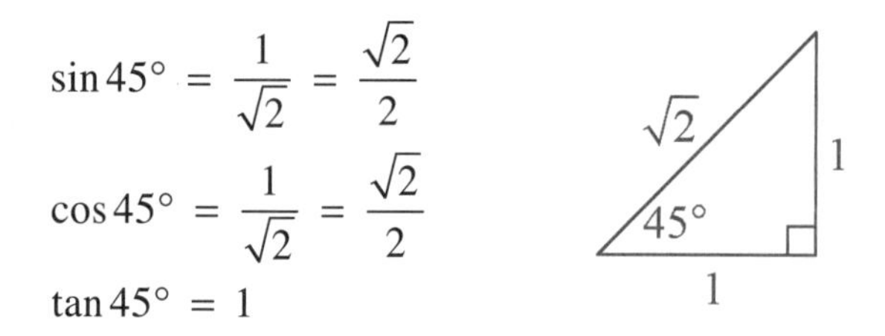

FIGURE 12.16

Note that some teachers or textbooks insist that denominators be rationalized; others do not.

Drawing an altitude in an equilateral triangle will create two 30°–60°–90° triangles. The important thing to remember about this triangle is that the hypotenuse is twice the length of the shorter leg (the leg opposite the 30° angle). It is convenient to let the shorter leg be 1 and the hypotenuse 2. Then either from memory or by using the Pythagorean theorem, you know that the longer leg is $\sqrt{3}$. This triangle, shown in Figure 12.17, can be used to find the trigonometric function values for both 30° ($\frac{\pi}{6}$ radians) and 60° ($\frac{\pi}{3}$ radians).

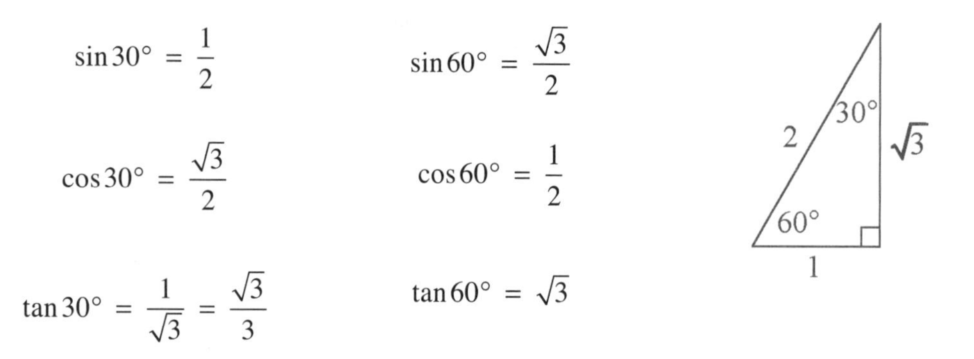

FIGURE 12.17

SUMMARY OF EXACT TRIGONOMETRIC VALUES FOR QUADRANT I

Table 12.2 shows a convenient way to remember the trigonometric function values for 0°, 30°, 45°, 60°, and 90°. They are not in simplest form but, instead, are written so you can see and remember the pattern for each function. Note that the numbers inside the radicals either count up 0, 1, 2, 3, 4 for sine or down 4, 3, 2, 1, 0 for cosine. For tangent, we count up in the numerator and down in the denominator.

TABLE 12.2
EXACT TRIGONOMETRIC VALUES FOR SPECIAL ANGLES

	0° (0)	**30°** $\left(\frac{\pi}{6}\right)$	**45°** $\left(\frac{\pi}{4}\right)$	**60°** $\left(\frac{\pi}{3}\right)$	**90°** $\left(\frac{\pi}{2}\right)$
Sin θ	$\frac{\sqrt{0}}{2}$	$\frac{\sqrt{1}}{2}$	$\frac{\sqrt{2}}{2}$	$\frac{\sqrt{3}}{2}$	$\frac{\sqrt{4}}{2}$
Cos θ	$\frac{\sqrt{4}}{2}$	$\frac{\sqrt{3}}{2}$	$\frac{\sqrt{2}}{2}$	$\frac{\sqrt{1}}{2}$	$\frac{\sqrt{0}}{2}$
Tan θ	$\frac{\sqrt{0}}{\sqrt{4}}$	$\frac{\sqrt{1}}{\sqrt{3}}$	$\frac{\sqrt{2}}{\sqrt{2}}$	$\frac{\sqrt{3}}{\sqrt{1}}$	$\frac{\sqrt{4}}{\sqrt{0}}$

Table 12.3 shows the same values simplified and rationalized.

TABLE 12.3
SIMPLIFIED TRIGONOMETRIC VALUES FOR SPECIAL ANGLES

	0° (0)	**30°** $\left(\frac{\pi}{6}\right)$	**45°** $\left(\frac{\pi}{4}\right)$	**60°** $\left(\frac{\pi}{3}\right)$	**90°** $\left(\frac{\pi}{2}\right)$
Sin θ	0	$\frac{1}{2}$	$\frac{\sqrt{2}}{2}$	$\frac{\sqrt{3}}{2}$	1
Cos θ	1	$\frac{\sqrt{3}}{2}$	$\frac{\sqrt{2}}{2}$	$\frac{1}{2}$	0
Tan θ	0	$\frac{\sqrt{3}}{3}$	1	$\sqrt{3}$	Undefined

TRIGONOMETRIC VALUES FOR SPECIAL ANGLES IN OTHER QUADRANTS

Knowing the exact values for the trigonometric functions of 30°, 45°, and 60° angles allows us to determine the exact values for angles in the other three quadrants that have a reference angle of 30°, 45°, or 60°. Keep in mind that only the sign (positive or negative) changes in the other quadrants; the absolute value remains the same. Figure 12.18 shows the three special angles reflected over the y-axis and x-axis forming four angles, each with the same reference angle.

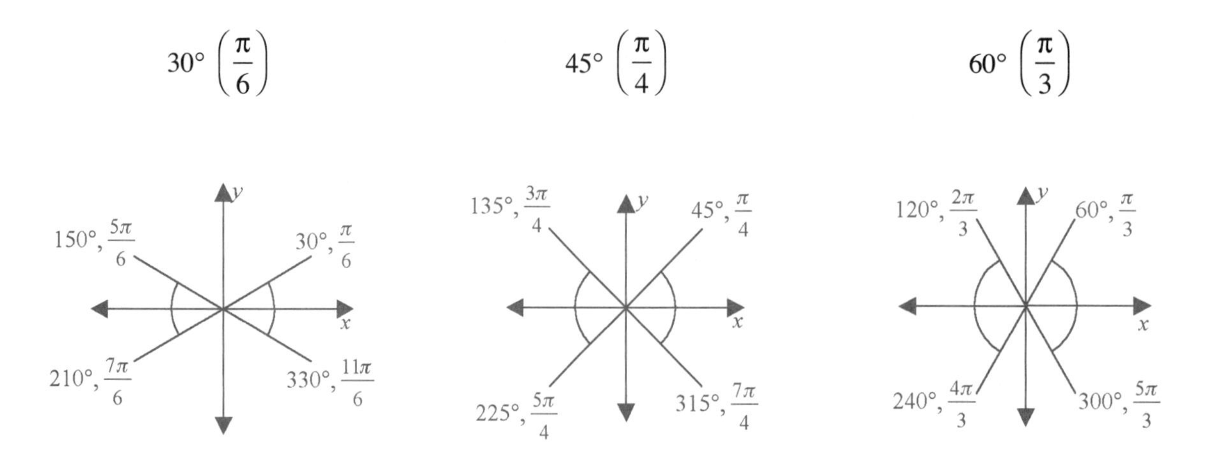

FIGURE 12.18

EXAMPLE 12.20 Write the exact values for sin 150°, cos 150°, and tan 150°.

SOLUTION

Determine the reference angle for 150°. Since 150° is in Quadrant II, subtract it from 180°: 180° – 150° = 30°. The reference angle is 30°; this will determine the values of the trigonometric functions.

$$\sin 30^\circ = \frac{1}{2}$$

$$\cos 30^\circ = \frac{\sqrt{3}}{2}$$

$$\tan 30^\circ = \frac{1}{\sqrt{3}} = \frac{\sqrt{3}}{3}$$

Since 150° is located in Quadrant II, sine will be positive but cosine and tangent will be negative.

$$\sin 150^\circ = \frac{1}{2}$$

$$\cos 150^\circ = -\frac{\sqrt{3}}{2}$$

$$\tan 150^\circ = -\frac{\sqrt{3}}{3}$$

SEE CALC TIP 12C

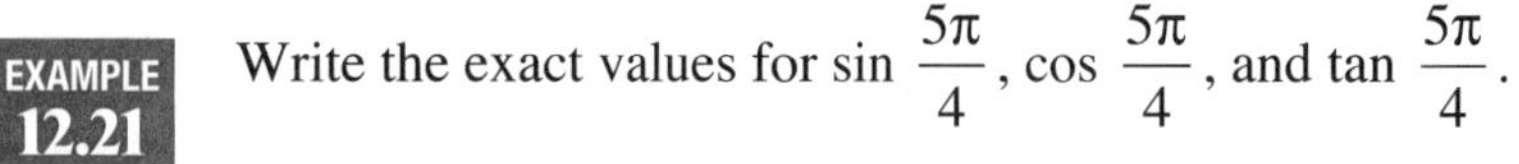

EXAMPLE 12.21 Write the exact values for $\sin \frac{5\pi}{4}$, $\cos \frac{5\pi}{4}$, and $\tan \frac{5\pi}{4}$.

SOLUTION

Determine the reference angle for $\frac{5\pi}{4}$. Since $\frac{5\pi}{4}$ is in Quadrant III, subtract π from it to find its reference angle: $\frac{5\pi}{4} - \pi = \frac{\pi}{4}$. The reference angle is $\frac{\pi}{4}$; this will determine the values of the trigonometric functions.

$$\sin\frac{\pi}{4} = \frac{1}{\sqrt{2}} = \frac{\sqrt{2}}{2}$$

$$\cos\frac{\pi}{4} = \frac{1}{\sqrt{2}} = \frac{\sqrt{2}}{2}$$

$$\tan\frac{\pi}{4} = 1$$

Since $\frac{5\pi}{4}$ is located in Quadrant III, tangent will be positive but sine and cosine will be negative.

$$\sin\frac{5\pi}{4} = -\frac{\sqrt{2}}{2}$$

$$\cos\frac{5\pi}{4} = -\frac{\sqrt{2}}{2}$$

$$\tan\frac{5\pi}{4} = 1$$

SEE CALC TIP 12D

SECTION EXERCISES

For 17–27, you should be able to do this without using your calculator. Find the exact value of each of the following.

12–17. $\sin 30°$

12–18. $2\tan 315°$

12–19. $\sin 225°$

12–20. $\cos\left(-\frac{\pi}{3}\right)$

12–21. $\sin\frac{2\pi}{3}$

12–22. $\frac{1}{2}\tan\pi$

12–23. $3\cos 2\pi$

12–24. $2\sin\frac{3\pi}{2}$

12–25. $\cos(5\pi)$

12–26. $\tan\frac{5\pi}{6}$

12–27. $4\sin\frac{5\pi}{4}$

For 28–30, evaluate the function and give the exact value.

12–28. $f(\pi)$ for $f(x) = \cos x + 2\sin \frac{1}{2}x$.

12–29. $f\left(\frac{3\pi}{2}\right)$ for $f(x) = \sin x + \cos \frac{1}{3}x$.

12–30. If $f(x) = 2\sin x + \cos 2x$, evaluate $f\left(\frac{\pi}{3}\right)$.

Reciprocal Trigonometric Functions

There are three other trigonometric functions that are not as frequently used as sine, cosine, and tangent. The functions are called *reciprocal trigonometric functions* because each is the reciprocal of one of the three basic trigonometric functions. *Cosecant* is the reciprocal of sine, *secant* is the reciprocal of cosine, and *cotangent* is the reciprocal of tangent.

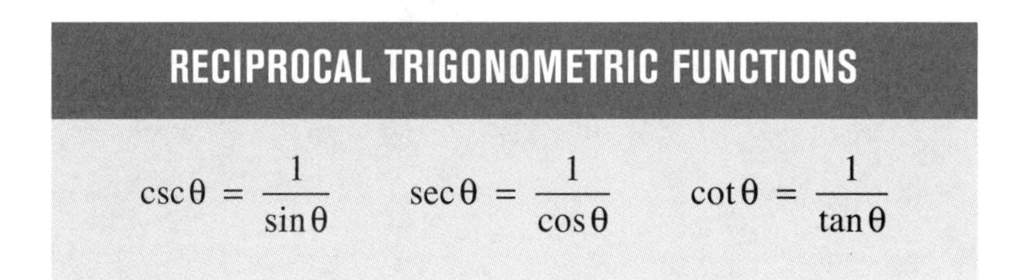

RECIPROCAL TRIGONOMETRIC FUNCTIONS

$$\csc\theta = \frac{1}{\sin\theta} \qquad \sec\theta = \frac{1}{\cos\theta} \qquad \cot\theta = \frac{1}{\tan\theta}$$

COMMON ERROR

A lot of students want the *co*secent to be the reciprocal of *co*sine (and secant to be the reciprocal of sine). However, these functions were not defined that way.

EXAMPLE 12.22 A cell phone tower on level ground is stabilized by wires anchored in the ground 50 feet from the base of the tower. Express the length, l, of the wire as a function of the angle, θ, it makes with the ground.

SOLUTION

Draw a diagram. Label the variables and known quantities.
In the triangle formed by the tower, the ground and the wires, we see that l is the hypotenuse and, for angle θ, 50 is the length of the adjacent side. Cosine is the ratio of the adjacent and hypotenuse sides.

$$\cos\theta = \frac{50}{l}$$

By solving for l, we get:

$$l = \frac{50}{\cos\theta} \text{ or } l = 50 \sec\theta$$

Trigonometric Identities

An *identity* is an equation that is true for all values of the variable in some domain. A very simple example is $x + x = 2x$. This is not an equation that is to be solved for x. Rather, it is true for all real values of x. A somewhat more interesting example is $\frac{x^2 - 1}{x - 1} = x + 1$, which is true for all real $x \neq 1$. There are many useful identities in trigonometry. Here we will cover only a couple of the most important ones.

QUOTIENT IDENTITIES

From the unit circle definitions of the trigonometric functions, we know the following:

$$\sin\theta = y$$
$$\cos\theta = x$$
$$\tan\theta = \frac{y}{x},\ x \neq 0$$

Substitute the trigonometric expressions for x and y into the ratio for tangent to get $\tan\theta = \frac{\sin\theta}{\cos\theta}$. Rewriting this ratio as its reciprocal gives $\cot\theta = \frac{\cos\theta}{\sin\theta}$.

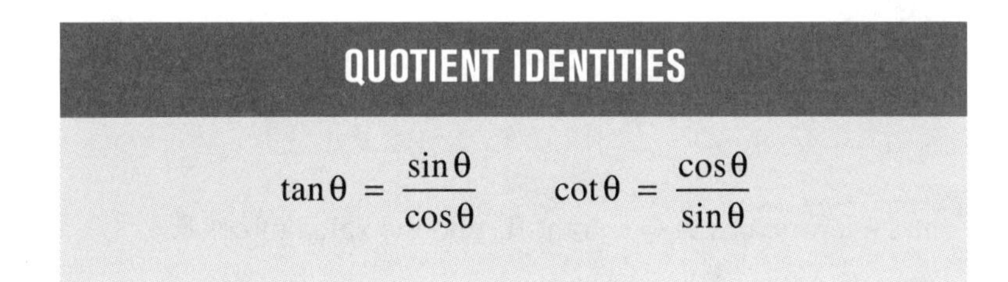
QUOTIENT IDENTITIES

$$\tan\theta = \frac{\sin\theta}{\cos\theta} \qquad \cot\theta = \frac{\cos\theta}{\sin\theta}$$

PYTHAGOREAN IDENTITIES

Arguably the most important trigonometric identities are the Pythagorean identities. We again use the unit circle definitions. The unit circle has radius 1 and is centered at the origin; its equation is $x^2 + y^2 = 1$. (The equation for a circle is derived from the Pythagorean theorem.) If θ is an angle in standard position that intersects the unit circle at the point (x, y), then by the definitions of sine and cosine we have $x = \cos\theta$ and $y = \sin\theta$.

Substituting into the equation of the unit circle, $x^2 + y^2 = 1$, gives $(\cos\theta)^2 + (\sin\theta)^2 = 1$.

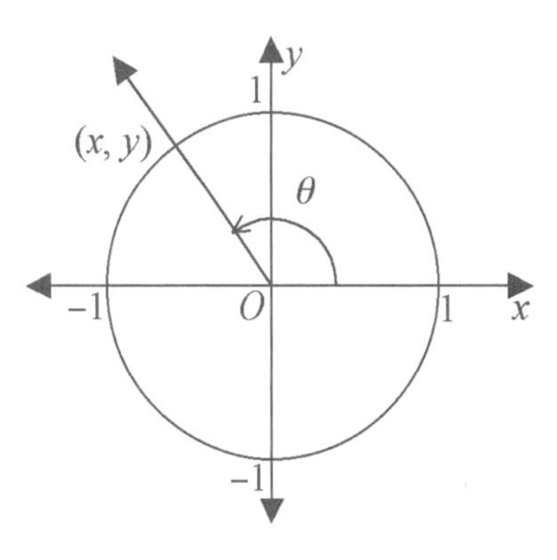

FIGURE 12.19

PYTHAGOREAN IDENTITY

For any angle θ, $\cos^2\theta + \sin^2\theta = 1$.

The notation $\cos^2\theta$ is commonly used to represent $(\cos\theta)^2$. This notation bothers some mathematicians, with good reason. However, it has been around for several centuries now, so we are probably stuck with it.

Verify the Pythagorean identity for $\theta = 60°$.

SOLUTION

The angle 60° is one for which we know exact trigonometric function values:

$\cos 60° = \frac{1}{2}$ and $\sin 60° = \frac{\sqrt{3}}{2}$. Substituting into the Pythagorean identity gives the following:

$$\cos^2 60° + \sin^2 60° = 1$$

$$\left(\frac{1}{2}\right)^2 + \left(\frac{\sqrt{3}}{2}\right)^2 = 1$$

$$\frac{1}{4} + \frac{3}{4} = 1$$

$$1 = 1$$

EXAMPLE 12.24

If $\sin\theta = \frac{3}{4}$ and θ terminates in Quadrant II, find the value of $\cos\theta$.

SOLUTION

Use the Pythagorean identity:

$$\cos^2\theta + \sin^2\theta = 1$$

$$\cos^2\theta + \left(\frac{3}{4}\right)^2 = 1$$

$$\cos^2\theta + \frac{9}{16} = 1$$

$$\cos^2\theta = 1 - \frac{9}{16}$$

$$\cos^2\theta = \frac{7}{16}$$

$$\cos\theta = \pm\sqrt{\frac{7}{16}}$$

$$\cos\theta = \pm\frac{\sqrt{7}}{4}$$

So far, this is good. However, we seem to have two possible answers. How do we know which is right, or are they both right? The problem said that θ is a Quadrant II angle. In Quadrant II, sine is positive but cosine is negative. For this problem, we must pick the negative value. The answer is $\cos\theta = -\frac{\sqrt{7}}{4}$.

OTHER PYTHAGOREAN IDENTITIES

The identity $\sin^2\theta + \cos^2\theta = 1$ can be rearranged to give $\sin^2\theta = 1 - \cos^2\theta$ and $\cos^2\theta = 1 - \sin^2\theta$. The identity $\sin^2\theta + \cos^2\theta = 1$ can also be divided by either $\sin^2\theta$ or $\cos^2\theta$ to yield new identities.

Start with the fundamental Pythagorean identity.

$$\sin^2\theta + \cos^2\theta = 1$$

Divide each term by $\sin^2\theta$.

$$\frac{\sin^2\theta}{\sin^2\theta} + \frac{\cos^2\theta}{\sin^2\theta} = \frac{1}{\sin^2\theta}$$

$$1 + \cot^2\theta = \csc^2\theta$$

Divide each term by $\cos^2\theta$.

$$\frac{\sin^2\theta}{\cos^2\theta} + \frac{\cos^2\theta}{\cos^2\theta} = \frac{1}{\cos^2\theta}$$

$$\tan^2\theta + 1 = \sec^2\theta$$

PYTHAGOREAN IDENTITIES

$\sin^2\theta + \cos^2\theta = 1$ $\qquad$ $\sin^2\theta = 1 - \cos^2\theta$ $\qquad$ $\cos^2\theta = 1 - \sin^2\theta$

$1 + \cot^2\theta = \csc^2\theta$ $\qquad$ $\sin\theta = \pm\sqrt{1 - \cos^2\theta}$ $\qquad$ $\cos\theta = \pm\sqrt{1 - \sin^2\theta}$

$\tan^2\theta + 1 = \sec^2\theta$

SECTION EXERCISES

For 31–32, draw an appropriate triangle and find the exact value for each of the following.

12–31. If θ is an acute angle and $\sin\theta = \frac{2}{5}$, find the values of cos θ and tan θ.

12–32. If $\cos\theta = -\frac{7}{25}$ and θ is in Quadrant III, find the value of sin θ.

12–33. What is the exact value of csc 60°?

12–34. Use the triangle at the right to evaluate the following:

a. sec A
b. cot B
c. csc A
d. csc B
e. sec B
f. cot A

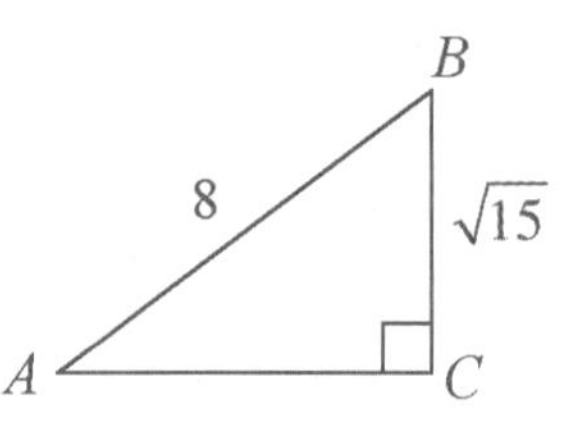

The Sum and Difference Formulas

It would be nice if $\sin(A + B) = \sin A + \sin B$. However, this is not true (and wishing does not make it so). The formulas below are used to find trigonometric values of the sum or difference of two angles. These formulas can be memorized, but frequently these and other trigonometric formulas are given on a reference sheet.

FUNCTIONS OF THE SUM OF TWO ANGLES

$$\sin(A + B) = \sin A \cos B + \cos A \sin B$$

$$\cos(A + B) = \cos A \cos B - \sin A \sin B$$

$$\tan(A + B) = \frac{\tan A + \tan B}{1 - \tan A \tan B}$$

FUNCTIONS OF THE DIFFERENCE OF TWO ANGLES

$$\sin(A - B) = \sin A \cos B - \cos A \sin B$$

$$\cos(A - B) = \cos A \cos B + \sin A \sin B$$

$$\tan(A - B) = \frac{\tan A - \tan B}{1 + \tan A \tan B}$$

EXAMPLE 12.25

Use the sum formula to find the exact value of sin 75°.

SOLUTION

To use the sum formula, we must find two angles that add to 75°. Because the problem asks for the exact value, choose special angles for which we know the exact trigonometric function values: 30°, 45°, and 60°. In this case, we can rewrite sin 75° = sin (45° + 30°) and then apply the angle sum formula for sine.

The formula is $\sin(A + B) = \sin A \cos B + \cos A \sin B$. In this problem, we are using $A = 45°$ and $B = 30°$. Substitute these values into the formula. Then see Table 12.3 (or, better yet, reproduce it from memory!) to find the needed trigonometric function values.

$$\begin{aligned}\sin(45° + 30°) &= \sin 45° \cos 30° + \cos 45° \sin 30° \\ &= \left(\frac{\sqrt{2}}{2}\right)\left(\frac{\sqrt{3}}{2}\right) + \left(\frac{\sqrt{2}}{2}\right)\left(\frac{1}{2}\right) \\ &= \frac{\sqrt{6}}{4} + \frac{\sqrt{2}}{4} \\ &= \frac{\sqrt{6} + \sqrt{2}}{4}\end{aligned}$$

SEE CALC TIP 12E

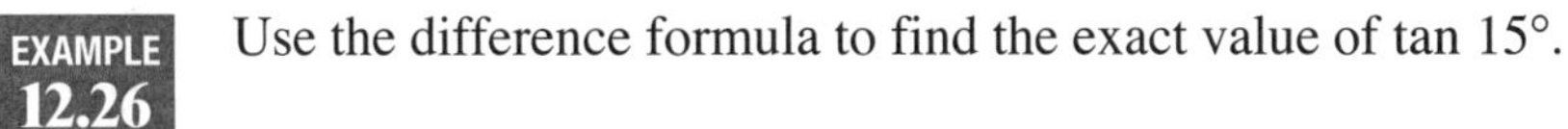

EXAMPLE 12.26 Use the difference formula to find the exact value of tan 15°.

SOLUTION

To use the difference formula, we must find two angles that subtract to 15°. Again, in order to find the exact value, we want to choose from among the special angles for which we know exact trigonometric function values. Rewrite $\tan 15° = \tan(45° - 30°)$ and then use the angle difference formula for tangent.

The formula is $\tan(A - B) = \dfrac{\tan A - \tan B}{1 + \tan A \tan B}$ where we are using $A = 45°$ and $B = 30°$. Substituting into the formula, we get the following:

$$\begin{aligned}\tan(45° - 30°) &= \frac{\tan 45° - \tan 30°}{1 + \tan 45° \tan 30°} \\ &= \frac{1 - \frac{\sqrt{3}}{3}}{1 + (1)\frac{\sqrt{3}}{3}}\end{aligned}$$

This is correct but is a far cry from simplest form; it is a complex fraction containing radicals. The steps for simplifying it are shown below. If you need to, see Chapters 8 and 9 to review how to simplify complex fractions and perform operations with radicals.

$$\tan(45° - 30°) = \frac{1 - \frac{\sqrt{3}}{3}}{1 + (1)\frac{\sqrt{3}}{3}}$$
$$= \frac{\frac{3 - \sqrt{3}}{3}}{\frac{3 + \sqrt{3}}{3}}$$
$$= \frac{3 - \sqrt{3}}{3 + \sqrt{3}}$$
$$= \frac{3 - \sqrt{3}}{3 + \sqrt{3}} \cdot \frac{3 - \sqrt{3}}{3 - \sqrt{3}}$$
$$= \frac{9 - 6\sqrt{3} + 3}{6}$$
$$= \frac{12 - 6\sqrt{3}}{6}$$
$$= 2 - \sqrt{3}$$

Simplifying and rationalizing the answer has a lot of steps and therefore a lot of opportunities to make a mistake. Double-check by comparing the decimal values for the trigonometric function and the answer on your calculator: tan 15° = 0.2679491924 and $2 - \sqrt{3}$ = 0.2679491924.

EXAMPLE 12.27

Find the exact value of cos 280° cos 50° – sin 280° sin 50°.

SOLUTION

The two angles given, 280° and 50°, are not special angles with known trigonometric values. If you use a calculator to answer this question, you will get a decimal approximation, not an exact value. Instead, try to match this expression with one of the sum or difference formulas. Pay attention to addition or subtraction and the position of the sine and cosines.

$$\cos(A + B) = \cos A \cos B - \sin A \sin B$$
$$= \cos 280° \cos 50° - \sin 280° \sin 50°$$

Matching up the formulas, we have $A = 280°$ and $B = 50°$ in the formula for $\cos(A + B)$. Substituting in the values for A and B, we get $\cos(280° + 50°) = \cos 330°$.

Remember that 330° is a Quadrant IV angle with a reference angle of 30°. Cosine is positive in Quadrant IV, and 30° is a special angle with a known trigonometric value. So we get $\cos 330° = \cos 30° = \frac{\sqrt{3}}{2}$.

EXAMPLE 12.28 If $90° < A < 180°$, $\sin A = \frac{\sqrt{5}}{3}$, $270° < B < 360°$, and $\cos B = \frac{5}{6}$, find the value of $\cos(A - B)$.

SOLUTION

To find $\cos(A - B)$, you need to use the formula $\cos(A - B) = \cos A \cos B + \sin A \sin B$. You know $\sin A$ but not $\sin B$. You know $\cos B$ but not $\cos A$. You need to find the missing values before using the formula.

$$A: \quad \sin A = \frac{\sqrt{5}}{3}, \quad \cos A = ?$$

$$B: \quad \cos B = \frac{5}{6}, \quad \sin B = ?$$

There is more than one way to find other trigonometric values if you know one. One way was shown in Example 12.16. Another way is to use one of the Pythagorean identities. Either method requires that you know the quadrant where the angle is located so you can determine the sign of the trigonometric value. In this example, $90° < A < 180°$, so A is in Quadrant II and $270° < B < 360°$, so B is in Quadrant IV.

$$A: \quad \text{Quadrant II}, \quad \sin A = \frac{\sqrt{5}}{3}, \quad \cos A = ?$$

$$B: \quad \text{Quadrant IV}, \quad \cos B = \frac{5}{6}, \quad \sin B = ?$$

One of the identities, $\cos\theta = \pm\sqrt{1 - \sin^2\theta}$ lets you use $\sin A$ to find $\cos A$.

$$\begin{aligned}
\cos A &= \pm\sqrt{1 - \sin^2 A} \\
&= \pm\sqrt{1 - \left(\frac{\sqrt{5}}{3}\right)^2} \\
&= \pm\sqrt{1 - \frac{5}{9}} \\
&= \pm\sqrt{\frac{4}{9}} \\
&= \pm\frac{2}{3}
\end{aligned}$$

Since angle A is in Quadrant II, cosine is negative and $\cos A = -\frac{2}{3}$.

Another identity, $\sin\theta = \pm\sqrt{1 - \cos^2\theta}$, lets you use $\cos B$ to find $\sin B$.

$$\begin{aligned}\sin B &= \pm\sqrt{1-\cos^2 B}\\ &= \pm\sqrt{1-\left(\frac{5}{6}\right)^2}\\ &= \pm\sqrt{1-\frac{25}{36}}\\ &= \pm\sqrt{\frac{11}{36}}\\ &= \pm\frac{\sqrt{11}}{6}\end{aligned}$$

Since angle B is in Quadrant IV, sine is negative and $\sin B = -\frac{\sqrt{11}}{6}$.

$$A: \quad \sin A = \frac{\sqrt{5}}{3}, \quad \cos A = -\frac{2}{3}$$

$$B: \quad \cos B = \frac{5}{6}, \quad \sin B = -\frac{\sqrt{11}}{6}$$

The last step is to substitute the values into the formula and simplify.

$$\begin{aligned}\cos(A-B) &= \cos A\cos B + \sin A\sin B\\ &= \left(-\frac{2}{3}\right)\left(\frac{5}{6}\right)+\left(\frac{\sqrt{5}}{3}\right)\left(-\frac{\sqrt{11}}{6}\right).\\ &= -\frac{10}{18}-\frac{\sqrt{55}}{18}\end{aligned}$$

So $\cos(A-B) = -\frac{10}{18} - \frac{\sqrt{55}}{18}$.

EXAMPLE 12.29 For $f(x) = \cos x$, evaluate $f(-x)$ and determine if $f(x) = \cos x$ is an even or odd function.

SOLUTION

We want to evaluate $f(-x) = \cos(-x)$. What does this have to do with sum or difference formulas? Remember that 0° is a special angle for which we know the exact values of the trigonometric functions. Then write $\cos(-x) = \cos(0 - x)$ and use the difference formula for cosine with $A = 0$ and $B = x$.

$$\begin{aligned}\cos(0^\circ - x) &= \cos 0^\circ \cos x + \sin 0^\circ \sin x\\ &= (1)\cos x + (0)\sin x\\ &= \cos x\end{aligned}$$

Since $\cos(-x) = \cos(x)$, $f(x) = \cos x$ is an even function. You will see the y-axis symmetry of $f(x) = \cos x$ in the next chapter.

SECTION EXERCISES

12–35. If $\cos\theta = \frac{4}{5}$ and θ is an acute angle, find the exact value of $\cos(\theta - 45°)$.

For 36–38, find the exact value of the following.

12–36. $\sin 80° \cos 20° - \cos 80° \sin 20°$

12–37. $\cos 60° \cos 15° + \sin 60° \sin 15°$

12–38. $\sin \frac{11\pi}{12} \cos \frac{\pi}{3} + \cos \frac{11\pi}{12} \sin \frac{\pi}{3}$

12–39. If A and B are both acute angles, $\tan A = \frac{4}{7}$, and $\tan B = 8$, find the value of $\tan (A + B)$.

12–40. If A and B are acute angles, $\cos A = \frac{24}{25}$, and $\tan B = \frac{15}{8}$, find the exact value of $\sin (A - B)$.

CHAPTER EXERCISES

C12–1. Convert the following angle into radians: 330º.

C12–2. Convert the following angle into degrees: $\frac{7}{8}\pi$ rad.

C12–3. Find the measure of the unknown.

$$r = 12.5 \text{ feet},\ s = 15.5 \text{ feet},\ \theta = ? \text{ radians}$$

C12–4. A boat is winched up a ramp on a cable that winds around a 6″ diameter drum. What total angle, to the nearest degree, must the winch turn through to pull the boat 8 feet up the ramp?

C12–5. Sketch an angle of $\frac{7}{5}\pi$ in standard position. Then name two other angles, one positive and one negative, that are coterminal with it.

C12–6. In what quadrant does θ terminate if $\tan\theta < 0$ and $\cos\theta > 0$?

C12–7. Find the measure of the reference angle for the following angle: 115°.

C12–8. An angle θ in standard position has a reference angle measuring 35°. Find the measure of θ if θ terminates in Quadrant III.

C12–9. Find the measure of the reference angle for the following angle: $\frac{17}{10}\pi$.

For 10–14, you should be able to do this without using your calculator. Evaluate the following and give the exact values.

C12–10. $\cos 150°$

C12–11. $\tan\left(\frac{5\pi}{6}\right)$

C12–12. $\tan\left(\frac{1}{2}\pi\right)$

C12–13. $\sin\left(\frac{3\pi}{2}\right)$

C12–14. $\cos\left(\frac{7\pi}{6}\right)$

For 15–16, evaluate the functions and give the exact values.

C12–15. $f\left(\frac{\pi}{2}\right)$ for $f(x) = \sin 2x + \cos x$.

C12–16. If $f(x) = 2\cos 2x - \sin^2\left(\frac{1}{2}x\right)$, evaluate $f\left(\frac{\pi}{2}\right)$.

C12–17. If $\tan\theta = -\frac{4}{3}$ and $\sin\theta > 0$, find the exact value of $\cos\theta$.

C12–18. Given that $75 = 45 + 30$, find the exact value of $\cos(75°)$.

C12–19. What is the expression $\cos 4x \cos 2x + \sin 4x \sin 2x$ equivalent to?

(1) $\cos 2x$ (2) $\sin 2x$ (3) $\cos 6x$ (4) $\sin 6x$

C12–20. Find the exact value of $\sin 75° \cos 30° - \cos 75° \sin 30°$

C12–21. If A and B are both acute angles, $\tan A = \frac{1}{3}$ and $\tan B = \frac{3}{4}$, find the value of $\tan(A - B)$.

C12–22. If $\sin A = \frac{3}{5}$, $\cos B = \frac{12}{13}$, and A and B are both positive acute angles, find the value of $\cos(A + B)$.

Trigonometric Graphs and Equations

WHAT YOU WILL LEARN

Graphs of trigonometric functions can be used to model any event that repeats in a periodic way. In order to understand these models, you need to understand the different features of trigonometric graphs and how changes to the equation are reflected in changes to the graph. In this chapter, you will learn to:

- Graph basic trigonometric functions;
- Identify features of trigonometric graphs;
- Graph transformations of trigonometric functions;
- Write equations of trigonometric graphs;
- Model real-life situations with trigonometric graphs and equations;
- Understand and use inverse trigonometric functions;
- Solve trigonometric equations.

SECTIONS IN THIS CHAPTER

Basic Graphs of Trigonometric Functions

In the previous chapter, you learned about sine, cosine, and tangent functions. In the next two sections, you will learn their graphs.

THE GRAPHS OF $y = \sin x$ AND $y = \cos x$

Which would you rather do, go for a ride on a Ferris wheel or read about graphs of trigonometric functions? Let's imagine that you are going on a Ferris wheel ride. This particular Ferris wheel has a radius of 1 and has been set up in a trench so that exactly half of it is above ground level (the x-axis) and half of it is below. You board at point A shown in Figure 13.1, and the wheel turns counterclockwise.

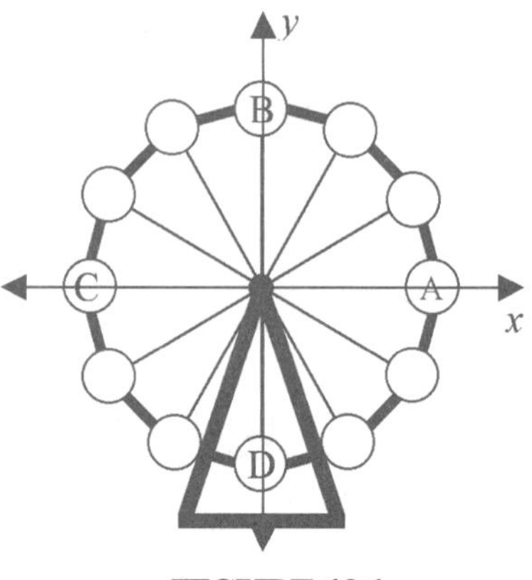

FIGURE 13.1

When the wheel has turned through an angle, θ, you are at the point $P(x, y)$ as shown in Figure 13.2. Because this is an educational Ferris wheel ride, your assignment is to make a graph of your height, y, above the ground as a function of the angle, θ. On the unit circle, $\sin \theta = y$. This will be a graph of $y = \sin \theta$.

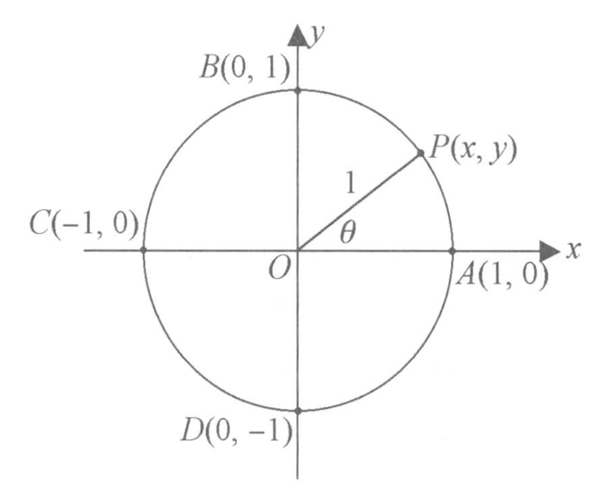

FIGURE 13.2

Trigonometric functions can be graphed using either degrees or radians, but radian measure is by far the more common choice. When you board the Ferris wheel at point A, $\theta = 0$ and you are at a height of 0. When the wheel has turned so that you are at point B, $\theta = \frac{\pi}{2}$ (90°), you have reached your maximum height of 1. What happens between those points? It should be clear that the graph is not a straight line. When θ is close to 0, you are rising relatively quickly. As θ approaches $\frac{\pi}{2}$, you are rising rather slowly. We can approximate some points on the graph using the special angles discussed in the last chapter.

$$\text{At } \theta = \frac{\pi}{6},\ y = \sin\frac{\pi}{6} = \frac{1}{2} \text{ or } 0.5.$$

$$\text{At } \theta = \frac{\pi}{4},\ y = \sin\frac{\pi}{4} = \frac{1}{\sqrt{2}} \approx .707.$$

$$\text{At } \theta = \frac{\pi}{3},\ y = \sin\frac{\pi}{3} = \frac{\sqrt{3}}{2} \approx 0.866.$$

By plotting these points and drawing a smooth curve, we find the first part of the ride looks like Figure 13.3.

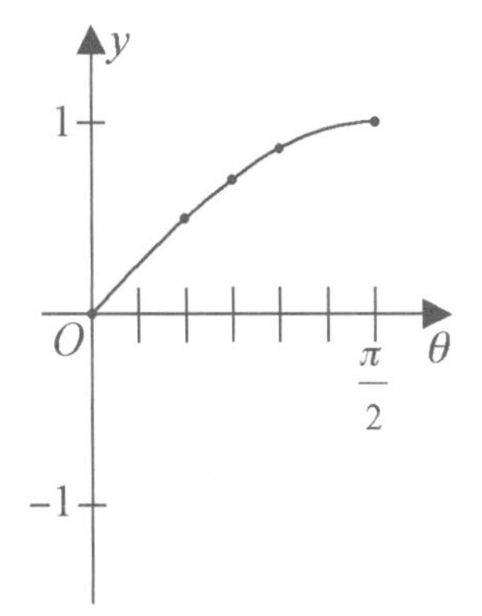

FIGURE 13.3

From $\theta = \frac{\pi}{2}$ to $\theta = \pi$, you are going back down. By symmetry, this part of the graph will be a reflection of the first part over the line $\theta = \frac{\pi}{2}$, leading to Figure 13.4.

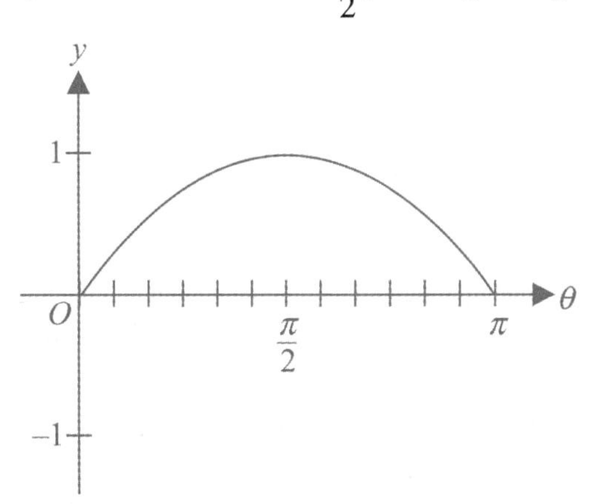

FIGURE 13.4

At this point, you drop below ground level until $\theta = \frac{3\pi}{2}$, when you are at a "height" of −1. Then you start going back up. When $\theta = 2\pi$, you are back to your starting point at ground level. Again, by symmetry, this part of the graph will be a reflection of the first part over the θ-axis (and translated to the right of course). One complete trip around the wheel looks like Figure 13.5.

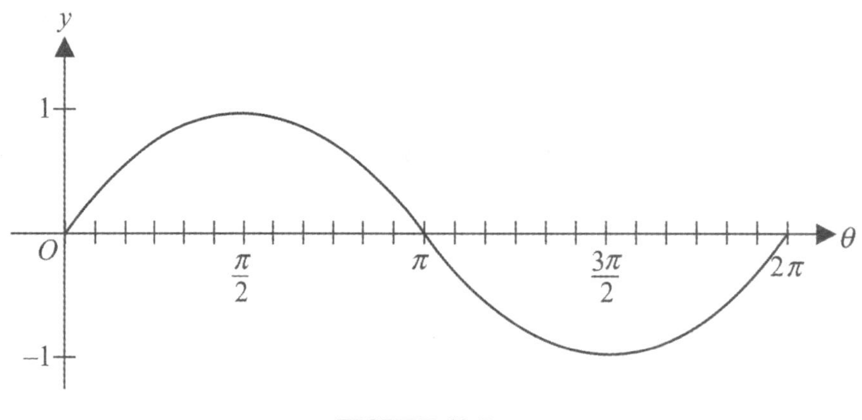

FIGURE 13.5

This is one complete cycle, or *period*, of the graph of $y = \sin\theta$. From now on, as θ continues to increase (as the wheel continues to turn), the graph will simply repeat itself, over and over. By imagining the wheel turning clockwise (backward), you can extend the graph to negative values of θ. The variable θ is usually renamed x. (This is not the same x as in Figure 13.2. This x represents the angle in the unit circle.) A graph of $y = \sin x$ appears in Figure 13.6.

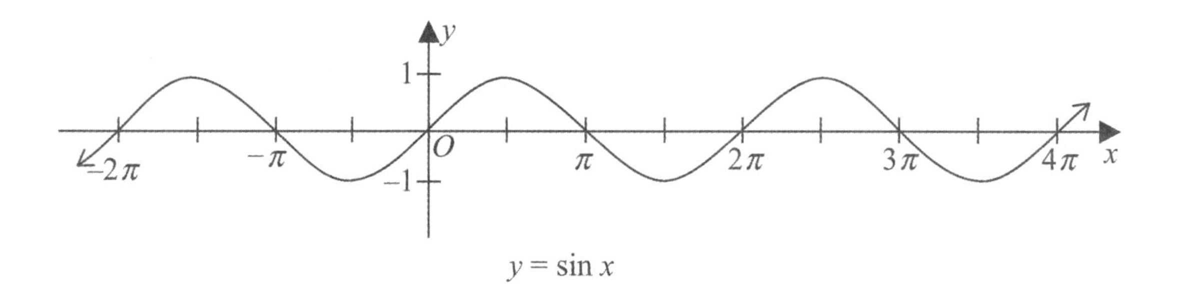

FIGURE 13.6

The Ferris wheel analogy can also be used to develop the graph of the cosine function. Instead of graphing height, y, above the ground (x-axis) as a function of θ, we graph the horizontal displacement, x, from the vertical axis as a function of θ. When you board at point A, $\theta = 0$ and you are 1 unit to the right of the axis. As θ increases, you move closer to the vertical axis. At point B, $\theta = \frac{\pi}{2}$, you are on the y-axis, so your horizontal displacement is 0 units. This part of the graph is shown in Figure 13.7a. As θ continues to increase, you move to the left of the axis, reaching a maximum distance of 1 unit to the left when $\theta = \pi$ as shown in Figure 13.7b.

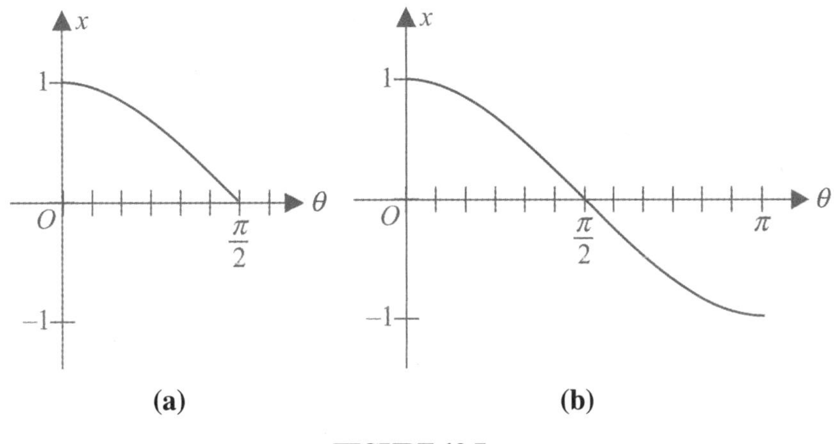

FIGURE 13.7

On the second half of the rotation (the "underground" portion) you repeat the process in reverse. One complete period is shown in Figure 13.8.

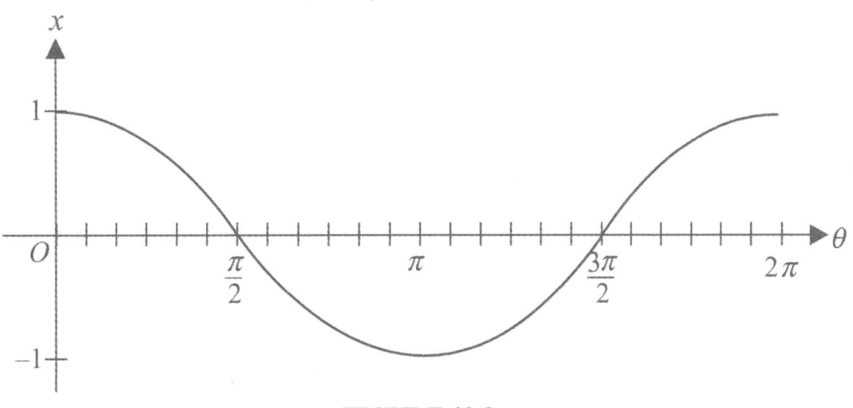

FIGURE 13.8

Again, the graph is periodic. As θ continues to increase, the curve shown above is repeated on every interval of 2π. This time we will rename both variables, x to y and θ to x. (This is not the same x and y as in Figure 13.2. Here x is the angle and y is the horizontal displacement.) We get the graph of $y = \cos x$ shown in Figure 13.9.

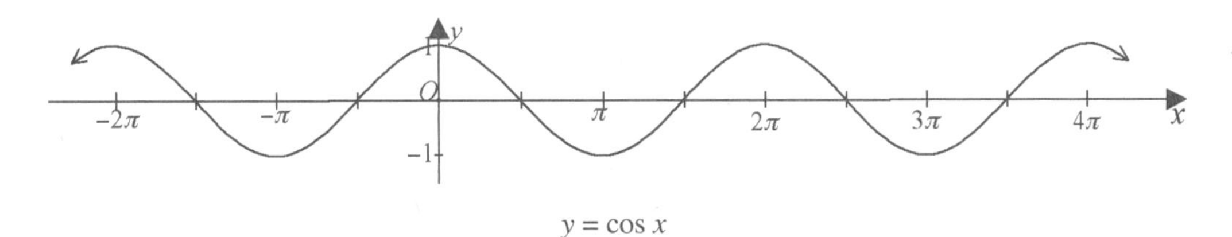

$y = \cos x$

FIGURE 13.9

Both graphs have several important things in common.

- Both $y = \sin x$ and $y = \cos x$ describe functions. The graphs pass the vertical line test. For each value of x, there is exactly one value of y.
- The domain for both functions is all real numbers. The range for both is $-1 \le y \le 1$.

- Both functions are *periodic* with period of 2π. This means that the part of the graph from $x = 0$ to $x = 2\pi$ repeats itself on $x = 2\pi$ to $x = 4\pi$ and on every subsequent interval of length 2π. In general, a function is periodic with period p if $f(x + p) = f(x)$ for all x. (This says that the graph remains unchanged after a horizontal translation of p units.) For the sine and cosine functions, we have $\sin(x + 2\pi) = \sin x$ and $\cos(x + 2\pi) = \cos x$ for all values of x.

GRAPH OF $y = \tan x$

The Ferris wheel analogy is not particularly helpful for graphing the tangent function, so we will do it the old-fashioned way instead with a table of values. Recall from the section on trigonometric identities in Chapter 12 that $\tan\theta = \dfrac{\sin\theta}{\cos\theta}$ for all values of θ except where $\cos\theta = 0$. We will use this identity and the special angles to help make a table of values for $y = \tan x$.

TABLE 13.1
SINE, COSINE, AND TANGENT VALUES

x	0	$\frac{\pi}{6}$	$\frac{\pi}{4}$	$\frac{\pi}{3}$	$\frac{\pi}{2}$
$\sin x$	0	$\frac{1}{2}$	$\frac{\sqrt{2}}{2}$	$\frac{\sqrt{3}}{2}$	1
$\cos x$	1	$\frac{\sqrt{3}}{2}$	$\frac{\sqrt{2}}{2}$	$\frac{1}{2}$	0
$\tan x = \frac{\sin x}{\cos x}$	0	$\frac{1}{\sqrt{3}} \approx 0.58$	1	$\sqrt{3} \approx 1.73$	Undefined

x	$\frac{2\pi}{3}$	$\frac{3\pi}{4}$	$\frac{5\pi}{6}$	π
$\sin x$	$\frac{\sqrt{3}}{2}$	$\frac{\sqrt{2}}{2}$	$\frac{1}{2}$	0
$\cos x$	$-\frac{1}{2}$	$-\frac{\sqrt{2}}{2}$	$-\frac{\sqrt{3}}{2}$	-1
$\tan x = \frac{\sin x}{\cos x}$	$-\sqrt{3} \approx -1.73$	-1	$-\frac{1}{\sqrt{3}} \approx -0.58$	0

What happens at $x = \frac{\pi}{2}$? As x gets close to $\frac{\pi}{2}$ from the left (staying just less than $\frac{\pi}{2}$), $y = \sin x$ gets close to 1 while $y = \cos x$ goes to 0 through positive values. The ratio increases without bound (goes to $+\infty$). As x gets close to $\frac{\pi}{2}$ from the right, $y = \sin x$ is still very close to 1 and $y = \cos x$ is getting close to 0 but through negative values. On this side of $x = \frac{\pi}{2}$, the ratio decreases without bound (goes to $-\infty$). The graph of $y = \tan x$ has a vertical asymptote at $x = \frac{\pi}{2}$. (A vertical asymptote is the same feature that some rational functions have, as discussed in Chapter 9.) One complete period of $y = \tan x$ is shown in Figure 13.10.

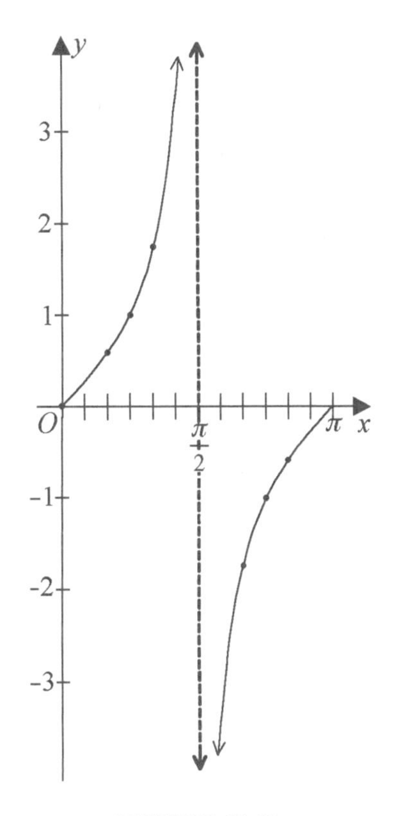

FIGURE 13.10

Like $y = \sin x$ and $y = \cos x$, $y = \tan x$ is periodic. Unlike the other two, its period is π rather than 2π. Also unlike $y = \sin x$ and $y = \cos x$, $y = \tan x$ has vertical asymptotes at every odd multiple of $\frac{\pi}{2}$: $-\frac{3\pi}{2}$, $-\frac{\pi}{2}$, $\frac{\pi}{2}$, $\frac{3\pi}{2}$, $\frac{5\pi}{2}$, and so on. Those values are therefore excluded from the domain of $y = \tan x$. The range of $y = \tan x$ is all real numbers. The graph of $y = \tan x$ is shown in Figure 13.11.

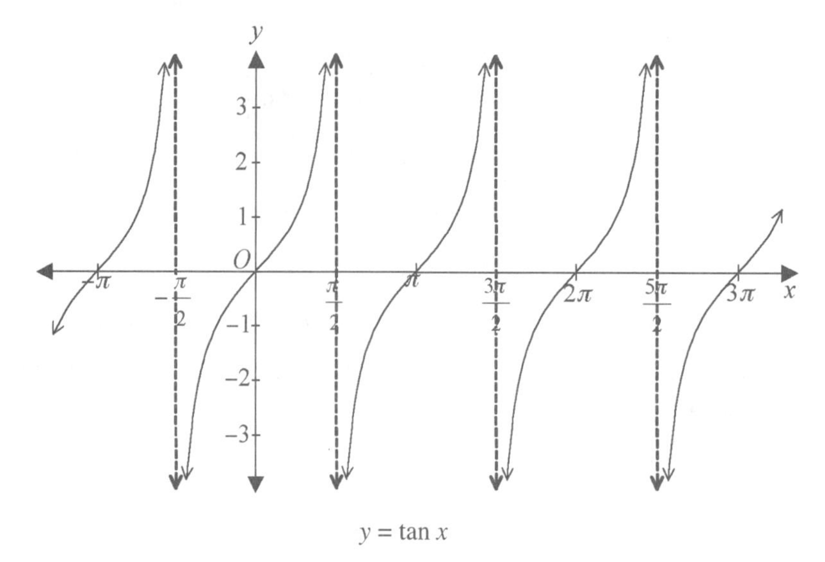

$y = \tan x$

FIGURE 13.11

SUMMARY OF BASIC TRIGONOMETRIC GRAPHS

$y = \sin x$

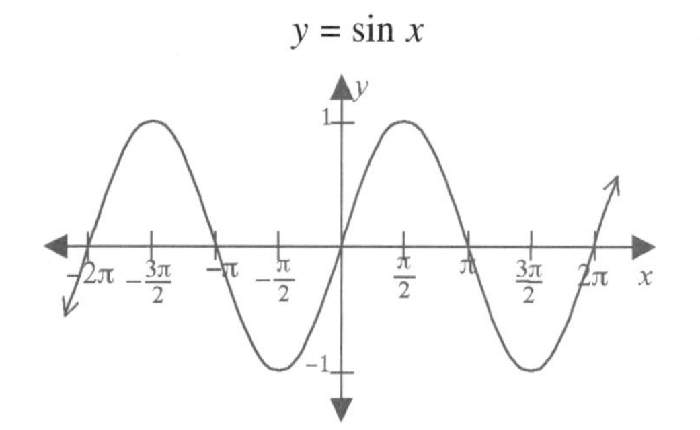

domain: all real numbers
range: $-1 \le y \le 1$

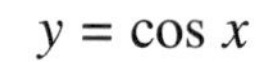
$y = \cos x$

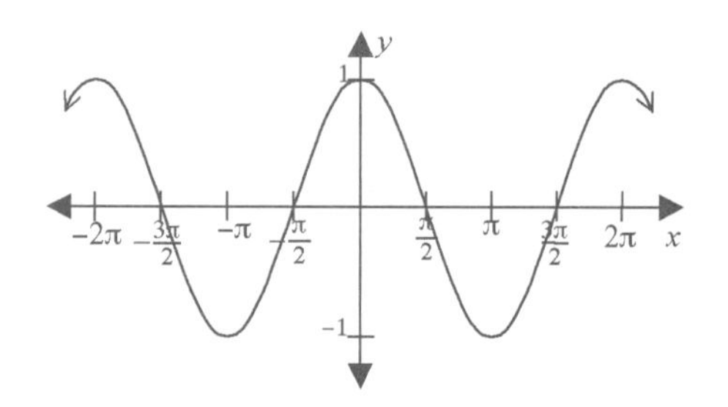

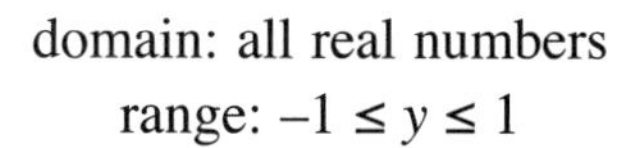
domain: all real numbers
range: $-1 \le y \le 1$

$y = \tan x$

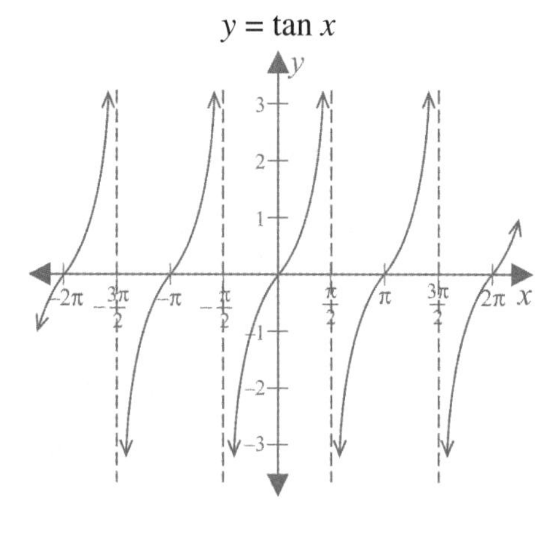

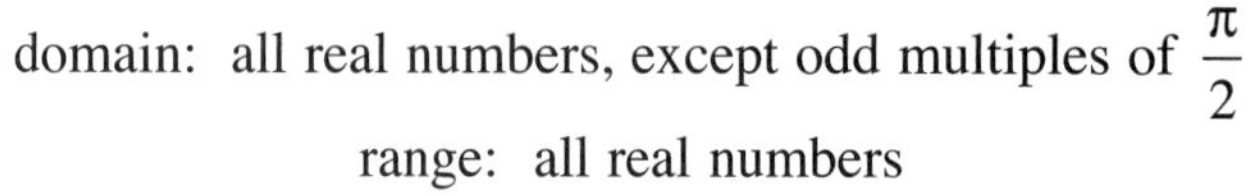
domain: all real numbers, except odd multiples of $\frac{\pi}{2}$

range: all real numbers

FIGURE 13.12

Transformations of Trigonometric Graphs

The graphs of $y = \sin x$ and $y = \cos x$ can be transformed in exactly the same ways as any other function. These transformations are very important when modeling with trigonometric functions.

REFLECTIONS OVER THE AXES

The graph of $y = -f(x)$ is the reflection of the graph of $y = f(x)$ over the x-axis. Therefore, $y = -\sin x$ and $y = -\cos x$ are the reflections of the graphs of $y = \sin x$ and $y = \cos x$ over the x-axis. One period of each is shown in Figure 13.13

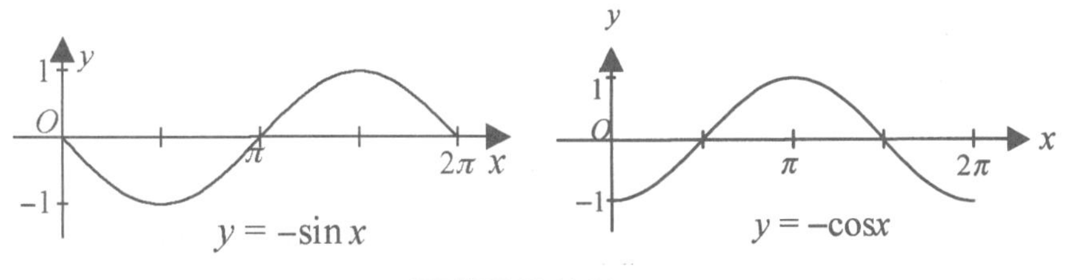

FIGURE 13.13

Reflections of sine and cosine graphs over the y-axis are not very interesting. Since cosine has y-axis symmetry, a reflection over the y-axis leaves the graph unchanged, $\cos(-x) = \cos(x)$. This means $y = \cos x$ is an even function. Because sine has origin symmetry, reflecting its graph over the y-axis is equivalent to reflecting it over the x-axis, $\sin(-x) = -\sin(x)$. This means $y = \sin x$ is an odd function.

VERTICAL DILATION

The graphs of $y = a \sin x$ and $y = a \cos x$ are vertical dilations of the graphs of $y = \sin x$ and $y = \cos x$ by a factor of a. Recall that the range of both sine and cosine is $-1 \le y \le 1$. A vertical dilation by a factor of a changes the range to $-|a| \le y \le |a|$. The quantity $|a|$ is called the *amplitude* of the graph. The amplitude tells how far above and below the *midline* (the x-axis if there is no vertical translation) the graph extends, as shown in Figure 13.14. This example assumes $a > 0$. If a is negative, the graph will be reflected over the x-axis but the amplitude, $|a|$, will still be positive.

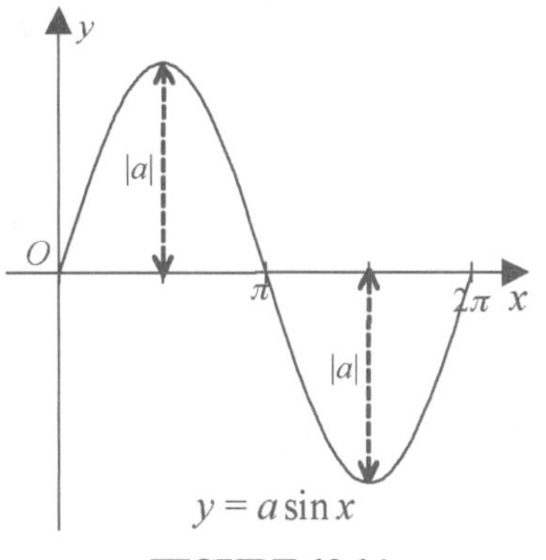

FIGURE 13.14

Part of the graph of $y = a \sin x$ is shown to the right. What is the amplitude of the graph, and what is the value of a?

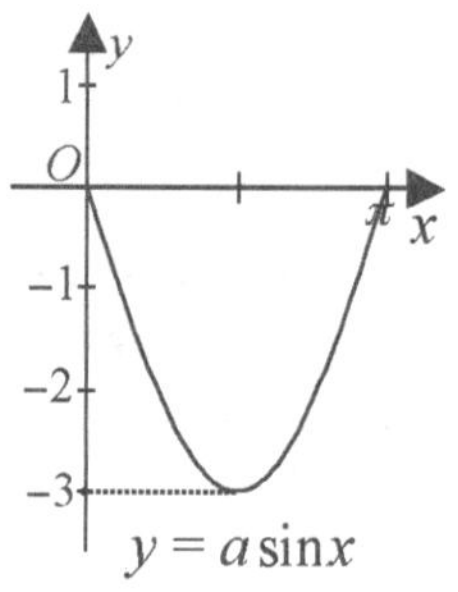

SOLUTION

The graph extends 3 units below its midline (the x-axis). If it were continued, it would also extend 3 units above the axis. The amplitude is 3. A basic sine graph starts at the origin and increases to a maximum at $x = \frac{\pi}{2}$. This one starts at the origin and decreases to a minimum at $x = \frac{\pi}{2}$. It has been reflected over the x-axis. This means a must be negative. Since the amplitude $|a| = 3$, a must be -3.

EXAMPLE 13.2

Sketch one period of the graph of $y = -2 \cos x$.

SOLUTION

A basic cosine graph starts at (0, 1) and passes through $\left(\frac{\pi}{2}, 0\right)$, $(\pi, -1)$, $\left(\frac{3\pi}{2}, 0\right)$, and $(2\pi, 1)$ as shown below left. The graph of $y = -2 \cos x$ has an amplitude of $|-2| = 2$, so it will be vertically dilated by a factor of 2. In addition, because $a = -2$ is negative, the graph will be reflected over the x-axis. It will start at $(0, -2)$ and pass through $\left(\frac{\pi}{2}, 0\right)$, $(\pi, 2)$, $\left(\frac{3\pi}{2}, 0\right)$, and $(2\pi, -2)$. The graph is sketched below right.

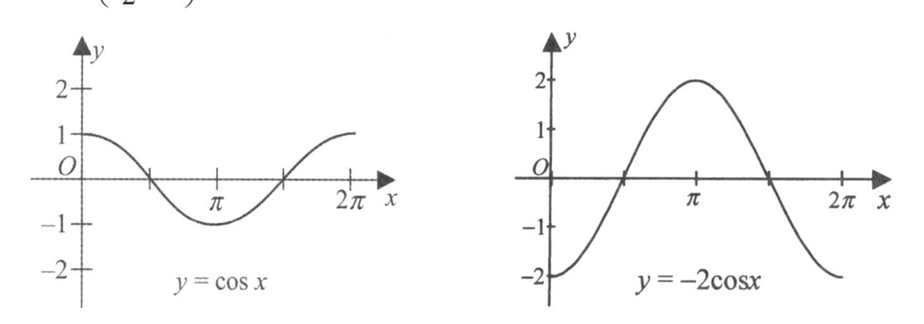

HORIZONTAL DILATION

The graphs of $y = a \sin bx$ and $y = a \cos bx$, $b > 0$, are horizontal dilations of the graphs of $y = a \sin x$ and $y = a \cos x$ by a factor of $\frac{1}{b}$. This has the effect of changing the period of the graph from 2π to $\frac{2\pi}{b}$.

The quantity b, called the *angular frequency*, tells the number of full periods that occur in an interval of 2π. Figure 13.15 shows the graph of $y = \cos 4x$. Note that four full periods of the graph occur in the interval $0 \le x \le 2\pi$ and that one period is $\frac{2\pi}{4} = \frac{\pi}{2}$.

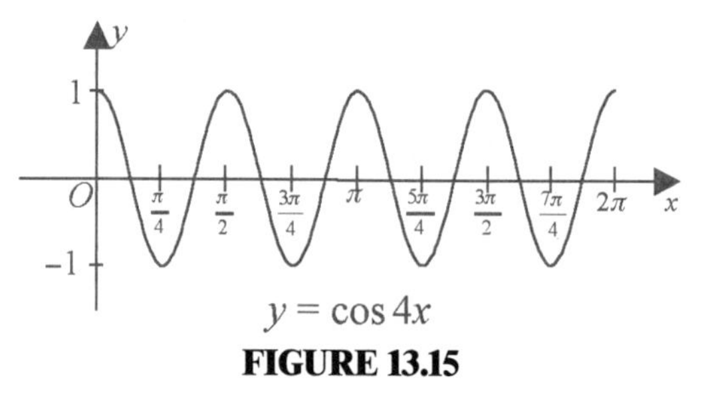

FIGURE 13.15

EXAMPLE 13.3 Find the period of the graph shown below, and determine the value of b.

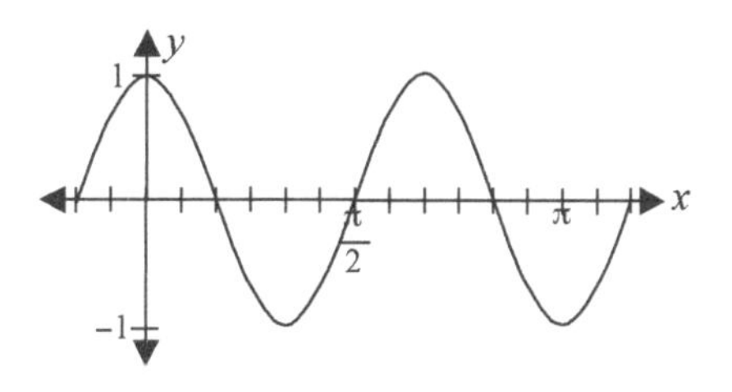

SOLUTION

To answer this question, we need to know the scale on the horizontal axis. Since six tick marks make $\frac{\pi}{2}$, each tick mark is $\frac{\frac{\pi}{2}}{6} = \frac{\pi}{12}$.

There are several ways to find the period from the graph. One way is to find the distance between two consecutive maxima. This graph has a maximum at $x = 0$ and its next maximum at $x = \frac{8\pi}{12} = \frac{2\pi}{3}$, so its period is $\frac{2\pi}{3} - 0 = \frac{2\pi}{3}$. You can also find the distance between consecutive minima or you can find the distance between a maximum and a minimum and double it. For graphs whose midline is the x-axis, as in this case, you can find the distance between every other zero (note that the distance between two consecutive zeros is only half a period). This graph has a zero at $x = \frac{2\pi}{12} = \frac{\pi}{6}$. The second zero after that is at $x = \frac{10\pi}{12} = \frac{5\pi}{6}$. The period is $\frac{5\pi}{6} - \frac{\pi}{6} = \frac{4\pi}{6} = \frac{2\pi}{3}$, as before.

If the graph had shown all the cycles between 0 and 2π, we could count the number of cycles and, if it were a whole number, we would have b. However, the graph does not show the entire interval, $0 \le x \le 2\pi$, so we will find b using the formula

$$\text{Period} = \frac{2\pi}{b}.$$

The formula can be solved for b to give $b = \frac{2\pi}{\text{Period}}$. For this example, we have

$$b = \frac{2\pi}{\frac{2\pi}{3}} = \frac{2\pi}{1} \cdot \frac{3}{2\pi} = 3.$$

EXAMPLE 13.4 Sketch one full period of the graph of $y = 3\sin\frac{1}{2}x$.

SOLUTION

Each complete period of a sine graph is divided into four equal intervals by certain key points on the graph. On the first interval, the graph goes from a zero to a maximum. On succeeding intervals, it goes from maximum to zero, zero to minimum, and finally minimum back to zero. The cosine graph is very similar except that the first interval starts at a maximum and the last interval ends at the next maximum.

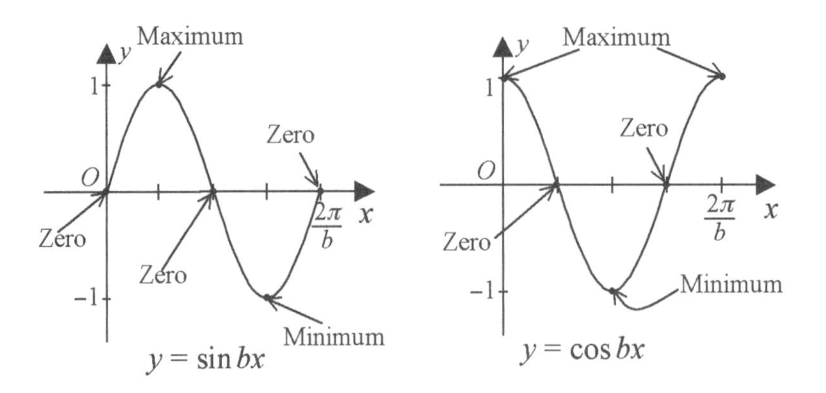

To graph a sine or cosine function, find the period and then divide the period by 4 to find the distance between consecutive key points. In this example, $b = \frac{1}{2}$ so the period is $\frac{2\pi}{\frac{1}{2}} = 4\pi$. The distance between each two consecutive key points on the graph is $\frac{4\pi}{4} = \pi$. Mark off intervals of π on the x-axis until you reach one full period.

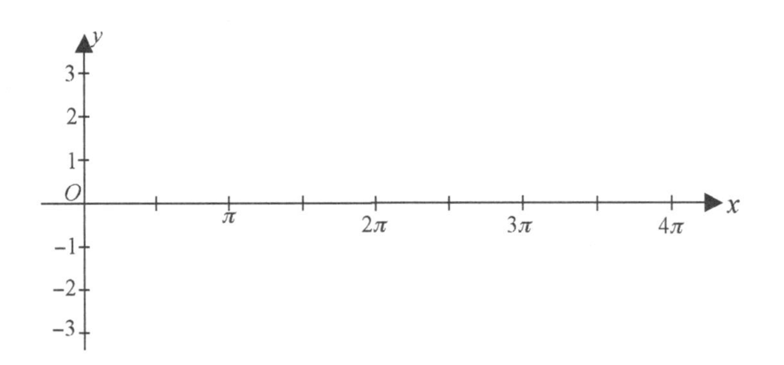

The amplitude is 3, so the graph will go up and down 3 units from the x-axis. A sine graph starts on its midline (the x-axis in this case) and then goes to its maximum, to its midline, to its minimum, and back to its midline. In this example, the maximum will be 3 units above the midline and the minimum will be 3 units below. Plot these points and then connect them with a smooth curve.

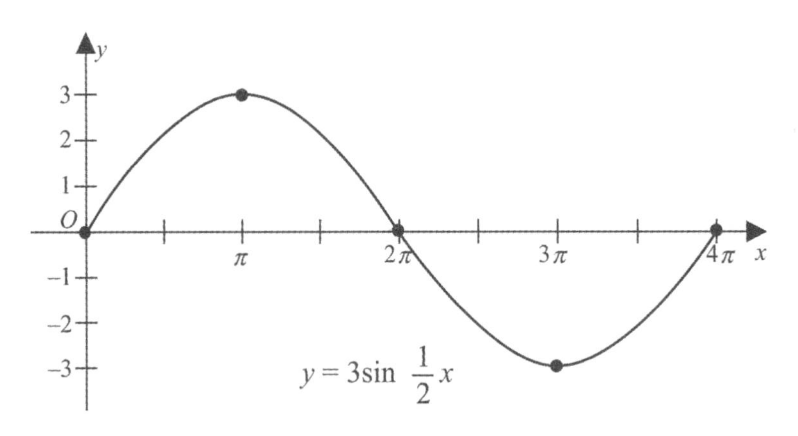

These transformations apply to $y = \tan x$ in a similar fashion. One difference is the period of $y = \tan x$ is π. So horizontal dilations of $\frac{1}{b}$ change the period to $\frac{\pi}{b}$.

EXAMPLE 13.5

Sketch the graph of $f(x) = \tan\left(\frac{1}{2}x\right)$ over the interval $-2\pi \le x \le 2\pi$.

SOLUTION

The graph of $f(x) = \tan\left(\frac{1}{2}x\right)$ is the graph of $f(x) = \tan(x)$ after a horizontal stretch by 2. After this stretch, the vertical asymptotes that were at $x = \pm\frac{\pi}{2}$ will now be at $x = \pm\pi$. The period has been changed from π to 2π.

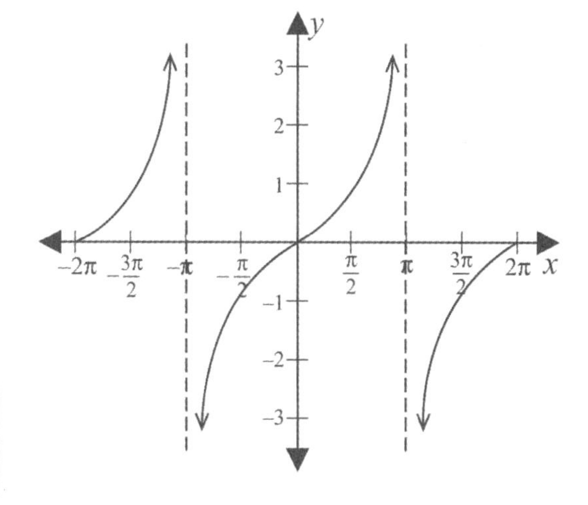

VERTICAL TRANSLATIONS

The graphs of $y = a \sin bx + d$ and $y = a \cos bx + d$ are vertical translations of the graphs of $y = a \sin bx$ and $y = a \cos bx$ by d units. This has the effect of shifting the midline of the graph from the x-axis to the line $y = d$. Remember that the amplitude, $|a|$, tells how far above and below the midline the graph extends. The range of the translated graph is $d - |a| \le y \le d + |a|$. The graph of $y = 2 \sin x + 3$ is shown in Figure 13.16. It has a midline (shown dashed on the graph) at $y = 3$, and the curve extends 2 units above and below that line. The range is $1 \le y \le 5$.

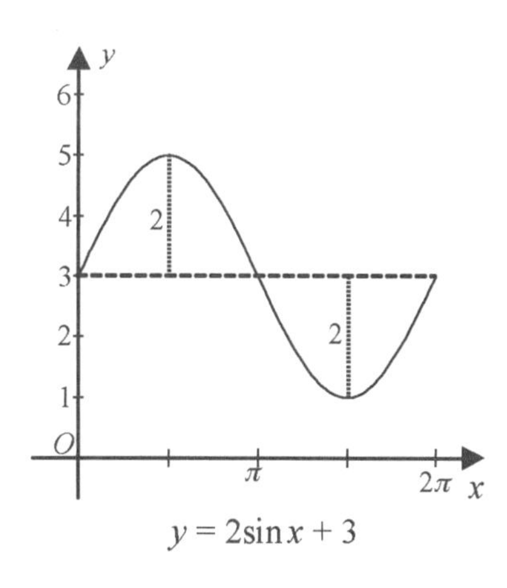

FIGURE 13.16

EXAMPLE 13.6

How does the graph of $y = -5 + 3\cos(4x)$ compare with the graph of $y = \cos(4x)$? Identify the minimum and maximum values of $y = -5 + 3\cos(4x)$.

SOLUTION

Note that $y = -5 + 3\cos(4x)$ is written with the vertical shift before the trigonometric function: $y = d + a\cos(bx)$ instead of after the function as in $y = a\cos(bx) + d$. This is rather common; don't let it confuse you. With that in mind, the vertical shift is easy. Since $d = -5$, the graph is shifted 5 units down.

The amplitude has changed from 1 to 3, so the graph of $y = -5 + 3\cos(4x)$ has been vertically stretched by 3.

Putting it together, the graph of $y = -5 + 3\cos(4x)$ is vertically stretched by a factor of 3 and shifted 5 units down compared with the graph of $y = \cos(4x)$. The new graph will have a minimum value of $-5 + -3 = -8$ and a maximum value of $-5 + 3 = -2$.

SUMMARY OF TRANSFORMATIONS OF SINE AND COSINE GRAPHS

The general form of a sine or cosine function is $y = a\sin(bx) + d$ or $y = a\cos(bx) + d$. Table 13.2 summarizes their characteristics.

TABLE 13.2
CHARACTERISTICS OF SIN AND COSINE GRAPHS

Characteristics	Symbols or Formula	Description in $y = a \sin(bx) + d$ or $y = a \cos(bx) + d$
Amplitude	$\lvert a\rvert$	How far above and below the midline the graph extends (if $a < 0$, the graph is reflected over its midline)
Angular frequency	b	How many complete cycles of the graph occur in any interval of length 2π
Period	$\frac{2\pi}{b}$	The length of one complete cycle (period) of the graph
Interval between key points	$\frac{\text{Period}}{4}$	The horizontal distance between the maximum, midline, and minimum points
Vertical shift	d	How far the midline of the graph has been shifted up or down
Range	$[d - \lvert a\rvert, d + \lvert a\rvert]$	The minimum y-value to the maximum y-value

Both sine and cosine graphs are shown in Figure 13.17. Note that the main difference is the location of the starting value on the y-axis. In the sine graph, the curve starts on the y-axis at its midline. In the cosine graph, the curve starts on the y-axis at its maximum or minimum.

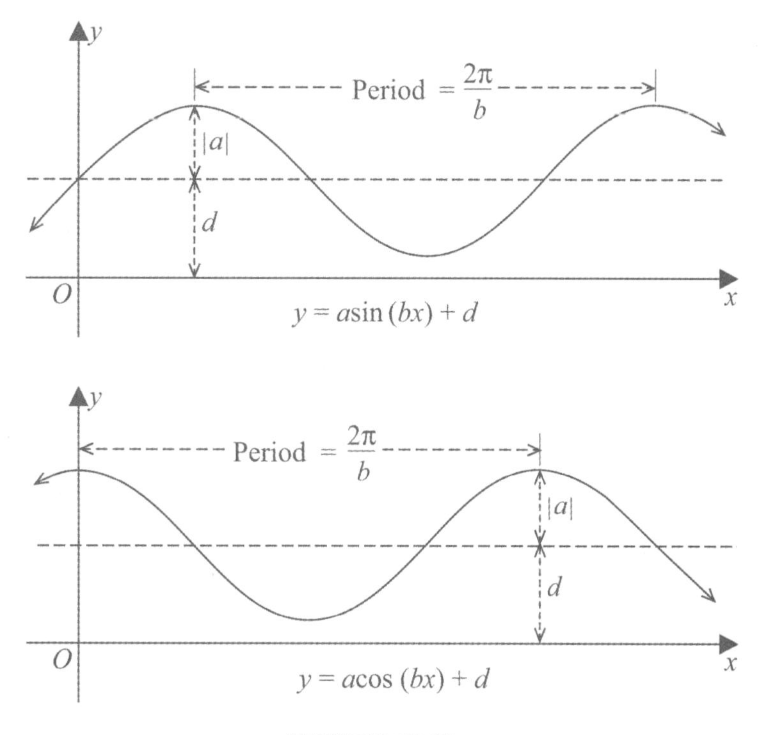

FIGURE 13.17

EXAMPLE 13.7

Sketch two cycles of the graph of $y = -2 \cos 4x + 1$.

SOLUTION

First find the period, $\frac{2\pi}{b} = \frac{2\pi}{4} = \frac{\pi}{2}$. Divide by 4 to find the interval between key points, $\frac{\text{Period}}{4} = \frac{\frac{\pi}{2}}{4} = \frac{\pi}{8}$. We want our first key point to be on the y-axis with succeeding key points every $\frac{\pi}{8}$. To graph one complete period, we need five consecutive key points: 0, $\frac{\pi}{8}$, $\frac{2\pi}{8}$, $\frac{3\pi}{8}$, and $\frac{4\pi}{8}$. Notice how the numerator increases by π and the denominator stays 8. We need to graph two cycles of the curve, so continue marking the x-axis at intervals of $\frac{\pi}{8}$ to $x = \frac{8\pi}{8} = \pi$.

Now find the range. The vertical shift is $d = 1$, so the midline of the graph will be shifted up one unit to $y = 1$. (Some people like to draw a dashed line at $y = 1$ to show the centerline.) The amplitude is $|a| = |-2| = 2$, so the graph will go both 2 units above and 2 units below $y = 1$. This means the range will be $-1 \leq y \leq 3$. Mark the y-axis appropriately. The axes are as shown below.

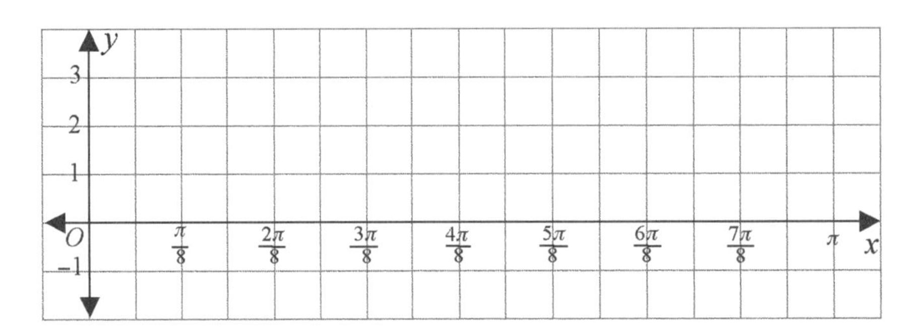

Now we are ready to plot points. This is a cosine graph with $a < 0$. It "starts" on the y-axis at its minimum. At succeeding key points, it goes up 2 units to the midline, up 2 to the maximum, returns to the midline, and returns to the minimum. Graph these points, and connect them with a smooth curve.

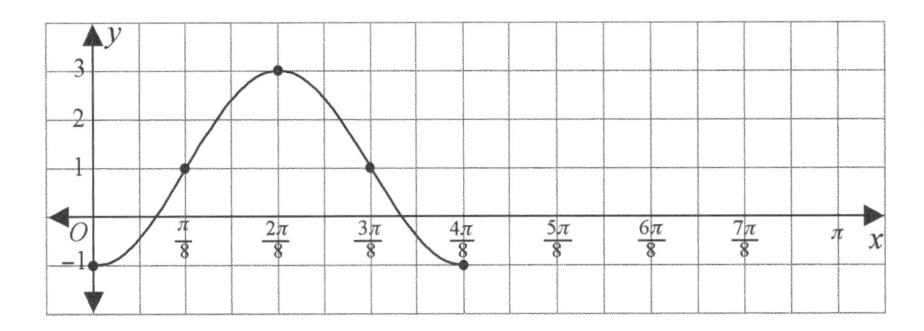

Now that one complete period is graphed, it is easy to extend the graph as far as you need in either direction. Just keep plotting new key points at intervals of $\frac{\pi}{8}$.

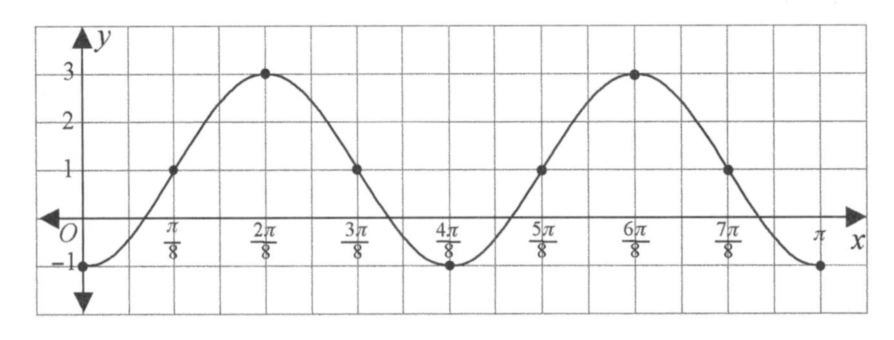

$y = -2\cos 4x + 1$

EXAMPLE 13.8

Sketch a graph of two full cycles of $y = 5 + 3\sin(2x)$.

SOLUTION

First find the period, $\frac{2\pi}{b} = \frac{2\pi}{2} = \pi$. Divide by 4 to find the interval between key points, $\frac{\text{Period}}{4} = \frac{\pi}{4}$. Mark the x-axis in intervals of $\frac{\pi}{4}$ out to two full periods (2π).

Now find the range. The vertical shift is $d = 5$ and the amplitude is $|a| = 3$. The graph will extend 3 units above and below the line $y = 5$, so the range will be $2 \le y \le 8$. The axes are shown below.

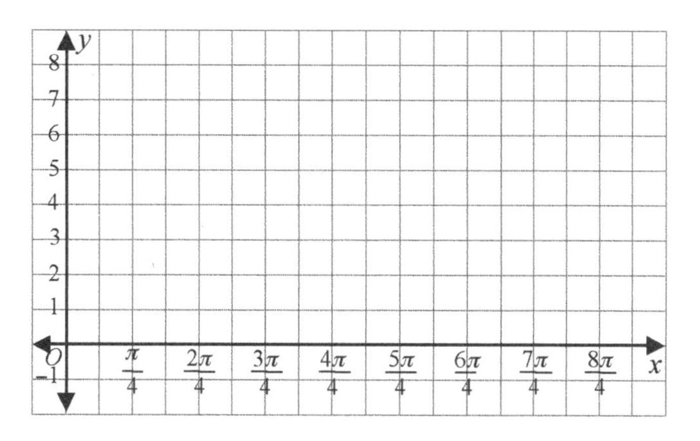

Now we are ready to plot points. This is a sine graph with $a > 0$. It "starts" at the y-axis on the midline, $y = 5$. At succeeding key points, it goes up 3, back to the midline, down 3, and returns to the midline. Graph these points, and connect them with a smooth curve.

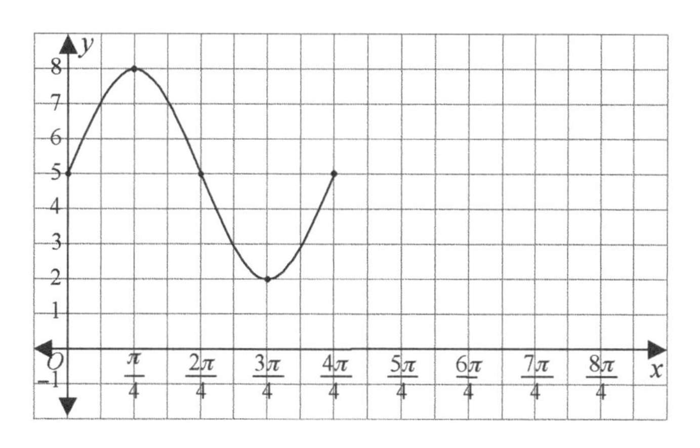

Now that one complete period is graphed, it is easy to extend the graph as far as you need in either direction. Just keep plotting new key points at intervals of $\frac{\pi}{4}$.

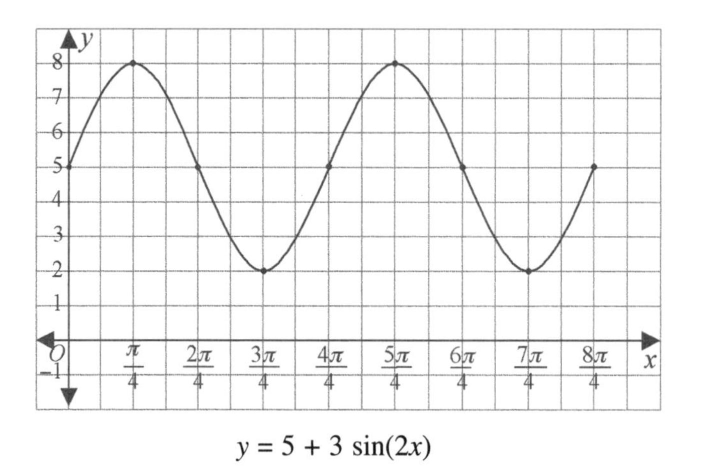

$y = 5 + 3\sin(2x)$

WRITING THE EQUATION FOR A SINE OR COSINE CURVE FROM A GRAPH

Given the graph of a sine or cosine curve, you should be able to write the equation of the function as $y = a\sin(bx) + d$ or as $y = a\cos(bx) + d$. The next two examples show how.

EXAMPLE 13.9 Write an equation of the function shown in the graph to the right.

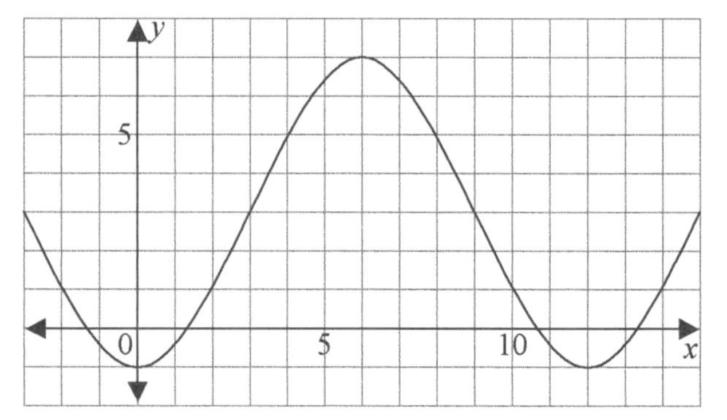

SOLUTION

You need to find the amplitude, the angular frequency, and the vertical shift. (You do not have to find them in that order.) You also need to decide whether to use sine or cosine.

Identify the maximum and minimum values of the function. On the graph, the maximum is 7 and the minimum is −1. The midline, or vertical shift, will be the average of the two values. The amplitude will be half the difference.

$$\text{Midline: } d = \frac{\text{maximum} + \text{minimum}}{2} = \frac{7 + (-1)}{2} = 3$$

$$\text{Amplitude: } |a| = \frac{\text{maximum} - \text{minimum}}{2} = \frac{7 - (-1)}{2} = 4$$

The amplitude could also be found by either $|a| = \text{maximum} - d$ or $|a| = d - \text{minimum}$. The graph has two consecutive minima at $x = 0$ and $x = 12$, so the period is $12 - 0 = 12$.

From this, you can find the angular frequency: $b = \frac{2\pi}{\text{Period}} = \frac{2\pi}{12} = \frac{\pi}{6}$.

The choice of sine or cosine depends on the starting point of the graph, what the y-value is at the y-axis. If the graph is at its midline at the x-axis, choose sine. This graph is at an extreme value at the y-axis, so you should choose cosine. Because the graph is at a minimum at the y-axis, a will be negative, $a = -4$.

Now put all the information together into one equation.

$$y = -4\cos\left(\frac{\pi}{6}x\right) + 3$$

EXAMPLE 13.10

Write an equation of the function shown in the graph below.

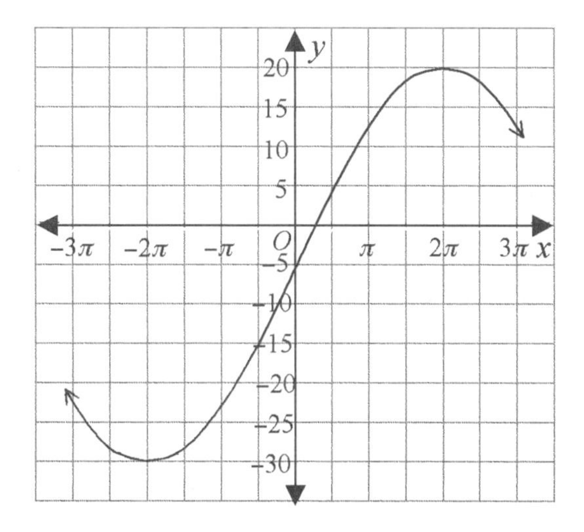

SOLUTION

Note the scales on the axes. The y-axis has a scale of 5. The maximum value of the function is 20 and the minimum is −30:

$$\text{Midline: } d = \frac{\text{maximum} + \text{minimum}}{2} = \frac{20 + (-30)}{2} = -5$$

$$\text{Amplitude: } |a| = \frac{\text{maximum} - \text{minimum}}{2} = \frac{20 - (-30)}{2} = 25$$

The graph does not show one full period. However, the distance between the minimum and the maximum is 4π, so one full period must be $2(4\pi) = 8\pi$. From this, we find the angular frequency, $b = \frac{2\pi}{\text{Period}} = \frac{2\pi}{8\pi} = \frac{1}{4}$.

This graph "starts" on the y-axis at its midline, so we choose to model the curve with the sine function. Because the graph increases from its midline at $x = 0$, a will be positive, $a = 25$. Putting all this together, we get:

$$y = 25\sin\left(\frac{1}{4}x\right) - 5$$

In later courses, you will learn about horizontal translations, called phase shifts, of trigonometric functions. This will let you graph and write equations for graphs of sine and cosine functions that are not at an extreme value or at the midline when $x = 0$.

SECTION EXERCISES

13–1. If $f(x) = 3\sin x$ and $g(x) = \sin 3x$, which of the following is true?
(1) $f(x)$ has a greater amplitude, and a greater period.
(2) $f(x)$ has a greater amplitude, and $g(x)$ has a greater period.
(3) $g(x)$ has a greater amplitude, and $f(x)$ has a greater period.
(4) $g(x)$ has a greater amplitude and a greater period.

13–2. Find the y-intercept of the function $f(x) = 3\cos(2x) + 1$.

13–3. What is the amplitude of the function $y = 4\sin 2x + 5$?

13–4. What is the amplitude of the function $y = -3\cos 5x + 8$?

13–5. What is the period of the function $y = -4\cos 2x + 10$?

13–6. What is the period of the graph of $y = 25 + 15\sin\frac{\pi}{2}x$?

13–7. What is the range of the function $y = 5\cos 2x$?

13–8. What is the range of the function $f(t) = -25\cos\frac{2\pi}{365}t + 75$?

13–9. The function $y = a\sin bx + d$ has range $-5 \le y \le 13$. Find the values of a and d.

13–10. **a.** Sketch the graph of $y = 3\sin x$ over the interval $0 \le x \le 2\pi$.
b. How does the graph of $y = 3\sin x$ compare with the graph of $y = \sin x$?
c. Sketch the graph of $y = \sin 2x$ over the interval $0 \le x \le 2\pi$.
d. How does the graph of $y = \sin 2x$ compare with the graph of $y = \sin x$?

13–11. For the function $f(x) = -5\cos 2x$,
a. What is the amplitude of the graph of $f(x)$?
b. What is the period of the graph of $f(x)$?
c. Sketch the graph of $f(x)$ over the interval $0 \le x \le 2\pi$.

13–12. Describe, in terms of transformations, how the graph of $y = -2\cos(x) + 3$ compares with the graph of $y = \cos x$.

For 13–15, graph one full period of the following.

13–13. $y = 2\cos\frac{1}{2}x$

13–14. $y = 5\sin 2x + 3$

13–15. $y = 3\sin\left(\frac{\pi}{4}x\right)$

For 16–17, answer the following.

a. Which is the better model for the graph, sine or cosine?

b. What is the amplitude?

c. What is the period?

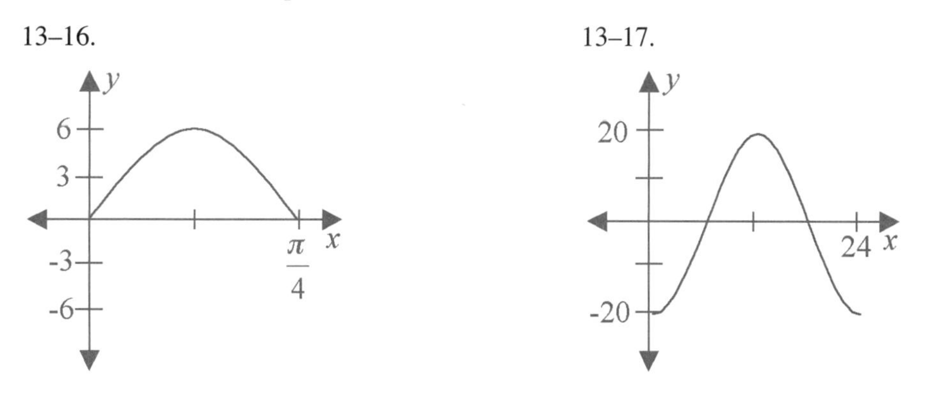

For 18–20, for each graph below, choose a model y = a *sin* (bx) + d *or* y = a *cos* (bx) + d, *state the values of* a, b, *and* d, *and then write an equation for the graph.*

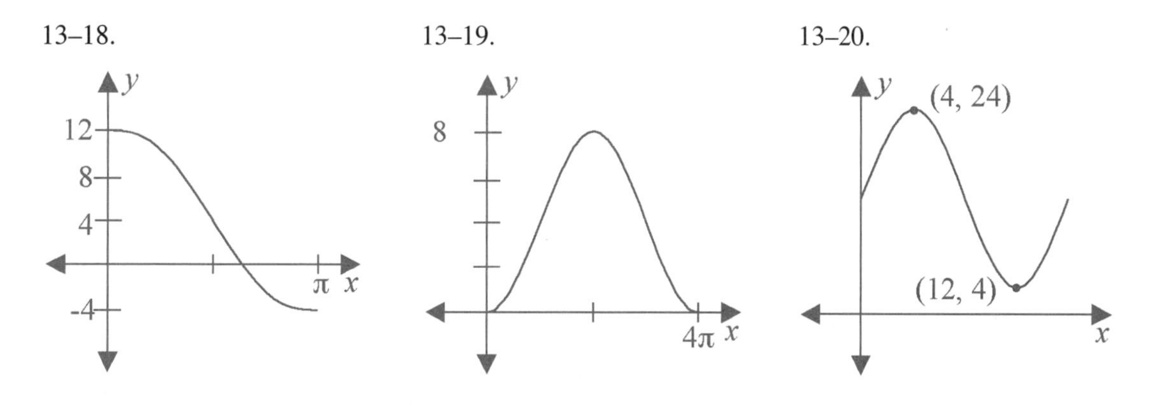

Modeling with Sine and Cosine Graphs

Many phenomena in real life repeat in a periodic way. Swinging pendulums, the number of hours of daylight in a day, the phases of the moon, and many other things can be modeled with sine or cosine functions.

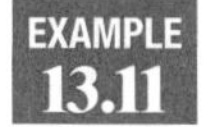

When a person's heart contracts, the blood pressure rises. When the heart relaxes, the blood pressure goes back down. A blood pressure reading is a ratio of the maximum pressure to the minimum; 120/80 is considered typical. Suppose a particular person's blood pressure is modeled by $P = 95 + 25 \cos 7.54t$, where P is the pressure in millimeters of mercury and t is time in seconds.

a. What is the person's maximum blood pressure? The minimum?

b. How long does it take the person's heart to beat once?

c. What is this person's pulse rate in beats per minute?

SOLUTION

a. The midline of the pressure model is 95. The pressure goes up and down 25 from that value. Thus the maximum is 95 + 25 = 120 and the minimum is 95 – 25 = 70. This person's blood pressure is 120/70.

b. One heartbeat is equivalent to one complete period of the function. In this equation $b = 7.54$:

$$\text{Period} = \frac{2\pi}{b} = \frac{2\pi}{7.54} \approx 0.833 \text{ seconds}$$

c. Period is the length of one complete cycle in seconds. The reciprocal of the period, $\frac{b}{2\pi}$, is called the *frequency* and tells how many complete cycles occur in one second. In this problem, the frequency is $\frac{7.54}{2\pi} \approx 1.2$ beats per second. That gives a pulse rate of $1.2 \times 60 = 72$ beats per minute.

EXAMPLE 13.12 Suppose you lived at a base in the northern hemisphere of Mars. The number of hours of daylight each day at the base is modeled by $D = 12.5 - 3.2\sin 0.0094t$, where t is time measured in Martian days from the beginning of the year.

a. How many hours of daylight are on the longest day of the year? On the shortest day?

b. How many Martian days are in one Martian year?

c. Which day has the most hours of daylight? The least?

SOLUTION

a. The greatest number of hours will be $d + |a| = 12.5 + 3.2 = 15.7$ hours. The least number will be $d - |a| = 12.5 - 3.2 = 9.3$ hours.

b. The number of hours of daylight should make one complete cycle in one year. In other words, one year is the period of this model. In the equation, $b = 0.0094$:

$$\text{Period} = \frac{2\pi}{b} = \frac{2\pi}{0.0094} \approx 668.4 \text{ days in one Martian year}$$

c. A sketch of the graph is helpful. The model is a sine curve with a period of 668 (rounded to the nearest whole day), a vertical shift of 12.5, and an amplitude of 3.2. The graph is reflected over its midline because a is negative. A sketch is shown below.

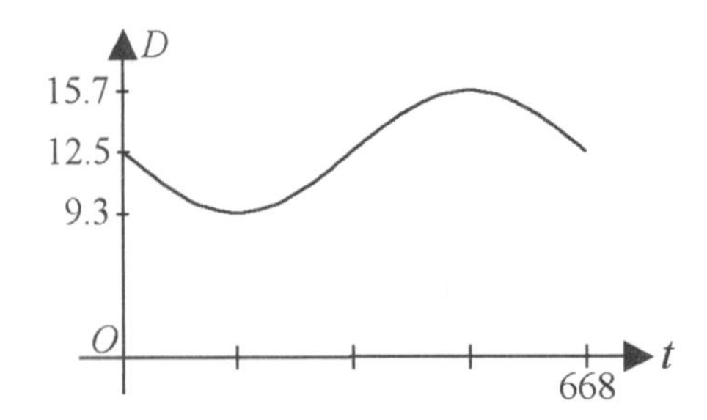

The shortest day of the year, in terms of hours of daylight, occurs one-quarter of the way through the year, or day $\frac{1}{4} \times 668 = 167$. The longest day of the year occurs three-quarters of the way through the year on day $\frac{3}{4} \times 668 = 501$.

EXAMPLE 13.13 On July 4, the afternoon high tide in Boston Harbor was 9.6 feet at 2:12 P.M. The next low tide was –0.8 feet at 8:24 P.M. Write a trigonometric equation to model the height of the tide as a function of time on that day where $t = 0$ means 2:12 P.M.

SOLUTION

A sketch of the graph is very helpful. We have two data points representing the endpoints of half of a full period. Note that we need to convert 2:12 P.M. and 8:24 P.M. to hours in decimal form. We know 12 minutes $= \frac{12}{60} = 0.2$ hours, so 2:12 P.M. is equivalent to 2.2 hours after noon and 8:24 P.M. is equivalent to 8.4 hours after noon. Since $t = 0$ means 2:12 P.M., then 8:24 P.M. is represented by $t = 8.4 - 2.2 = 6.2$.

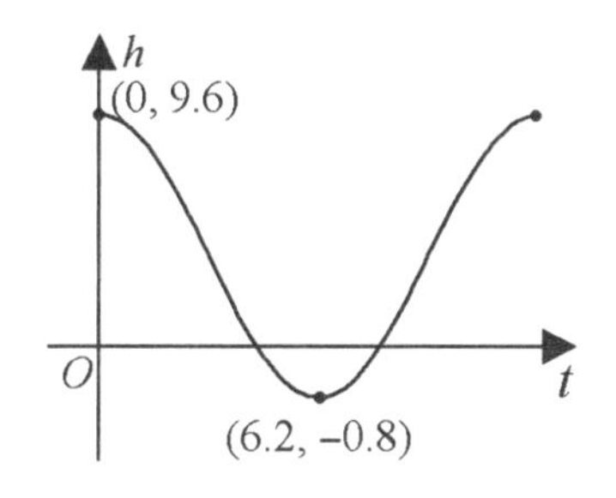

The vertical shift is $d = \frac{9.6 + (-0.8)}{2} = 4.4$ feet.

The amplitude is $|a| = 9.6 - 4.4 = 5.2$ feet.

The period is $2(6.2 - 0) = 12.4$ hours, so $b = \frac{2\pi}{12.4} \approx 0.507$.

Again we have a choice of sine or cosine. Use cosine because the maximum of the graph occurs at $t = 0$. So $a = 5.2$; we get the equation $h = 5.2\cos(0.507t) + 4.4$.

EXAMPLE 13.14 The average monthly high temperatures in St. Louis are shown in the table below. Write a trigonometric model to represent the data.

Month	Jan.	Feb.	Mar.	Apr.	May	Jun.	Jul.	Aug.	Sep.	Oct.	Nov.	Dec.
Average High Temperature	38°	44°	55°	67°	76°	85°	90°	88°	80°	68°	54°	42°

SOLUTION

A scatter plot of the data is shown to the right.

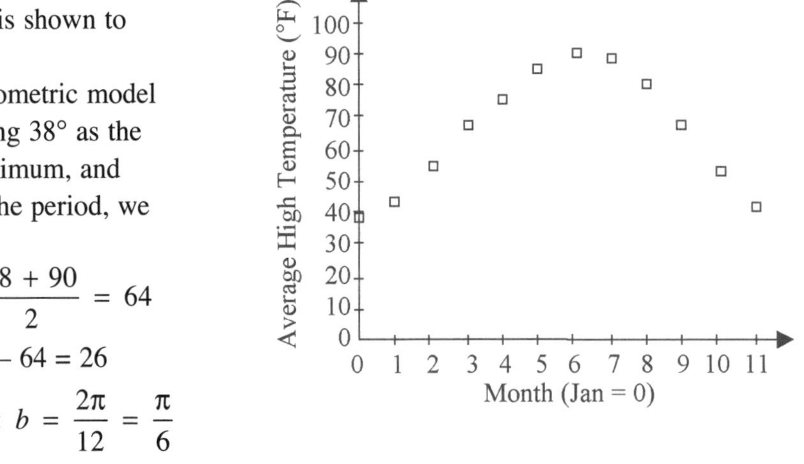

It appears that a trigonometric model would be reasonable. Using 38° as the minimum, 90° as the maximum, and 12 months (one year) as the period, we can find the following:

Vertical shift: $d = \dfrac{38 + 90}{2} = 64$

Amplitude: $|a| = 90 - 64 = 26$

Period is 12 months: $b = \dfrac{2\pi}{12} = \dfrac{\pi}{6}$

Since a minimum value occurs at $t = 0$, we will use a cosine model and $a = -26$.

$$T = -26\cos\left(\frac{\pi}{6}t\right) + 64$$

The figure below shows one period of our model superimposed on the data. You can see that it is not a perfect fit but does a very good job representing the data.

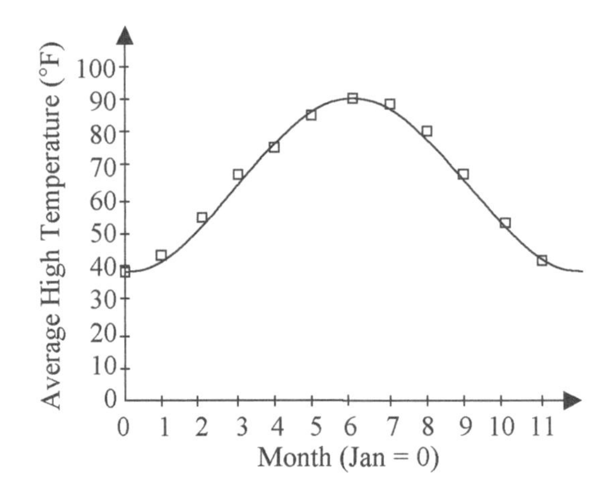

SECTION EXERCISES

13–21. Sunspot activity is cyclical. Suppose the number of sunspots per year is modeled by $N(t) = 60 + 50\sin 0.57t$, where t is in years. According to this model:

a. What is the maximum number of sunspots you would expect in one year?

b. What is the minimum number of sunspots you would expect in one year?

c. How many years is one complete sunspot cycle?

d. In this model, which year corresponds to $t = 0$?
(1) A year of minimum sunspot activity
(2) A year of maximum sunspot activity
(3) A year of average sunspot activity

13–22. Harvey the hamster is running in his exercise wheel. Lily the ladybug is clinging to the edge of the wheel. As the wheel turns, Lily's height above the floor of Harvey's cage is given by $y = 5 + 4\sin 4\pi t$, with y in inches and t in seconds.
a. How long does Harvey's wheel take to go around once?
b. What is Lily's greatest height above the floor of the cage? Her least height?
c. What is the diameter of Harvey's exercise wheel?

13–23. A weight hangs from a spring in the physics lab next to a vertical meter stick. At rest, the weight is at the 50 cm mark on the meter stick. Phyllis pulls the weight down a certain amount and lets go. Phyllis observes the weight's position at various times and records the data in a table.

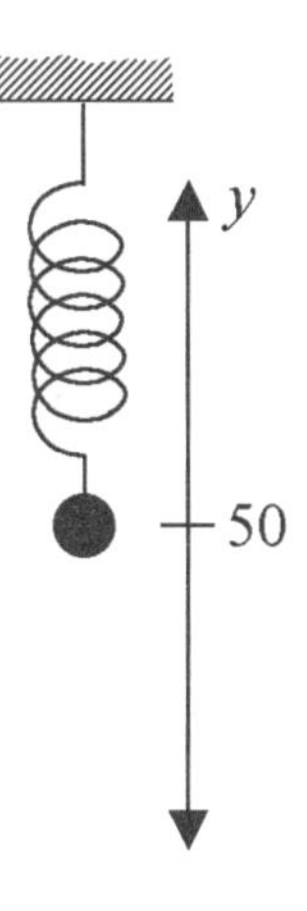

Time (s)	0.0	0.5	1.0	1.5	2.0	2.5	3.0
Position (cm)	42	50	58	50	42	50	58

Write an equation for the weight's position, y, as a function of time, t.

13–24. The graph shows a rough approximation of the average high temperature in Phoenix, AZ, over the course of a year. Write an equation for temperature as a function of time in months where $t = 0$ represents the beginning of the year.

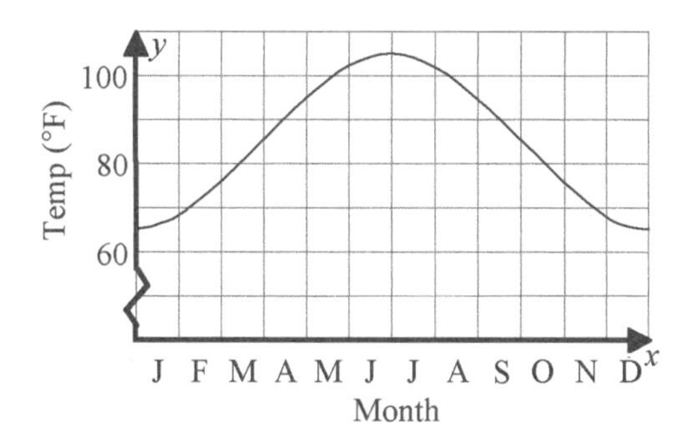

(Use a scientific or graphing calculator.)

13–25. A satellite goes around Earth in an orbit that takes it alternately 4,500 km north of the equator and then half an orbit later 4,500 km south of the equator. One complete orbit takes 90 minutes. The satellite's position relative to the equator can be modeled by $y = a\cos bt$, where y is the distance in kilometers north (+) or south (–) of the equator and t is time in hours.
a. Find the period of the function in hours.
b. Write the equation of the function (substituting correct values for a and b).
c. Where is the satellite relative to the equator when $t = 100$ hours?

Inverse Trigonometric Functions

You learned about inverse functions in Chapter 3. Inverse functions are used to "undo" a function and are helpful when solving equations. For example, to solve $x^3 = 5$, we take the cube root of both sides because taking the cube root is the inverse of cubing. In this section, we learn about the inverses of trigonometric functions, which will help solve equations such as $\sin x = \frac{1}{2}$.

Recall that for the inverse of a function to also be a function, the original function must be one-to-one; every output must come from only one input. Graphically, the original function must pass the horizontal line test. It is clear from their graphs that none of the trigonometric functions are one-to-one. For example, in one period of the function $y = \sin x$, the y-value $\frac{1}{2}$ comes from both $x = \frac{\pi}{6}$ and $x = \frac{5\pi}{6}$ as shown in Figure 13.18. If we extend the graph in both directions, it is clear that any y-value in the range of $\sin x$ can come from an infinite number of x-values. The same is true for $y = \cos x$ and $y = \tan x$.

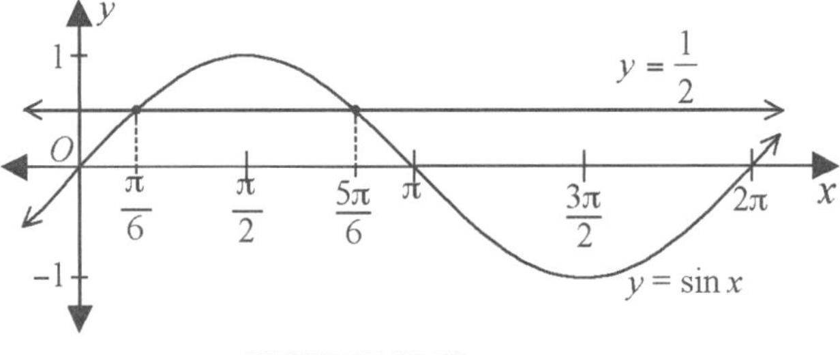

FIGURE 13.18

To define an inverse function for $\sin x$, we must first restrict its domain to make it one-to-one. In other words, we must choose an interval of x over which the graph of $y = \sin x$ passes the horizontal line test. Further, we want the interval we choose to include the entire range of $\sin x$, $[-1, 1]$. There are an infinite number of such intervals along the x-axis, so we can add one more requirement; we'd like to include the origin. This narrows it down to one possible interval. To define the inverse function for $y = \sin x$, we restrict its domain to the interval $-\frac{\pi}{2} \leq x \leq \frac{\pi}{2}$. This is shown in Figure 13.19.

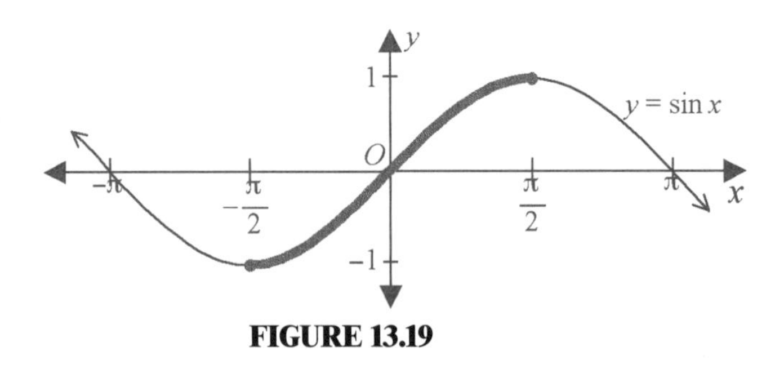

FIGURE 13.19

Recall that to find the inverse of a function algebraically, we switch x and y and then solve for y. For the sine function, we get the following.

$$\text{Original: } y = \sin x,\ -\frac{\pi}{2} \le x \le \frac{\pi}{2}$$

$$\text{Inverse: } x = \sin y,\ -\frac{\pi}{2} \le y \le \frac{\pi}{2}$$

$$\text{then } y = ?$$

You'll recall we had the same problem when finding the inverse of an exponential function in Chapter 11. Again, we need a new function. The inverse of the sine function is simply called the *inverse sine function*. There are two common notations for it, $y = \sin^{-1}x$ and $y = \arcsin x$. Both mean the same thing as shown in Figure 13.20.

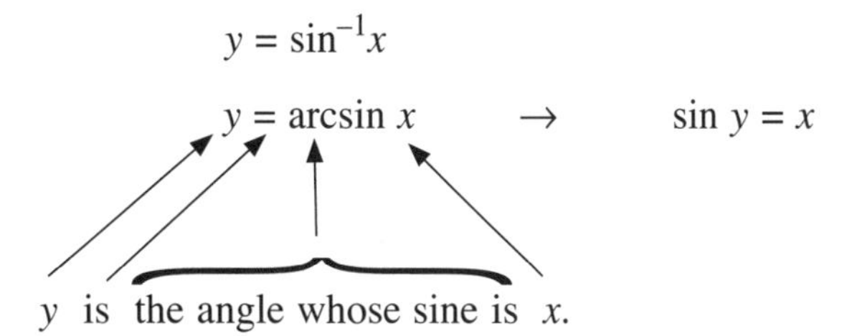

FIGURE 13.20

If you are curious where "arc" came from, recall that in radians, the measure of an angle is the same as the length of its intercepted arc on the unit circle. We will use the $\sin^{-1}x$ notation because it is common on graphing calculators. Remember that the -1 is the symbol for inverse function, as in $f^{-1}(x)$, not an exponent meaning reciprocal. The graph of $y = \sin^{-1}x$ is the reflection of the graph of $y = \sin x$ over the line $y = x$. The domain of $y = \sin^{-1}x$ is the range of $y = \sin x$, $[-1, 1]$. The range of $y = \sin^{-1}x$ is the restricted domain of $y = \sin x$, $\left[-\frac{\pi}{2}, \frac{\pi}{2}\right]$. Note that this means the inverse sine function will always return an angle in either Quadrant I or Quadrant IV.

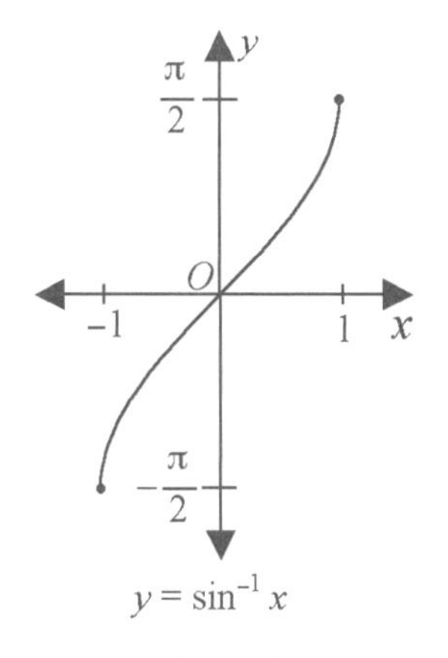

FIGURE 13.21

Remember that the basic idea of an inverse function is that it will "undo" the original function. For sine, this means the following:

$$\sin^{-1}(\sin x) = x \text{ for } -\frac{\pi}{2} \le x \le \frac{\pi}{2}$$

$$\sin\left(\sin^{-1} x\right) = x \text{ for } -1 \le x \le 1$$

The cosine and tangent functions likewise must have their domains restricted in order to have inverse functions. For $y = \cos x$, the appropriate restricted domain is $[0, \pi]$ as shown in Figure 13.22. Note that on this domain, $y = \cos x$ is one-to-one and takes on all values in its range, $[-1, 1]$. The domain also includes the origin.

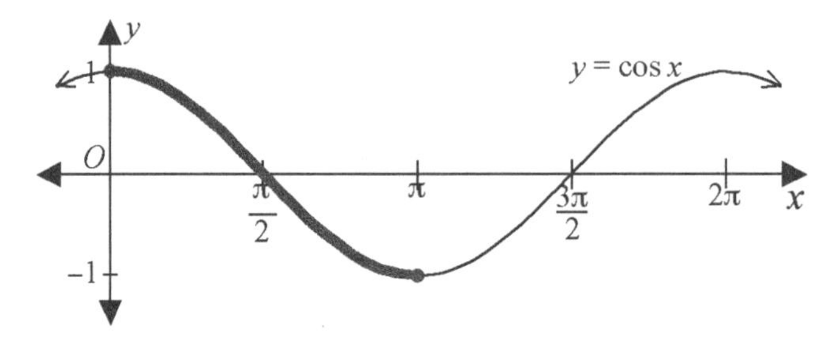

FIGURE 13.22

On this restricted domain, we can define the *inverse cosine function*, $y = \cos^{-1} x$ (or $y = \arccos x$), having domain $[-1, 1]$ and range $[0, \pi]$. The inverse cosine function always returns an angle in either Quadrant I or Quadrant II.

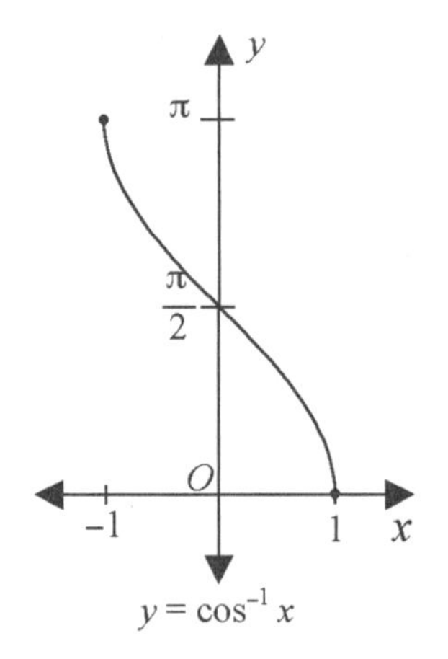

FIGURE 13.23

The restricted domain for $y = \tan x$ is $\left(-\frac{\pi}{2}, \frac{\pi}{2}\right)$, almost the same as for $y = \sin x$.

The endpoints are not included because $y = \tan x$ is not defined at $x = \pm\frac{\pi}{2}$.

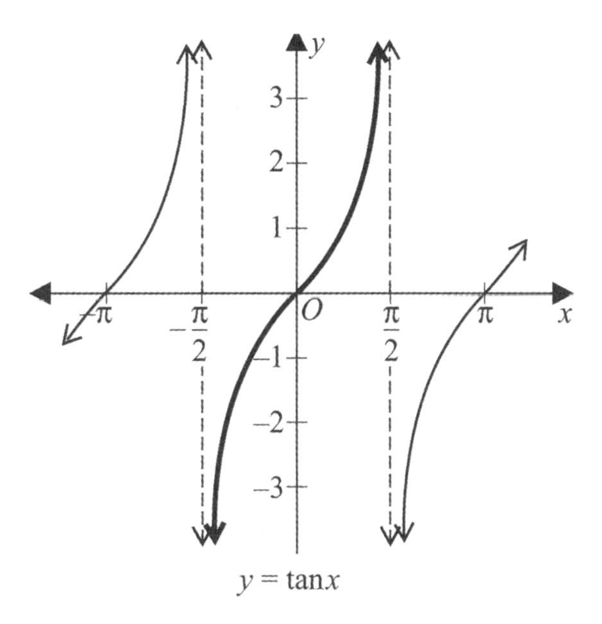

FIGURE 13.24

The *inverse tangent function* $y = \tan^{-1} x$ (or $y = \arctan x$) has domain $(-\infty, \infty)$ and range $\left(-\frac{\pi}{2}, \frac{\pi}{2}\right)$. Like inverse sine, it always returns an angle in Quadrant I or Quadrant IV.

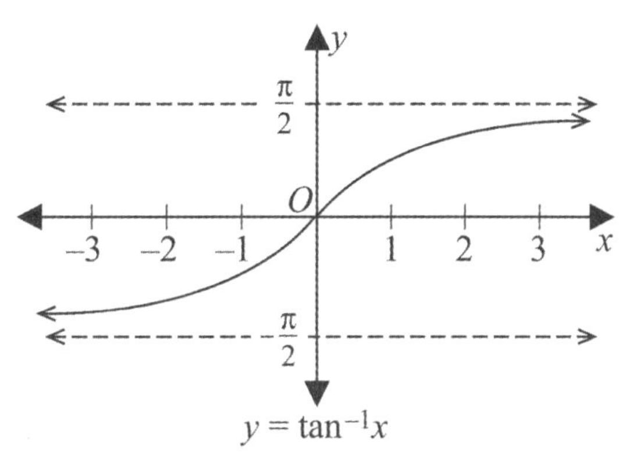

FIGURE 13.25

The inverse trigonometric functions are summarized in Table 13.3.

TABLE 13.3
INVERSE TRIGONOMETRIC FUNCTIONS

Function	Domain	Range	Graph	Properties
$y = \sin^{-1} x$	$[-1, 1]$	$\left[-\frac{\pi}{2}, \frac{\pi}{2}\right]$ Quadrants I and IV	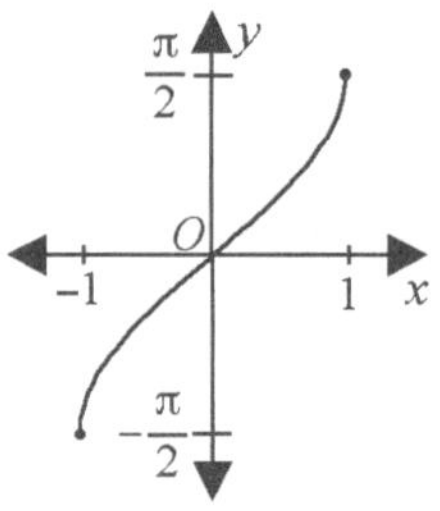	$\sin^{-1}(\sin x) = x$ for $-\frac{\pi}{2} \le x \le \frac{\pi}{2}$ and $\sin(\sin^{-1} x) = x$ for $-1 \le x \le 1$
$y = \cos^{-1} x$	$[-1, 1]$	$[0, \pi]$ Quadrants I and II	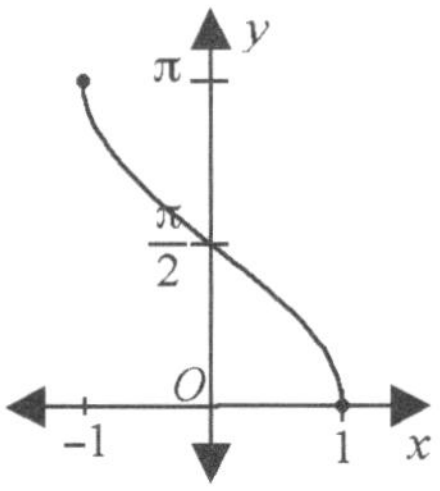	$\cos^{-1}(\cos x) = x$ for $0 \le x \le \pi$ and $\cos(\cos^{-1} x) = x$ for $-1 \le x \le 1$
$y = \tan^{-1} x$	$(-\infty, \infty)$	$\left(-\frac{\pi}{2}, \frac{\pi}{2}\right)$ Quadrants I and IV	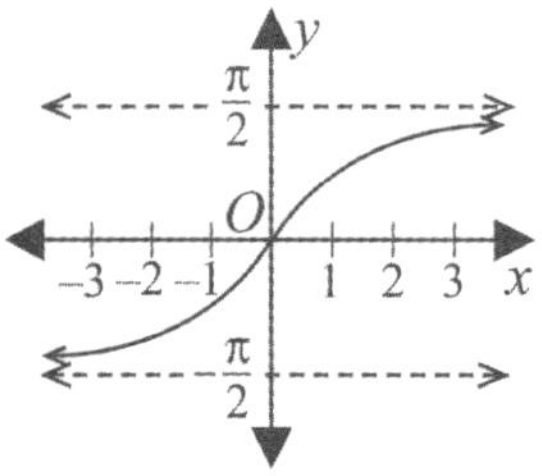	$\tan^{-1}(\tan x) = x$ for $-\frac{\pi}{2} < x < \frac{\pi}{2}$ and $\tan(\tan^{-1} x) = x$ for all real numbers

You have probably noticed that so far in this section, we have worked exclusively in radians. This is the usual mode when dealing with inverse trigonometric functions. However, there are problems where we prefer our answers in degrees. Scientific and graphing calculators let you choose the mode for your answers. In degrees, the restricted domain of the sine function is $-90° \le x \le 90°$. When taking an inverse sine in degree mode, you will always get an answer between $-90°$ and $90°$. The restricted domain for cosine in degrees is $0° \le x \le 180°$, and the restricted domain for tangent is $-90° < x < 90°$.

For your use in the next sections, Table 13.4 gives the trigonometric function values for special angles.

TABLE 13.4
VALUES FOR SPECIAL ANGLES

	0° (0)	**30°** $\left(\frac{\pi}{6}\right)$	**45°** $\left(\frac{\pi}{4}\right)$	**60°** $\left(\frac{\pi}{3}\right)$	**90°** $\left(\frac{\pi}{2}\right)$
sin θ	0	$\frac{1}{2}$	$\frac{\sqrt{2}}{2}$	$\frac{\sqrt{3}}{2}$	1
cos θ	1	$\frac{\sqrt{3}}{2}$	$\frac{\sqrt{2}}{2}$	$\frac{1}{2}$	0
tan θ	0	$\frac{\sqrt{3}}{3}$	1	$\sqrt{3}$	Undefined

Evaluate the following expressions.

a. $\sin^{-1}\left(\frac{\sqrt{3}}{2}\right)$

b. $\sin^{-1}\left(-\frac{1}{\sqrt{2}}\right)$

c. $\cos^{-1}\left(-\frac{\sqrt{3}}{2}\right)$

d. $\tan^{-1}\left(-\sqrt{3}\right)$

SOLUTION

All of these can be done quickly and easily on a calculator as long as you don't mind decimal approximations for the answers. Some teachers want their students to be able to do calculations involving special angles (see Chapter 12) without a calculator. We'll do them without a calculator.

a. To evaluate $\sin^{-1}\left(\frac{\sqrt{3}}{2}\right)$ means to find the angle within the restricted domain of sine whose sine is $\frac{\sqrt{3}}{2}$. In other words, find θ such that $-\frac{\pi}{2} \le \theta \le \frac{\pi}{2}$ and $\sin\theta = \frac{\sqrt{3}}{2}$. First figure out what quadrant the angle is in. The restricted domain of sine includes only Quadrants I and IV. Of those two, sine is positive only in Quadrant I. We want a Quadrant I angle. Next figure out the reference angle. From Table 13.4 or your memory, you find that $\sin\theta = \frac{\sqrt{3}}{2}$ when $\theta = \frac{\pi}{3}$. Alternatively, you can sketch a reference triangle in standard position. If $\sin\theta = \frac{\sqrt{3}}{2}$, then the opposite side is $\sqrt{3}$ and the hypotenuse is 2. From memory or using the Pythagorean theorem, you find the missing side to be 1. You should recognize the resulting triangle as a 30°–60°–90° triangle with the 60° angle across from $\sqrt{3}$. This means the reference angle we want is 60° or $\frac{\pi}{3}$ radians.

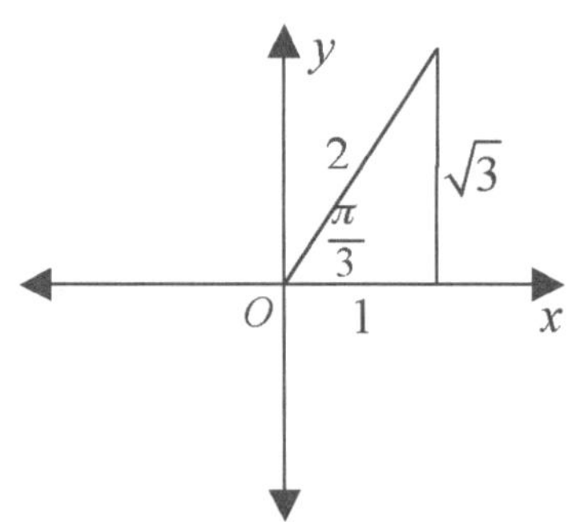

In Quadrant I, the reference angle and the actual angle are the same, so we are done; $\sin^{-1}\left(\frac{\sqrt{3}}{2}\right) = \frac{\pi}{3}$ or $\sin^{-1}\left(\frac{\sqrt{3}}{2}\right) = 60°$.

b. This time we want to find θ such that $-\frac{\pi}{2} \leq \theta \leq \frac{\pi}{2}$ and $\sin\theta = -\frac{1}{\sqrt{2}}$. Of the restricted domain of the sine function, Quadrants I and IV, sine is negative only in Quadrant IV. The reference triangle is shown below. You should recognize it as a 45°–45°–90° triangle, so the reference angle marked on the diagram is 45° or $\frac{\pi}{4}$ radians.

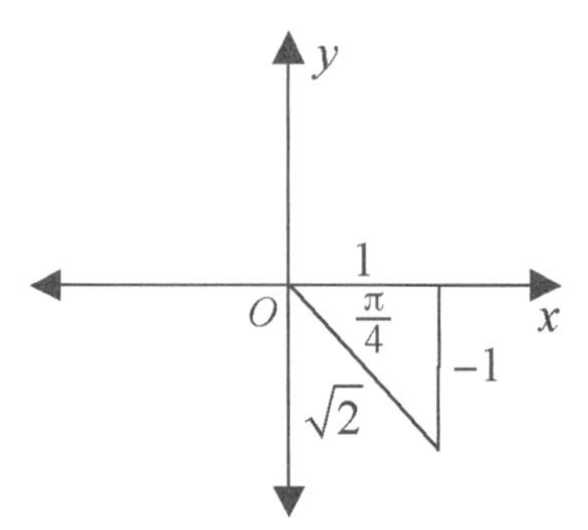

We want a Quadrant IV angle between $-\frac{\pi}{2}$ and 0. The angle is $\theta = -\frac{\pi}{4}$. Note that the answer is not $\theta = \frac{7\pi}{4}$. Although $\frac{7\pi}{4}$ is coterminal with $-\frac{\pi}{4}$, it is not in the restricted domain of the sine function.

The answer is $\sin^{-1}\left(-\frac{1}{\sqrt{2}}\right) = -\frac{\pi}{4}$ or $\sin^{-1}\left(-\frac{1}{\sqrt{2}}\right) = -45°$.

c. Now we want an angle in the restricted domain of cosine that satisfies $\cos\theta = -\frac{\sqrt{3}}{2}$. Remember that the restricted domain for cosine is different from that for sine. For this problem, we need an angle in the range $[0, \pi]$ or Quadrants I and II. Of those, cosine is negative only in Quadrant II. The reference triangle is shown below. Again you should recognize the 30°–60°–90° triangle. This time the desired reference angle is 30° or $\frac{\pi}{6}$ radians.

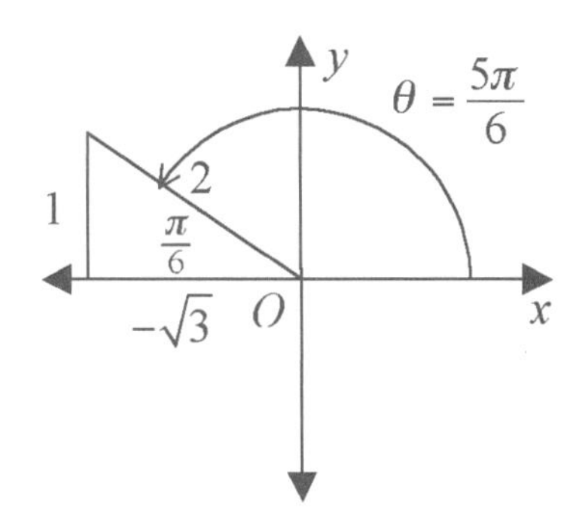

We need a Quadrant II angle having a reference angle of $\frac{\pi}{6}$ or 30°. The desired angle is $\theta = \pi - \frac{\pi}{6} = \frac{5\pi}{6}$ or $\theta = 180° - 30 = 150°$.

The answer is $\cos^{-1}\left(-\frac{\sqrt{3}}{2}\right) = \frac{5\pi}{6}$ or $\cos^{-1}\left(-\frac{\sqrt{3}}{2}\right) = 150°$.

d. Like sine, tangent's restricted domain is limited to angles in Quadrants I and IV. Of those, tangent is negative in only Quadrant IV. The reference triangle is shown below. Again it is a 30°–60°–90° triangle, and the reference angle is 60° or $\frac{\pi}{3}$ radians.

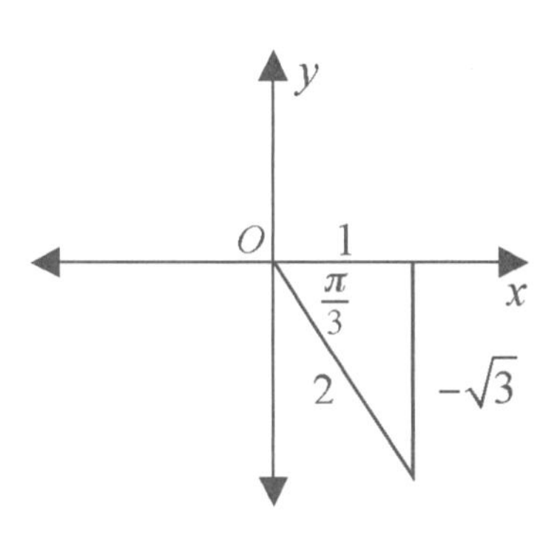

Because we need a Quadrant IV angle, the final answer is $\tan^{-1}\left(-\sqrt{3}\right) = -\frac{\pi}{3}$ or $\tan^{-1}\left(-\sqrt{3}\right) = -60°$.

Evaluate the following:

a. $f(x) = \cos^{-1} x$ at $x = 0$

b. $g(x) = \sin^{-1} x$ at $x = -1$

SOLUTION

When evaluating inverse sine or inverse cosine at $x = 0$ or $x = \pm 1$, reference triangles won't work. Instead think about the unit circle.

a. We want to find $\cos^{-1}(0)$. In other words, we want to find where in the restricted domain of cosine that $\cos\theta = 0$. On the unit circle, $\cos\theta$ corresponds to x. We know $x = 0$ on the y-axis. Possible choices for the angle are $\theta = \frac{\pi}{2}$, $\theta = \frac{3\pi}{2}$, or $\theta = -\frac{\pi}{2}$.

The only one in the restricted domain of cosine is $\theta = \frac{\pi}{2}$. So $f(0) = \cos^{-1}(0) = \frac{\pi}{2}$ or $f(0) = \cos^{-1}(0) = 90°$.

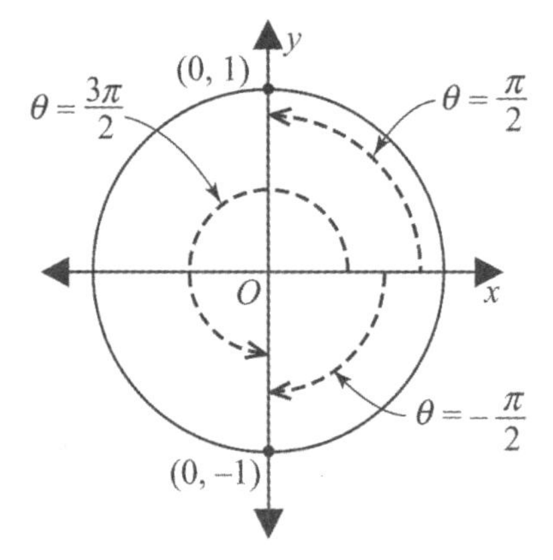

b. Now we want $\sin^{-1}(-1)$, the angle in the restricted domain of sine where $\sin\theta = -1$. On the unit circle, sine corresponds to y. We know $y = -1$ only on the negative side of the y-axis. Using the same diagram as above, we see two choices, $\theta = \frac{3\pi}{2}$ or $\theta = -\frac{\pi}{2}$. However, the restricted domain of sine is $\left[-\frac{\pi}{2}, \frac{\pi}{2}\right]$, so we want $\theta = -\frac{\pi}{2}$.

The answer is $g(-1) = \sin^{-1}(-1) = -\frac{\pi}{2}$ or $g(-1) = \sin^{-1}(-1) = -90°$.

SECTION EXERCISES

13–26. Evaluate the following expressions.

a. $\sin^{-1}\left(-\frac{\sqrt{2}}{2}\right)$

b. $\cos^{-1}\left(-\frac{1}{2}\right)$

c. $\tan^{-1}\left(\frac{\sqrt{3}}{3}\right)$

13–27. Evaluate the following.

a. $f(x) = \cos^{-1} x$ at $x = -1$

b. $g(x) = \sin^{-1} x$ at $x = 1$

13–28. Bertrand was solving the equation $\sin x = k$. He found $\sin^{-1} k = 65°$ on his calculator. Find two values of x in the interval $0 \le x < 360°$ that solve Bert's equation.

13–29. Norman was solving the equation $\cos x = -k$. When he found $\cos^{-1} k$ on his calculator, he got 35°. Find two values of x in the interval $0 \le x < 360°$ that will solve Norman's equation.

Solving Trigonometric Equations

The use of trigonometry in applications and modeling naturally leads to the solving of trigonometric equations. Suppose a ball has an initial velocity of 100 feet/second. At what angle should it leave the ground so that it lands 250 feet away? To find out, we need to solve the equation $\frac{(100)^2 \sin(2\theta)}{32} = 250$. This looks pretty hard, so we'll do some simpler examples first.

Solving a trigonometric equation algebraically is done in two main steps. First we use appropriate algebra techniques to simplify the equation to a single trigonometric function that equals a value, for example, $\cos x = -\frac{1}{2}$. Then we use the appropriate inverse trigonometric function and our knowledge of the graphs of trigonometric functions to finish solving for the variable. Because of the periodic nature of trigonometric functions, trigonometric equations have an infinite number of solutions. In Algebra 2, you will most likely be asked to find only the solutions in a particular interval, often one period, or only a particular solution such as the smallest positive one. Read the question carefully so you know what you need to find.

To solve trigonometric equations algebraically, you need to understand the unit circle definitions of the trigonometric functions well. In particular, you need to be very familiar with the signs of the trigonometric functions by quadrant and comfortable with the ideas of reference angles. Also, if you need exact answers to some problems, you need to know the trigonometric function values for special angles such as the quadrantal angles and $\frac{\pi}{6}$, $\frac{\pi}{4}$, and $\frac{\pi}{3}$. If you are not confident with those topics, this would be a good time to review the relevant sections in Chapter 12.

Trigonometric equations can be solved in either radians or degrees. Read the question or understand the context to know which one to use. As a general rule, unless something in the problem explicitly suggests the use of degrees, trigonometric equations should be solved in radians.

As with many things, solving trigonometric equations is best learned through examples. We'll start easy and work our way up.

EXAMPLE 13.17 Find all the solutions to $\sin x = \frac{\sqrt{2}}{2}$ in the interval $0 \leq x < 2\pi$.

SOLUTION

Note that the interval $0 \leq x < 2\pi$ implies that we want our answers in radians. In this problem, the trigonometric function is already isolated. Finding the appropriate angles involves three steps.

1. Identify the quadrants where the solutions may be.
2. Find the reference angle.
3. Find the actual solutions.

Sine is positive in Quadrants I and II (remember "all students take calculus" from Chapter 12). This problem will have two solutions, one in each of those quadrants. It is helpful to draw reference triangles in both quadrants.

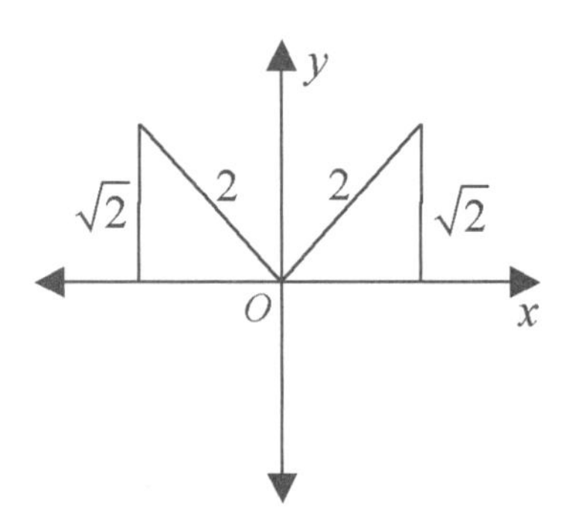

For both solutions, the reference angle will be $x_{ref} = \sin^{-1}\left(\frac{\sqrt{2}}{2}\right)$. Either by recognizing the special triangle on your diagram (using the Pythagorean theorem to find the missing side, $\sqrt{2}$, may help) or by referring to Table 13.4 (which you are working on memorizing), we find the reference angle to be $x_{ref} = \frac{\pi}{4}$. Add it to your diagram.

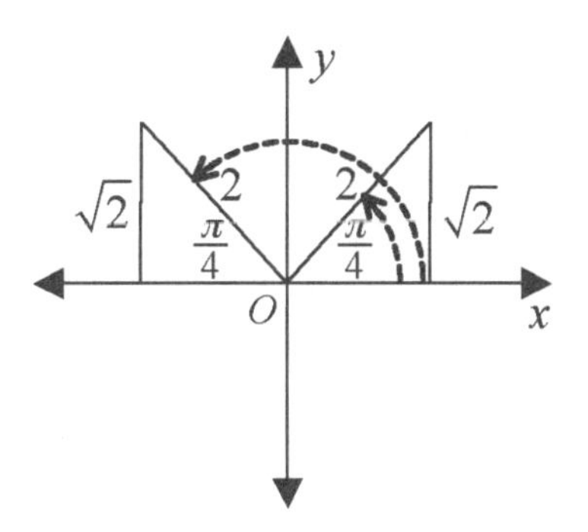

Now we can find the two solutions; they are marked with dashed arcs on the diagram. The Quadrant I solution is the same as the reference angle, $x = \frac{\pi}{4}$. The Quadrant II solution is $x = \pi - \frac{\pi}{4} = \frac{3\pi}{4}$.

The solutions are $x = \frac{\pi}{4}$ and $x = \frac{3\pi}{4}$ or, as a solution set, $\left\{\frac{\pi}{4}, \frac{3\pi}{4}\right\}$.

Frequently you need to isolate the trigonometric function before undoing with an inverse trigonometric function.

EXAMPLE 13.18

Solve for all x in $0° \leq x < 360°$ that satisfies $2\cos x + \sqrt{3} = 0$.

SOLUTION

Note that this problem asks for solutions in degrees. First isolate the trigonometric function.

$$2\cos x + \sqrt{3} = 0$$
$$2\cos x = -\sqrt{3}$$
$$\cos x = -\frac{\sqrt{3}}{2}$$

Cosine is negative in Quadrants II and III. Sketch reference triangles in those quadrants. In the diagram to the right, the missing sides have been found using the Pythagorean Theorem. This is optional but may help you recognize the special triangle as a 30°–60°–90° triangle.

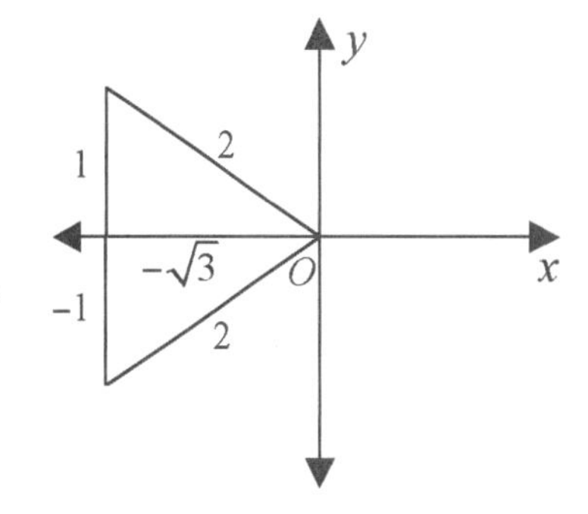

When finding the reference angle, don't use the negative sign. In this problem, the reference angle is $x_{ref} = \cos^{-1}\left(\frac{\sqrt{3}}{2}\right)$, not $\cos^{-1}\left(-\frac{\sqrt{3}}{2}\right)$. From the diagram or from Table 13.4, you will find the reference angle to be 30°. Add it to the diagram.

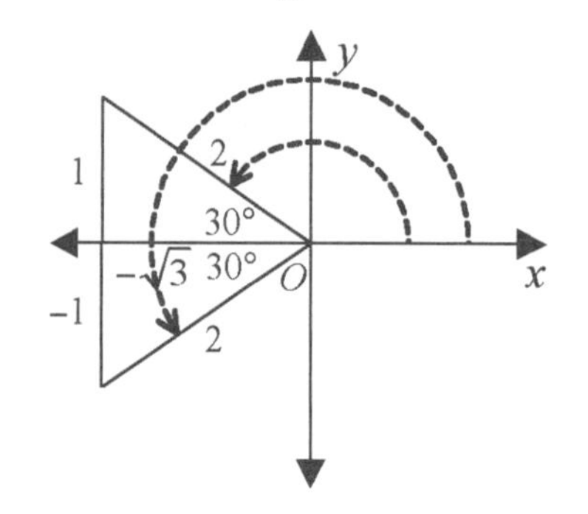

The Quadrant II solution is 180° − 30° = 150°. The Quadrant III solution is 180° + 30° = 210°. The solution set is {150°, 210°}.

SEE CALC TIP 13C

EXAMPLE 13.19

Solve for all x in $0 \le x < 2\pi$ that satisfy $5 \sin x - 2 = 3$.

SOLUTION

First isolate the trigonometric function.

$$\begin{aligned} 5 \sin x - 2 &= 3 \\ 5 \sin x &= 5 \\ \sin x &= 1 \end{aligned}$$

This is one of the special cases mentioned in the last section where reference triangles won't help. If a problem reduces to either $\sin x$ or $\cos x$ equal to 0, 1, or −1 or to $\tan x = 0$, use the unit circle. On the unit circle, sine is represented by y. So solving $\sin x = 1$ is equivalent to finding points on the unit circle where $y = 1$. There is only one such point, (0, 1), which corresponds to an angle of $\frac{\pi}{2}$.

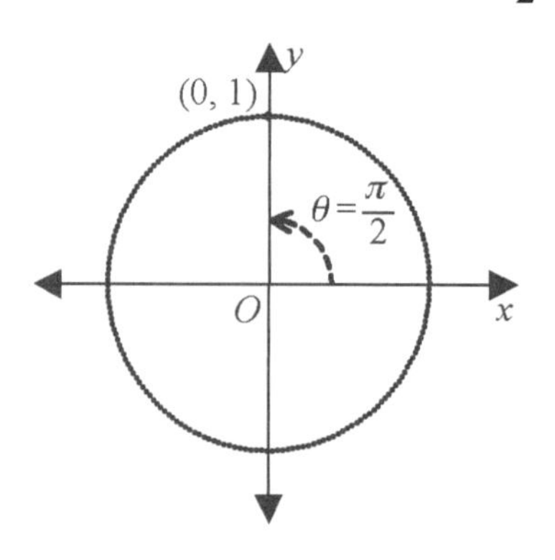

The only solution to this problem in the given interval is $x = \frac{\pi}{2}$.

EXAMPLE 13.20 Solve for all x in $0° \leq x < 360°$ that satisfies $\sqrt{3}\tan x + 1 = 0$.

SOLUTION

First isolate the trigonometric function.

$$\sqrt{3}\tan x + 1 = 0$$
$$\sqrt{3}\tan x = -1$$
$$\tan x = -\frac{1}{\sqrt{3}} = -\frac{\sqrt{3}}{3}$$

Tangent is negative in Quadrants II and IV. Sketch reference triangles in those quadrants. The unknown sides can be found using the Pythagorean theorem if you need to. You should recognize the special triangle as a 30°–60°–90° triangle.

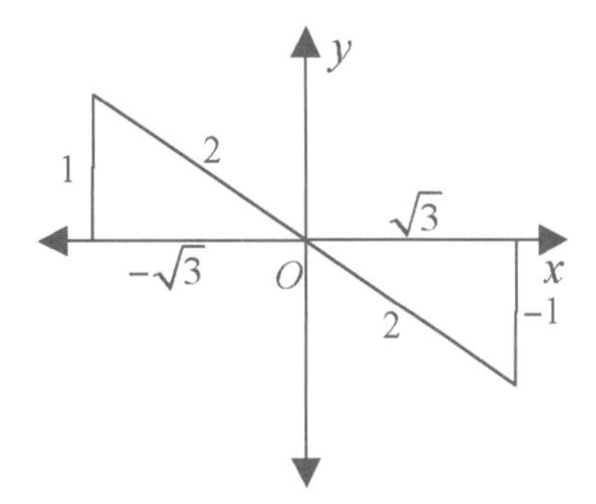

Again, when finding the reference angle, don't use the negative sign. In this problem, the reference angle is $x = \tan^{-1}\left(\frac{\sqrt{3}}{3}\right)$. From the diagram or from Table 13.4, you will find the reference angle to be 30°. Add it to the diagram.

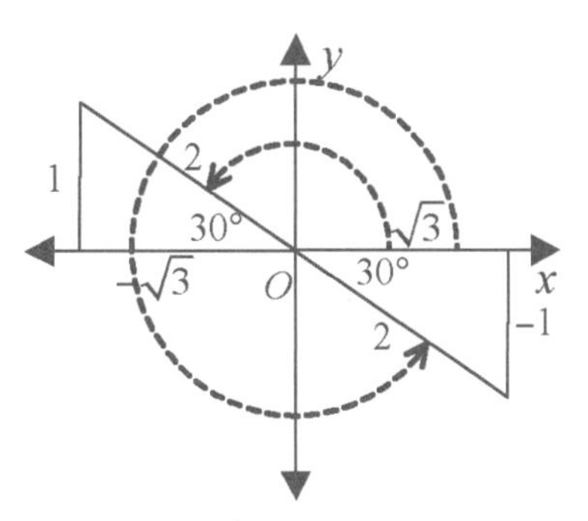

The Quadrant II solution is 180° – 30° = 150°. The Quadrant IV solution is 360° – 30° = 330°. The solution set is {150°, 330°}.

You will need a scientific or graphing calculator for the next example.

Solve for all x in $0° \leq x < 360°$ that satisfies $3 \cos x - 1 = 0$. Round your answers to the nearest tenth of a degree.

SOLUTION

First isolate the trigonometric function.

$$3\cos x - 1 = 0$$
$$3\cos x = 1$$
$$\cos x = \frac{1}{3}$$

Cosine is positive in Quadrants I and IV; draw appropriate reference triangles. Since $\frac{1}{3}$ is not a trigonometric function value from one of the special angles (in other words, it is not a value that shows up in Table 13.4), you do not need to label the sides of the triangles. Instead, use your calculator (make sure you are in degree mode for this problem) to find the reference angle $x_{ref} = \cos^{-1}\left(\frac{1}{3}\right) = 70.5°$ to the nearest tenth. Add this to the diagram.

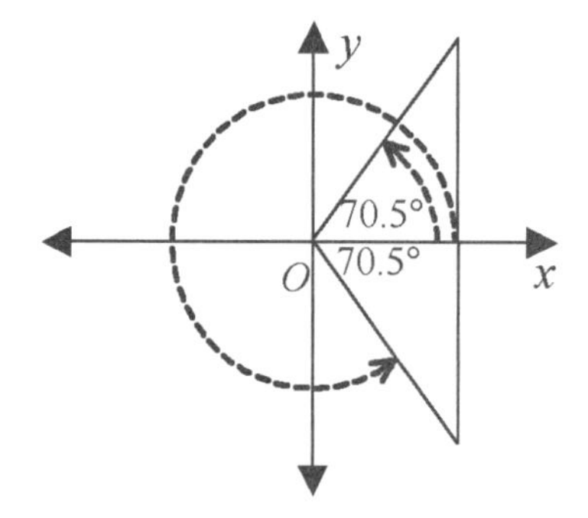

The Quadrant I angle is 70.5°. The Quadrant IV angle is $360° - 70.5° = 289.5°$. The solution set is {70.5°, 289.5°}.

We started this section with an example about hitting a ball a specific distance. If a projectile has an initial velocity v_0 in feet per second and angle θ with the horizontal, its range, which is the horizontal distance in feet it travels in the air, can be modeled by $R = \frac{v_0^2 \sin(2\theta)}{32}$. (This model ignores the effects of air resistance.) Suppose a ball has an initial velocity of 100 feet/second. At what angle should it leave the ground so that it lands 250 feet away?

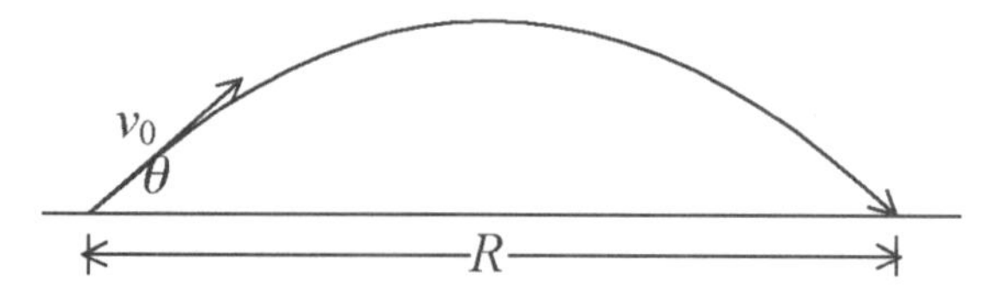

SOLUTION

After substituting 100 for v_0 and 250 for R, we have the equation $\frac{(100)^2 \sin(2\theta)}{32} = 250$.
This looks hard at first. After isolating the trigonometric function, it reduces to:

$$\sin(2\theta) = 0.8$$

This equation has a new twist, 2θ instead of just θ. Solve for 2θ first and then divide by 2 at the end. (Some students find it helpful to define a temporary variable to represent 2θ. If you let $A = 2\theta$, then the first part of the problem is simply to solve $\sin A = 0.8$. After finding A, divide by 2 to find θ. This extra step is optional. It doesn't change the work you do to solve the problem.)

Sine is positive in Quadrants I and II. The reference angle is $\sin^{-1}(0.8)$. The problem did not specify degrees or radians, but degrees seem reasonable in the context of the problem. The reference angle is 53.13010235°. Draw two reference triangles, and label the reference angles. (On the diagram, we rounded the reference angle to the nearest tenth of a degree. In your actual calculations you should keep all the decimals and round appropriately at the end of the problem.)

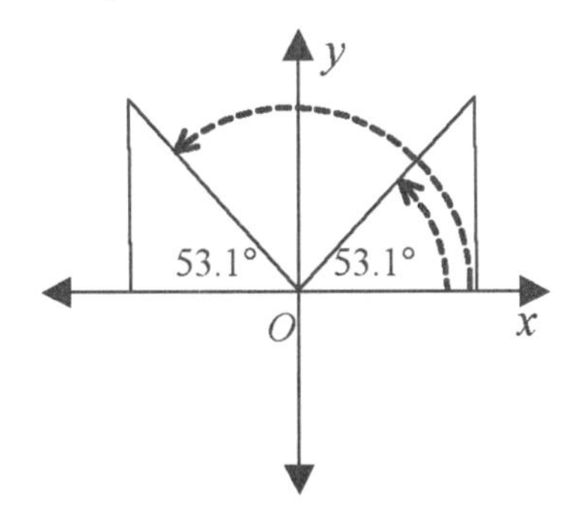

Remember, right now we are solving for 2θ. The two angles shown with dashed arcs on the diagram each represent 2θ, not θ. In Quadrant I we have $2\theta = 53.13010235°$. In Quadrant II, we have $2\theta = 180 - 53.13010235° = 126.8698976°$. We can now divide by 2 to finish solving the problem.

$$2\theta = 53.13010235° \rightarrow \theta \approx 26.6°$$
$$2\theta = 126.8698976° \rightarrow \theta \approx 63.4°$$

Some people are surprised that there are two valid solutions to this problem. The two trajectories both start and end in the same place. However, one is much lower and "flatter" than the other, a lot like the difference between a high fly ball and a line drive in baseball.

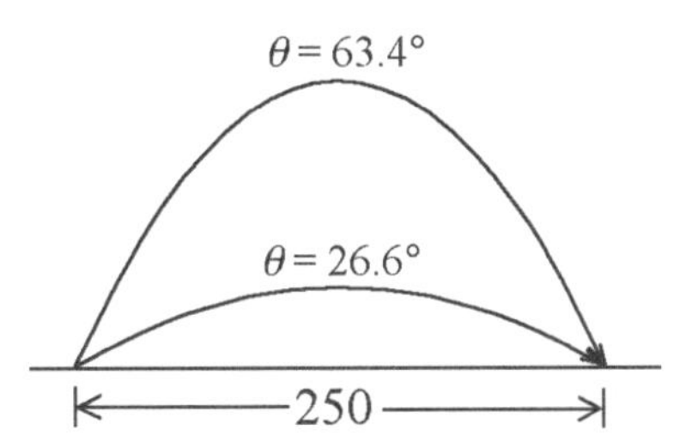

SECTION EXERCISES

13–30. What is the smallest positive value of x, in degrees, for which $\sin x = -\frac{1}{2}$?

13–31. Without using your calculator, find two angles, $0° \le x < 360°$, such that

a. $\cos x = -\frac{1}{\sqrt{2}}$

b. $\tan x = -\sqrt{3}$

13–32. Using your calculator, find two angles rounded to the nearest degree, $0° \le x < 360°$, such that

a. $\sin x = -0.6$

b. $\tan x = 20$

For 33–34, without using your calculator, solve the following for all values of x *in the interval* $0° \le x < 360°$.

13–33. $2\cos x + 1 = 0$

13–34. $4(1 + \sin x) = 6$

For 35–36, using your calculator, solve the following for all values of x, *rounded to the nearest tenth of a degree, in the interval* $0 \le x < 360°$.

13–35. $3 \sin x + 1 = 0$

13–36. $4(\tan x - 1) = \tan x$

For 37–38, without using your calculator, find, in terms of π, *all the values of* θ *in the interval* $[0, 2\pi)$ *that satisfy each equation.*

13–37. $-2\cos\ x = \sqrt{2}$

13–38. $\tan x = \frac{-1}{\sqrt{3}}$

13–39. Without using your calculator, find the first two positive solutions, in degrees, for the equation $\tan \frac{1}{2}x = 1$.

13–40. The distance Chuck Armstrong throws a baseball depends on the angle θ up from the horizontal at which he throws it. Suppose the distance in feet is given by $d = 222 \sin 2\theta$. Using your calculator, find two angles, rounded to the nearest degree, in the interval $0 \le x < 90°$ at which Chuck could throw the ball for it to travel 165 feet.

CHAPTER EXERCISES

C13–1. Which of the following could not be reasonably modeled by a periodic function?

(1) Average monthly temperatures
(2) The phases of the moon
(3) The height off the ground of the right foot of an athlete running the 110-meter hurdles
(4) The population of the United States

C13–2. Parts of two trigonometric graphs, labeled A and B, are shown in the graph below. One of the graphs has no vertical shift.

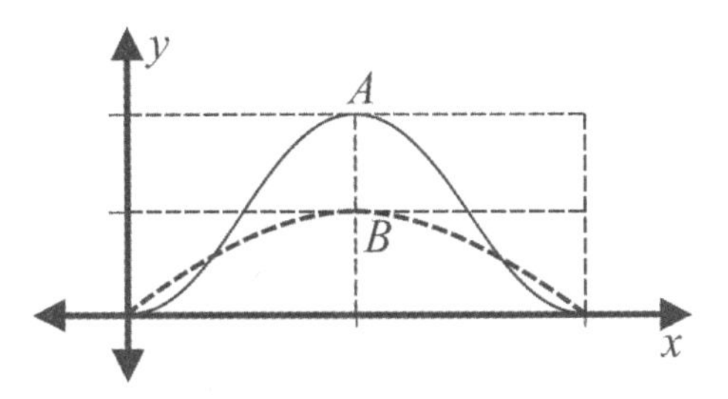

a. Which graph has the larger amplitude?

(1) A (2) B (3) The amplitudes are equal.

b. Which graph has the larger period?

(1) A (2) B (3) The periods are equal.

C13–3. If $y = \cos kx$, $k > 0$, as k increases:

(1) The amplitude increases (2) The amplitude decreases
(3) The period increases (4) The period decreases

C13–4. What is the amplitude of the function $y = 15 - 6\cos 2x$?

C13–5. What is the period of the function $y = 6 + 2\cos\frac{1}{4}x$?

C13–6. What is the range of the function $y = 5\sin\frac{1}{2}x + 3$?

C13–7. If the amplitude of $y = a \sin bx$ is $\frac{3}{2}$ and the period is $\frac{2\pi}{3}$, find the values of a and b.

For 8–9, graph one full period of the following.

C13–8. $y = 1 - 2\cos x$

C13–9. $y = 4\sin\left(\frac{\pi}{2}x\right) + 3$

For 10–12, write an equation for each graph below.

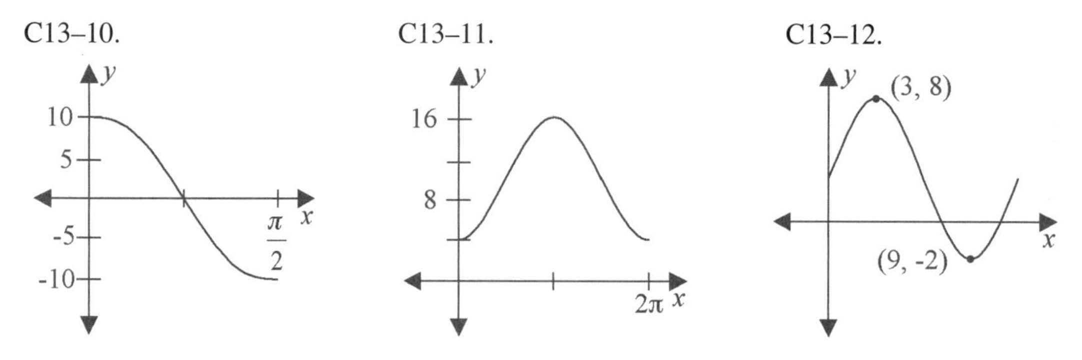

C13–13. The number of hours of daylight in Anchorage, Alaska, can be modeled by

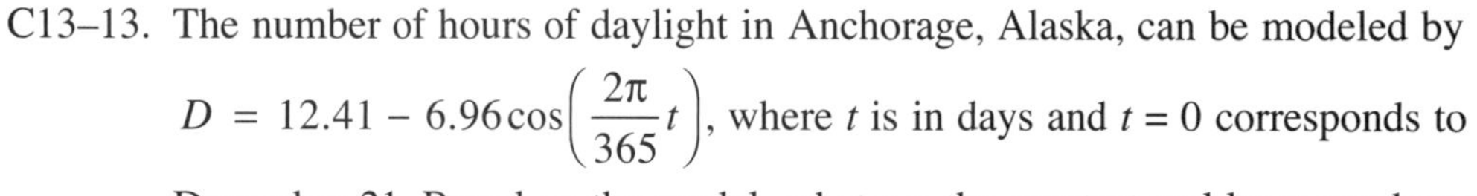

$D = 12.41 - 6.96\cos\left(\frac{2\pi}{365}t\right)$, where t is in days and $t = 0$ corresponds to December 21. Based on the model, what are the greatest and least number of hours of daylight in Anchorage?

C13–14. Normal electrical outlets in the United States provide alternating current. The voltage varies sinusoidally with a maximum of about 115 volts and a minimum of –115 volts (measured from a 0 level called ground). The period of the current is 1/60 of a second. If the voltage is represented by $V = A\sin Bt$, find the values of A and B.

C13–15. The difference between low tide and high tide at Backwash Bay is 6 feet. The tide comes in and goes out twice in a day.

a. What is the amplitude of the tides in feet?

b. What is the period of the tides in hours?

c. Write a trigonometric equation to model the water depth if the depth is at a maximum of 8 feet at time $t = 0$ hours.

C13–16. Without using your calculator, find two angles, $0° \le x < 360°$, such that $\sin x = \frac{1}{2}$.

C13–17. Using your calculator, find two angles rounded to the nearest degree, $0° \le x < 360°$, such that $\cos x = 0.9$.

C13–18. Without using your calculator, solve the following for all values of x in the interval $0° \le x < 360°$ for the equation $\tan \theta = -1$.

C13–19. Using your calculator, solve for all values of x, rounded to the nearest tenth of a degree, in the interval $0 \le x < 360°$, such that $2\sin x = \frac{1}{2}$.

C13–20. Without using your calculator, find all the values of θ in the interval $[0, 2\pi)$, in terms of π, that satisfy the equation $\sin x = 0$.

C13–21. Without using your calculator, find the first two positive solutions, in degrees, for $\sin 3x = \frac{\sqrt{3}}{2}$.

Sequences and Series

WHAT YOU WILL LEARN

Often we receive information as a list of numbers in order. Lots of situations are described with discrete data, which are individual numbers, instead of with continuous data. Working with lists of ordered numbers and finding their totals occurs often in life. In this chapter, you will learn to:

- Evaluate sequences given the explicit or recursive formulas;
- Write the explicit or recursive formula for an arithmetic or geometric sequence;
- Evaluate a series given in summation notation;
- Determine the sum of a finite arithmetic or geometric sequence;
- Determine the sum of an infinite geometric sequence.

SECTIONS IN THIS CHAPTER

Introduction to Sequences

Suppose your slightly eccentric Uncle Albert gave you \$5 on your first birthday. On each succeeding birthday, Uncle Al gave you \$2 more than he had given you the previous year.

Birthday	Uncle Al's Gift
1	\$5
2	\$7
3	\$9
4	\$11
5	\$13
6	\$15

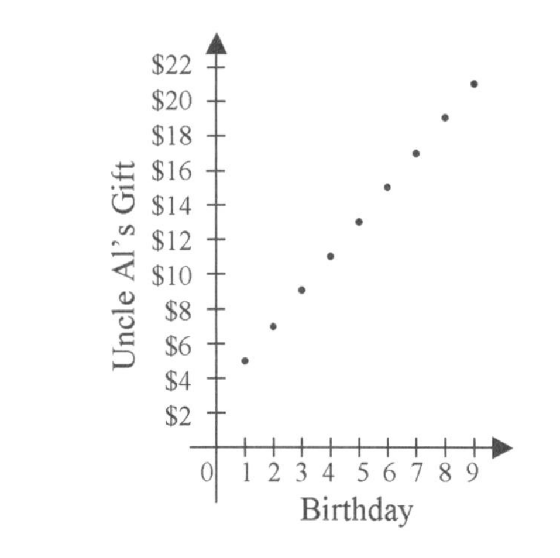

FIGURE 14.1

From the table and the graph, this example looks a lot like a linear function in that it has a constant rate of change, \$2 per year. However, the graph makes clear that there is a big difference in the domain between this and a typical linear function. The dots are not connected; this is not a mistake. Uncle Al gives you money only on your birthday. He gives you no money on any of the days between birthdays. Also, Uncle Al did not give you money before you were born. He did not even give you money the day you were born. The first money from Uncle Al came on your first birthday. So instead of being a continuous line, this graph is a series of isolated points.

This is an example of a *sequence*. A sequence can be thought of as simply a list of numbers, called *terms*, in a particular order. This sequence is 5, 7, 9, 11, 13, ... A sequence can be either finite or infinite. If Uncle Al stopped giving you money after your 18th birthday, this sequence would end with the number 39 and be a finite sequence. If you and Uncle Al both live forever and he keeps giving you money (which seems unlikely), the sequence would continue indefinitely and be an infinite sequence.

Another way to think of a sequence is as a function, $f(n)$, whose domain is the positive integers (or just some limited number of the positive integers if the sequence is finite). The sequence in this example could be written as $f(n) = 3 + 2n$, $n = 1, 2, 3, \ldots$. Note that in math, it is common to use n for a variable that can take on only integral values (other common choices are i, j, k, and m). Thus we use n in this function instead of x, which often signifies a continuous variable. Also in sequences, the dependent variable is often a (or b or c) instead of y or f. Additionally, function notation, $f(n)$, is usually replaced with subscript notation, a_n. Therefore, instead of $y = 3 + 2n$ or $f(n) = 3 + 2n$, we usually write $a_n = 3 + 2n$, $n = 1, 2, 3, \ldots$; a_n is called the "nth term" of the sequence. The subscript, n, gives the location of the number in the list. For example, a_{10} means the value of the tenth term of the sequence, in this case 23.

EXAMPLE 14.1 Write the first four terms of the sequence $a_n = n^2 - 2n + 2$, $n = 1, 2, 3, \ldots$.

SOLUTION

We evaluate terms of this sequence by substituting in values of n.

	n	$a_n = n^2 - 2n + 2$	a_n
First term	1	$a_1 = (1)^2 - 2(1) + 2 = 1$	1
Second term	2	$a_2 = (2)^2 - 2(2) + 2 = 2$	2
Third term	3	$a_3 = (3)^2 - 2(3) + 2 = 5$	5
Fourth term	4	$a_4 = (4)^2 - 2(4) + 2 = 10$	10

The first four terms of this sequence are 1, 2, 5, and 10.

RECURSIVE FORMULAS FOR SEQUENCES

Let's make the birthday example more profitable. Your slightly eccentric, rich Aunt Beatrice gives you \$5 on your first birthday and then doubles the amount of money she gives you on each successive birthday. What is a formula for the amount of money you get from Aunt Bea on your nth birthday?

In this example, you get \$5 on your first birthday. On each successive birthday, you receive twice as much as you got on the previous year. If b_n represents the amount you receive on your nth birthday, then b_{n-1} represents the amount you got on your preceding birthday. With this notation, the sequence can be described by $b_1 = 5$, $b_n = 2b_{n-1}$ for $n \geq 2$. Some people find the notation $b_n = 2b_{n-1}$ confusing at first. Try it with a specific value of n. For $n = 5$, b_5 is the amount you got on your fifth birthday, $b_{5-1} = b_4$ is the amount you received on your fourth birthday, and $b_5 = 2b_4$ means you received twice as much on your fifth birthday as on your fourth.

This is an example of a *recursive* definition of a sequence. In a recursive definition, one or more of the initial terms is given and then each new term is defined as a function of one or more of the preceding terms. In this example, $b_1 = 5$ says you received \$5 on your first birthday. And, $b_n = 2b_{n-1}$ for $n \geq 2$ means that on each birthday starting with your second, you received twice as much as the year before. Note that with a recursive formula, you must have an initial value (or sometimes values). Without knowing that $b_1 = 5$, there is no way to figure out b_2 or any succeeding values.

Many students find recursive relationships confusing at first. Take the time to make sure you understand the notation. Remember that a_n (or b_n) simply represents the nth term in a sequence. No matter what n is, a_{n+1} represents the next term in the sequence, the one after a_n. As long as a_n is not the first term of the sequence, a_{n-1} represents the preceding term, the one right before a_n in the sequence. The sequence looks like:

$$a_1, a_2, a_3, \ldots, a_{n-1}, a_n, a_{n+1}, \ldots$$

Be very careful to make sure you don't mix up a_{n-1} and $a_n - 1$; they have very different meanings. The expression a_{n-1} represents a particular term in the sequence, the one immediately before the nth term. In contrast, $a_n - 1$ means one less than the value of the nth term. In Example 14.1, if $n = 4$, $a_4 = 10$, $a_{4-1} = a_3 = 5$, and $a_4 - 1 = 10 - 1 = 9$.

Suppose the first term of a sequence is 3 and each term after the first is 4 less than 3 times the preceding term.

a. Write a recursive definition of the sequence.

b. Write the first four terms of the sequence.

c. If one term of the sequence is 6,563, what is the next term? What was the preceding term?

d. What is the 25th term of the sequence?

SOLUTION

a. To write a recursive definition, we need to specify an initial value and an equation showing how to get new values from previous ones. The problem says the first term is 3, so we write $a_1 = 3$. For the second and later terms, each term is "4 less than 3 times the preceding term." If a_n represents the nth term for some $n \geq 2$, then the preceding term is a_{n-1} and the recursive relation is $a_n = 3a_{n-1} - 4$ for $n \geq 2$. Putting them together, our answer is:

$$a_1 = 3,\ a_n = 3a_{n-1} - 4 \text{ for } n \geq 2$$

Before we go on, it is worth noting that the relationship can be written in a slightly different way. If we know a particular term, a_n, then the next term, a_{n+1}, must be 4 less than 3 times a_n : $a_{n+1} = 3a_n - 4$. So we could write:

$$a_1 = 3,\ a_{n+1} = 3a_n - 4 \text{ for } n \geq 1$$

Both these answers represent the exact same sequence. Sometimes one may have a small advantage over the other (see part c below), but usually the choice is a matter of personal preference.

b. To find the first four terms of the sequence, it is helpful to set up a table.

n	$a_n = 3a_{n-1} - 4$ for $n \geq 2$	a_n
1	given	3
2	$a_2 = 3a_1 - 4 = 3(3) - 4 = 5$	5
3	$a_3 = 3a_2 - 4 = 3(5) - 4 = 11$	11
4	$a_4 = 3a_3 - 4 = 3(11) - 4 = 29$	29

The first four terms of the sequence are 3, 5, 11, and 29.

c. Let $a_n = 6{,}563$. From the definition of the sequence, the next term should be 4 less than 3 times this value: $a_{n+1} = 3a_n - 4$. By substituting, we find $a_{n+1} = 3(6{,}563) - 4 = 19{,}685$.

To find the previous term, we use the same definition. The term 6,563 should be 4 less than 3 times the preceding term: $a_n = 3a_{n-1} - 4$. By substituting and solving, we get:

$$6{,}563 = 3a_{n-1} - 4$$
$$6{,}567 = 3a_{n-1}$$
$$2{,}189 = a_{n-1}$$

Again, be sure you understand the notation (and read it carefully). In solving the equation $6{,}563 = 3a_{n-1} - 4$ above, it makes perfect sense to add 4 to both sides in the first line. However, it makes no sense at all to add 1 to both sides of the answer, $2{,}189 = a_{n-1}$. If you don't see the difference, reread the last paragraph before this example.

The term that follows 6,563 is 19,685. The term before it is 2,189.

d. Finding the 25th term of the sequence is a problem. With a recursively defined sequence, to find a_{25} we first need to know a_{24}. To find a_{24}, though, we need a_{23}. To find a_{23} . . . ; you see the problem. The only way to find a_{25} is to start with $a_1 = 3$ and apply the formula repeatedly 24 times. In other words, keep extending the table in part *a* until you get to $n = 25$. This is where technology becomes very helpful. With a graphing calculator, this answer is found, with a little effort, to be $a_{25} = 282{,}429{,}536{,}483$.

SEE CALC TIP 14A

Example 14.2 illustrates both the advantage and the disadvantage of recursively defined sequences. Many sequences arise naturally where each term depends on the term before it. It is often easy to write a recursive relationship for such sequences. With a recursive sequence, though, there is no way to directly calculate a particular term without first calculating all the preceding terms.

EXAMPLE 14.3

A famous sequence called the Fibonacci sequence is given by the recursive relation $a_1 = 1, a_2 = 1, a_n = a_{n-1} + a_{n-2}$ for $n > 2$. Find the first seven terms of the sequence.

SOLUTION

Once more, it is important to understand the notation: a_n represents a term (the *n*th term) of the sequence, a_{n-1} represents the term before it, and a_{n-2} represents the term before that (two terms before a_n). In this problem, the first two terms are both 1. After that, the recursive definition $a_n = a_{n-1} + a_{n-2}$ says that each new term is the sum of the two terms preceding it.

n	$a_n = a_{n-1} + a_{n-2}$	a_n
1	Given	1
2	Given	1
3	1 + 1 = 2	2
4	2 + 1 = 3	3
5	3 + 2 = 5	5
6	5 + 3 = 8	8
7	8 + 5 = 13	13

SEE CALC TIP 14B

The first seven terms are {1, 1, 2, 3, 5, 8, 13}.

EXPLICIT FORMULAS FOR SEQUENCES

In the first birthday example (Uncle Al), we had an *explicit formula* for the nth term of the sequence: $a_n = 3 + 2n$ for $n = 1, 2, 3, \ldots$ In an explicitly defined sequence, each term a_n is a function of only n rather than of the preceding terms. The big advantage is that we can calculate any term directly. If you want to know how much Uncle Al will give you on your 18th birthday, you can find it quickly and easily without having to first figure out the previous 17 years of gifts: $a_{18} = 3 + 2(18) = \$39$.

EXAMPLE 14.4 An explicit formula for the amount of money Aunt Beatrice gives you on each birthday is $a_n = 2.5(2)^n$. How much would Aunt Bea give you on your 18th birthday?

SOLUTION

Simply substitute $n = 18$ into the formula: $a_{18} = 2.5(2)^{18} = \$655,360$.
You'd better write Aunt Bea a really nice thank you card.

SEE CALC TIP 14C

Since explicit formulas have such an advantage for calculation, why would we ever use a recursive formula? Unfortunately, it is not yet known how to find an explicit formula for every recursion relation. Formulas are known for some special ones, like both versions of the birthday gift problem, but not for many others.

SECTION EXERCISES

For 1, the terms do not need to be reduced.

14–1. Write the first four terms of each sequence.

a. $a_n = \dfrac{2n}{n+3}, n \geq 1$

b. $a_n = \dfrac{(-1)^n}{n+1}, n \geq 1$

For 2–5, write the next four terms, after the given term(s), of each sequence.

14–2. $a_1 = 20,000,\ a_n = 0.8a_{n-1}$

14–3. $a_1 = 100,\ a_n = a_{n-1} - 5$

14–4. $a_1 = 5,\ a_n = \dfrac{a_{n-1} + 1}{2}$

14–5. $a_1 = 1,\ a_2 = 1;\ a_n = a_{n-1} + 2a_{n-2}$

Summation Notation

Let's return once more to the birthday examples. A very natural question might be "How much total money has Uncle Albert or Aunt Beatrice given you after your nth birthday?" To find out, we need to *sum* the first n terms in the sequence. After your nth birthday, Uncle Al will have given you $S_n = a_1 + a_2 + a_3 + \ldots + a_n$ dollars. This is an example of a *series*. A series is the sum of the terms in a sequence. Like sequences, series can bc either finite or infinite.

Remember that Uncle Al gives you $a_n = 3 + 2n$ dollars on your nth birthday. We can now find out how much total money he has given you after your fifth birthday. By using S_n to represent the sum of the first n terms of a sequence, we have:

$$\begin{aligned} S_5 &= a_1 + a_2 + a_3 + a_4 + a_5 \\ &= (3 + 2(1)) + (3 + 2(2)) + (3 + 2(3)) + (3 + 2(4)) + (3 + 2(5)) \end{aligned}$$

The answer is \$45. Of more interest here is learning a compact notation, called *summation notation*, for writing sums such as those above. Summation notation uses the capital Greek letter sigma, Σ, as the symbol for summation:

$$\sum_{i=1}^{n} a_i = a_1 + a_2 + \cdots + a_n$$

The expression on the left is read "The sum from i equals 1 to n of a_i." In the expression:

- Σ means sum
- i is the *index*, an integer variable. Other common indices are j, k, m, and n.
- 1 (as in $i = 1$) is the initial value of i; i starts at 1 in this example.
- n is the final value of i; i starts at 1 and increases in steps of 1 until it reaches n.
- a_i are the terms being added, starting with a_1 and ending at a_n.

In the birthday example, the total amount of money Uncle Al has given you after 5 years can be expressed as:

$$\begin{aligned} S_5 &= \sum_{i=1}^{5} a_i \\ &= \sum_{i=1}^{5} (3 + 2i) \\ &= (3 + 2(1)) + (3 + 2(2)) + (3 + 2(3)) + (3 + 2(4)) + (3 + 2(5)) \\ &= 45 \end{aligned}$$

As you can see, summation notation does not change the amount of work we have to do, in terms of number of calculations, to work out a sum. What it does is give us a compact way to write long sums.

Students are sometimes bothered by the choice of i for the index. Earlier we defined $a_n = 3 + 2n$. Why not use n for the index and write $S_5 = \sum_{n=1}^{5} a_n = \sum_{n=1}^{5} (3 + 2n)$? We can.

The index is a dummy variable. It is used to keep track of the terms in the sum but does not appear in the final answer. We can use any variable we want for the index as long as it is not being used for something else in the problem. The answer will still be 45. A reason for not using n is that we started this section by asking how much total birthday money you have received after n years. It would not be correct to write $S_n = \sum_{n=1}^{n} a_n$. This is using n for two different things in the same problem (and it says n starts at 1 and increases until it equals itself, which makes no sense). Use a different letter for the index.

EXAMPLE 14.5 Evaluate: $\sum_{k=1}^{4} k^2$

SOLUTION

This says to add up values of k^2 with k starting at 1, increasing by steps of 1 and ending at 4.

$$\begin{aligned}\sum_{k=1}^{4} k^2 &= 1^2 + 2^2 + 3^2 + 4^2 \\ &= 1 + 4 + 9 + 16 \\ &= 30\end{aligned}$$

EXAMPLE 14.6 Evaluate: $\sum_{m=3}^{5} m(m-1)$

SOLUTION

Although it is very common for the index to start at 1, it does not have to. In this problem, the sum starts at $m = 3$ and goes to $m = 5$. Note that 5 is the final value of the index, m. It does not represent the number of terms in the series.

$$\begin{aligned}\sum_{m=3}^{5} m(m-1) &= 3(3-1) + 4(4-1) + 5(5-1) \\ &= 6 + 12 + 20 \\ &= 38\end{aligned}$$

EXAMPLE 14.7 In a certain year, 54 total inches of rain fell in Miami, FL. Let x_i represent the number of inches of rainfall for the ith month that year where $i = 1$ corresponds to January.

a. Explain in words precisely what $x_2 = 6$ means in this problem.

b. Find the value of $\frac{1}{12}\sum_{i=1}^{12} x_i$ and explain in words what it represents.

SOLUTION

a. 2 is the index and 6 is the value. This means that in the 2nd month, February, Miami had 6 inches of rain.

b. $\sum_{i=1}^{12} x_i$ is sigma notation for adding up all the rain in each month from month 1, which is January, to month 12, which is December. So $\sum_{i=1}^{12} x_i$ gives the total amount of rain for the year, which is given in the problem as 54 inches. From that, $\frac{1}{12}\sum_{i=1}^{12} x_i = \frac{1}{12}(54) = 4.5$. So this represents the average rainfall per month over the year.

EXAMPLE 14.8

Express $S = \frac{1}{1} + \frac{2}{2} + \frac{3}{4} + \frac{4}{8} + \frac{5}{16} + \frac{6}{32}$ using summation notation.

SOLUTION

Write a general expression for the terms first. Look for patterns that will allow you to express the terms in the series as a function of a single integer variable.

The pattern in the numerator is easy; the numerator increases by one each term. What is going on in the denominator? Observe that the denominators are all powers of 2. The series can be rewritten as:

$$S = \frac{1}{2^0} + \frac{2}{2^1} + \frac{3}{2^2} + \frac{4}{2^3} + \frac{5}{2^4} + \frac{6}{2^5}$$

Now we have a choice. If we think "The exponent in the denominator is always 1 less than the numerator," we can let the numerator be the index, say k, and write each term in the form $\frac{k}{2^{k-1}}$:

$$S = \frac{1}{2^{1-1}} + \frac{2}{2^{2-1}} + \frac{3}{2^{3-1}} + \frac{4}{2^{4-1}} + \frac{5}{2^{5-1}} + \frac{6}{2^{6-1}}$$

Clearly, k goes from 1 to 6, so we get $S = \sum_{k=1}^{6} \frac{k}{2^{k-1}}$.

Alternatively, we could think "The numerator is always one more than the exponent in the denominator." Using the exponent in the denominator as the index, say n this time, each term can be written in the form $\frac{n+1}{2^n}$:

$$S = \frac{0+1}{2^0} + \frac{1+1}{2^1} + \frac{2+1}{2^2} + \frac{3+1}{2^3} + \frac{4+1}{2^4} + \frac{5+1}{2^5}$$

In this case, n goes from 0 to 5. We have $S = \sum_{n=0}^{5} \frac{n+1}{2^n}$.

Either $S = \sum_{k=1}^{6} \frac{k}{2^{k-1}}$ or $S = \sum_{n=0}^{5} \frac{n+1}{2^n}$ represents the given sum.

SECTION EXERCISES

For 6–9, write out each summation and find its value.

14–6. $\sum_{k=0}^{4} 2^k$ 14–7. $\sum_{n=1}^{4} n^2$ 14–8. $12\sum_{j=1}^{5} \frac{1}{j}$ 14–9. $\sum_{k=2}^{5} \frac{k+3}{2k}$

For 10–12, write each of the following using sigma notation.

14–10. $1 + \frac{1}{4} + \frac{1}{9} + \frac{1}{16} + \frac{1}{25}$

14–11. $\frac{12}{1} + \frac{12}{2} + \frac{12}{3} + \frac{12}{4} + \ldots + \frac{12}{20}$

14–12. $\frac{1}{1} + \frac{3}{2} + \frac{5}{4} + \frac{7}{8} + \frac{9}{16} + \ldots + a_n$

14–13. Philo Farnsworth recorded the number of hours he spent watching TV each day for the month of February. Suppose x_i represents the number of hours of TV Philo watched on the ith day of the month.

a. What does $x_5 = 2$ mean?

b. What does $\sum_{i=1}^{28} x_i$ represent?

c. Suppose $\frac{1}{28}\sum_{i=1}^{28} x_i = 1.5$. What exactly does this mean?

Arithmetic Sequences

Remember Uncle Albert? Starting with \$5 on your first birthday, Uncle Al gave you \$2 more each year than he did the year before. The sequence was 5, 7, 9, 11, 13, ...

This is an example of a particular kind of sequence called an *arithmetic sequence*. In an arithmetic sequence, the difference between any two consecutive terms is a constant, usually called d for difference.

Arithmetic sequence: $a_n - a_{n-1} = d$ for all $n \geq 2$

Note that the order of the subtraction is important. It is always term − preceding term = difference. In the Uncle Al example, the common difference is 2.

EXAMPLE 14.9 Consider the series 4, 10, 16, 22, 28, ...

a. Determine if the series is arithmetic. If so, find the common difference.
b. Find the next two terms of the series.
c. Write a recursive definition for the series.
d. Write an explicit definition for the series.
e. Find the value of the 50th term of the sequence.

SOLUTION

a. By checking differences between consecutive terms, we see $10 - 4 = 6$, $16 - 10 = 6$, $22 - 16 = 6$, and $28 - 22 = 6$. The differences are all the same. This is an arithmetic series with the common difference $d = 6$.

b. To find successive terms in an arithmetic sequence, just keep adding the difference d. The next two terms are $28 + 6 = 34$ and $34 + 6 = 40$.

c. For any arithmetic sequence, the recursion relation is $a_n = a_{n-1} + d$ for $n \geq 2$. For this example, and remembering we need to specify a_1, we have $a_1 = 4$; $a_n = a_{n-1} + 6$ for $n \geq 2$.

d. Let's write out a few terms using the recursive definition but without simplifying.

$a_1 = 4$
$a_2 = 4 + 6$
$a_3 = 4 + 6 + 6$
$a_4 = 4 + 6 + 6 + 6$

There is a pattern here. Each term consists of a 4 with some number of 6s added to it. Specifically, the number of 6s added is one less than the term number, n. The third term has one 4 with $3 - 1 = 2$ 6s added to it. The nth term will have one 4 with $(n - 1)$ 6s added. Repeated addition is the same as multiplication:

$$a_n = 4 + \overbrace{6 + 6 + \ldots + 6}^{n-1 \text{ sixes}}$$
$$= 4 + 6(n - 1)$$

An explicit formula for this sequence is $a_n = 4 + 6(n - 1)$.

e. Use the explicit formula to find a specific term. For the 50th term, substitute $n = 50$ into $a_n = 4 + 6(n - 1)$. So $a_{50} = 4 + 6(50 - 1) = 298$. The 50th term is 298.

We can easily generalize the answer to part *d* to an explicit function for any arithmetic sequence. Note that 4 was the first term, a_1, and 6 was the common difference, d. So for any arithmetic sequence, the explicit formula is the following:

$$a_n = a_1 + d(n - 1)$$

EXAMPLE 14.10 Consider the series 32, 27, 22, 17, 12, . . .

a. Determine if the series is arithmetic. If so, find the common difference.
b. Write a recursive definition for the series.
c. Write an explicit definition for the series.

SOLUTION

a. By checking differences between common terms, we see $27 - 32 = -5$, $22 - 27 = -5$, $17 - 22 = -5$, and $12 - 17 = -5$. The differences are all the same. This is an arithmetic series with the common difference $d = -5$.

b. For any arithmetic sequence, the recursion relation is $a_n = a_{n-1} + d$ for $n \geq 2$. For this example and remembering we need to specify a_1, we have $a_1 = 32$; $a_n = a_{n-1} - 5$ for $n \geq 2$.

c. For any arithmetic sequence, the explicit formula is $a_n = a_1 + d(n - 1)$. In this example, $a_1 = 32$ and $d = -5$, which gives $a_n = 32 - 5(n - 1)$.

SUMMARY FOR ARITHMETIC SEQUENCES

The important ideas of arithmetic sequences are summarized in Table 14.1. Whether you have to memorize the explicit formula is up to your teacher (we think it's worthwhile). However, you definitely need to understand it thoroughly.

TABLE 14.1
ARITHMETIC SEQUENCE SUMMARY

Common Difference	$d = a_n - a_{n-1}$
Recursive Formula	State a_1 $a_n = a_{n-1} + d$ for $n \geq 2$
Explicit Formula	$a_n = a_1 + d(n - 1)$

SECTION EXERCISES

14–14. Determine whether the following could be an arithmetic sequence.

a. 10, 7, 4, 1, –2, . . .

b. 1, 3, 6, 10, . . .

c. $\frac{1}{2}, \frac{3}{4}, 1, \frac{5}{4}, \frac{3}{2}, \ldots$

d. 1, 1.1, 1.11, 1.111, . . .

14–15. Find a formula for a_n as a function of n of the arithmetic sequence.

a. $a_1 = -7$, $d = 4$

b. $a_1 = 6$, $a_2 = 20$

c. $a_2 = 6$, $a_3 = 4$

14–16. Write the first five terms of the arithmetic sequence. Then write the nth term of the sequence as a function of n.

a. $a_1 = 2$, $a_k = a_{k-1} - 3$

b. $a_1 = 200$, $a_k = a_{k-1} - 25$

Arithmetic Series

A *finite arithmetic series* is the sum of the terms in a finite arithmetic sequence.

Arithmetic sequence: $a_1, a_2, a_3, \ldots, a_n$
where $a_k = a_1 + d(k-1)$ for $k = 1, 2, 3, \ldots, n$

Arithmetic series: $S_n = \sum_{k=1}^{n} a_k = a_1 + a_2 + a_3 + \ldots + a_n$

$$= a_1 + (a_1 + d) + (a_1 + 2d) + (a_1 + 3d) + \cdots + (a_1 + (n-1)d)$$

There is a formula for the sum of a finite arithmetic sequence. To see where it comes from, we return one last time to Uncle Al and his birthday gifts. What total amount of money has Uncle Al given you after your 18th birthday? Obviously, we could grab a calculator and add up the 18 separate years (the answer is \$396). However, we want to find a formula and that involves a bit of a trick. Remember, he gave you \$5 on your first birthday and \$2 more each succeeding year, leading to you receiving \$39 on your 18th birthday.

Write the sum: $S_{18} = 5 + 7 + 9 + \cdots + 35 + 37 + 39$

Write the sum in reverse: $S_{18} = 39 + 37 + 35 + \cdots + 9 + 7 + 5$

Add the two sums: $2S_{18} = 44 + 44 + 44 + \cdots + 44 + 44 + 44$

Simplify: $2S_{18} = 18(44)$

Solve for S_{18}: $S_{18} = \frac{18(44)}{2} = 396$

Notice that the 18 represents the number of terms we added and 44 is the sum of the first and last (18th) terms. With a little bit of algebra, you can show this trick works for any finite arithmetic sequence and so we have the following formula:

$$\text{Sum of a finite arithmetic sequence: } S_n = \sum_{k=1}^{n} a_k = \frac{n(a_1 + a_n)}{2}$$

One way to remember it is as the average of the first and last terms, multiplied by the number of terms in the sequence.

EXAMPLE 14.11 Consider the arithmetic series 4, 10, 16, 22, 28, . . . Find the sum of the first 50 terms.

SOLUTION

In Example 14.9, we found the 50th term in this sequence to be 298. By using the formula for an arithmetic series, $S_n = \sum_{k=1}^{n} a_k = \frac{n(a_1 + a_n)}{2}$ with $n = 50$, $a_1 = 4$, and $a_{50} = 298$, we get $S_{50} = \frac{50(4 + 298)}{2} = 7{,}550$.

SECTION EXERCISES

14–17. Each row in a theater has two more seats than the row in front of it. If the first row seats 25 people and there are 20 rows in the theater, what is the total seating capacity?

14–18. On his first birthday, Rufus's Uncle Wheezer gave him $25. On every succeeding birthday, Uncle Wheezer gave Rufus $10 more than he did the year before.

a. In terms of n, how much money did Uncle Wheezer give Rufus on his nth birthday?

b. How much total birthday money has Uncle Wheezer given Rufus up to and including his 18th birthday?

Geometric Sequences

It is time to revisit Aunt Beatrice. Like Uncle Al, she gave you $5 on your first birthday. Unlike Uncle Al, she doubled the amount she gave you on each succeeding birthday. Her gifts followed the sequence 5, 10, 20, 40, 80, . . .

This is another important kind of sequence called a *geometric sequence*. In a geometric sequence, the ratio of any two consecutive terms is a constant, usually called r for ratio.

$$\text{Geometric sequence: } \frac{a_n}{a_{n-1}} = r \text{ for all } n \geq 2$$

Again, the order in which the division is done is important. It is always $\frac{\text{term}}{\text{previous term}}$ = ratio. In the Aunt Bea example, the common ratio is 2.

EXAMPLE 14.12

Consider the series 8, 12, 18, 27, 40.5, . . .

a. Determine if the series is geometric. If so, find the common ratio.

b. Find the next two terms of the series.

c. Write a recursive definition for the series.

d. Write an explicit definition for the series.

e. Find the value, to the nearest whole number, of the 50th term of the sequence.

SOLUTION

a. By checking ratios between consecutive terms, we see $\frac{12}{8} = 1.5$, $\frac{18}{12} = 1.5$, $\frac{27}{18} = 1.5$, and $\frac{40.5}{27} = 1.5$. The ratios are all the same. This is a geometric series with common ratio $r = 1.5$.

b. To find successive terms in a geometric sequence, just keep multiplying by the ratio r. The next two terms are $40.5(1.5) = 60.75$ and $60.75(1.5) = 91.125$.

c. For any geometric sequence, the recursion relation is $a_n = ra_{n-1}$ for $n \geq 2$. For this example, remembering to specify a_1, we have $a_1 = 8$; $a_n = 1.5a_{n-1}$ for $n \geq 2$.

d. Write out a few terms using the recursive definition but without simplifying.

$a_1 = 8$
$a_2 = 8(1.5)$
$a_3 = 8(1.5)(1.5)$
$a_4 = 8(1.5)(1.5)(1.5)$

We see a pattern here. Each term consists of an 8 multiplied by some quantity of 1.5s. Specifically, the quantity of 1.5s is one less than the term number, n. The third term has one 8 multiplied by $3 - 1 = 2$ factors of 1.5. The nth term will have an 8 multiplied by $n - 1$ factors of 1.5. Repeated multiplication is represented by an exponent:

$$a_n = 8\overbrace{(1.5)(1.5)\ldots(1.5)}^{n-1 \text{ factors of } 1.5}$$
$$= 8(1.5)^{n-1}$$

The explicit formula is $a_n = 8(1.5)^{n-1}$.

e. Use the explicit formula to find a specific term. For the 50th term, substitute $n = 50$ into $a_n = 8(1.5)^{n-1}$, getting $a_{50} = 8(1.5)^{n-1} = 8(1.5)^{49} \approx 3{,}400{,}648{,}001$. To the nearest whole number, the 50th term is 3,400,648,001.

Again, we can generalize the answer to part d to find the explicit formula for any geometric sequence. The 8 was the first term, a_1, and the 1.5 was the common ratio, r. So for any geometric sequence the explicit formula is the following:

$$a_n = a_1 r^{n-1}$$

EXAMPLE 14.13

Consider the series 28, –14, 7, –3.5, 1.75, . . .

a. Determine if the series is geometric. If so, find the common ratio.

b. Write a recursive definition for the series.

c. Write an explicit definition for the series.

SOLUTION

a. By checking ratios between common terms, we see $\frac{-14}{28} = -\frac{1}{2}$, $\frac{7}{-14} = -\frac{1}{2}$, $\frac{-3.5}{7} = -\frac{1}{2}$, and $\frac{1.75}{-3.5} = -\frac{1}{2}$. The ratios are all the same. This is a geometric series with common ratio $r = -\frac{1}{2}$.

b. For any geometric sequence, the recursion relation is $a_n = ra_{n-1}$ for $n \geq 2$. For this example, remembering to specify a_1, we have $a_1 = 28$; $a_n = -\frac{1}{2}a_{n-1}$ for $n \geq 2$.

c. For any geometric sequence, the explicit formula is $a_n = a_1r^{n-1}$. In this example, $a_1 = 28$ and $r = -\frac{1}{2}$, which gives $a_n = 28\left(-\frac{1}{2}\right)^{n-1}$.

SUMMARY FOR GEOMETRIC SEQUENCES

Table 14.2 shows a summary of the main ideas of geometric sequences.

TABLE 14.2
GEOMETRIC SEQUENCE SUMMARY

Common Ratio	$\frac{a_n}{a_{n-1}} = r$ for all $n \geq 2$
Recursive Formula	State a_1 $a_n = ra_{n-1}$ for $n \geq 2$
Explicit Formula	$a_n = a_1r^{n-1}$

EXAMPLE 14.14

Determine if the following sequences are arithmetic, geometric, or neither.

a. 1, 4, 9, 16, 25, . . .
b. 2, 6, 18, 54, 162, . . .
c. 5, 8, 11, 14, 17, . . .

SOLUTION

For each sequence, check for a common difference by subtracting consecutive terms (arithmetic). If there is no common difference, check for a common ratio by dividing consecutive terms (geometric). If all pairs of consecutive terms fail to have either a common difference or a common ratio, the sequence is neither arithmetic nor geometric.

a. Since $4 - 1 = 3$ and $9 - 4 = 5$, the sequence is not arithmetic.

Since $\frac{4}{1} = 4$ and $\frac{9}{4} = 2.25$, the sequence is not geometric either.

This sequence is neither arithmetic nor geometric.

b. The sequence does not have a common difference. However, $\frac{6}{2} = 3$, $\frac{18}{6} = 3$, $\frac{54}{18} = 3$, and $\frac{162}{54} = 3$, so it has a common ratio. The sequence is geometric.

c. This sequence has a common difference: $8 - 5 = 3$, $11 - 8 = 3$, $14 - 11 = 3$, and $17 - 14 = 3$. The sequence is arithmetic.

SECTION EXERCISES

14–19. Determine whether each sequence could be geometric.

a. 2, 6, 24, 120, . . .

b. 2, 6, 18, 54, . . .

c. 12, 8, $\frac{16}{3}$, $\frac{32}{9}$, . . .

d. 1, –1, 1, –1, 1, –1, . . .

14–20. Write the first five terms of the sequence. Then write the nth term of the sequence as a function of n.

a. $a_1 = 16,\ a_{k+1} = \frac{1}{2}a_k$

b. $a_1 = 4,\ a_{k+1} = -3a_k$

Geometric Series

A *finite geometric series* is the sum of a finite geometric sequence.

Geometric sequence: $a_1, a_2, a_3, \ldots, a_n$
where $a_k = a_1 r^{k-1}$ for $k = 1, 2, 3, \ldots, n$

Geometric series:
$$S_n = \sum_{k=1}^{n} a_k = a_1 + a_2 + a_3 + \ldots + a_n$$
$$= a_1 + a_1 r + a_1 r^2 + a_1 r^3 + \cdots + a_1 r^{n-1}$$

There is a formula for the sum of the terms in a geometric sequence. Like the formula for the arithmetic series, this one involves a trick to derive it.

Write the sum: $S_n = a_1 + a_1 r + a_1 r^2 + a_1 r^3 + \cdots + a_1 r^{n-1}$

Multiply by r: $rS_n = a_1 r + a_1 r^2 + a_1 r^3 + \cdots + a_1 r^{n-1} + a_1 r^n$

Subtract the two sums: $S_n - rS_n = a_1 - a_1 r^n$

Factor out common factors: $S_n(1 - r) = a_1\left(1 - r^n\right)$

Solve for S_n: $S_n = a_1 \frac{\left(1 - r^n\right)}{(1 - r)}$ assuming $r \neq 1$

So for any finite geometric sequence with $r \neq 1$, we have:

Sum of a finite geometric sequence: $S_n = a_1 \frac{\left(1 - r^n\right)}{(1 - r)}$

EXAMPLE 14.15 Determine how much total birthday money Aunt Beatrice has given you after your 18th birthday.

SOLUTION

Each year, Aunt Bea gave you $b_n = 5(2)^{n-1}$ dollars on your nth birthday. We have $b_1 = 5$ and $r = 2$. So after $n = 18$ years, Aunt Bea has given you a total of:

$$S_{18} = (5)\frac{\left(1 - 2^{18}\right)}{(1 - 2)} = \$1,310,715$$

EXAMPLE 14.16 Find the sum of the first twelve terms of the geometric sequence –3.5, 7, –14 , 28, . . .

SOLUTION

To use the formula for the series, $S_n = a_1\dfrac{\left(1 - r^n\right)}{(1 - r)}$, you need to know the common ratio; $r = \dfrac{7}{-3.5} = -2$. We were told the sequence is geometric, so we do not need to test the rest of the ratios. We know $n = 12$ and $a_1 = -3.5$.

$$S_{12} = (-3.5)\frac{\left(1 - (-2)^{12}\right)}{(1 - (-2))} = 4,777.5$$

The sum is 4,777.5.

SECTION EXERCISES

14–21. Mortimer Moneybags agrees to give his son Marvin the following allowance: 5 cents on the first of January, 15 cents on the second, 45 cents on the third, and so on.

a. Write a summation to represent the total amount of money Marvin has received after n days.

b. How much total allowance will Marvin receive in the first week of January?

c. How much total allowance would Marvin get in the month of January?

Infinite Series

Conceptually, a finite series is not a big deal. You have added finite lists of numbers as far back as the early elementary grades. In more advanced mathematics, there are many very important *infinite* series. A thorough discussion of what it means to add up an infinite number of terms is well beyond the scope of Algebra 2. However, to complete this chapter on sequences and series, we include a brief discussion of some special infinite series.

With a little thought, you can see that an infinite arithmetic series is not very interesting. There are two main possibilities. If the common difference is positive, then the individual terms increase without bound. We add up an infinite number of ever larger numbers and the sum increases without bound. In other words, the sum goes to $+\infty$. If the common difference is negative, the terms decrease without bound (becoming larger and larger negative numbers). The sum goes to $-\infty$. The only other possibility is if a_1 and d are both 0 and we add up an infinite number of 0s. This case is particularly uninteresting.

So is it even possible to add up an infinite number of nonzero quantities and get a finite answer? Here is an example.

A very hungry bunny is crouched with the tip of his nose 2 yards away from a large, juicy carrot. The bunny hops exactly half the distance to the carrot, which is 1 yard. Still hungry but now a little tired, the bunny makes a second hop that carries him half the remaining distance to the carrot, or 0.5 yard. Now even more tired, the bunny makes a third hop that again takes him half the remaining distance, or 0.25 yard. The bunny continues hopping, with each hop only half the length of the one before it. Does he ever get to the carrot?

The bunny's hops are of lengths $1, \frac{1}{2}, \frac{1}{4}, \frac{1}{8}, \ldots$. This is a geometric sequence with $a_1 = 1$ and $r = \frac{1}{2}$. The length of the kth hop is $a_k = \left(\frac{1}{2}\right)^{k-1}$ for $k = 1, 2, 3, \ldots$. The total distance the bunny has moved after n hops is the sum of the first n terms of this sequence: $S_n = 1 + \frac{1}{2} + \frac{1}{4} + \cdots + \left(\frac{1}{2}\right)^{n-1}$, which by our formula for the sum of a *finite* geometric sequence is $S_n = \dfrac{1 - \left(\frac{1}{2}\right)^n}{1 - \frac{1}{2}}$. A table of these sums is shown below (values for $n > 5$ have been rounded).

n	S_n
1	1
2	1.5
3	1.75
4	1.875
5	1.9375
10	1.9980
15	1.99994
20	1.999998
25	1.99999994

This table probably confirms your intuition. As the bunny continues to hop, he gets closer and closer to the carrot (which was 2 yards away). However, no finite number of hops will ever actually get him to the carrot (let alone past it). On the other hand, by taking enough hops, the bunny can get within any nonzero distance, no matter how small, of the carrot. What if the bunny wants to be less than 0.000,001 of a yard (less than the diameter of a whisker) away? No problem, that will take 21 hops.

Now it is just one step (although a very big step) to talking about an infinite sum. What total distance does the bunny travel as the number of hops increases without bound or "goes to ∞"? More precisely, what happens to S_n as $n \to \infty$? Saving the details for a more advanced class, we say that if the bunny could somehow make an infinite number of hops, he would have traveled exactly 2 yards. Mathematically, we are saying that we can add up an infinite number of positive quantities and get a finite answer:

$$\begin{aligned} S &= 1 + \frac{1}{2} + \frac{1}{4} + \cdots + \left(\frac{1}{2}\right)^{k-1} + \cdots \\ &= 2 \end{aligned}$$

Let's find a general formula for the sum of an infinite geometric sequence. For a finite geometric sequence, we have $S_n = a_1 \frac{\left(1 - r^n\right)}{(1 - r)}$. The only thing in this formula that depends on n is r^n. What happens to r^n as $n \to \infty$? If $r \geq 1$, then as $n \to \infty$, $r^n \to \infty$ and the sum of the sequence grows without bound. Something similar happens if $r \leq -1$. In this case, the signs of the sequence S_n alternate, but $|S_n|$ still increases without bound. What if $-1 < r < 1$ (usually written compactly as $|r| < 1$)? In this case, as $n \to \infty$, $r^n \to 0$, and $S_n \to a_1\left(\frac{1 - 0}{1 - r}\right) = \frac{a_1}{1 - r}$. Therefore, for an infinite geometric series, we say:

$$S = \sum_{k=1}^{\infty} a_k = \frac{a_1}{1 - r} \text{ provided } |r| < 1$$

In the case of the bunny and the carrot, we have $a_1 = 1$ and $r = \frac{1}{2}$: $S = \dfrac{1}{1 - \frac{1}{2}} = 2$.

If you are not completely convinced by the preceding arguments, don't feel bad; you are in good company. The idea of infinity leads to some surprising and counterintuitive results and has challenged even the best of mathematicians for more than 2,000 years. If you are really curious, keep studying math.

EXAMPLE 14.17

Find the sum of the geometric sequence 1,000; 200; 40; 8; 1.6; . . . .

SOLUTION

First find r and check that $-1 < r < 1$. Here $r = \dfrac{200}{1{,}000} = 0.2$; this satisfies $|r| < 1$. By using the formula for an infinite geometric series, $S = \sum_{k=1}^{\infty} a_k = \dfrac{a_1}{1 - r} = \dfrac{1{,}000}{1 - 0.2} = 1{,}250$

It is interesting to note that the sum of only the first five terms of the sequence is 1,249.6. The remaining infinite number of terms are needed just to get that last 0.4.

Find the sum of the geometric sequence 10, 25, 62.5, 156.25, ...

SOLUTION

First find r and check that $|r| < 1$. Here $r = \dfrac{25}{10} = 2.5$. This does not satisfy $|r| < 1$. This series is unbounded and has no sum.

SUMMARY OF ARITHMETIC AND GEOMETRIC SERIES

Table 14.3 shows the formulas for the sums of arithmetic and geometric series. Whether you need to memorize them is up to your teacher, but you must know how to use them.

TABLE 14.3
FORMULAS FOR SERIES

Finite Arithmetic Series	$S_n = \dfrac{n(a_1 + a_n)}{2}$
Finite Geometric Series	$S_n = a_1 \dfrac{(1 - r^n)}{(1 - r)}$
Infinite Geometric Series	$S = \dfrac{a_1}{1 - r}$ provided $\lvert r \rvert < 1$

Sometimes part of a problem is determining which formula to use as well as finding out values that are needed for the formula but are not directly given.

Find the sum of 3, 7, 11, . . . , 59.

SOLUTION

Determine the type of sequence first. 7 – 3 = 4 and 11 – 7 = 4. This is an arithmetic sequence with $d = 4$. To use the formula for a finite arithmetic series, we need to know the value of n, the number of terms. You could keep adding 4 on your calculator and count terms until you got to 59. A better solution is to use the explicit formula for arithmetic sequences: $a_n = a_1 + d(n - 1)$ to find n. Here $a_n = 59$, $a_1 = 3$, and $d = 4$. Substitute and solve for n.

$$59 = 3 + 4(n - 1)$$
$$56 = 4(n - 1)$$
$$\frac{56}{4} = n - 1$$
$$14 = n - 1$$
$$15 = n$$

Now use the formula for a finite arithmetic series, $S_n = \frac{n(a_1 + a_n)}{2}$.

$$S_{15} = \frac{15(3 + 59)}{2} = 465$$

The sequence sums to 465.

EXAMPLE 14.20

Find the sum of 54, 18, 6, 2, . . .

SOLUTION

This is an infinite sequence. Among the sequences covered here, only a geometric sequence with $|r| < 1$ will have a finite sum. Check if it has a common ratio: $\frac{18}{54} = \frac{1}{3}$, $\frac{6}{18} = \frac{1}{3}$, and $\frac{2}{6} = \frac{1}{3}$. It has a common ratio of $r = \frac{1}{3}$. It is a geometric sequence with $|r| < 1$, so you can use the formula for an infinite geometric series, $S = \frac{a_1}{1 - r}$.

$$S = \frac{54}{1 - \left(\frac{1}{3}\right)}$$
$$S = \frac{54}{\frac{2}{3}}$$
$$S = 81$$

The sum of the sequence is 81.

EXAMPLE 14.21 Norman Nerdling throws a ball hard against the gym floor and watches as it rebounds straight up to a height of 12 feet. On each successive bounce, the ball rebounds to two-thirds the height of its previous bounce.

a. Write out the heights of the first 4 bounces.

b. Find an expression for the total vertical distance the ball has traveled at the end of its 4th bounce (not counting the initial distance from Norman's hand to the floor).

c. Find an expression for the height of the nth bounce.

d. If the ball is allowed to bounce forever in this fashion, what total vertical distance will it travel (again not counting the initial hand-to-floor distance)?

SOLUTION

a. The first bounce is 12 feet. The second bounce is $\frac{2}{3}(12) = 8$ feet, the third bounce $= \frac{2}{3}(8) = \frac{16}{3}$ feet (judging by his name, Norman wants you to be precise and not round the fractions), and the fourth bounce is $\frac{2}{3}\left(\frac{16}{3}\right) = \frac{32}{9}$ feet.

b. You might think you could find the total vertical distance of the bounces by finding $12 + 8 + \frac{16}{3} + \frac{32}{9} = \frac{260}{9}$. However, the ball traveled up this distance and then fell back down the same distance, so you will need to double the sum to find the total distance.

The total distance traveled by the ball is $2\left(12 + 8 + \frac{16}{3} + \frac{32}{9}\right) = \frac{520}{9} \approx 57.8$ feet.

c. This is a geometric sequence with $a_1 = 12$ and $r = \frac{2}{3}$. The explicit formula is

$a_n = 12\left(\frac{2}{3}\right)^{n-1}$.

d. Bouncing forever means an infinite series. Since $r = \frac{2}{3}$, this series does have a sum. Use the infinite geometric series formula, $S = \frac{a_1}{1-r}$, with $a_1 = 12$ and $r = \frac{2}{3}$.

$S = \frac{12}{1 - \frac{2}{3}} = 36$ feet. Again, as in part b, this is the distance in only one direction, up or down. We have to double the sum to find the total distance to count both directions. If the ball bounces forever, it will travel a total distance of 72 feet.

A finite arithmetic series adds to 10,098. If $a_1 = 23$ and $a_n = 373$, find the number of terms, n, in the series and find the common difference.

SOLUTION

To find the number of terms in the series, n, use $S_n = \frac{n(a_1 + a_n)}{2}$. Substitute $S_n = 10{,}098$, $a_1 = 23$, and $a_n = 373$ into the formula. Solve for n.

$$\begin{aligned} 10{,}098 &= \frac{n(23 + 373)}{2} \\ 10{,}098 &= 198n \\ 51 &= n \end{aligned}$$

There are 51 terms in the series.

To find the common difference, d, use $a_n = a_1 + d(n - 1)$. With $n = 51$, $a_1 = 23$, and $a_{51} = 373$, we get:

$$\begin{aligned} 373 &= 23 + d(51 - 1) \\ &= 23 + 50d \\ 350 &= 50d \\ d &= 7 \end{aligned}$$

The common difference for the terms in this series is 7.

SECTION EXERCISES

14–22. Old Uncle Albert has finally gone around the bend. One day he gives \$25 to a total stranger on the street. Each succeeding day, Al gives money to another stranger. The amounts follow an arithmetic sequence. The 50th lucky stranger gets \$760.

a. How much extra money does Uncle Al give to each new stranger?

b. How much does lucky stranger number 120 get?

c. Assuming Uncle Al's money lasts, which lucky stranger will receive exactly \$4,000?

d. How much total money has Al given away after one (nonleap) year?

14–23. When dropped, a certain ball always returns to 80% of its original height. Suppose it was originally dropped from a height of 20 feet.

a. How high does the ball bounce on its first bounce?

b. How high does the ball bounce on its 6th bounce?

c. How high does the ball bounce on its nth bounce?

14–24. Sam goes bungee jumping. On her first downward fall, she travels 100 feet. She rebounds only 90 feet before falling again 81 feet. Suppose this sequence continues. How much total distance (up and down) does Sam travel?

14–25. For each infinite sequence, classify the sequence as arithmetic or geometric and determine if its sum exists.

a. $\frac{2}{3}, \frac{2}{9}, \frac{2}{27}, \frac{2}{81}, \frac{2}{243}, \ldots$

b. $\frac{1}{2}, 1, 2, 4, 8, \ldots$

c. $2, 6, 10, 14, 18, \ldots$

d. $\frac{1}{1}, \frac{-1}{2}, \frac{1}{4}, \frac{-1}{8}, \frac{1}{16}, \ldots$

For 26–29, for each infinite geometric series, find the sum if it exists or justify why the sum does not exist.

14–26. $8 - 4 + 2 - 1 + \frac{1}{2} - \cdots$

14–27. $18 + 6 + 2 + \frac{2}{3} + \cdots$

14–28. $\frac{1}{64} + \frac{1}{32} + \frac{1}{16} + \frac{1}{8} + \cdots$

14–29. $\sum_{k=0}^{\infty} 30(0.9)^k$

CHAPTER EXERCISES

C14–1. A particular sequence is defined explicitly by $a_n = \frac{n}{2^n}$ for $n = 1, 2, 3, \ldots$. Write the first six terms of this sequence. The terms do not need to be reduced.

C14–2. A particular sequence is defined recursively by $a_1 = 4$, $a_n = 2a_{n-1} - 3$. Write the first six terms of this sequence.

C14–3. Write out the terms of the summation and find its value: $3\sum_{k=3}^{6} \frac{1}{k-1}$

C14–4. Write the following using sigma notation: $\frac{1}{4} + \frac{2}{9} + \frac{3}{16} + \frac{4}{25} + \frac{5}{36} + \frac{6}{49}$

C14–5. Hoop Shooter played in 15 basketball games this season. Let x_i represent the number of points Hoop scored in the ith game, for $i = 1$ to $i = 15$.

a. What does $x_{10} = 7$ mean?

b. What does $\sum_{i=1}^{15} x_i = 126$ mean?

c. Write an expression involving a summation to represent Hoop's average points per game.

C14–6. The numbers 2, 5, 8, 11, 14, . . . represent an arithmetic sequence. If a_n represents the nth number in the sequence:

a. Find a recursive definition for the sequence.

b. Find an explicit definition for the sequence.

C14–7. Find an explicit formula for the nth term of the arithmetic sequence, and find the sum of the first n terms.

a. 8, 20, 32, 44, . . . , $n = 10$

b. $a_1 = 50$, $a_2 = 38$, $n = 25$

C14–8. The numbers 2, 6, 18, 54, 162, . . . represent a geometric sequence. If a_n represents the nth number in the sequence:

a. Find a recursive definition of the sequence.

b. Find an explicit definition of the sequence.

C14–9. Find a formula for a_n as a function of n of the geometric sequence, and find the sum of the first n terms.

a. $a_1 = 2$, $r = 5$, $n = 8$

b. $a_1 = 8$, $r = -\frac{1}{2}$, $n = 6$

C14–10. Logs at a paper mill are stacked in a pile with 25 logs in the bottom layer, 24 logs in the next layer, and so on. The pile is 12 layers high. How many logs does it contain?

C14–11. A theater has 15 seats in the front row. The number of seats in each row follows an arithmetic series with 72 seats in the last row. The theater has a total of 870 seats. How many rows does the theater have? By how many seats does each row increase?

For 12–13, for each infinite geometric series, find the sum if it exists or justify why the sum does not exist.

C14–12. $1 - \frac{2}{3} + \frac{4}{9} - \frac{8}{27} + \cdots$

C14–13. $\sum_{k=0}^{\infty} 25(1.02)^k$

Statistics

WHAT YOU WILL LEARN

We use statistics to organize and make sense out of the large amount of data that routinely appear in our daily lives. Statistics can also help us use all that information to make better, more informed decisions. In this chapter, you will learn to:

- Analyze different methods of collecting data;
- Use measures of central tendency and variance to describe data;
- Use the normal distribution to describe data;
- Model data with regression equations;
- Make predictions and interpret your model.

SECTIONS IN THIS CHAPTER

- Introduction to Statistics
- Data Collection
- Statistical Measures
- Normal Distribution
- Modeling with Functions

Introduction to Statistics

Statistics is the collection, organization, and analysis of data. Statistics can be divided into two broad categories. Descriptive statistics are mathematical techniques for organizing, describing, and interpreting data. Familiar examples are averages and proportions such as your average grade in math class or the proportion (percent) of free throws you make in basketball.

Inferential statistics use data to make inferences or predictions that, in turn, can help make decisions. If you want to know which of three candidates will win the election for junior class president, you might poll a sample of the class and use the results to estimate who is most likely to win. You will also want to know how certain your estimates are. If your poll shows candidate A with 38% support, candidate B with 36%, and candidate C with 22%, you may believe that candidate C is unlikely to win. However, just how sure are you that A will beat B? Meanwhile, you have heard shooting free throws underhand may lead to a higher success rate than shooting them overhand. After practicing for a few days, you try an experiment. You shoot 40 free throws, alternating between overhand and underhand. You make 12 of 20 of the overhand ones and 16 of 20 of the underhand ones. Is this convincing evidence that you should throw underhand?

Data Collection

Descriptive statistics are methods of organizing, summarizing, displaying, and interpreting data. Before using statistics with data, we should briefly discuss the different types of data we may be describing.

TYPES OF DATA

In statistics, we distinguish between *quantitative* and *qualitative* data. Quantitative data are numbers that represent quantities that can be measured on a numerical scale. Examples include the numbers of students in different high school classes, the cost of your school lunch in the cafeteria, the heights of players on the basketball team, or the time a student spends on the bus to school. On the other hand, qualitative data, also called categorical data, can be put into categories but cannot be measured on a scale. Examples are gender (male or female), eye color (blue, green, brown, etc.), and preferences (like, dislike, no opinion). Some qualitative data may be numerical but the numbers are used to describe, not measure. For example, telephone numbers and zip codes are descriptions, not measurements, and are types of qualitative data. Many different mathematical models and tests can be applied to qualitative data. In this chapter, though, we will focus on quantitative data.

Quantitative data can be further broken down into *discrete data* and *continuous data.* Discrete data can take on only certain values. These values are often, though not always, whole numbers. For example, the number of students in a high school class will always be a whole number. You will never see a class with 17.5 students. Another way of thinking about it is that discrete data has gaps between possible values. A cafeteria lunch might cost $2.78 or $2.79, but it cannot be any number between those values, such as $2.784. In contrast, continuous data can take on any value within an appropriate range; there are no gaps. The height of a basketball player could be any number within the range of possible human heights. The time spent on the school bus might be any number between 0 and the longest bus ride, say, 60 minutes.

Data can also be classified by how many variables are measured. *Univariate data* consists of measurements of just one variable. For example, we might select a sample of adults in a certain city and find out how many years of formal education each one has. Instead, we might find out each one's annual income. Such data are descriptive. We can use the data to estimate the average number of years of education among adults in the city or the proportion of people in the city with incomes above a certain level. The information cannot be used to study relationships between variables.

Bivariate data consists of *paired* measurements of two different variables. For example, we could collect people's years of education and annual salaries as *ordered pairs* instead of two separate data sets. Bivariate data is useful for studying relationships between variables such as whether and how additional years of education affect one's annual income.

POPULATIONS AND SAMPLES

In statistics, a *population* consists of all members of some group that we are trying to study. A *sample* is just part of the whole population. For example, suppose you are interested in knowing about the heights of the girls in a high school. The population would be all the girls in the high school. In a large school, measuring the height of every girl might be impractical. So you would measure the heights of just a sample of the girls.

Measures, such as mean or standard deviation, can be calculated for either the population or the sample. To distinguish between the two measures, different symbols are used for the population and the sample. The mean (average) height of all the girls in your high school is called the *population mean*, designated μ. The mean height of the girls in your sample is the *sample mean*, $\bar{x}$. Finding the exact value of the population mean is often impractical. One purpose of inferential statistics is to use measures from the sample to estimate measures for the population.

RANDOM SAMPLES

To get meaningful estimates for the population, you cannot use just any convenient sample. You would not get a good estimate of the average height of the girls in a high school by measuring all the girls on the varsity basketball team. The girls on the varsity basketball team are likely to have an average height that is taller than the average height of all the girls in the school. In statistical sampling, we try to find a sample that is representative of the population. The best way to get a representative sample is to select a *simple random sample*. In a simple random sample, each individual in the population has an equal chance of being selected for the sample and each individual's chance of being selected is independent of everyone else's chance. The old-fashioned way of getting a simple random sample would be to write the name of each individual in the population on separate slips of paper, mix all the slips in a container (traditionally a hat), and then randomly (without looking) select the desired number of names for the sample. Today, computers can be used instead of papers in hats, but the idea is the same.

SECTION EXERCISES

15–1. Categorize each of the following as quantitative or qualitative data.
- **a.** Favorite pizza toppings
- **b.** Age of children in a family
- **c.** Cell phone numbers
- **d.** Scores on a biology test

15–2. Categorize each of the following as univariate or bivariate data.
- **a.** Odometer readings on a sample of used cars
- **b.** Miles driven on one tank of gas and the size of the gas tank
- **c.** Height and shoe size for buyers in a shoe store
- **d.** Age of buyers in the shoe store

15–3. Categorize each of the following as the population or a sample.
- **a.** Voting preference of every tenth person in the telephone book to estimate the voting preference for the town.
- **b.** The number of turkey sandwiches sold for the year in the cafeteria to estimate the yearly sales of turkey sandwiches in the school.
- **c.** The grade point average of students in the honor society to estimate the honor society GPA.
- **d.** The grade point average of students in the honor society to estimate the GPA for the school.

15–4. Categorize each of the following as a random sample or not a random sample.
- **a.** Voting preference of every tenth person in the telephone book to estimate the voting preference for the town.
- **b.** The grade point average of students in the honor society to estimate the GPA for the school.
- **c.** The age of students in the kindergarten class to estimate the age of students in the elementary school.
- **d.** The opinion of each shopper at the north end of the mall on the convenience of the mall hours.

Statistical Measures

One of the goals of statistics is to describe or summarize data. Today, data sets may consist of many thousands of numbers. It would be difficult to see patterns if you looked at all the numbers separately. Instead, we would like ways to display or summarize the most important features of the data. Three things of interest about a data set are the shape or distribution of the data, the location of the center of the data, and how spread out the data are from the center.

The shape of a data set is often shown graphically in a histogram. Distributions can have many different shapes; three examples are shown in Figure 15.1.

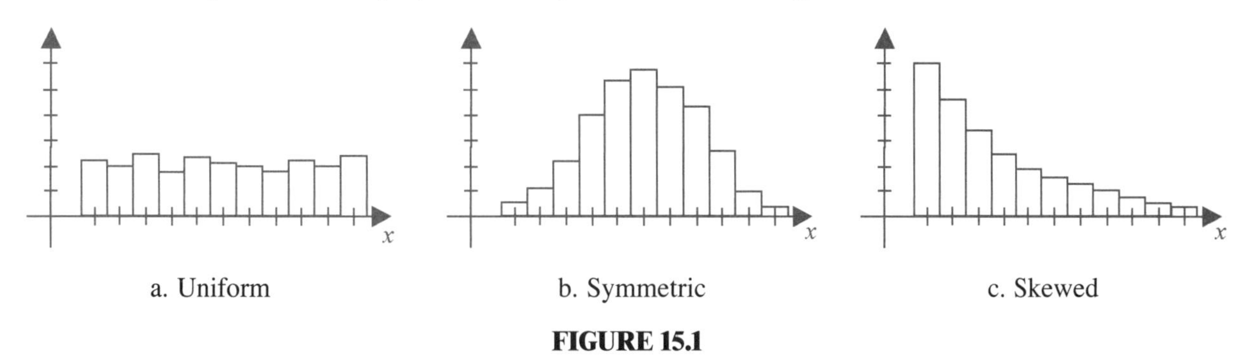

a. Uniform b. Symmetric c. Skewed

FIGURE 15.1

The distribution in Figure 15.1a is approximately uniform. In a uniform distribution, all possible x-values would be equally likely to occur. The distribution in Figure 15.1b is symmetric or bell shaped. Values of x near the center are the most likely. The likelihood of particular x-values decreases as you move away from the center. The distribution in Figure 15.1c is an example of a distribution that is skewed right. In a skewed right histogram, most of the observations are on the left side and there is a tail that goes off to the right. In a distribution that was skewed left, the tail would go to the left and most of the observations would be on the right side.

To describe the distribution further, we would like a number that tells where the center of the data is. We would like another number that somehow describes how spread out the data is around its center.

MEASURES OF CENTRAL TENDENCY

A *measure of central tendency* is a summary statistic that describes where the middle of a data set is. The three most common measures of central tendency are the mean, the median, and the mode.

The *arithmetic mean*, usually shortened to just mean, is the average you are already familiar with. The mean of a data set is the sum of the values in the data set divided by the number of values in the data set. Remember that in statistics, we distinguish between measures for the population and the sample. The symbol for a population mean is μ (mu); the symbol for a sample mean is $\bar{x}$ (x-bar). Both are calculated exactly the same way. Add up all the values then divide by the number of values. If the data consists of n numbers represented by $x_1, x_2, \ldots, x_n$:

$$\text{Mean} = \mu \text{ or } \bar{x} = \frac{\sum_{i=1}^{n} x_i}{n}$$

The *median* of a set of data is the middle value when the data are arranged in order from least to greatest. If there is an even number of data points, the median is the average of the middle two values. There is no one generally accepted symbol for median. You may see $\mu_{1/2}$, $\tilde{x}$, M, m, or simply the abbreviation med.

The *mode* of a data set is the value that occurs most often. If no value occurs more than once, there is no mode. There can be more than one value for mode if there is a tie for the greatest frequency.

EXAMPLE 15.1 A college baseball team played 30 games in its season. The following data show the numbers of runs scored in a simple random sample of 10 of their games.

2, 4, 1, 5, 0, 3, 7, 10, 2, 5

Find the mean, median, and mode of this sample.

SOLUTION

To find the mean, we add up all the values and divide by 10, the number of games. Note that we cannot leave out the 0 value. Although it does not change the sum in the numerator, it needs to be included in the total count for the denominator.

$$\bar{x} = \frac{2 + 4 + 1 + 5 + 0 + 3 + 7 + 10 + 2 + 5}{10} = 3.9$$

This says that a typical number of runs scored in a game by this team is 3.9. We do not round this to 4. It is true that a baseball team can never score 3.9 runs in a game. However, the mean of a data set does not have to be a value that actually appears in the data set.

Note that because this is a sample from a larger population, we use the symbol $\bar{x}$ for sample mean. The mean number of runs scored in all 30 games, population mean, would be designated μ. We use $\bar{x}$ to estimate μ.

To find the median, we must first put the data in order:

0, 1, 2, 2, 3, 4, 5, 5, 7, 10

Because this data set contains an even number of values, the median will be the average of the middle two values, 3 and 4:

$$\text{Median} = \frac{3 + 4}{2} = 3.5$$

The median represents a value that divides the data set into two equal sized parts. This team scored less than 3.5 runs in half its games and more than 3.5 runs in half its games. (This will not always be precisely the case since it is possible that some of the data values could equal the median.)

The mode is the value that occurs most often. In the problem, both 2 and 5 occur twice. So this data has two modes, 2 and 5. (A data set with two modes is sometimes called *bimodal*.) Unlike the mean and the median, there can be more than one mode or can be no mode. The mode, if there is one, must be a value in the data set.

The mode is the most commonly used measure of central tendency for qualitative data. You cannot average eye colors or put them in order and pick the middle one. However, you can find which one occurs most frequently. For quantitative data, the mean and the median are by far the most commonly used measures of central tendency. For many data sets, the mean and median will be close to each other. This won't always be the case, though, as the following example illustrates.

EXAMPLE 15.2

Suppose twelve employees of a large company were sampled to find their annual salaries. The salaries were $28,000; $33,000; $37,000; $42,000; $45,000; $47,000; $51,000; $54,000; $57,000; $63,000; $115,000; and $1,252,000. Find the mean and the median of this data set. Discuss which one better represents the data.

SOLUTION

The mean of this sample is $152,000; the median is $49,000.

The mean of $152,000 does not seem like a typical salary at this company. Eleven out of twelve people sampled make less than that; ten of them make well under half of that. Only one person makes more, but she makes a lot more, over eight times the mean. The median salary of $49,000 seems more representative of the middle of the data. Half the people in the sample make less than that while half make more than that.

Example 15.2 illustrates one advantage the median has over the mean; the median is less sensitive to extreme values in the data. Note what happens if the largest salary is deleted from the data set. The mean drops by $100,000, from $152,000 to $52,000. The median drops only $2,000 to $47,000. The $1,252,000 data point is an *outlier*, a value that is far away from the rest of the data. In this case, the outlier is more than ten times the size of the next largest value. It has a much greater impact on the mean than it does on the median. (Note that just because a data point is an outlier does not make it wrong. It is entirely possible that a high-level executive in the company makes over a million dollars a year. However, that probably does not represent a typical salary.)

EXAMPLE 15.3

A survey of local gas prices found the following cost, in dollars per gallon, at a number of gas stations. Find the mean, median, and mode of the price of gas on this day.

Price of gas ($ per gallon)	Frequency (number of gas stations)
3.89	7
3.92	14
3.93	9
3.95	6
3.98	3
4.04	1

SOLUTION

In this problem, we are given a frequency column with the number of gas stations offering to sell gas at each price. For example, 7 stations sell gas at \$3.89. By summing the frequencies, we find that a total of 7 + 14 + 9 + 6 + 3 + 1 = 40 gas stations were surveyed. To find the average price, we must add up all 40 individual prices. We could add \$3.89 seven times, then \$3.92 thirteen times, and so on. It is quicker to multiply each price by its frequency and then add to get the total.

$$7(\$3.89) + 14(\$3.92) + 9(\$3.93) + 6(\$3.95) + 3(\$3.98) + 1(\$4.04) = \$157.16$$

To find the mean, divide this total by the total number of stations, 40.

The mean is $\bar{x} = \dfrac{\$157.16}{40} \approx \$3.93$.

In general, if the data values are represented by x_i and the frequencies, the number of times each x-value occurs, are represented by f_i:

$$\text{Mean} = \mu \text{ or } \bar{x} = \frac{\sum_{i=1}^{n} f_i x_i}{\sum_{i=1}^{n} f_i}$$

There is an even number of gas stations, 40. To find the median, you need to average the two prices in the middle in order. You could, with a lot of work, write out all 40 prices in order. Instead, you could think that the middle of 40 numbers must be the 20th and 21st numbers. The cumulative frequency column below counts the number of stations with that price or lower. For example, there are 30 stations offering gas priced at \$3.93 or less. According to this chart, stations 20 and 21 both offer gas priced at \$3.92, so the median is \$3.92.

Price of gas (\$ per gallon)	Frequency (number of gas stations)	Cumulative Frequency
3.89	7	7
3.92	14	21
3.93	9	30
3.95	6	36
3.98	3	39
4.04	1	40

Mode is easy to find in a frequency table. Just choose the price with the highest frequency. The highest frequency in the table is 14 for the price \$3.92, so the mode is \$3.92.

SEE CALC TIP 15A

COMMON ERROR

Some students forget to use the frequency of each value when computing the mean and median. If you find the mean or median using just the price column, you will get the correct answer only if all the frequencies are equal, which is a very unlikely situation.

MEASURES OF DISPERSION

Suppose an Algebra 2 teacher gives four tests to her class of 12 students. Her grading system results in all grades being multiples of 5. The distributions of the scores on the four tests are shown in Figure 15.2. The mean and the median were both 70 for all four tests. Clearly there are differences in the distributions that are not captured by those measures of central tendency. The distributions differ in how they are spread out around the mean. On Test 1, most students scored close to the mean. Test 2 was similar except for a couple of outliers. On Test 3, the scores were much more uniformly distributed between the low and high scores. On Test 4, there was no score at the mean, only a few even close to the mean, and many out near the extremes.

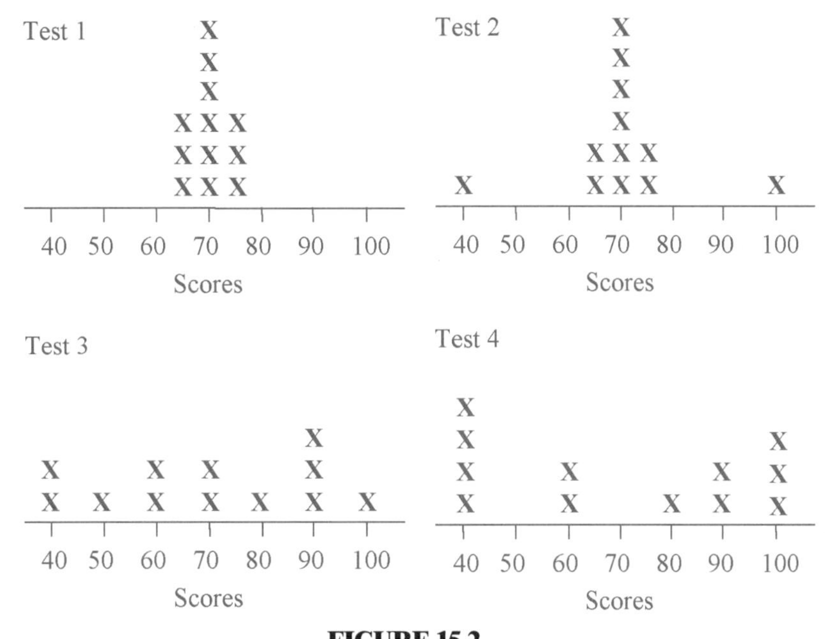

FIGURE 15.2

A *measure of dispersion* is a summary statistic that describes how spread out the data are around the mean or how much variability there is among the data. The simplest measure of dispersion is the *range*, which is the difference between the greatest and least data values.

$$\text{Range} = \text{Maximum value} - \text{Minimum value}$$

In the test scores example, the range correctly picks Test 1 as having the least dispersion or variability. The range of Test 1 is 75 – 65 = 10. Using only the range, all the other tests are equally variable; they each have a range of 100 – 40 = 60. This is not very descriptive. The range summarizes the variability in the whole data set using just two of its values. A statistic that uses all the data values would better describe the variability.

To keep the calculations manageable, let's look at a simple example. A statistics student named Sam has five test scores of 76, 83, 79, 72, and 90. Her average for the five tests was $\mu = 80$. We define the *deviation*, D_i, of a particular score to be the difference between that score, x_i, and the mean, $D_i = x_i - \mu$. The subscript, i, represents the order of the scores. The first test score is represented as x_1, and D_1 is its deviation from the mean. So on Sam's first test, her deviation was $D_1 = 76 - 80 = -4$. Sam's first test was 4 points below her average. On her second test, $D_2 = 83 - 80 = +3$, which means she scored 3 points above her average. The deviations for Sam's five tests are –4, +3, –1, –8, and +10.

If we average Sam's deviations, we get $\overline{D} = 0$. This will always be true because the mean describes the middle of the data. So the deviations of scores below the mean will always exactly cancel out the deviations of the scores above the mean. This tells us nothing about the variability of the data. We need to make all the deviations positive so that we find the average distance of the scores from the mean without regard to whether they are above or below the mean.

One way to do this is to find the *mean absolute deviation*, the mean of the absolute values of the deviations. For Sam, the mean absolute deviation is $\frac{|-4| + |3| + |-1| + |-8| + |10|}{5} = 5.2$. This says the average distance between Sam's test scores and her mean test score is 5.2 points.

The mean absolute deviation is easy to calculate and easy to understand. Unfortunately, the absolute values in it make studying its statistical properties rather complicated. Because of this, mean absolute deviation is rarely used. We use a different way of making the deviations positive: we square each deviation before averaging. The *variance* of a data set is the mean of the *squared* deviations:

$$\text{Variance} = \frac{\sum_{i=1}^{n}(x_i - \mu)^2}{n}$$

In words, the variance formula says to find each score's deviation from the mean, square each deviation, and then average all the squared values. Sam's test scores have a variance of:

$$\frac{(76 - 80)^2 + (83 - 80)^2 + (79 - 80)^2 + (72 - 80)^2 + (90 - 80)^2}{5} = 38$$

Clearly, Sam's typical test scores were not 38 points away from her mean. The number is large because we squared all the deviations. A natural solution is to take the square root of the variance. This is called the *standard deviation,* and its symbol is the lower case Greek letter sigma: σ.

$$\text{Standard deviation: } \sigma = \sqrt{\frac{\sum_{i=1}^{n}(x_i - \mu)^2}{n}}$$

The standard deviation of Sam's test scores is $\sigma = \sqrt{38} \approx 6.2$. This says that a typical distance between Sam's test scores and her average is about 6.2 points. Notice that this is a little larger than the mean absolute deviation, which was 5.2. Squaring the deviations gives the larger ones a greater influence in the standard deviation. Despite this bias toward larger deviations, the standard deviation has computational advantages that make it the most commonly used measure of dispersion for quantitative data.

The formula given above is for the *population standard deviation*, σ. In the example, we were treating Sam's five test scores as the entire population of her scores. If those scores had instead been a random sample from a larger population of Sam's test scores, we might want to estimate the population standard deviation σ using the *sample standard deviation*,

s. However, if we calculate the sample standard deviation, s, with the same formula as the population standard deviation, σ, s will tend to slightly underestimate the true value of σ. It turns out that the best estimate of σ from a sample is found by dividing by $n - 1$ instead of n. Thus we have two slightly different formulas for standard deviation, one for the population standard deviation, σ, and another for the sample standard deviation, s.

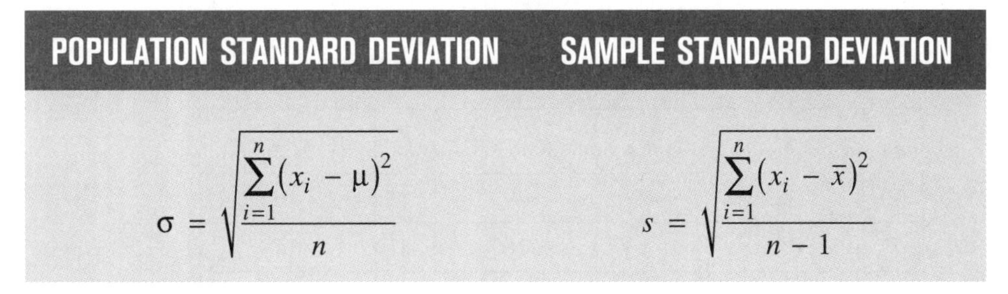

Note that in the sample standard deviation formula, the population mean, μ, has been replaced by the sample mean, $\bar{x}$. Usually when we are dealing with a sample, we don't know μ but $\bar{x}$ is the best estimate of it. Remember that μ and $\bar{x}$ are calculated exactly the same way. There is no change in the formula for mean when dealing with a sample instead of a population.

For both populations and samples, the variance is the square of the standard deviation. So the population variance is σ^2 and the sample variance is s^2. As you can see, calculating standard deviation by hand can be tedious even for relatively small data sets. Find out from your teacher whether you really need to memorize the standard deviation formula and be able to calculate it by hand. Most modern scientific and graphing calculators have built-in functions to find the mean and both σ and s. Your choice of σ or s depends on the problem. If it is clear that the data represent a sample of a larger population, use s. If the data represent the entire population, use σ.

Returning to the distributions of the Algebra 2 tests shown in Figure 15.2, the standard deviations of the tests were $\sigma_1 = 3.54$, $\sigma_2 = 12.58$, $\sigma_3 = 19.58$, and $\sigma_4 = 24.83$. As we would expect, the first test, with all the students grouped close to the mean, had the smallest standard deviation, while the last test, with many students having scores far from the mean, had the largest standard deviation.

EXAMPLE 15.4

Find the population variance and population standard deviation, each to the nearest thousandth, for the height of students given in the table to the right.

Height (inches)	Frequency (number of students)
59	1
62	4
65	12
67	7
70	4
72	2

SOLUTION

Find the population variance first, and then take its square root to find the population standard deviation. Computing the variance by hand is complicated. Here we break it into steps. The first step is to find the mean.

$$\bar{x} = \frac{1(59) + 4(62) + 12(65) + 7(67) + 4(70) + 2(72)}{1 + 4 + 12 + 7 + 4 + 2} = 66$$

Use a table to organize the rest of the steps. The first column is the height. The second column is the mean. The third column is the deviation of the height from the mean. The fourth column is the square of the deviation. The fifth column is the frequency. The sixth column is the frequency times the squared deviation.

x_i	$\bar{x}$	$x_i - \bar{x}$	$(x_i - \bar{x})^2$	f_i	$f_i(x_i - \bar{x})^2$
59	66	−7	49	1	49
62	66	−4	16	4	64
65	66	−1	1	12	12
67	66	1	1	7	7
70	66	4	16	4	64
72	66	6	36	2	72
			Total	30	268

The variance is the average found by totaling the last column and dividing by the number of students.

$$\text{Population variance} = \frac{268}{30} \approx 8.933$$

SEE CALC TIP 15B

The population standard deviation is the square root of the variance.

$$\sigma = \sqrt{8.933} \approx 2.989$$

SECTION EXERCISES

15–5. The table shows the grades received on the calculus test by students in the last eight years of Mr. Leibniz's calculus classes.

Grade	**Number of Students**
4.0	8
3.0	14
2.0	12
1.0	7
0	1

a. How many students have taken the test over the last eight years?
b. What is the average number of students to take the test each year?
c. What is the mean grade received by students over the last eight years?
d. What is the median grade received?
e. What is the modal grade?
f. If four students take the test this year, what would the grades have to be to get a unique new mode?
g. If four students take the test this year, what would the grades have to be to raise the overall average to at least 2.6?

15–6. The table below shows the distribution of grades on last year's Algebra 2 final exam for those students in Mr. Cardano's class who passed.

Algebra 2 Final Score	67	72	75	77	78	79	80	81	82
Number of Students	2	1	1	2	3	1	2	5	4
Algebra 2 Final Score	84	85	87	89	90	92	93	94	97
Number of Students	2	2	1	2	1	3	2	2	1

a. To the nearest tenth, what was the mean grade among those who passed?
b. What was the median grade?
c. To the nearest hundredth, what was the standard deviation of the grades?
d. What percent of the students scored above the average?
e. What percent of the students got scores within one standard deviation of the mean?

15–7. Ryan made a batch of chocolate chip cookies. As a math project, his sister Lara counted the number of chocolate chips in each cookie as she ate them. From her data, she created the table below.

Number of Chips	**Number of Cookies**
7	2
8	x
9	4
10	3
11	1
12	2
16	1

a. If the unique mode is 8, what is the least possible value of x?
b. Using your value of x, what is the mean number of chocolate chips per cookie?
c. Using your value of x, what is standard deviation of the number of chocolate chips per cookie to the nearest hundredth?

15–8. Norman Nerdling recorded the number of hours (rounded to the nearest whole hour) he played video games each day over a six-week period.

Video Game Hours	0	1	2	3	4	5	6	7	8	12
Number of Days	2	4	8	7	5	5	3	4	2	1

a. What was the mode of the number of hours of video games played?
b. What was the median of the number of hours of video games played?
c. What was the mean, to the nearest tenth, of the number of hours of video games played?
d. What was the standard deviation, to the nearest tenth, of the number of hours of video games played?
e. If Norman took data for exactly six weeks, how many days did he forget to write down his hours?
f. What percent of days is Norman's gaming time within one standard deviation of the mean?

15–9. Hoops Shooter and his twin sister Sharpe both finished their high school basketball careers with averages of exactly 16 points per game. Hoops's scores had a standard deviation of 5, while Sharpe's scores had a standard deviation of 2.
a. Which player most likely had the single highest scoring game?
b. Which player most likely had the single lowest scoring game?

Normal Distribution

Now we introduce an extremely important continuous distribution called the *normal distribution*. The normal distribution describes a random variable that, at least in principle, can take on any real number.

Suppose you pick an American adult man at random and measure his height. Is it equally likely that he would be 5′10″ as opposed to 6′10″? Of course not; 5′10″ is about average for adult American men while 6′10″ men are rare. Let's imagine that we collected a very large sample of heights of adult American men and made a histogram. What would we expect it to look like? We might reasonably expect a large proportion of the heights to be around average, say between 68 and 72 inches. Many men are somewhat above average, say between 72 and 76 inches but probably not as many as between 68 and 72 inches tall. We might expect a similar number of men a little below average, 64 to 68 inches. The farther away from average we get, the fewer men we expect to find at those heights. Men between 80 and 84 inches tall or between 56 and 60 inches tall certainly exist but are very uncommon. Our histogram might look like that shown in Figure 15.3.

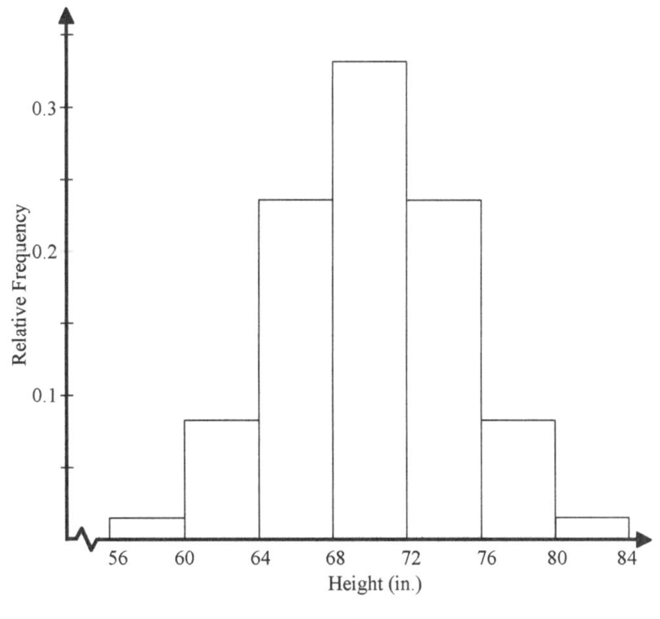

FIGURE 15.3

Now let us redraw the histogram with more, but narrower, intervals. Let's halve the width of each interval. The new histogram might look like Figure 15.4. The vertical scale has changed (if the scale had been unchanged, the heights of all the intervals would have approximately halved), but the shape of the histogram remains the same. Frequencies decrease as we move farther away from the mean in either direction.

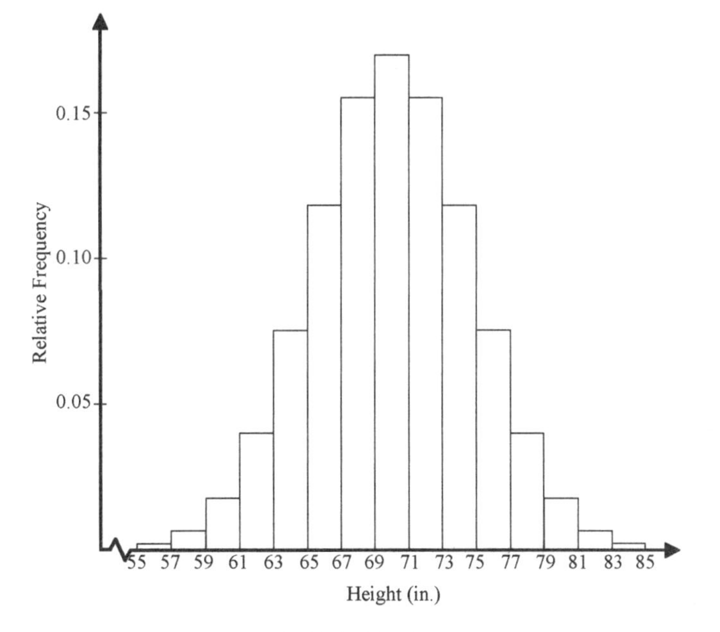

FIGURE 15.4

Halving the widths of the intervals another time and then yet another time would give the histogram in Figure 15.5. (Again, the vertical scale has changed each time.)

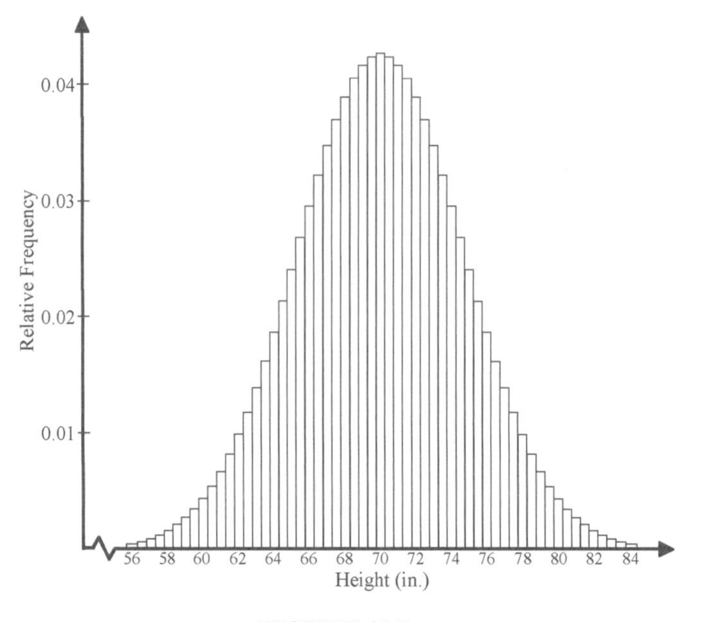

FIGURE 15.5

Imagine we could keep doing this, making histograms with ever more and ever narrower intervals. Our histogram would approach a smooth curve like the one shown in Figure 15.6. (An optional note for the curious. In the preceding histograms, the vertical scale is labeled Relative Frequency. When the distribution changed from discrete to continuous, the vertical scale no longer represents relative frequency. It is renamed "Density.")

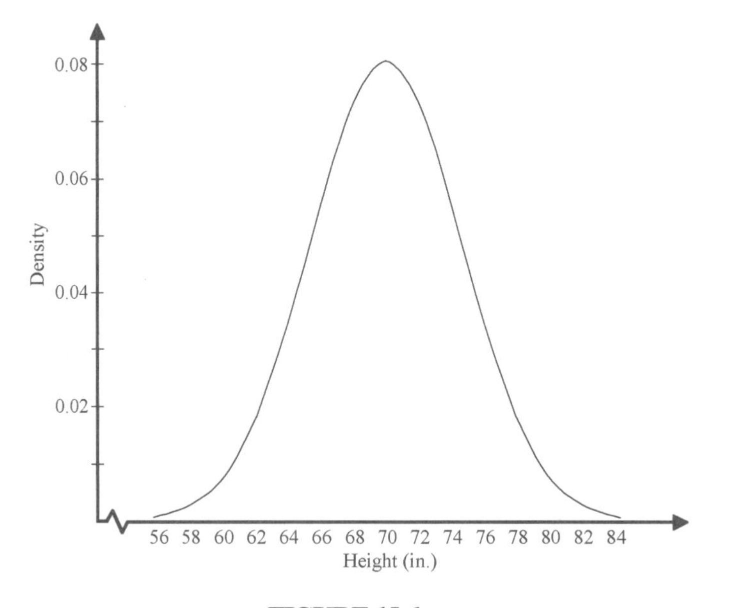

FIGURE 15.6

This curve is an example of a *normal distribution curve* and is possibly the most important distribution in statistics. It turns out that for large enough samples, a wide range of measurable phenomena are well modeled by normal distributions. Examples, in addition to heights, include weights, such as watermelons grown on a farm; volumes, such as the amount of soda actually in a serving at a fast food restaurant; standardized test scores, such as IQ tests and the SAT; and lifetimes, such as how long your cell phone will work on one battery charge.

A normal distribution is often referred to as a bell-shaped curve because its characteristic shape reminds many people of a bell. Mathematically, all normal curves are translations and/or dilations of the graph of $f(x) = e^{-x^2}$. A normal distribution is completely defined by two values, its mean, μ, and its standard deviation, σ. The mean is the location of the center of the curve, the point where the curve reaches its maximum height. The standard deviation determines how spread out the scores are from the mean. In a sense, the standard deviation describes the width of the curve. Small standard deviations have most scores closer to the mean; the graph appears tall and narrow. Larger standard deviations indicate more scores farther from the mean; the graph looks short and wide. For example, adult American women have an average height of 64.5 inches and the standard deviation of women's heights is slightly smaller than that for men. The graph of women's heights would be the same shape as the graph for men in Figure 15.6 but translated about 5.5 units to the left and horizontally dilated (narrowed) by a small amount.

You should know the following properties of normal distributions.

1. The mean is the x-value where the distribution reaches its maximum value. It is the same as the median and the mode.
2. The distribution is symmetric about the mean.
3. The likelihood of a particular value decreases the farther it is from the mean.

EXAMPLE 15.5 The diagram shows three normal curves sketched on the same axes.

a. Which has the largest mean? The smallest?
b. Which has the largest standard deviation? The smallest?

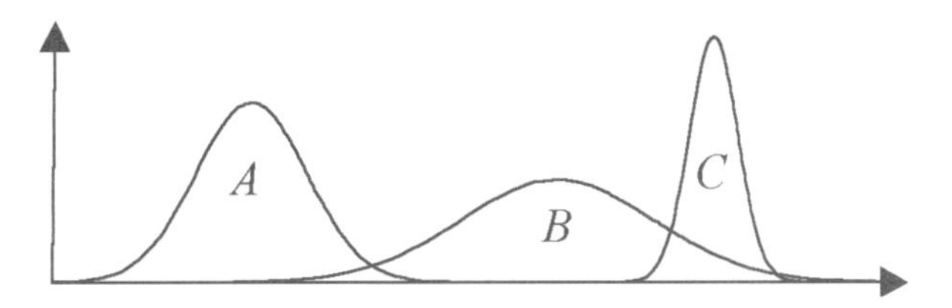

SOLUTION

a. The mean is at the center of the distribution. The distribution labeled C has the greatest mean; the one labeled A has the smallest mean.

b. A larger standard deviation means greater variability from the mean; the curve will appear wider or more spread out horizontally. Distribution B appears the most spread out; it has the largest standard deviation. Distribution C is the least spread out; it has the smallest standard deviation.

In addition to the basic properties listed previously, the scores in a normal distribution will obey the following rule, often called the *empirical rule*.

1. About 68.2% of all the scores will fall within one standard deviation of the mean.
2. About 95.4% of all the scores will fall within two standard deviations of the mean.
3. About 99.7% of all the scores will fall within three standard deviations of the mean.

For convenience, most people just use 68%, 95%, and "more than 99%." These values are good enough for many practical applications. The empirical rule is shown graphically in Figure 15.7.

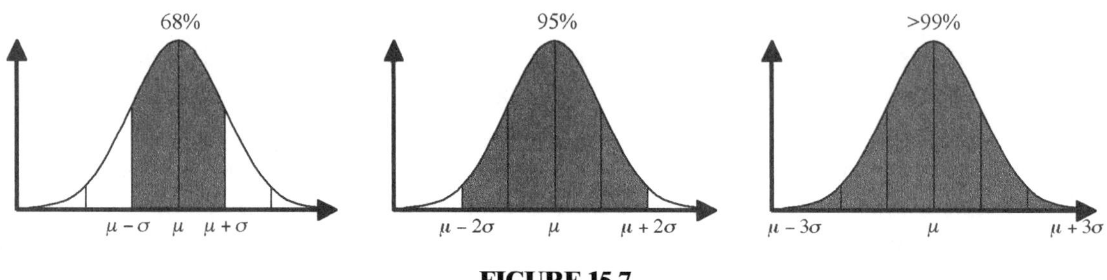

FIGURE 15.7

EXAMPLE 15.6

The average height of adult men in the United States is 70 inches. If the standard deviation is 2.5 inches, what interval, centered on the mean, will include the heights of about 95% of all adult men in the United States?

SOLUTION

By the empirical rule, about 95% of all scores should fall within two standard deviations of the mean:

$$\mu - 2\sigma \le x \le \mu + 2\sigma.$$

By substituting $\mu = 70$ and $\sigma = 2.5$, we get:

$$70 - 2(2.5) \le x \le 70 + 2(2.5)$$

$$65 \le x \le 75$$

So about 95% of all adult American men are between 65 and 75 inches tall.

EXAMPLE 15.7

A standardized math test was given to all the fourth graders in a large city school district. About 68% of the students scored between 88 and 104 on the test. Estimate the mean and standard deviation of the scores.

SOLUTION

We need to make two assumptions. We need to assume that the scores are normally distributed. This is a reasonable assumption provided the number of students who took the test is fairly large. We will also assume that the 68% referred to in the problem is centered on the mean.

If the 68% is centered on the mean, the mean should be the middle of the interval 88 to 104. This is found by averaging the endpoints of the interval.

$$\mu = \frac{88 + 104}{2} = 96$$

In a normal distribution, about 68% of all the scores will be within one standard deviation of the mean. For this problem, that means a score of 104 should be one standard deviation above the mean. Using our estimated mean of 96, we get $96 + \sigma = 104$ or $\sigma = 8$.

With the assumptions stated above, the scores should have a mean of 96 and a standard deviation of 8.

EXAMPLE 15.8

The prices of existing homes in a certain suburban neighborhood are approximately normally distributed with a mean of $187,500 and a standard deviation of $12,500. What percent of the homes in the neighborhood are priced between $150,000 and $225,000?

SOLUTION

To answer this question, we need to know how many standard deviations away from the mean are the values $150,000 and $225,000. The number of deviations from the mean is called the z-score and is found with the formula $z = \frac{x - \mu}{\sigma}$. For the lower value, we have $z = \frac{150{,}000 - 187{,}500}{12{,}500} = -3.0$. This means that $150,000 is 3 standard deviations below the mean. For the upper value, $z = \frac{225{,}000 - 187{,}500}{12{,}500} = 3.0$ and $225,000 is 3 standard deviations above the mean. By the empirical rule, about 99.7% of the homes, almost all of them, will be priced between $150,000 and $225,000.

SEE CALC TIP 15C

Clearly, the numbers in the previous example were chosen so that the answer could be found using only the empirical rule. In practice, we want to be able to answer questions that involve percents other than just 68, 95, and 99.7 and z-scores other than ±1, ±2, and ±3. Not too many years ago, you would have solved such problems by looking up values in a table of the *standard normal distribution*, the normal distribution having mean 0 and standard deviation 1. Today these problems can be answered quickly and easily with technology. Many graphing calculators have functions to calculate normal distribution percentages. For many, you do not even need to first compute z-scores; you can simply enter the actual data. This book does not go into detail on the various methods of finding normal distribution values. If you are taking a course that includes a unit on the normal distribution, find out how you are expected to find the values you need to solve the problems.

Normal distribution can be used to find probabilities of events if you know the mean and standard deviation for the data. Usually these probabilities are expressed in percent form.

EXAMPLE 15.9 The battery life of a certain laptop computer is normally distributed with a mean of 5.5 hours and a standard deviation of 0.4 hours. If you take this laptop on a 6-hour airline flight, what is the probability you will be able to use it for the whole flight?

SOLUTION

Six hours is $\frac{6 - 5.5}{0.4} = 1.25$ standard deviations above the mean. Using the empirical rule, the chance of a score at least one standard deviation above the mean is $\frac{1}{2}(100\% - 68\%) = 16\%$.

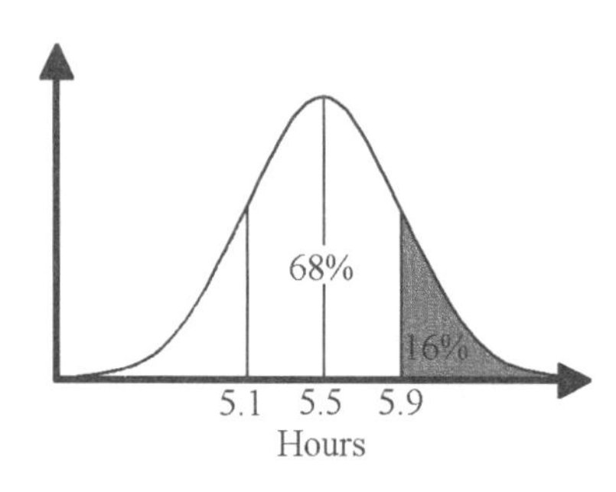

SEE CALC TIP 15D

The chance of a score more than 1.25 standard deviations above the mean will be somewhat less than that. Using either a table or a calculator gives an actual probability of about 0.106 or 10.6% that your laptop battery will last the whole six hours.

PERCENTILES

For large data sets (many hundreds of observations), a convenient way to describe how one data point compares with the whole data set is with its percentile. A data point at the *p*th percentile is greater than or equal to *p*% of all the scores. For example, when you were a child and had a medical checkup, the doctor would measure your height to compare you with other children of your age and sex. If your height was at the 50th percentile, then you were as tall as or taller than 50% of the children your age. Assuming heights are normally distributed, this would mean you were average height. If your height was at the 95th percentile, then 95% of children your age were your height or shorter than you; you were tall for your age.

Using the empirical rule, a normal distribution graph can be labeled with percentiles based on how many standard deviations a score is from the mean. These percentiles of the normal distribution, rounded to the nearest whole number, are shown in Figure 15.8.

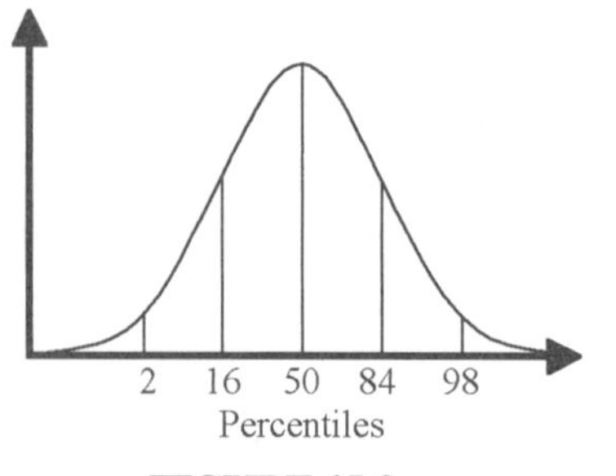

FIGURE 15.8

EXAMPLE 15.10

A standardized exam with a normal distribution has a mean score of 500 and a standard deviation of 100. If a student scores 700 on this exam, what is her percentile ranking? If 20,000 students took this exam, how many students have a lower score than this student?

SOLUTION

A score of 700 is $\frac{700-500}{100} = 2$ standard deviations above the mean. The percentile at 2 standard deviations above the mean is 98. This student would have a percentile ranking of 98.

This means about 98% of the students scored 700 or lower (we say "about" 98% because that percentile was rounded to the nearest whole number). If 20,000 students took the exam, then about 98% of 20,000 or 0.98 (20,000) = 19,600 students scored 700 or lower on the exam.

EXAMPLE 15.11

A two-year-old girl is getting a well-child checkup. The nurse notes that the girl is 35 inches tall, which places her taller than 84% of two-year-old girls. If she were 34 inches tall, then she would be average for her age. Based on this information, what is the mean and standard deviation for the height of two-year-old girls?

SOLUTION

If 34 inches is average, the mean = 34. Since the girl is taller than 84% of other two-year-old girls, she is in the 84th percentile. On the normal distribution curve, the 84th percentile is one standard deviation above the mean. The standard deviation is 35 – 34 = 1 inch.

EXAMPLE 15.12

The weights of the eggs laid by hens at a particular poultry farm are normally distributed with a mean of 2.1 ounces and a standard deviation of 0.25 ounces. Eggs weighing over 2.0 ounces are labeled large or bigger. What percent of the eggs from this farm are smaller than large?

SOLUTION

An egg weighing 2 ounces has a z-score of $\frac{2-2.1}{0.25} = -0.4$, meaning it is 0.4 standard deviations below the mean. The empirical rule is not very accurate for this problem. However, you can estimate the percent to be between 16% and 50%. A calculator gives a probability of getting an egg smaller than 2.0 ounces as 0.345 or 34.5%.

SEE CALC TIP 15E

EXAMPLE 15.13

The size of a single scoop of ice cream at the Frozen Cow is approximately normally distributed with a mean of 0.9 of a cup and a standard deviation of 0.1 of a cup. Dorothy orders a single scoop of ice cream.

a. Dorothy will eat up to 0.75 of a cup of ice cream. Any leftover she feeds to her dog, Toto. What is the probability of Toto getting some ice cream?

b. Suppose Dorothy's scoop of ice cream is in the 80th percentile for size. How much ice cream does Toto get?

SOLUTION

a. Toto gets ice cream if Dorothy's scoop is larger than 0.75 of a cup. A scoop of 0.75 of a cup has a z-score of $\frac{0.75 - 0.9}{0.1} = -1.5$, which makes it 1.5 standard deviations below average. From the graph of the percentiles of the normal distribution in Figure 15.8, we see that the probability of getting a scoop smaller than 0.75 of a cup is somewhere between 2% and 16% and somewhat closer to 2%. This means the chance of getting more than 0.75 of a cup of ice cream is between 84% and 98% and closer to 98%. With a calculator, we can find the probability of getting more than 0.75 of a cup is 0.9332. Toto has slightly better than a 93% chance of getting ice cream.

b. We need to find the actual size of Dorothy's scoop of ice cream given that it was at the 80th percentile. Again using the graph of the percentiles of the normal distribution in Figure 15.8, we see that the 80th percentile will be a little less than one standard deviation above the mean, which means Dorothy got somewhat less than 0.9 + 0.1 = 1 cup of ice cream. With a calculator, we can find that the 80th percentile is 0.984 of a cup. After Dorothy eats her 0.75 of a cup, Toto gets 0.984 – 0.75 = 0.234, or slightly less than a quarter of a cup.

SEE CALC TIP 15F

SECTION EXERCISES

15–10. The figures show three different normal distributions. Assume they all have the same scale on the x-axis.

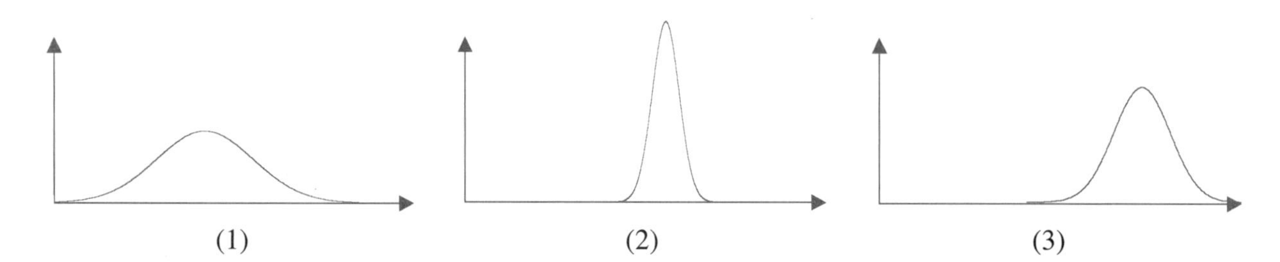

a. For which distribution is the mean the greatest?

b. For which distribution is the mean the least?

c. Which distribution has the largest standard deviation?

d. Which distribution has the smallest standard deviation?

e. Which distribution would require the widest interval to contain 95% of all its scores?

For 11–14, use a calculator with statistical features.

15–11. Rita Roller's average bowling score is 162, and the standard deviation of her scores is 12. Assume her scores are normally distributed.

a. What is the probability Rita bowls 180 or better on her next game?

b. In what percent of her games does Rita bowl less than 150?

c. On her next game, which of the following scores would Rita be most likely to get?

(1) 150 (2) 158 (3) 170 (4) 176

d. Of the choices above, which score would Rita be least likely to get?

e. Out of her next 100 games, about how many times can Rita expect to break 200?

15–12. The heights of American men ages 18 to 24 are normally distributed with mean 69.7 inches and a standard deviation of 3.0 inches.

a. The heights of approximately 95% of all American men in the age group will fall within what interval centered on the mean?

b. There are 1,250 men at Elsewhere Community College (ECC). About how many of them are between 66.7 and 72.7 inches tall?

c. Approximately what percent of men in the age group are taller than 64 inches?

d. Estimate the percentile height for a man who is exactly 6 feet tall.

15–13. On their final day at the Felix Golf Clinic, about 95% of the participants scored between 75 and 101.

a. Estimate the average score on the final day.

b. Estimate the standard deviation of the scores on the final day.

c. Estimate the score at or below which the top 10% of the class golfed on the last day.

15–14. In an ACME widget, a good set of alkaline batteries will last an average of 5 hours with a standard deviation of 40 minutes. What is the probability a widget with new alkaline batteries will run for at least 6 hours?

Modeling with Functions

Recall that bivariate data consist of paired measurements of two different variables and is useful for studying relationships between variables. Real-life data, whether collected in a lab or in the field, almost never perfectly fits some nice mathematical model or equation. There will always be measurement error when measuring quantities. Even if there were not, other variables will often affect the data that cannot be controlled. One of our goals is to find a mathematical model for bivariate data, an equation that approximates the data and can be used to study the relationship between the variables and possibly to make predictions. This is called curve fitting. At its most basic, it can be thought of as a four-step process.

CURVE FITTING

Curve fitting involves four separate steps. They are described briefly here and in more detail later.

1. Make a scatter plot of the data, which is a graph showing all the data points.
2. Select the general type of function to use to model the data. Some possibilities are linear, quadratic, power, exponential, logarithmic, and trigonometric.
3. Find the best-fit equation, which is the equation that best matches the data.
4. Analyze the equation to see how well it models the data.

Let us see how it works. Most people know that water pressure increases as one goes deeper beneath the surface. Suppose the pressure was measured at eight different depths in a lake resulting in the measurements shown in Table 15.1. We would like to find a mathematical equation that fits the data as closely as possible.

TABLE 15.1
UNDERWATER PRESSURES

Depth (ft.)	Pressure (psi)
5	16.1
10	20.2
15	20.9
20	22.6
25	26.3
30	27.0
40	33.4
50	35.6

STEP 1: SCATTER PLOT

A scatter plot of the data is shown in Figure 15.9. A scatter plot is just a graph of all the data. Points in the plot are not connected. It is possible that the points in the plot might not pass the vertical line test, for example if we had taken more than one pressure measurement at some or all of the depths. The scatter plot helps us visualize the data and select an appropriate function to model it with.

STEP 2: CHOOSING A FUNCTION

In practice, a certain amount of knowledge and experience goes into selecting the type of function to model the data. For our purposes, we will choose a function by examining the shape of the curve in the scatter plot. Looking at the data in this example suggests that pressure increases at a roughly constant rate with depth. True, the data does not make a perfect straight line. However, a general rule is to keep things simple. It does not appear that the data would be better modeled with another function such as a parabola or an exponential. So we will choose to model this data with a linear function.

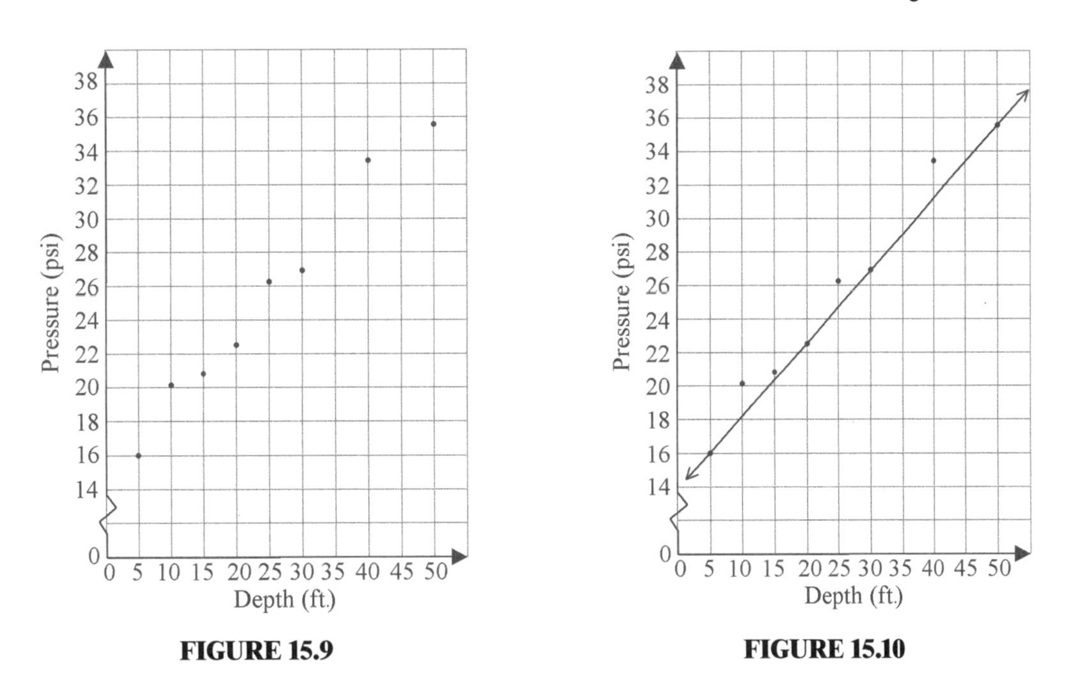

FIGURE 15.9

FIGURE 15.10

STEP 3: FINDING THE EQUATION OF BEST FIT

So how do we find an equation for the line of best fit, the line that comes closest to all the data points? One rather natural idea is to use just the first and last data points to write the equation of a line. In this example, the points (5, 16.1) and (50, 35.6) give the line $y = 0.433x + 13.933$. This line is graphed in Figure 15.10. It appears to be a reasonable fit to the data. Observe, though, that four of the data points are clearly above the line while none are noticeably below it (one is below, but by so little you can't tell at this scale). This suggests that maybe the line ought to be translated upward a bit. How far? Although the slope of our line looks good, is it really the best possible slope?

What exactly is meant by line of best fit and how to find it is the subject of entire textbooks. Whole college courses are devoted to *regression analysis*, the term for techniques for finding these models. For models with just one independent variable, such as the example here, the most common method for finding the line of best fit is *simple linear regression* (also known as *ordinary least squares*). The resulting line is called the *least squares regression line* or, often, simply the *regression line*. Working out a least squares regression line by hand for more than a small handful of data points is very tedious work. Many scientific and almost all graphing calculators will do it for you.

The idea of least squares linear regression is fairly simple. We want to somehow minimize the sum of all the errors, the vertical distances between actual measured y-values and the corresponding y-values on the line of best fit, as shown in Figure 15.11. (Figure 15.11 shows just three data points for example purposes. Typically, there are many more.)

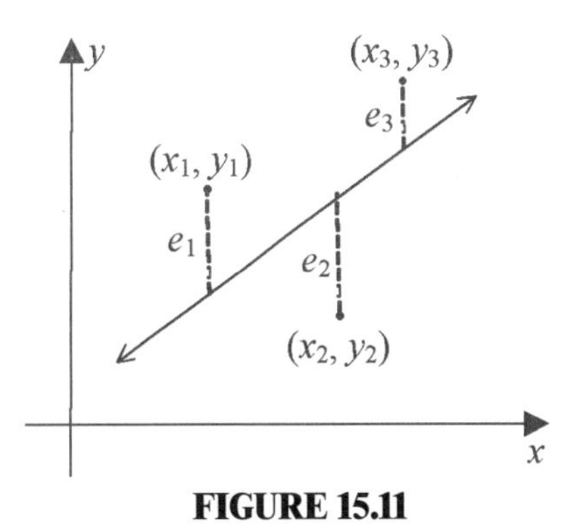

FIGURE 15.11

SEE CALC TIP 15G

Just as we needed to make the deviations positive when calculating standard deviation, we need to make the errors positive to find the line of best fit. Similar to computing standard deviation, squaring the errors turns out to be more convenient than using absolute value. The line of best fit is then the line that minimizes the sum of the squared errors. In Figure 15.12, we want the line that makes $e_1^2 + e_2^2 + e_3^2$ as small as possible. This is where the name "least squares" comes from. The computations involved in actually finding this line are beyond the scope of Algebra 2. In this book, we assume it is done on a calculator or with other technology.

So what is the line of best fit for the pressure data? With coefficients rounded to three decimal places, the result is $y = 0.435x + 14.662$.

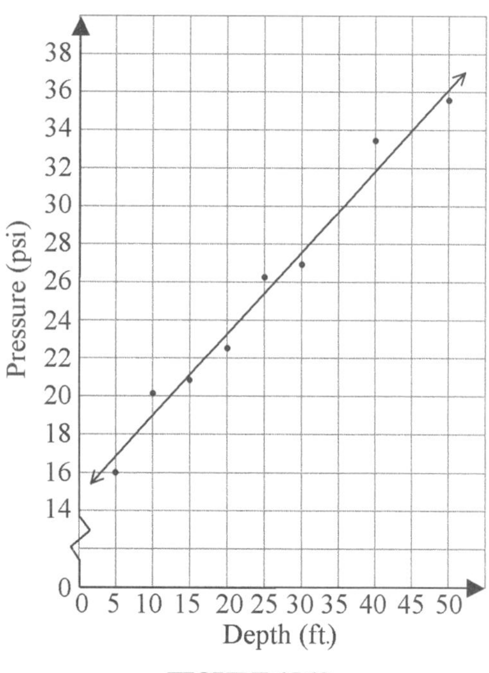

FIGURE 15.12

STEP 4: ANALYZE THE EQUATION

There are several considerations when deciding if we have an appropriate model for the data; they are beyond the scope of an Algebra 2 introduction to regression. Once again, we will compare the data to the line visually. The line of best fit is shown in Figure 15.12. It looks close to the data. Comparing it to the "first point–last point" equation we started with, we see that the slope is almost the same but the y-intercept is greater, as we expected.

Now that we have the equation, what is it good for? Here are three simple questions we might use it to answer.

1. For some reason, no measurement was made at $x = 35$. What does the model predict is the pressure at that depth?

When 35 is substituted for x, the model predicts $y = 0.435(35) + 14.662 \approx 29.9$ psi. This is called *interpolation*. It means using the model to find a new data point within the domain of the measured data. The original data was from $x = 5$ to $x = 50$; $x = 35$ is within this interval. In general, as long as the equation models the data well, estimates found by interpolation are expected to be reasonably accurate.

2. What is the pressure at a depth of 100 feet?

When $x = 100$, the model predicts $y = 0.435(100) + 14.662 \approx 58.2$ psi. This is an example of *extrapolation*, using the model outside the domain of the original data. Extrapolation should not be done with the same confidence as interpolation. We are assuming that the same equation that models pressure for depths from $x = 5$ to $x = 50$ feet will continue to be a good model at twice that depth. Without more information, we do not really have any choice but to make this assumption. However, we should be aware that this estimate has the potential for considerably greater error than our estimate for $x = 35$.

3. At what depth would the pressure be 40 psi?

In this case, we are trying to solve for the value of x that will make $y = 40$. Solving $40 = 0.435x + 14.662$ gives $x \approx 58.25$ feet. Again, this estimate should be viewed with caution since it is outside the domain of the measured data (although not nearly as far outside $x = 100$.)

The basic idea of ordinary least squares can be extended to find polynomial regressions such as quadratic, cubic, and even higher-order best-fit curves. With appropriate transformations of the data, the method can also be used to find other types of regressions including exponential, logarithmic, and power models. Most graphing calculators will do these regressions. Linear, quadratic, and exponential models are among the most common. You should choose the type of model from the shape of the scatter plot and knowledge of the data.

TABLE 15.2
REGRESSION EXAMPLES

Linear (LinReg)	**Quadratic (QuadReg)**	**Exponential (ExpReg)**
$y = ax + b$	$y = ax^2 + bx + c$	$y = ab^x$

EXAMPLE 15.14 Which type of model—linear, quadratic, or exponential—is the most appropriate for each of the following?

a. Three midsize cars from a particular manufacturer are tested to determine their fuel efficiency in miles per gallon as a function of speed in miles per hour.

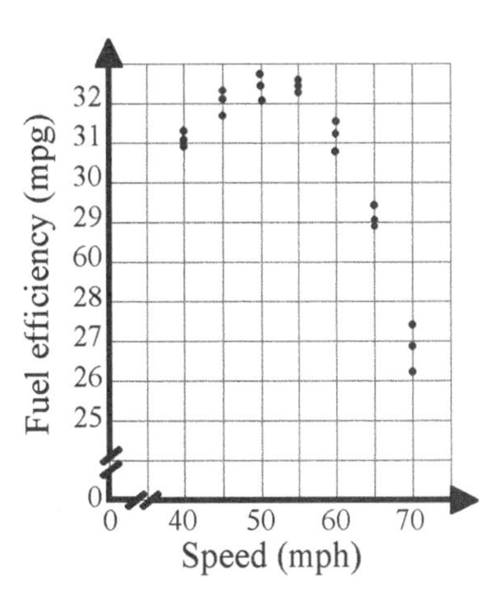

b. The ACME Widget Company has recorded its widget sales for each year since 2001. They want to model sales as a function of time to predict future sales.

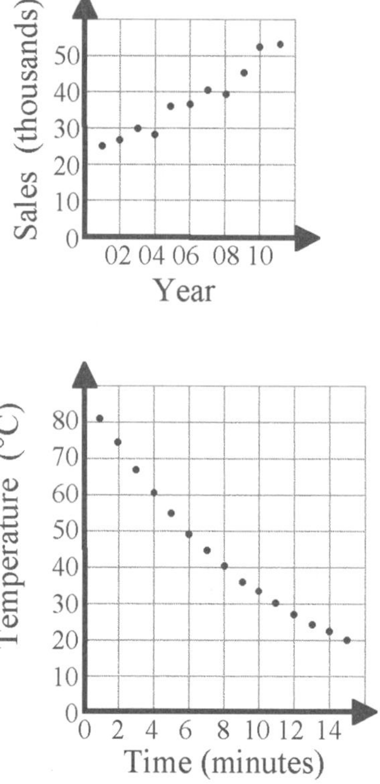

c. A hot cup of coffee cools outdoors on a day when the temperature is 0° Celsius. The temperature of the coffee is recorded at one-minute intervals.

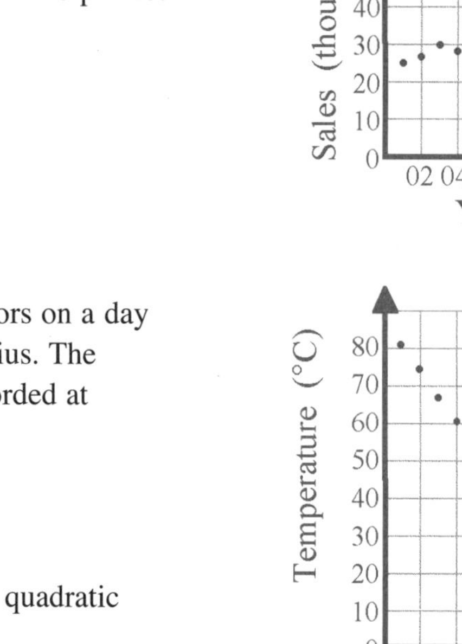

SOLUTION

a. This data is clearly best modeled by a quadratic function.

b. This scatter plot suggests a linear model. Although the data does not form a perfect line, average sales seem to be increasing at a constant rate.

c. This is an example where some understanding of the problem can help choose the best model. The data shown might be well modeled by either a quadratic or an exponential model. However, a quadratic model would suggest that the temperature of the coffee reaches some minimum and then starts to rise again. An exponential model would show the temperature gradually leveling off. The exponential model makes much more sense.

EXAMPLE 15.15

A baseball batter hits a high pop fly. The height of the ball is recorded every half second for the first two and a half seconds as shown in the table.

Time (seconds)	Height (feet)
0.5	47.2
1.0	83.6
1.5	112.5
2.0	134.1
2.5	148.3

a. Determine which equation—linear, quadratic, or exponential—best models this data.

b. Find the regression equation for your chosen model, rounding coefficients to the nearest thousandth.

c. According to your model, how high off the ground was the ball when it was hit?

d. Use your equation to predict the maximum height of the ball.

e. Use your equation to find when the ball hits the ground (assuming it is not caught).

SOLUTION

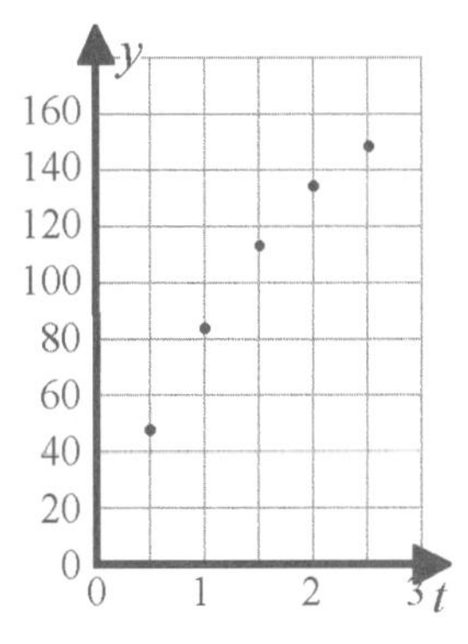

a. A scatter plot of the data is shown at the right.
A brief look might suggest we could get away with a linear model. A closer look shows we should consider a nonlinear model. We know the ball will eventually stop going up and then fall back to the ground. A quadratic model is appropriate.

b. Using a graphing calculator, the quadratic regression equation is
$y = -14.771t^2 + 94.854t + 3.480$.

c. According to the model, at time $t = 0$ the ball was 3.480 feet off the ground. This seems reasonable for a batted baseball.

d. Since a graphing calculator was used to find the regression equation, it makes sense to use it to graph the equation and find the maximum height, $y = 155.76$ at time $t = 3.211$ seconds. In the interest of reviewing quadratic functions (Chapter 5), we note the ball will reach its maximum height at $t = \frac{-b}{2a} = \frac{-94.854}{2(-14.771)} = 3.211$ seconds after it was hit. At this time, the model predicts the ball will be at a height of $y = -14.771(3.211)^2 + 94.854(3.211) + 3.480 = 155.759$ feet.

e. The ball will hit the ground when $y = 0$. The equation $0 = -14.771t^2 + 94.854t + 3.480$ can be solved with either technology or the quadratic formula. According to the model, the ball will hit the ground after 6.458 seconds.

EXAMPLE 15.16 Thousands of years ago, the last small herd of unicorns swam to a remote island and remained there. Over time, the population grew as shown in the table below.

Time (years since arrival)	Population (unicorns)
5	26
10	31
15	40
20	48
25	62
30	76

a. Determine which equation—linear, quadratic, or exponential—best models this data.

b. Find the regression equation for your chosen model, rounding coefficients to the nearest ten-thousandth.

c. Use your equation to predict the number of unicorns that originally went to the island.

d. Use your model to predict the number of unicorns on the island one century after the first ones arrived.

e. Use your model to estimate when the unicorn population reaches 2,500.

SOLUTION

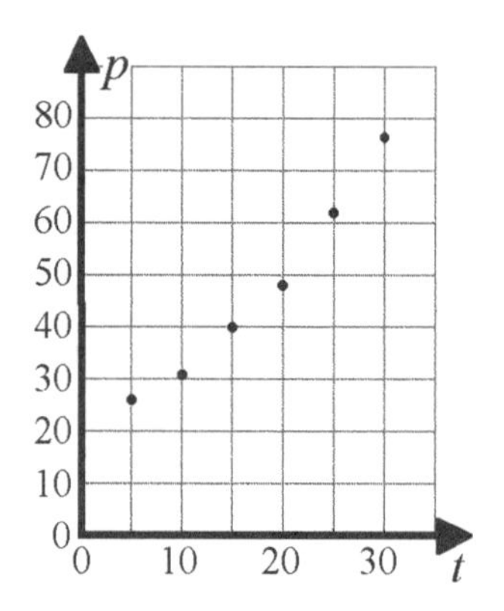

a. A scatter plot of the data is shown at the right. It is clearly not linear. By looking only at the scatter plot, it might be hard to choose between quadratic and an exponential model. This is where additional knowledge about what you are modeling can be helpful. Most simple population growth models are exponential.

b. Using a graphing calculator shows the regression equation is $p = 20.5428(1.0445)^t$.

c. At time $t = 0$, when the unicorns first arrived, the model predicts a population of 20.5428 unicorns. Either 20 or 21 would be reasonable estimates.

d. At time $t = 100$, the model predicts a population of $p = 20.5428(1.0445)^{100} \approx 1{,}598$ unicorns.

e. We need to solve $2{,}500 = 20.5428(1.0445)^t$. This can be solved using technology or logarithms (see Chapter 11). You get $t \approx 110.3$. So the population will reach 2,500 unicorns sometime during the 111th year after the unicorns first came to the island.

SECTION EXERCISES

15–15. A submersible has been resting on the ocean floor while its crew takes water and soil samples. The craft then begins to rise toward the surface. The table shows the water pressure on the hull of the submersible as a function of its height above the ocean floor.

Height (feet)	50	100	180	240	300
Pressure (pounds per square inch)	153	132	97	71	46

a. Draw a scatter plot for the data.

b. Find the equation of the line of best fit to the data. Round your coefficients to two decimal places.

c. Based on your equation, what was the pressure on the submersible when it was on the ocean floor, to the nearest hundredth?

d. Based on your equation, how much does water pressure decrease with each additional foot of height above the ocean floor, to the nearest hundredth?

e. The pressure at the surface of the ocean is 14.7 psi. Based on your equation, how high is the surface above the ocean floor at the point where the submersible was? Round your answer to the nearest tenth of a foot.

15–16. A car tire is slowly leaking air. The table below shows the pressure in the tire in pounds/in.^{2} for several times since it began leaking.

Time (minutes)	1	3	5	8	10
Pressure (psi)	31.5	29.7	27.8	25.5	24.0

a. Write the exponential regression equation for pressure as a function of time, rounding the calculated values to the nearest hundredth.

b. Based on your equation, what will the tire pressure be after 25 minutes?

15–17. For his own math project, Ryan's brother Sean selected some cookies, measured their diameters, and counted the number of chocolate chips in each one. He then created the table below.

Diameter (cm)	**Number of Chips**
7	8
8	10
7.5	8
8	9
8.5	12
7.5	9

a. Write the equation of the quadratic regression for the number of chips y as a function of the diameter x of the cookie. Round all coefficients to the nearest hundredth.

b. Use your equation to estimate, to the nearest tenth, the number of chips a 10 cm diameter cookie would contain.

c. Use your model to find, to the nearest tenth, the diameter of the cookie that would be expected to have 11 chocolate chips in it.

15–18. Having time on his hands, Grandpa selected some brownies, weighed them, and counted the number of nuts in each one. He created the table below.

Weight (oz)	**Number of Nuts**
4	8
4.5	8
3.5	7
4	7
3	4
3.5	7

a. Write the equation of the line of best fit for the number of nuts y as a function of the weight x of the brownie. Round all coefficients to the nearest hundredth.

b. Use your equation to estimate, to the nearest tenth, the number of nuts a 2-ounce brownie would contain.

c. Grandma loves nuts (she claims she married one) and wants a brownie with a dozen nuts in it. Use your model to estimate, to the nearest tenth, the weight of the brownie she would need to find to get that many nuts.

CHAPTER EXERCISES

C15–1. Mr. Napier recorded the time in minutes it took for each of his Algebra 2 students to finish their statistics test. His data are shown in the table.

Time in Minutes	28	30	31	32	33	34	35
Number of Students	1	2	1	2	2	3	5
Time in Minutes	36	37	38	39	40	41	45
Number of Students	6	8	5	5	3	2	1

a. Find the mean of the data.
b. Find the standard deviation of the data.
c. What percent of the students finished the test within one standard deviation of the mean time?
d. What percent of the students finished the test within two standard deviations of the mean time?
e. Is Mr. Napier justified in believing that the data are normally distributed?

C15–2. The salaries of the employees of the ACME Widget Company are shown below.

Title	**Number of Employees**	**Annual Salary ($)**
President	1	250,000
Chief financial officer	1	100,000
Personnel director	1	80,000
Sales representatives	4	32,000
Assembly line forepersons	3	28,000
Assembly line workers	15	24,000
Clerical staff	3	22,000
Custodial staff	2	18,000

a. What is the mean salary at ACME?
b. What is the median salary?
c. What is ACME's total payroll (total amount of money paid annually in salaries)?

d. ACME advertises that average annual salary of its workers is over $36,500 per year. Are these ads factually correct?

e. Do these ads accurately reflect the salary a typical worker might expect to earn at ACME?

C15–3. Dr. Pete E. Atrician recorded the weights of a random sample of babies born at St. Elsewhere Hospital during the first three months of this year. His data are recorded below.

Baby's Weight (lbs)	5.3	6.1	7.8	8.2	8.5	8.9	9.3	9.6
Number of Babies	1	1	2	1	3	2	1	1

a. What is the average weight of the babies in the sample?

b. What is the sample standard deviation of the weights to the nearest hundredth?

c. Suppose the data in the table accurately reflect the weights of all babies born at St. Elsewhere. What is the probability a baby born at St. Elsewhere weighs 8 pounds or more at birth?

C15–4. A class of students is said to be homogeneous if they are all very much alike on some measure. Below are statistics for the scores achieved on the Algebra 2 final exam by Mr. Cardano's last four classes. Which group appears to have been the most homogeneous in math? Why?

(1) Three years ago: $\bar{x} = 79$; $\sigma = 9.6$

(2) Two years ago: $\bar{x} = 83$; $\sigma = 3.7$

(3) Last year: $\bar{x} = 81$; $\sigma = 12.8$

(4) This year: $\bar{x} = 85$; $\sigma = 6.2$

C15–5. On a standardized test, the mean was 78. Gomer earned a 90, which put him in the 93rd percentile of all the students who took the test. Which of the following could be the standard deviation of the scores?

(1) 8 (2) 10 (3) 12 (4) 15 (5) 18

C15–6. There are 1,300 women ages 18–24 at Elsewhere Community College (ECC); 1,240 of them are between 59.1 and 69.9 inches tall.

a. What percent of the women are between 59.1 and 69.9 inches tall?

b. Assuming the women's heights are normally distributed, find the mean and standard deviation of the heights.

c. Estimate the number of women at ECC who are 5'10" or taller.

C15–7. The refills for a mechanical pencil are supposed to be 0.5 mm in diameter. Anything smaller than 0.485 mm will not stay in the pencil. Anything larger than 0.520 mm will not fit in the pencil. The Number 2 Pencil Company makes refills that are normally distributed with a mean of 0.5 mm and a standard deviation of 0.01 mm. What percent of the refills actually work in the pencils?

C15–8. About 95% of all the doohickeys manufactured by the Normal Doohickey Co. (NDC) measure between 2.3 and 2.7 inches long. The lengths are assumed to be normally distributed.

a. Estimate the mean and standard deviation of the length of doohickeys manufactured by NDC.

b. The ACME Widget Company buys doohickeys to use in manufacturing widgets. They need doohickeys that are within 0.05 inches of 2.5 inches long. In a batch of 1,000 doohickeys from NDC, about how many will be usable in widget making?

C15–9. The table below shows the latitude and the average daily maximum temperature in April for several major North American cities.

City	Acapulco, Mexico	Miami, FL	Dallas, TX	Washington, DC
Latitude	16.9	25.8	32.9	38.9
Temperature (°F)	87	81	75	64
City	Los Angeles, CA	New York, NY	Ottawa, Canada	Juneau, AK
Latitude	33.9	40.8	45.4	58.3
Temperature (°F)	69	61	51	39

a. Draw a scatter plot for the data.

b. Find the equation of the line of best fit to the data. Round your coefficients to two decimal places.

c. Using your equation, estimate the average daily high temperature for April in San Francisco (latitude 37.7).

d. Based on your equation, at what latitude would the average daily high temperature for April be 32°F?

e. What, exactly, does the slope of your line tell you? Be specific (use numbers).

f. Based on your equation, by how much will average daily high temperature in April change for each 5° farther north in latitude?

C15–10. The table below gives the price you might expect to pay for a used car based on its age.

Age (yr)	1	2	3	4	5
Price ($)	12,031	10,955	8,985	8,017	6,929

a. Find an exponential regression model, $y = ab^x$, for the price as a function of age.

b. According to your model, what would a 6-year-old car cost?

c. According to this model, if you had only $4,000 to spend, how old a car would you be looking at?

C15–11. The table below gives stopping distances for a car traveling at different speeds under the same road conditions.

Speed (mph)	25	25	30	35	45	45	55	55	65	65
Stopping Distance (ft)	63	56	84	107	153	164	204	220	285	303

a. Find the quadratic regression equation for stopping distance as a function of speed. Round coefficients to the nearest thousandth.
b. Use your equation to predict the stopping distance for the car traveling at 40 miles per hour.
c. Use your model to determine how fast the car could be going and still stop within 250 feet.

C15–12. Cars traveling at various speeds were stopped by slamming on the brakes. The lengths of the resulting skid marks were measured, and the data are shown in the table. We wish to help police determine the speed of a vehicle by measuring the skid marks it left.

Skid Marks (ft)	28	44	40	76	97	140	145	168	206	197
Speed (mph)	25	30	30	40	45	55	55	60	65	65

a. Find a power regression, $y = ax^b$ for speed as a function of skid mark length. Round coefficients to the nearest hundredth.
b. Use your equation to estimate the speed of a car that left skid marks 256 feet long.

Answers to Exercises

1 Linear Functions

1–1. $x = -2$

1–2. $x = 3$

1–3. $x = 7.5$

1–4. $x = \dfrac{a^2}{a - 3}$

1–5. $h = \dfrac{S - \pi r^2}{2\pi r}$

1–6. $x = \dfrac{c - b}{a - c}$

1–7. $(-4, 6)$

1–8. $(-\infty,-1] \cup [5,\infty)$

1–9. $[-3,\infty)$

1–10. $[-2,\infty)$

1–11. $(-\infty,-1] \cup (5,\infty)$

1–12. $(-10,10]$

1–13.

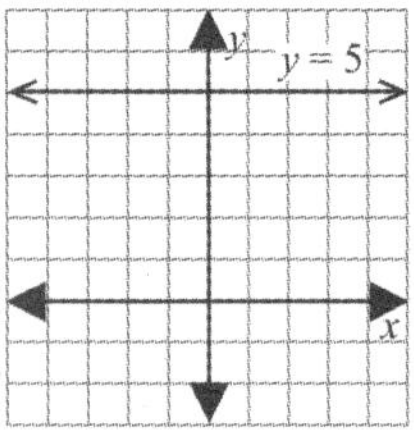

1–14.

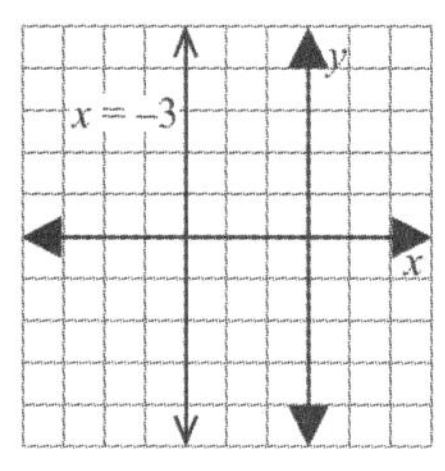

1–15.

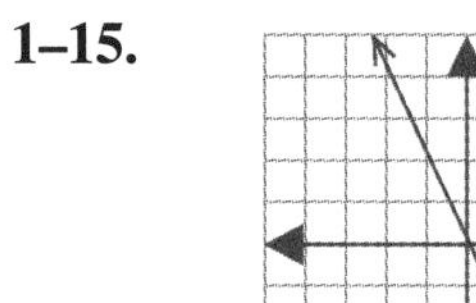
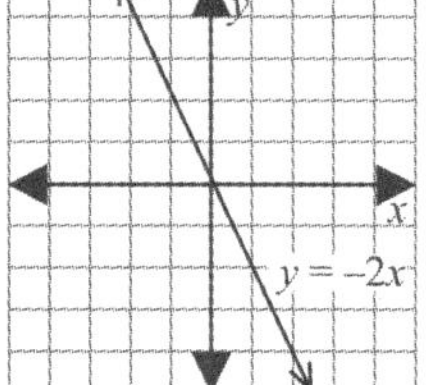

1–16.

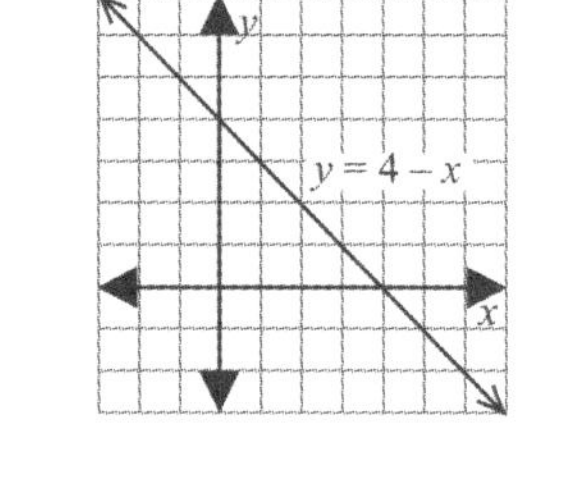

1–17. **a.** $y = -\dfrac{4}{3}x + 2$

b. slope $= -\dfrac{4}{3}$, y-intercept $= 2$

c.

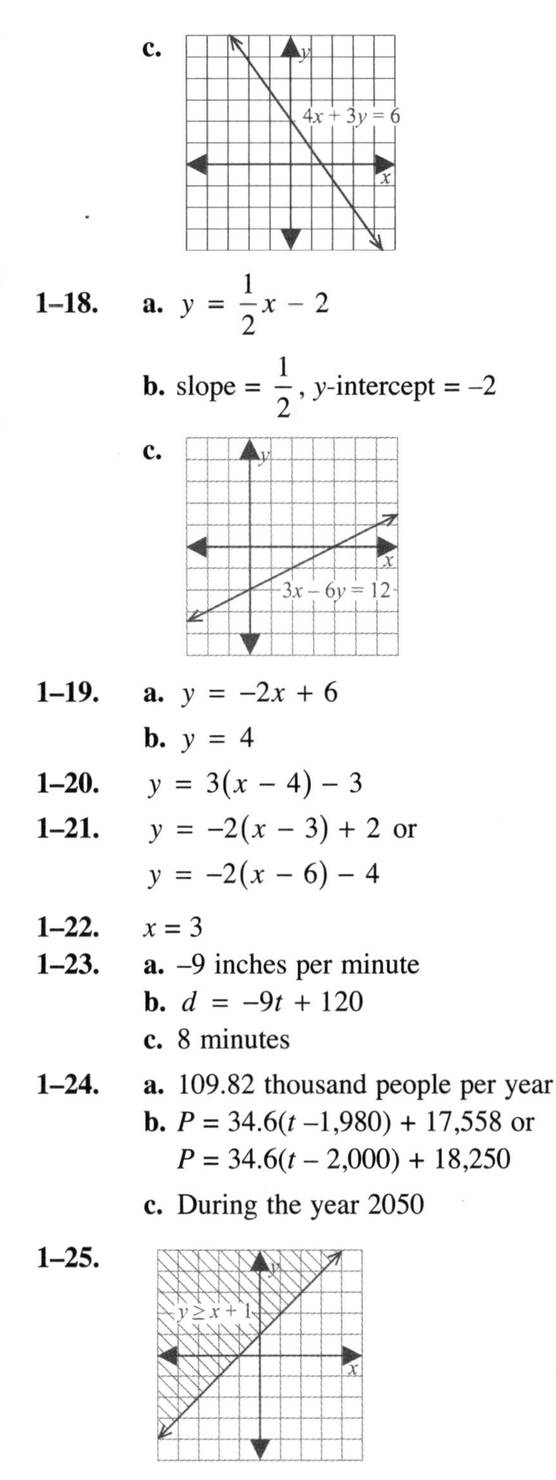

1–18. **a.** $y = \frac{1}{2}x - 2$

b. slope = $\frac{1}{2}$, y-intercept = –2

c.

1–19. **a.** $y = -2x + 6$

b. $y = 4$

1–20. $y = 3(x - 4) - 3$

1–21. $y = -2(x - 3) + 2$ or

$y = -2(x - 6) - 4$

1–22. $x = 3$

1–23. **a.** –9 inches per minute

b. $d = -9t + 120$

c. 8 minutes

1–24. **a.** 109.82 thousand people per year

b. $P = 34.6(t - 1{,}980) + 17{,}558$ or

$P = 34.6(t - 2{,}000) + 18{,}250$

c. During the year 2050

1–25.

1–26.

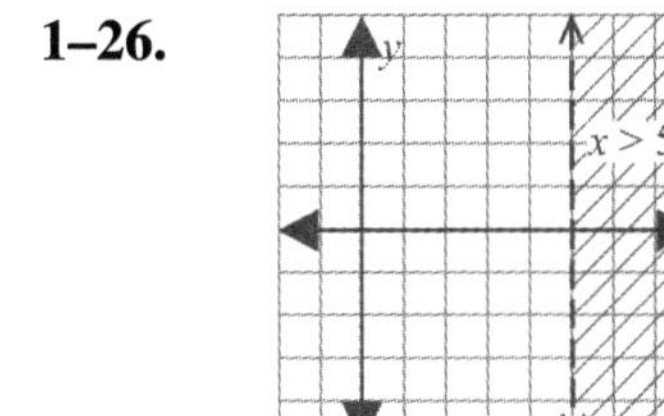

1–27.

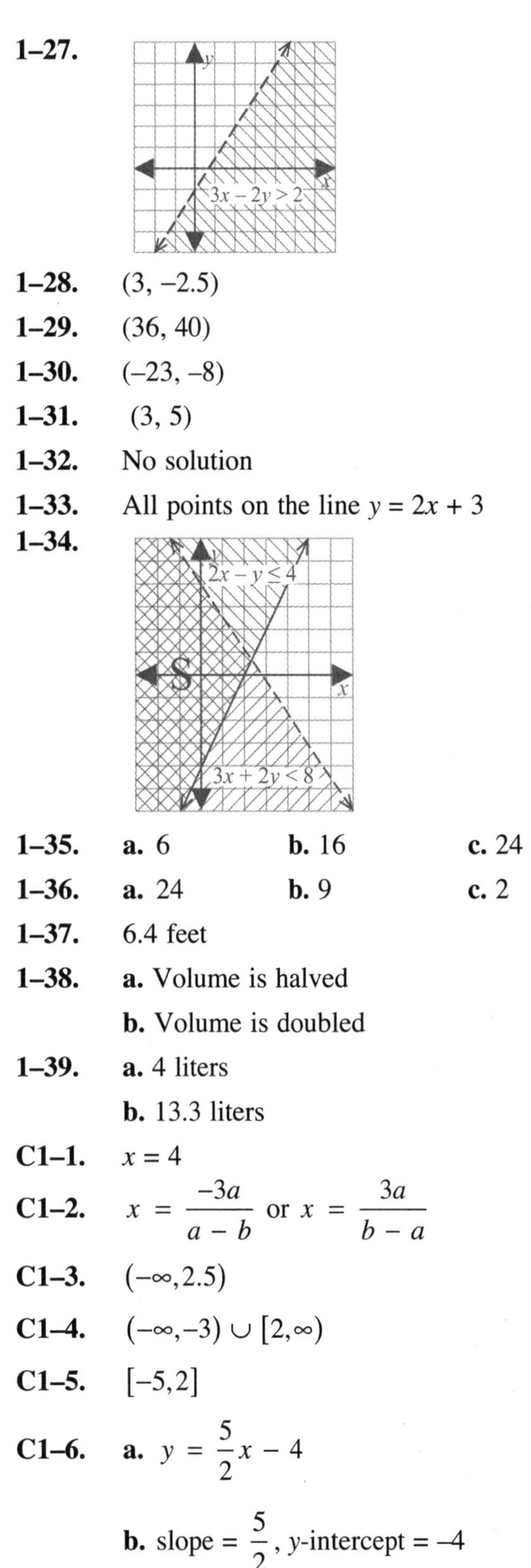

1–28. (3, –2.5)

1–29. (36, 40)

1–30. (–23, –8)

1–31. (3, 5)

1–32. No solution

1–33. All points on the line $y = 2x + 3$

1–34.

1–35. **a.** 6 **b.** 16 **c.** 24

1–36. **a.** 24 **b.** 9 **c.** 2

1–37. 6.4 feet

1–38. **a.** Volume is halved

b. Volume is doubled

1–39. **a.** 4 liters

b. 13.3 liters

C1–1. $x = 4$

C1–2. $x = \frac{-3a}{a - b}$ or $x = \frac{3a}{b - a}$

C1–3. $(-\infty, 2.5)$

C1–4. $(-\infty, -3) \cup [2, \infty)$

C1–5. $[-5, 2]$

C1–6. **a.** $y = \frac{5}{2}x - 4$

b. slope = $\frac{5}{2}$, y-intercept = –4

c.

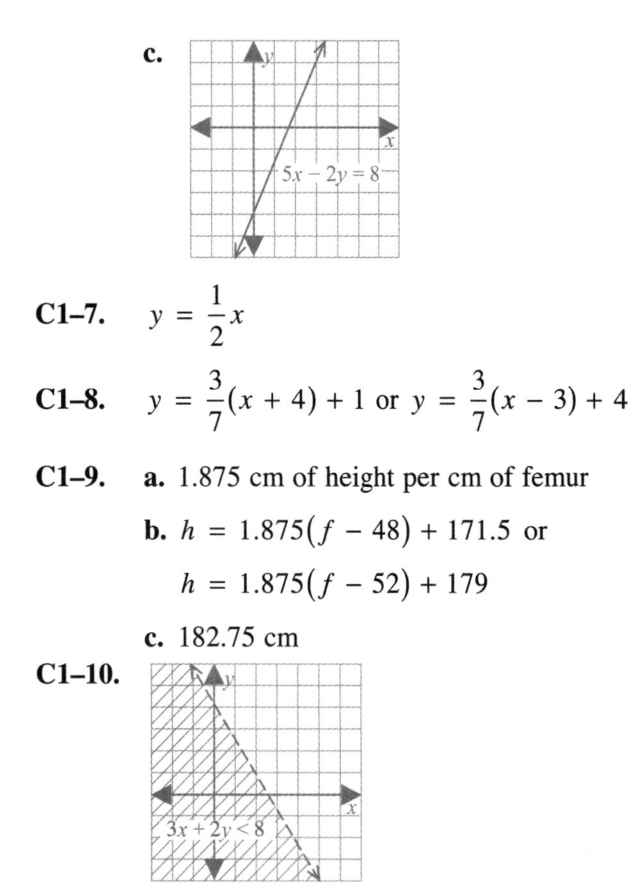

C1–7. $y = \frac{1}{2}x$

C1–8. $y = \frac{3}{7}(x + 4) + 1$ or $y = \frac{3}{7}(x - 3) + 4$

C1–9. **a.** 1.875 cm of height per cm of femur

b. $h = 1.875(f - 48) + 171.5$ or

$h = 1.875(f - 52) + 179$

c. 182.75 cm

C1–10.

C1–11. (48, 36)

C1–12. (–4, 0)

C1–13. The solution is (–2, 1).

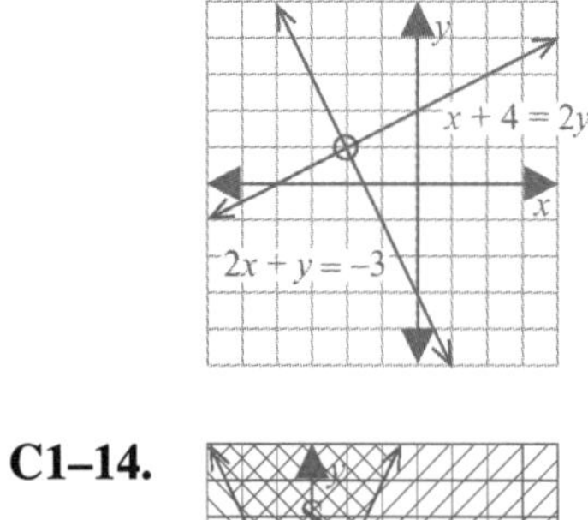

C1–14.

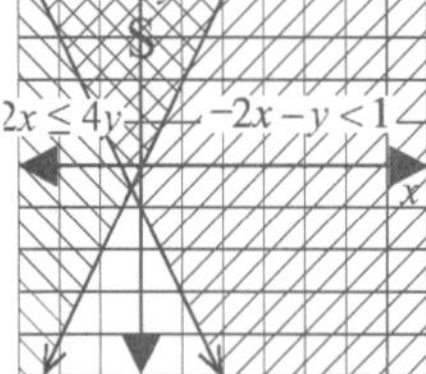

C1–15. 176 cm.

2 Polynomial Operations

2–1. $9x^8y^6$

2–2. $-6x^{4n}y^{9k}$

2–3. $\frac{3y^2}{2}$

2–4. **a.** 10^{x-1}

b. 10^{4x}

c. c^6

2–5. 0

2–6. $2x^3 - 6x^2 + 5x - 2$

2–7. $2x^3 - 13x^2 + 24x - 10$

2–8. $5y^2 - 2x^3$

2–9. $x^3 + 2x^2 - 5x - 6$

2–10. $x^2 - 3x - 5$

2–11. $2x^2 - 3x + 4$

2–12. $2x^3 + x^2 - 4x - 6 + \frac{-3}{x + 5}$

2–13. $81x^4 - 216x^3 + 216x^2 - 96x + 16$

2–14. **a.** 12
b. 125,970

2–15. $-112640x^9y^3$

2–16. $729y^6$

2–17. $6a^2b^3(6a^2 - 7ab + 3b^2)$

2–18. $(x + 3)(2x + 1)$

2–19. $(5x + 3y)(5x - 3y)$

2–20. $(x - 6y)(x + 2y)$

2–21. $5(x + 4)(x - 4)$

2–22. $(2x - 9)(x + 1)$

2–23. $3(2x - 1)(x + 5)$

2–24. $xy(x + y)(x - y)$

2–25. $(x^2 + 9)(x + 3)(x - 3)$

2–26. $(4x^2 + 3)(x + 1)(x - 1)$

2–27. $(2x + 5)(4x^2 - 10x + 25)$

2–28. $(4x - 3)(16x^2 + 12x + 9)$

2–29. $xy(5x - y)(25x^2 + 5xy + y^2)$

2–30. $(x - 2)(x + 2)(x^2 - 2x + 4)(x^2 + 2x + 4)$

2–31. $(x^2 + 4)(x - 3)$

2–32. $(3x + 1)(x - 2)(x + 2)$

C2–1. $\frac{-x^7}{y^2}$

C2–2. $1000c$

C2–3. $4x^5 + 2x^4 + 9$

C2–4. $-2x^3 + 15x^2 - 31x + 12$

C2–5. $x^2 - 2x + 3$

C2–6. $2x^3 + 10x^2 + 5x + 25$

C2–7. $x^2 - 5x + 6 + \frac{7}{x - 3}$

C2–8. $x^3 - 9x^2y + 27xy^2 - 27y^3$

C2–9. $-189x^2$

C2–10. $4x^2(2x^2 + 4x - 1)$

C2–11. $3(1 + 3x)(1 - 3x)$

C2–12. $(2x - 1)(x + 3)$

C2–13. $5x^2(3x + 2)(x - 4)$

C2–14. $(x^2 + 9)(x + 2)(x - 2)$

C2–15. $x(x + 2)(x^2 - 2x + 4)$

C2–16. $2x(x - 10)(x^2 + 10x + 100)$

C2–17. $4(x^2 + 1)(x^4 - x^2 + 1)$

C2–18. $(2x - 3)(x^2 + 5)$

C2–19. $x^2(2x^2 - 7)(x + 2)$

3 Functions and Relations

3–1. (4)

3–2. (1), (2), (3)

3–3. (3), (4)

3–4. (1), (2), (3)

3–5. **a.** 3 **b.** 15 **c.** $\sqrt{a}$ **d.** b **e.** $\sqrt{a + h}$ **f.** $x = 256$

3–6. **a.** -4 **b.** -39 **c.** $-27a^2 - 6a + 1$ **d.** $-3z^2 + 4z$ **e.** $\left\{-2, \frac{4}{3}\right\}$

3–7. **a.** 1 **b.** 2 **c.** -1.5 **d.** -2 **e.** $\{-2, 4\}$ **f.** $\{-1, 2\}$

3–8. **a.** 1 **b.** -1 **c.** 2 **d.** 1

3–9. Domain: $\{0, 1, 2, 3, 4\}$; Range: $\{3, 4, 7\}$

3–10. **a.** Domain: $(-\infty, \infty)$; Range: $[-4, \infty)$

b. Domain: $(-4, 4]$; Range: $(1, 7]$

3–11. $\{0, 1, 2, 3, 4, 5, 6, 7, 8, 9, 10, 11, 12\}$

3–12. **a.** no, $1 - 2 < 0$

b. yes

c. no, undefined

d. yes

3–13. all real numbers, $x \neq -1$, $x \neq 5$

3–14. $[1, 10) \cup (10, \infty)$

3–15. $\{y \mid 4 \leq y \leq 12\}$

3–16. **a.** y-axis; even

b. x-axis, y-axis, origin; not a function

c. line $y = x$; not a function

3–17. **a.** odd **b.** neither

3–18. **a.** even **b.** odd

3–19. **a.** $x^2 + 3x + 3$

b. $x^2 - 3x + 7$

c. $3x^3 - 2x^2 + 15x - 10$

d. $\frac{x^2 + 5}{3x - 2}$, $x \neq \frac{2}{3}$

e. $\frac{3x - 2}{x^2 + 5}$

3–20. **a.** 100
b. 28
c. $c(b(x)) = 3\sqrt{\frac{x}{200}} - 2$

3–21. **a.** reflection over the x-axis
b. reflection over the y-axis
c. shift up 3 units
d. shift right 5 units
e. shift left 4 units and down 6 units
f. horizontal shrink by factor of $\frac{1}{2}$

3–22. **a.** (1) **c.** (2)
b. (3) **d.** (4)

3–23. **a.** (2) **c.** (4)
b. (3) **d.** (1)

3–24. **a.** $y = f(-x)$
b. $y = -f(x)$
c. $y = f(x + 5)$
d. $y = f(x) + 3$
e. $y = f(x - 3) - 4$
f. $y = f\left(\frac{1}{2}\right)x$
g. $y = \frac{1}{2}f(x)$

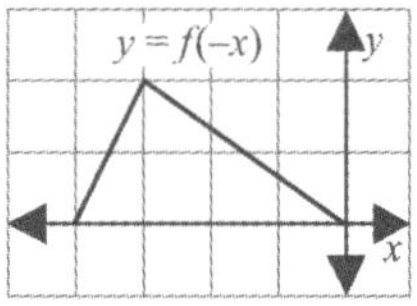

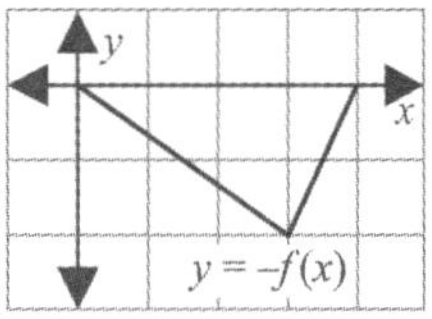

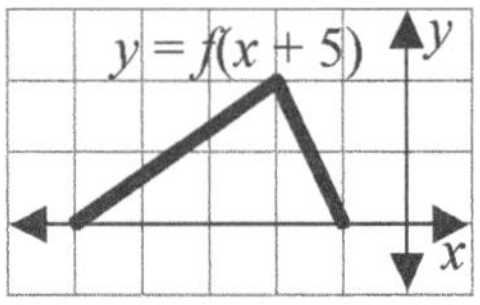

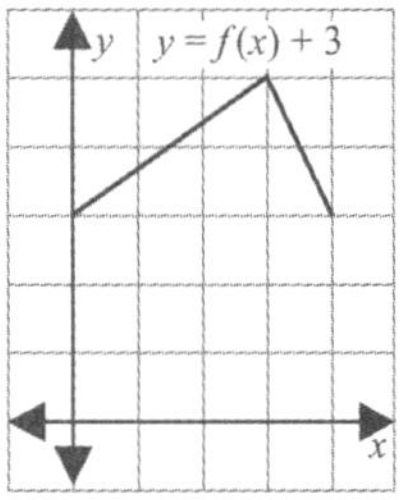

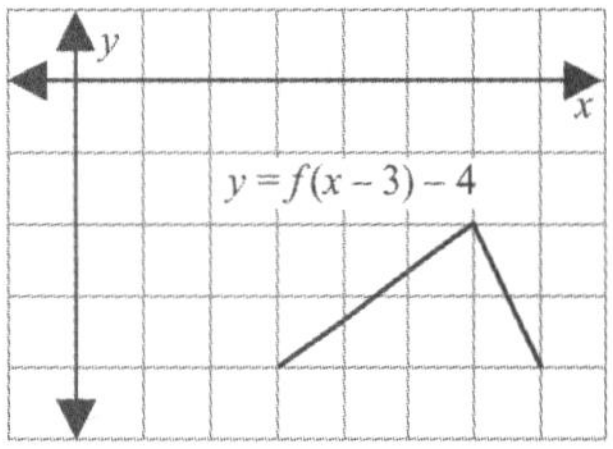

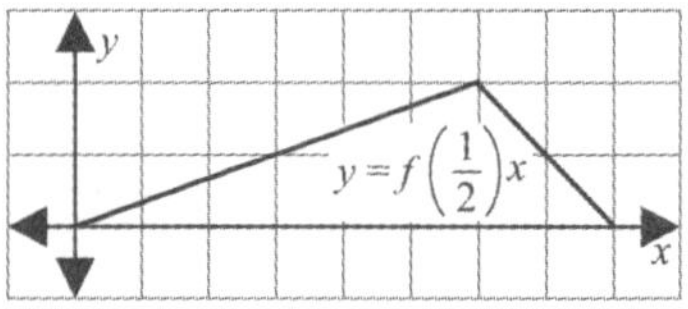

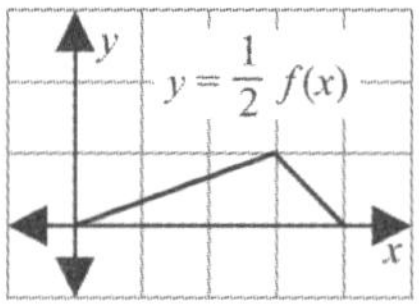

3–25. **a.** $y = (-x - 4)\sqrt{-x}$
b. $y = 2(x - 4)\sqrt{x}$
c. $y = \left(\frac{x}{3} - 4\right)\sqrt{\frac{x}{3}}$

3–26.

x	$f^{-1}(x)$
10	0
5	1
2	2
1	3

3–27. **a.** $f^{-1}(x) = 3x - 2$
b. $f^{-1}(x) = \frac{3}{2}(x + 12)$
c. $f^{-1}(x) = \frac{\sqrt{x} + 4}{2}, x \geq 0$
d. $f^{-1}(x) = (3 - x)^2 - 9, x \leq 3$
$= x^2 - 6x, x \leq 3$
e. $f^{-1}(x) = \frac{-2x}{x - 3}$ or $f^{-1}(x) = \frac{2x}{3 - 3x}$
$x \neq 3$
f. $f^{-1}(x) = \frac{36 - \sqrt{x}}{2}, x \geq 0$

3–28.

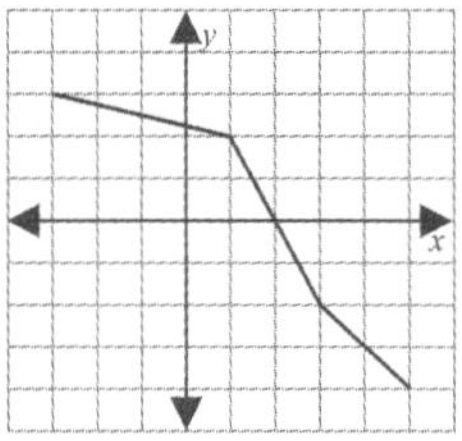

C3–1. **a.** (1), (3), (4) **b.** (3)

C3–2. **a.** When $x = 9$, $y = 3$.
b. (9, 3)
c. (3, 9)

C3–3. **a.** When $x = 5$, $f(x) = 7$.
b. 1
c. 3
d. –8
e. $x = -2$
f. all real numbers, $x \neq 1$
g. $\dfrac{4x}{x-3}$

C3–4. **a.** Domain: $(-\infty, 2] \cup [4, \infty)$; Range: $[0, \infty)$
b. Domain $[4, 5]$; Range: $(-2, 4]$

C3–5. $x \leq 8$

C3–6. $[-1,0) \cup (0,\infty)$

C3–7. $\{y \mid -4 \leq y < 12\}$

C3–8. **a.** x-axis; not a function
b. origin; odd
c. x-axis, y-axis, origin; not a function

C3–9. even

C3–10. **a.** even **c.** odd
b. odd

C3–11. **a.** 11 **c.** $2x^2 - 21$
b. 16 **d.** $4x^2 - 12x$

C3–12. **a.** reflection over the origin
b. shift left 2 units
c. vertical stretch by a factor of 3

C3–13. **a.** (3) **c.** (4)
b. (2) **d.** (1)

C3–14.

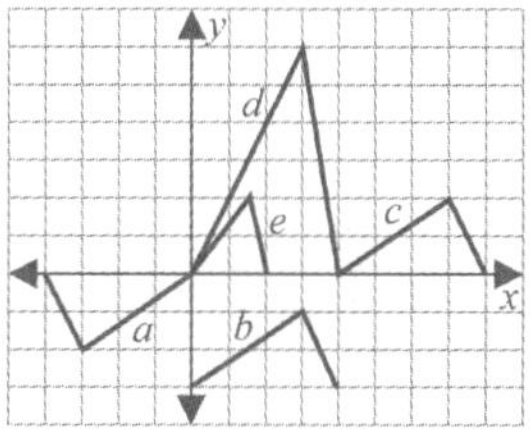

C3–15. **a.** $y = \dfrac{-2x}{x^2+1}$
b. $y = \dfrac{2(x+4)}{(x+4)^2+1} + 3$

C3–16. (–5, 2)

C3–17. **a.** $f^{-1}(x) = \dfrac{3}{x} - 2, x \neq 0$
b. $f^{-1}(x) = \left(\dfrac{3x+12}{2}\right)^2, x \geq -4$

C3–18.

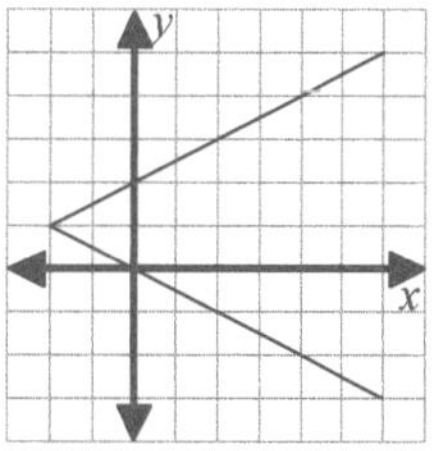

C3–19. **a.** Yes, each number of people has a unique cost.
b. {1, 2, 3, 4, 5, 6, 7, 8}
c. {30, 34.5, 40, 48, 60, 80, 120, 240}

C3–20. **a.** 5
b. –0.5
c. 2
d. (–2, 5]
e. (–2, 6]
f. Yes, each y-value comes from only one x-value (the function passes the horizontal line test).
g. 4

4 Absolute Value Functions

4–1. {–9, 4}

4–2. no solution

4–3. $\left\{\dfrac{2}{3}, 4\right\}$

4–4. 1

4–5. $-\dfrac{2}{3}$

4–6. no solution

4–7. $-8 \leq x \leq 2$

4–8. no solution

4–9. $x < -1$ or $x > 7$

4–10. $-3 \leq x \leq 1$

4–11. $x < \dfrac{4}{3}$ or $x > 18$

4–12. $x > 2$

4–13. (2)

4–14. $|T - 20| \le 2$

4–15.

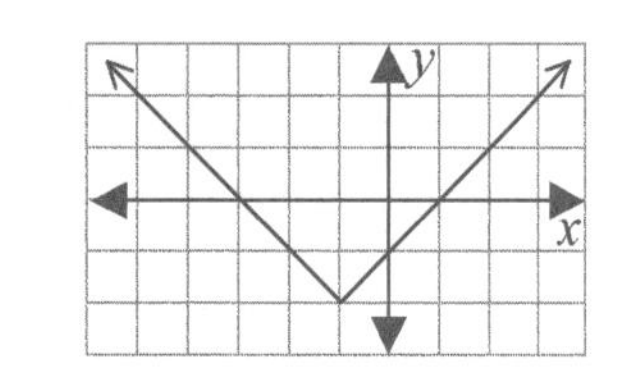

4–16.

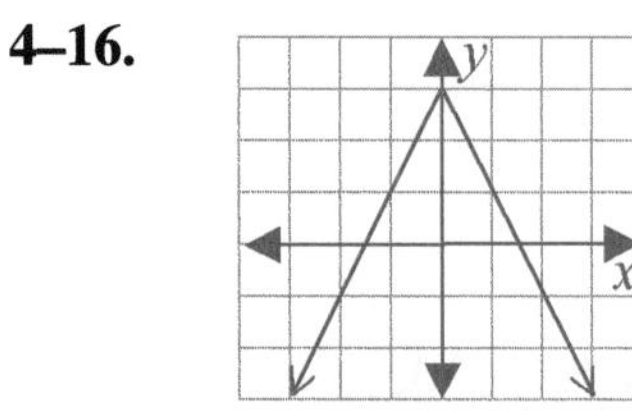

4–17.

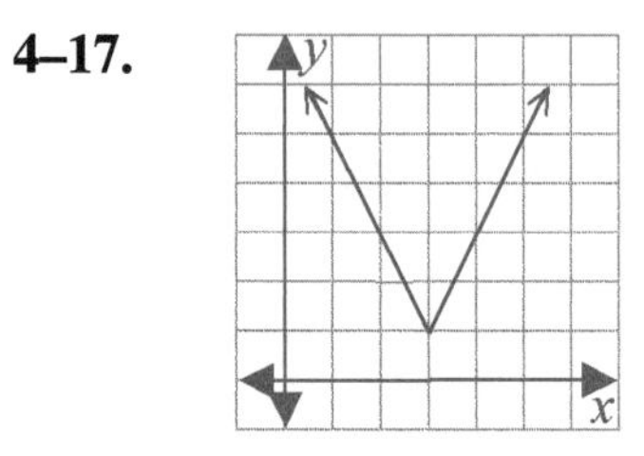

4–18.

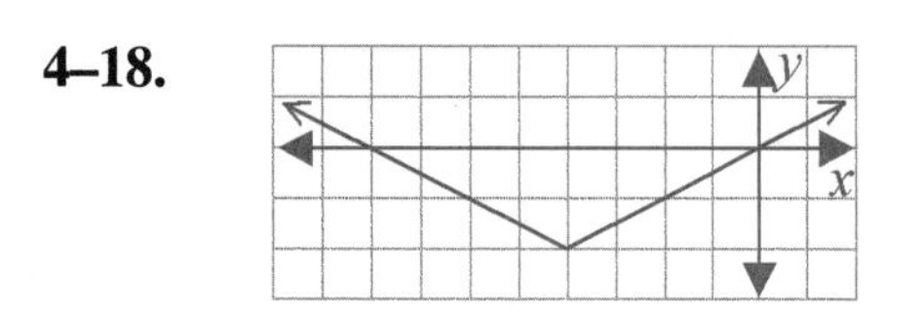

4–19. $\{-7, 1\}$

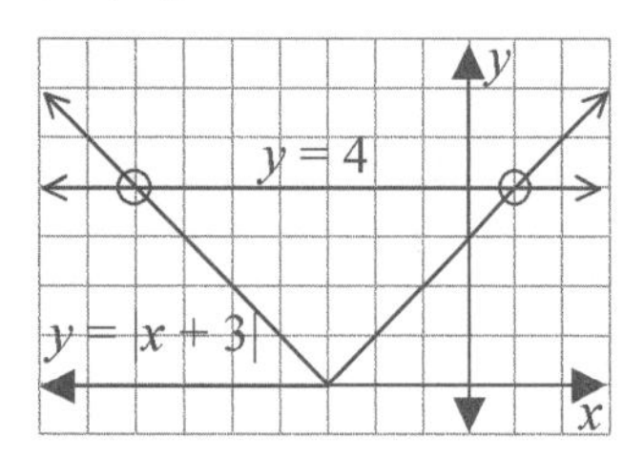

4–20. $\{-2, 4\}$

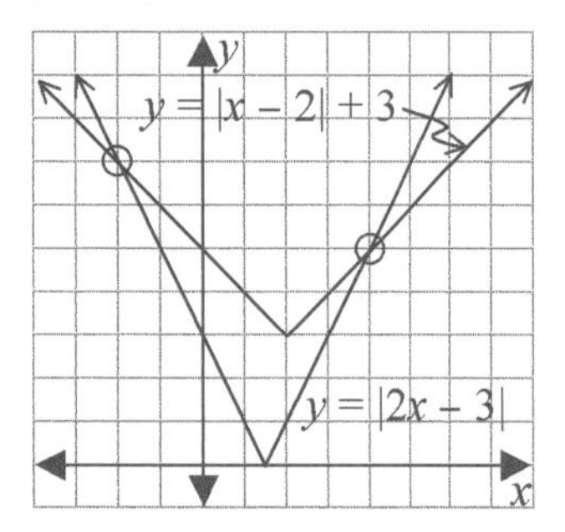

4–21. $x < -2 \text{ or } x > \dfrac{2}{3}$

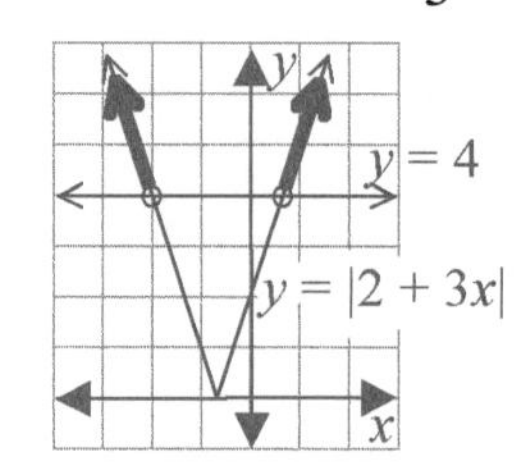

4–22. $-1.5 \le x \le 3$

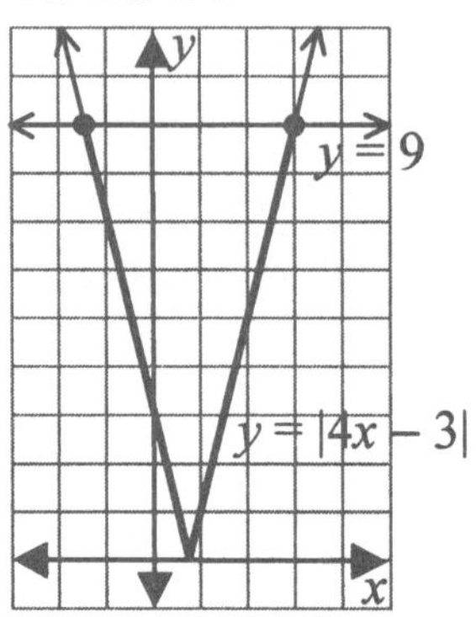

4–23. **a.** 4 **b.** 6 **c.** 2

4–24.

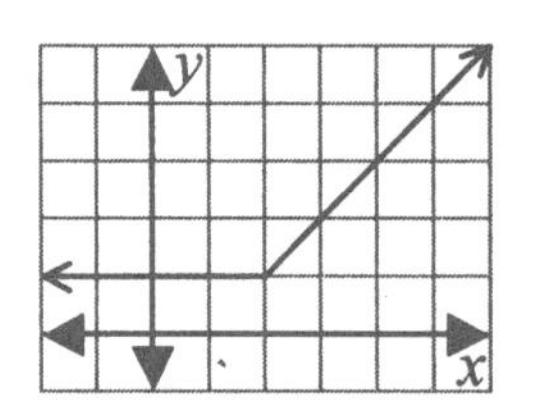

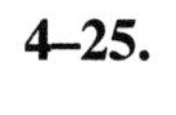

4–25.

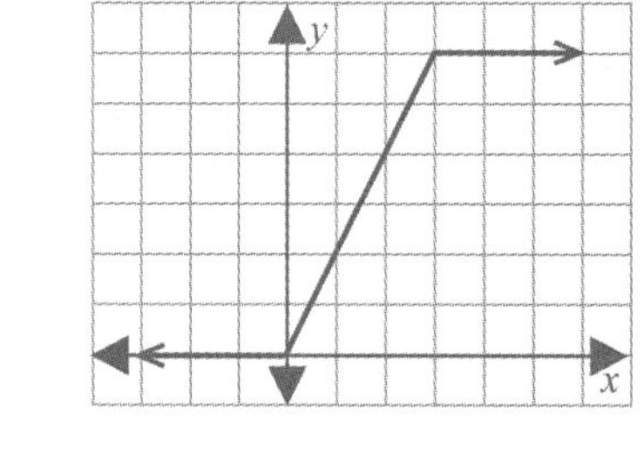

4–26. **a.** $y = \begin{cases} 15 & \text{if } 0 < x \le 300 \\ 15 + 0.05(x - 300) & \text{if } x > 300 \end{cases}$

b.

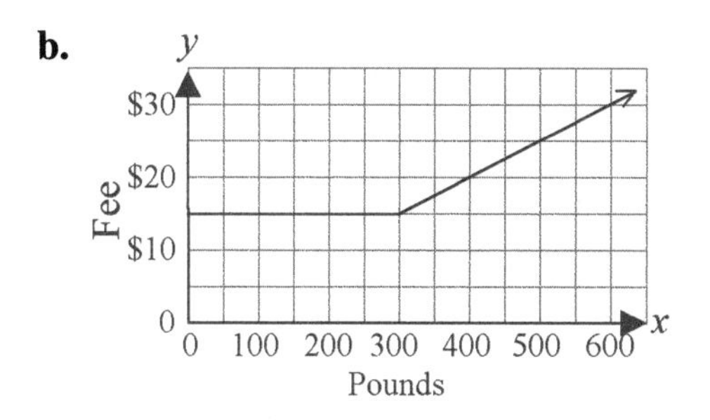

4–27.

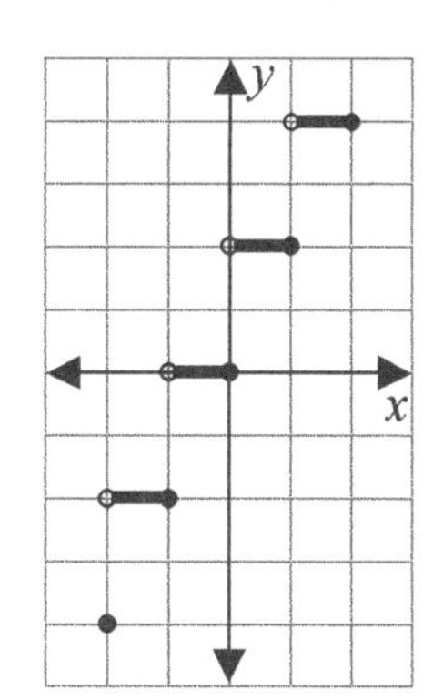

4–28.

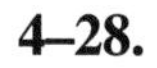

C4–1. {–5, 13}

C4–2. 15

C4–3. $x \leq -6$ or $x \geq 1$

C4–4. $2 < x < 10$

C4–5. (3)

C4–6. $|D - 15| > 0.25$

C4–7.

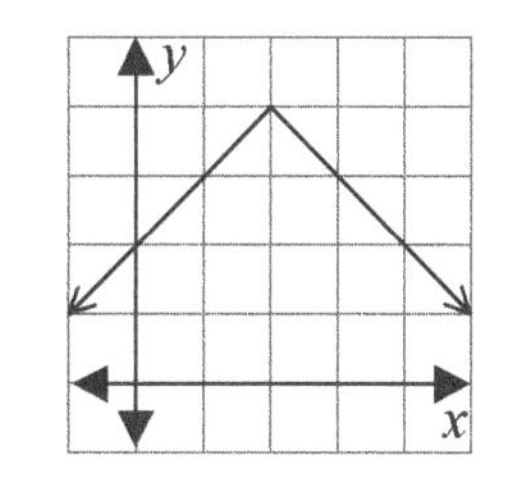

C4–8.

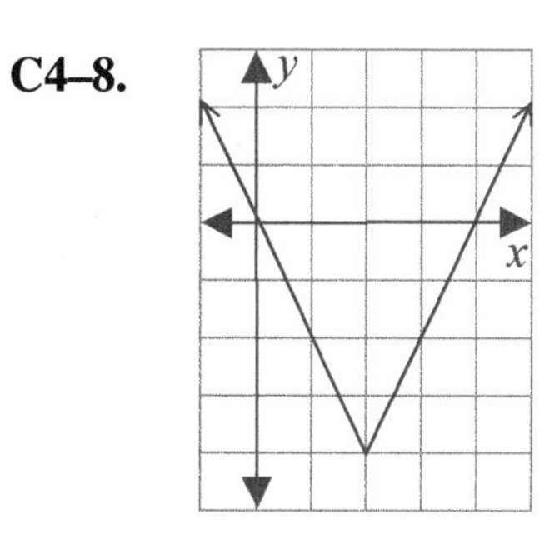

C4–9. {–4, 2}

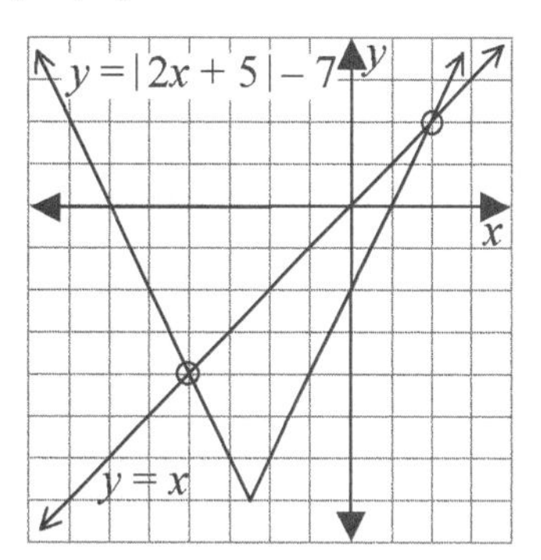

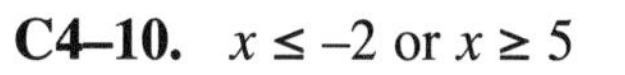

C4–10. $x \leq -2$ or $x \geq 5$

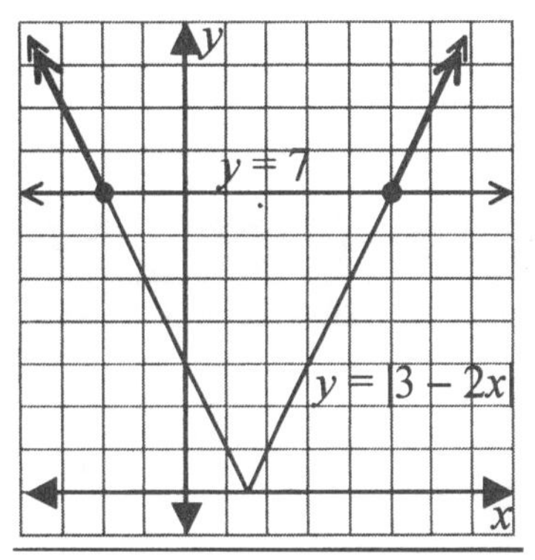

C4–11.

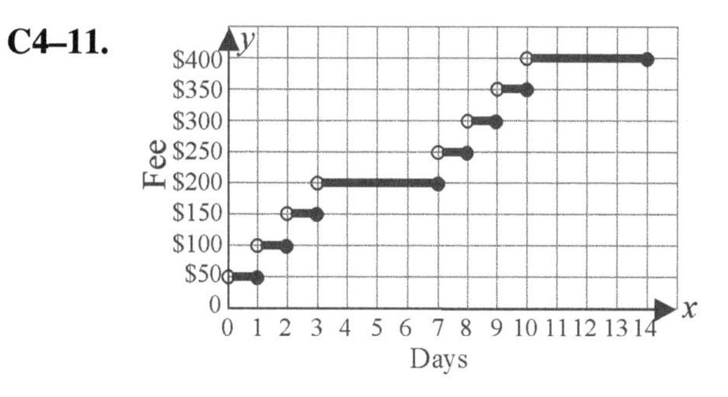

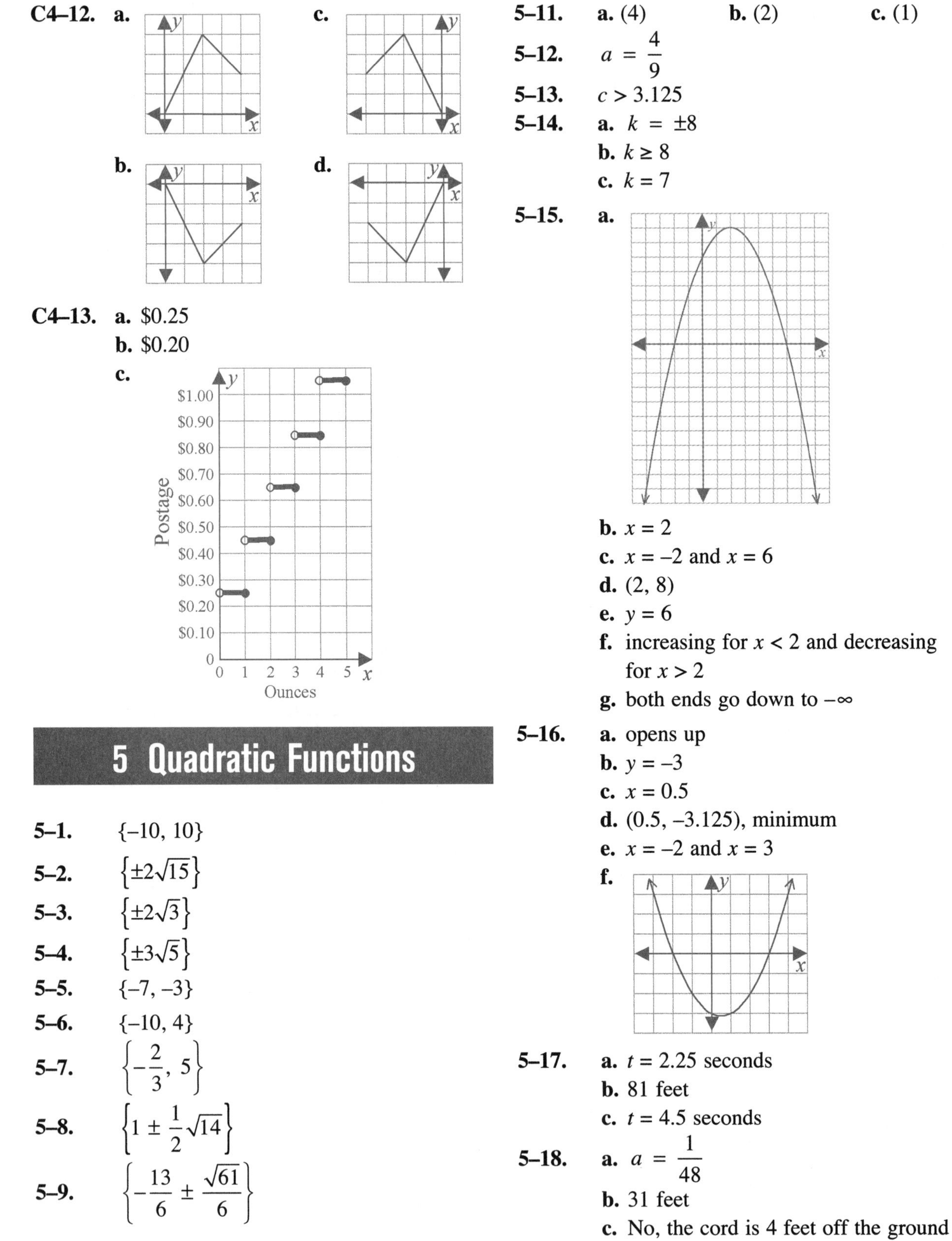

C4–12. **a.** **b.** **c.** **d.**

C4–13. **a.** \$0.25

b. \$0.20

c.

5 Quadratic Functions

5–1. $\{-10, 10\}$

5–2. $\{\pm 2\sqrt{15}\}$

5–3. $\{\pm 2\sqrt{3}\}$

5–4. $\{\pm 3\sqrt{5}\}$

5–5. $\{-7, -3\}$

5–6. $\{-10, 4\}$

5–7. $\left\{-\frac{2}{3}, 5\right\}$

5–8. $\left\{1 \pm \frac{1}{2}\sqrt{14}\right\}$

5–9. $\left\{-\frac{13}{6} \pm \frac{\sqrt{61}}{6}\right\}$

5–10. $b = 14$ and $h = 4$

5–11. **a.** (4) **b.** (2) **c.** (1)

5–12. $a = \frac{4}{9}$

5–13. $c > 3.125$

5–14. **a.** $k = \pm 8$

b. $k \geq 8$

c. $k = 7$

5–15. **a.**

b. $x = 2$

c. $x = -2$ and $x = 6$

d. (2, 8)

e. $y = 6$

f. increasing for $x < 2$ and decreasing for $x > 2$

g. both ends go down to $-\infty$

5–16. **a.** opens up

b. $y = -3$

c. $x = 0.5$

d. (0.5, –3.125), minimum

e. $x = -2$ and $x = 3$

f.

5–17. **a.** $t = 2.25$ seconds

b. 81 feet

c. $t = 4.5$ seconds

5–18. **a.** $a = \frac{1}{48}$

b. 31 feet

c. No, the cord is 4 feet off the ground at its lowest point.

5–19. **a.** 42 feet **b.** 2.5 seconds **c.** 75.75 feet **d** 6.2 seconds **e.** 5 seconds

5–20. $-\frac{25}{9}$ mph per second

5–21. 6.65 ft per mph. For each extra mph in speed, on average an extra 6.65 feet are needed to stop.

5–22. **a.** (5, –3) **b.** $x = 5$

5–23. Reflect over the x-axis; vertically shrink by a factor of $\frac{1}{2}$; and move right 5 and up 4 units.

5–24.

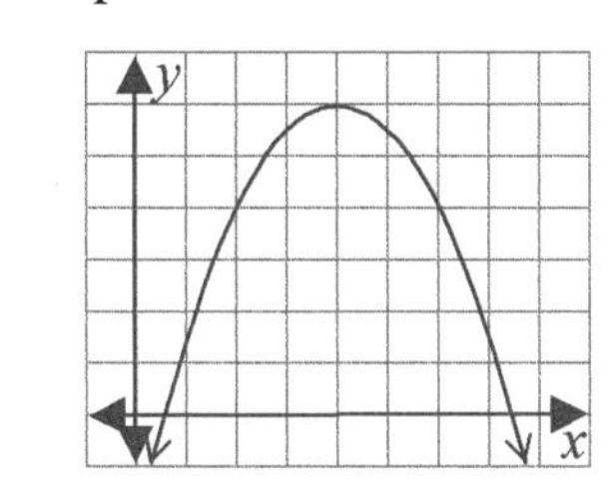

5–25. **a.** 7 feet
b. 16 ft high, 12 ft away
c. $y = -\frac{1}{16}(x - 12)^2 + 16$
d. No, at 23 feet away, the ball is only 8.4-feet high, not 10-feet high.

5–26. $y = (x - 4)^2 + 7$

5–27. $y = -3(x + 9)^2 - 13$

5–28. $y = -4x^2 + 40x - 97$

5–29. $-4 < x < 4$

5–30. $x \leq -\sqrt{10}$ or $x \geq \sqrt{10}$

5–31. All real numbers

5–32. $x = 8$

5–33. no solution

5–34. $x < -5$ or $x > 3$

5–35. $0.75 < t < 3$

5–36. $y = -\frac{1}{4}(x + 2)^2 + 6$

5–37. $y = -6x^2 + 26x + 20$

5–38. $y = -25x^2 - 70x - 31$

5–39. $y = -0.02(x - 20)^2 + 10$

5–40. {(–2, 1), (1, 4)}

5–41. {(–2, –7), (2, 1)}

5–42. {(0, 6), (4, –6)}

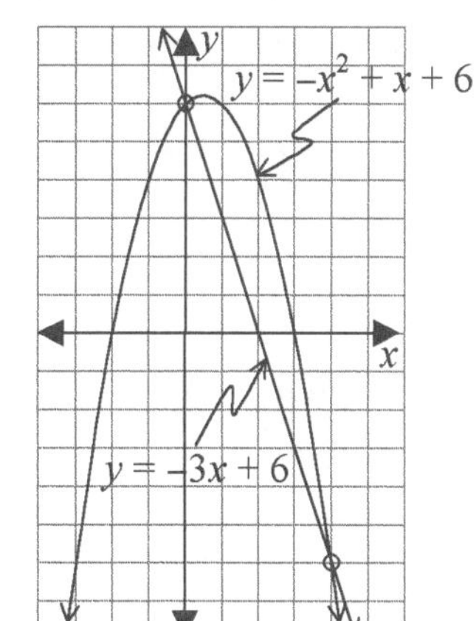

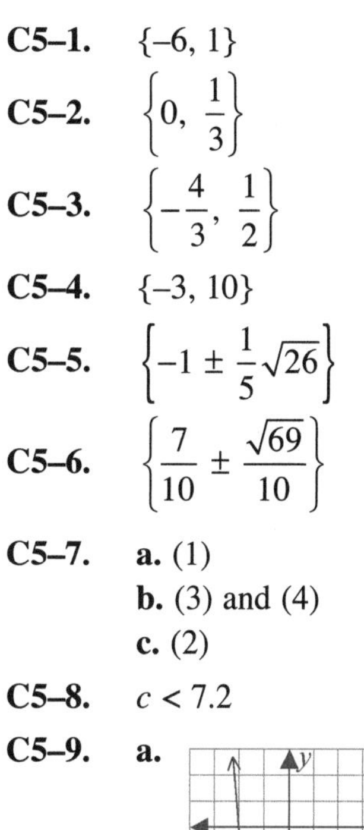

C5–1. {–6, 1}

C5–2. $\left\{0, \frac{1}{3}\right\}$

C5–3. $\left\{-\frac{4}{3}, \frac{1}{2}\right\}$

C5–4. {–3, 10}

C5–5. $\left\{-1 \pm \frac{1}{5}\sqrt{26}\right\}$

C5–6. $\left\{\frac{7}{10} \pm \frac{\sqrt{69}}{10}\right\}$

C5–7. **a.** (1)
b. (3) and (4)
c. (2)

C5–8. $c < 7.2$

C5–9. **a.**

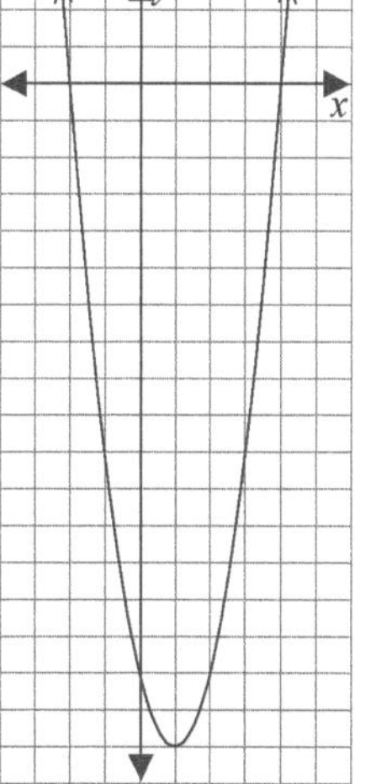

b. $x = 1$
c. $x = -2$ and $x = 4$
d. $(1, -18)$
e. $y = -16$
f. decreasing for $x < 1$ and increasing for $x > 1$
g. both ends go up to $+\infty$

C5–10. **a.** opens down
b. $y = 12$
c. $x = -0.5$
d. $(-0.5, 12.5)$ maximum
e. $x = -3$ and $x = 2$
f.

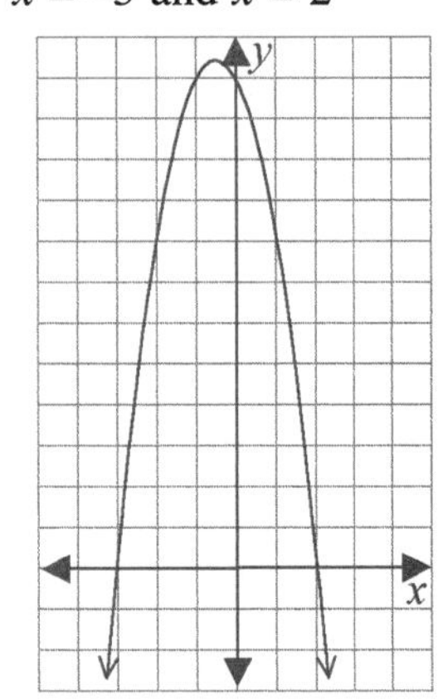

C5–11. **a.** 2 feet **c.** 25 feet
b. 27 feet **d.** 51 feet

C5–12. **a.** 2 meters per second
b. −4 meters per second
c. $y = -4(x - 1) + 10$ or $y = -4(x - 9) - 22$

C5–13. Vertically stretch by a factor of 3; move left 2 and down 7 units.

C5–14.

C5–15. $y = (x + 5)^2 - 1$

C5–16. $y = 2(x - 1)^2 - 6$

C5–17. $-3 \leq x \leq 8$

C5–18. $x < -5$ or $x > 1$

C5–19. **a.** $-1 \leq x \leq 5$
b. $x < 0$ or $x > 4$

C5–20. $y = 3(x - 10)^2 + 15$

C5–21. $y = 0.25(x + 7)(2x - 5)$

C5–22. $y = -8x^2 + 56x - 10$

C5–23. $\{(-4, -3), (6, 2)\}$

C5–24. $\{(-2, 0), (5, 7)\}$

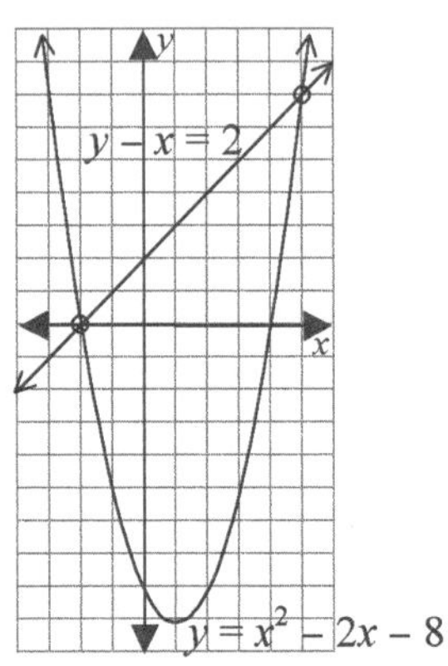

6 Complex Numbers

6–1. $9i$

6–2. $-2i\sqrt{6}$

6–3. $8i$

6–4. $10i\sqrt{3}$

6–5. 1

6–6. −6

6–7. $-i$

6–8. 0

6–9. $14i$

6–10. −6

6–11. $2i$

6–12. $7 - 4i$

6–13. $-7 + 7i\sqrt{2}$

6–14. $13 + 8i$

6–15. $7i\sqrt{3}$

6–16. 5

6–17. 36

6–18. $4 + i$

6–19. $-\frac{1}{2} - \frac{7}{4}i$

6–20. $-\frac{5}{13} - \frac{12}{13}i$

6–21. **a.** $-4 + 3i$ **b.** $4 + 3i$
c. $\frac{4}{25} + \frac{3}{25}i$

6–22. $42i$

6–23. $-7 - 2i$

6–24. $2 - 11i$

6–25. $-8i$

6–26. $\pm 2i$

6–27. $\pm \frac{9}{2}i$

6–28. $2i\sqrt{2}$

6–29. $\pm 2i$

6–30. $3 \pm 2i$

6–31. $\frac{1}{3} \pm \frac{2}{3}i$

6–32. $\frac{7}{2} \pm \frac{1}{2}i\sqrt{31}$

6–33. $1 \pm \frac{1}{2}i\sqrt{2}$

6–34. $x^2 + 25 = 0$

6–35. **a.** $2 + i\sqrt{6}$

b. $x^2 - 4x + 10 = 0$

6–36. A: $3 + 6i$, B: $4 + 0i$, C: $0 - 3i$, D: $-1 + 4i$

6–37.

C6–1. $\frac{3}{4}i\sqrt{5}$

C6–2. $2i\sqrt{5}$

C6–3. $2\sqrt{3}$

C6–4. 0

C6–5. $1 + 15i$

C6–6. $14 + 5i$

C6–7. 7

C6–8. $\frac{7}{10} + \frac{1}{10}i$

C6–9. $-1 + 2i$

C6–10. -15

C6–11. $-119 - 120i$

C6–12. $\pm 8i$

C6–13. $\pm 2i\sqrt{3}$

C6–14. $-2 \pm \frac{1}{2}i$

C6–15. $4 \pm 2i$

C6–16. $2 \pm \frac{1}{2}i\sqrt{3}$

C6–17. $x^2 - 8x + 25 = 0$

C6–18. C: $4 + 5i$, A: $0 + 3i$, T: $-1 + 0i$, S: $-2 - 3i$

C6–19.

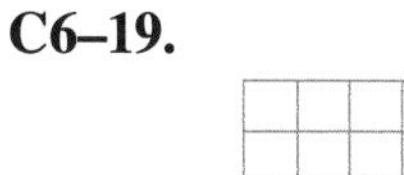

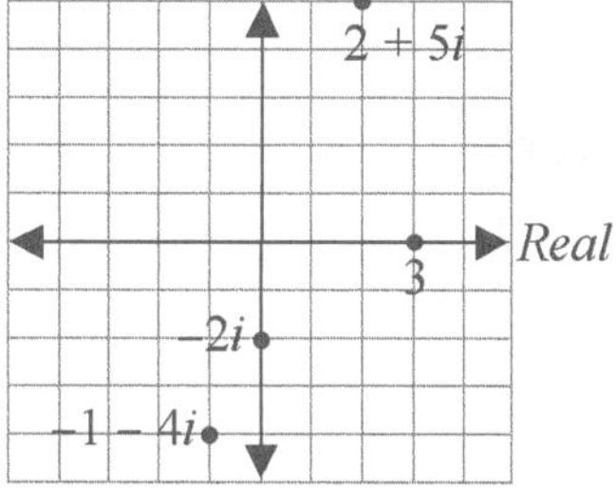

7 Polynomial Functions

7–1. **a.** 4

b. $a_n = -2$, $a_0 = -3$, $a_3 = 0$

c. 4

d. 3

e. $y = -3$

f. both ends go down to $-\infty$

7–2. y-intercept = 6, degree = 5

7–3. **a.** even

b. neither

c. odd

7–4. -10

7–5. $\{-5, -3, 0, 3\}$

7–6. $\left\{-3, -1, \frac{2}{3}, 2\right\}$;

$y = (x - 2)(x + 1)(x + 3)(3x - 2)$

7–7. $y = 2(x + 2)(x - 1)(2x - 7)$

7–8. $\left\{-\frac{1}{2}, 4, 1 \pm \sqrt{3}\right\}$;

$y = (x - 4)(2x + 1)\left(x^2 - 2x - 2\right)$

7–9. $y = (x + 4)\left(x^2 + 2x - 5\right)$

7–10. $y = (x - 2)(x + 3)\left(2x^2 + 3x - 4\right)$; $\left\{-3,\ 2,\ \dfrac{-3 \pm \sqrt{41}}{4}\right\}$

7–11. **a.** 3 **b.** $\{3, \pm 2i\}$

7–12. $y = (x - 4)\left(2x^2 - 3x + 4\right)$; $\left\{4, \dfrac{3}{4} \pm \dfrac{\sqrt{23}}{4}i\right\}$

7–13. $y = (x - 3)(x + 1)\left(x^2 + 2x + 3\right)$; $\left\{-1, 3, -1 \pm i\sqrt{2}\right\}$

7–14. $y = x(x + 5)(2x - 1)\left(2x^2 + 5\right)$; $\left\{-5,\ 0,\ \dfrac{1}{2},\ \pm\dfrac{1}{2}i\sqrt{10}\right\}$

7–15. 2

7–16.

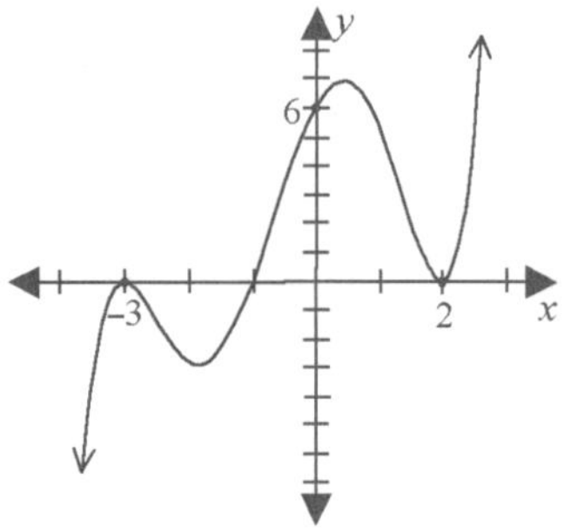

7–17. $y = \dfrac{1}{8}(x + 1)^3(x - 2)(x - 4)^2$

7–18. $\left\{\dfrac{2}{5}, \pm\sqrt{7}\right\}$

7–19. $\left\{-4,\ \dfrac{5 \pm \sqrt{17}}{2}\right\}$

7–20. $\{-3, \pm i\}$

7–21. $\left\{-2,\ 5,\ -\dfrac{1}{2} \pm \dfrac{1}{2}i\sqrt{3}\right\}$

7–22. $(-3, 0)$

7–23. $\left(-\infty, -\dfrac{5}{2}\right) \cup (4, \infty)$

7–24. $[-3, 0]$

7–25. $(-\infty, -5) \cup \left(\dfrac{1}{2}, 4\right) \cup (4, \infty)$

C7–1. **a.** 5
b. $a_n = -2$, $a_0 = -5$, $a_4 = 0$
c. 5 **d.** 4
e. $y = -5$
f. right end goes down to $-\infty$, left end goes up to ∞

C7–2. y-intercept = –60, degree = 6

C7–3. 30

C7–4. $\{-1, 0, 1, 5\}$

C7–5. $\left\{-5,\ \dfrac{1}{2},\ 1,\ \dfrac{4}{3}\right\}$; $y = (x - 1)(x + 5)(2x - 1)(3x - 4)$

C7–6. $y = -2(x + 3)(x - 4)(4x - 3)$

C7–7. $\dfrac{3 \pm \sqrt{29}}{2}$

C7–8. **a.** 6
b. 6 real and 0 complex, 4 real and 2 complex, 2 real and 4 complex, or 0 real and 6 complex
c. 5 **d.** $-1 - 3i$
e. both ends go down to $-\infty$

C7–9. **a.** 9 **b.** once

C7–10. **a.** 4 **b.** 4

C7–11. $y = -4(3x + 2)(x - 1)^2$

C7–12.

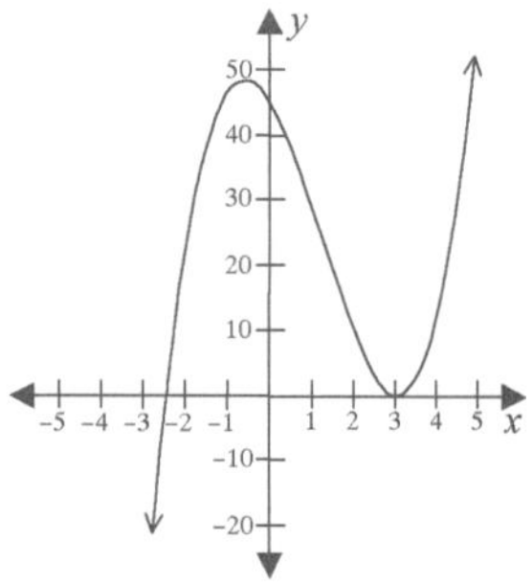

C7–13. **a.** 6 **b.** 7 **c.** 1

C7–14. $y = (x - 4)(x - 2)^2(2x + 5)$; $\left\{-\dfrac{5}{2}, 2, 4\right\}$

C7–15. $y = (x + 2)(2x - 5)\left(x^2 + x + 3\right)$; $\left\{-2,\ \dfrac{5}{2},\ -\dfrac{1}{2} \pm \dfrac{1}{2}i\sqrt{11}\right\}$

C7–16. $y = -\frac{1}{25}x^3(x+4)^2(x-3)$

C7–17. $\{-5, -3, 3\}$

C7–18. $\left\{-1, \frac{1}{2}, \pm\sqrt{2}\right\}$

C7–19. $\left\{-5, 1, 2, \frac{5}{6} \pm \frac{1}{6}i\sqrt{35}\right\}$

C7–20. $(-\infty, 0] \cup [4, \infty)$

C7–21. $(0, 1) \cup \left(\frac{3}{2}, \infty\right)$

C7–22. $(-8, 1) \cup (1, 3) \cup (7, \infty)$

8 Radical Functions

8–1. $2x\sqrt{5}$

8–2. $9a^5b^8$

8–3. $3x^2y^2\sqrt[3]{2y^2}$

8–4. $2x^2y^2\sqrt[4]{3y}$

8–5. 0

8–6. $12x^3$

8–7. $8 - 15y + 10\sqrt{5y}$

8–8. $6x^2 + 6x\sqrt{3}$

8–9. $5x\sqrt{2}$

8–10. $13x^2\sqrt[3]{2x^2}$

8–11. $30x^2y^2\sqrt[3]{9x^2}$

8–12. $3y\sqrt[3]{y}$

8–13. $-9x\sqrt[4]{2x}$

8–14. $\frac{\sqrt{30}}{15}$

8–15. $\sqrt{14}$

8–16. $\frac{3xy^3}{2}$

8–17. $\frac{\sqrt{3}+1}{2}$

8–18. $\frac{3\sqrt{5}+5}{2}$

8–19. $-16 - 7\sqrt{7}$

8–20. 1,000

8–21. 25

8–22. $8\sqrt[3]{x^2}$

8–23. $\sqrt[3]{x^3 - 8}$

8–24. $-2x^{\frac{2}{3}} + 9y^{\frac{5}{4}}$

8–25. $12\left(x^2 - 5\right)^{\frac{1}{3}}$

8–26. $27x^9y^3$

8–27. $\frac{y}{3x}$

8–28. shift left 4 and down 1 unit

8–29. reflect over the y-axis

8–30. reflect over the x-axis and a vertical dilation by a factor of $\frac{1}{2}$

8–31.

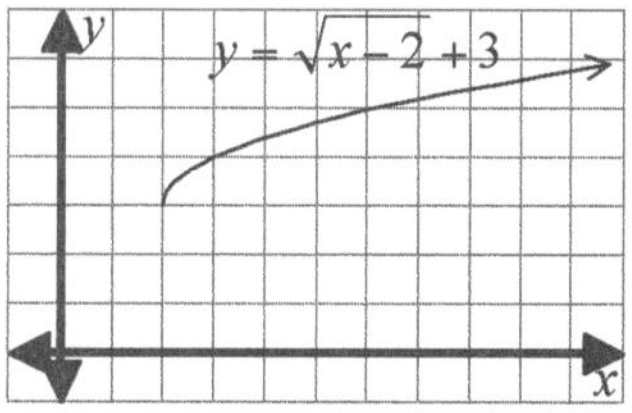

$y = \sqrt{x-2} + 3$

8–32.

$y = \sqrt[3]{x+4} - 1$

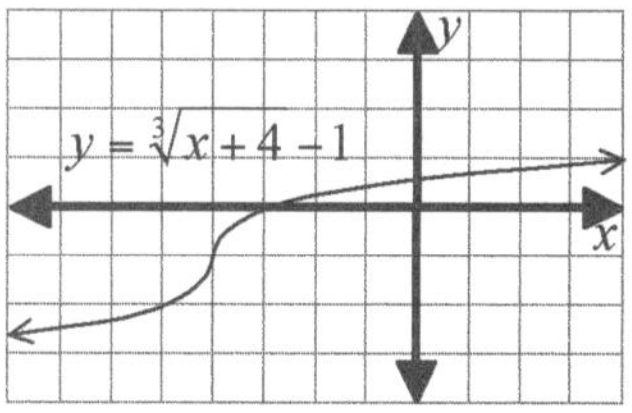

8–33. $y = -\sqrt{x+6} + 5$

8–34. $x = -13$

8–35. no solution

8–36. $x = 5$ only

8–37. $x = -4$ only

8–38. $x = 81$

8–39. $x = 27$

8–40. 893.7 meters

8–41. $f^{-1}(x) = 4x^2 - 5, x \geq 0$

8–42. $f^{-1}(x) = \dfrac{(x+1)^2}{25}, x \geq -1$

8–43. $f^{-1}(x) = (x+1)^3 - 4$

C8–1. $6xy^3\sqrt{3}$

C8–2. $2x^2y^3\sqrt[3]{6x^2}$

C8–3. $25 + 12x - 20\sqrt{3x}$

C8–4. $-15x^4y^3\sqrt{2}$

C8–5. $-3x^3\sqrt{6y}$

C8–6. $3y\sqrt{5y}$

C8–7. $4x^2\sqrt[3]{2x}$

C8–8. $5x\sqrt[3]{6}$

C8–9. $\dfrac{\sqrt{10}}{4}$

C8–10. $5x^4y^2$

C8–11. $21 - 7\sqrt{10}$

C8–12. $\dfrac{5 + 3\sqrt{3}}{2}$

C8–13. $5\sqrt{xy}$

C8–14. $5\sqrt{\left(x^2 - 4\right)^3}$

C8–15. $3(1 - 2x)^{\frac{1}{2}}$

C8–16. $10x^{\frac{3}{4}}$

C8–17. $\dfrac{2\sqrt[3]{r^2}}{s}$

C8–18. reflect over the x-axis and vertical dilation by a factor of 3

C8–19. shift right 5 and up 2 units

C8–20.

$y = \sqrt{x+1} - 4$

C8–21.

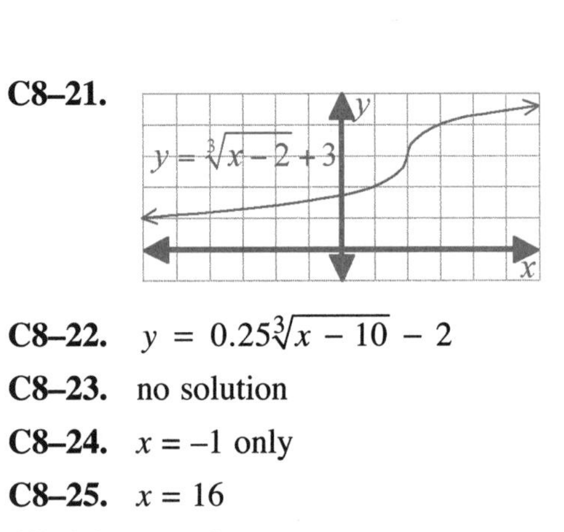

C8–22. $y = 0.25\sqrt[3]{x - 10} - 2$

C8–23. no solution

C8–24. $x = -1$ only

C8–25. $x = 16$

C8–26. 20.3 feet

C8–27. $f^{-1}(x) = (x - 3)^2 + 2, \; x \geq 3$

C8–28. $f^{-1}(x) = \dfrac{(x-4)^3}{8} + 3$

9 Rational Functions

9–1. $\{-3, 4\}$

9–2. $\{0, -2\}$

9–3. $\dfrac{-(y+3)}{4}$

9–4. $\dfrac{4x}{x - 4}$

9–5. $\dfrac{-1}{a^2}$

9–6. $\dfrac{-2}{x}$

9–7. $\dfrac{7}{3(x-1)}$

9–8. $\dfrac{3(a+2)}{2a}$

9–9. $\dfrac{y + x}{2y}$

9–10. $\dfrac{x + 2}{x - 2}$

9–11. $a - b$

9–12. **a.** $\dfrac{10x^2 + 55x - 56}{2x(x-1)}$ **b.** $\dfrac{18}{x - 1}$

9–13. $\dfrac{y^5}{4x^3}$

9–14. $\dfrac{2}{3x^5}$

9–15. $\dfrac{3y^2z^{\frac{1}{2}}}{x}$

9–16. $\dfrac{9x^2}{4}$

9–17. $\dfrac{3}{8}$

9–18. -1

9–19. $\dfrac{x}{x-1}$

9–20. $\dfrac{x-1}{x^2}$

9–21. $f(x) = \dfrac{24(x-3)}{(x-2)(x+3)^2}$

9–22. $x = a$ is a root; $x = b$ and $x = d$ are vertical asymptotes; $x = c$ and $x = e$ are neither.

9–23. $g(x) = 2x - 3 + \dfrac{4x-7}{x^2-4}$

9–24. y-intercept: $y = -\dfrac{1}{4}$
root: $x = 1$
vertical asymptote: $x = 2$, odd; $x = -2$, odd
end behavior: $y = 0$ is a horizontal asymptote; on the right, the graph is below; on the left, the graph is above.

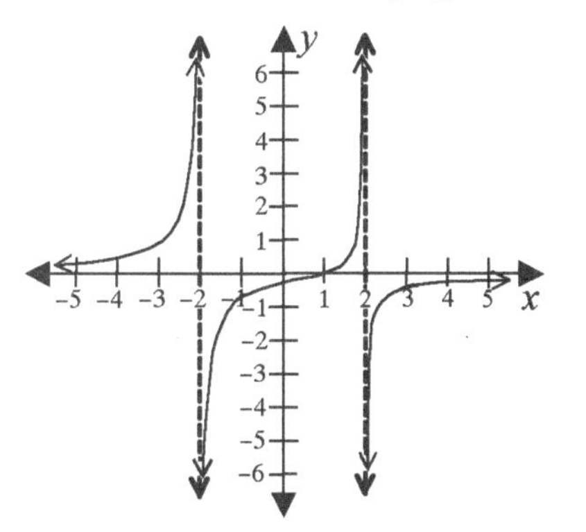

9–25. y-intercept: $y = -1$
root: $x = -2$
vertical asymptote: $x = 2$, odd; $x = -5$, even
end behavior: $y = 0$ is a horizontal asymptote; both ends of the graph, left and right, are above the asymptote.

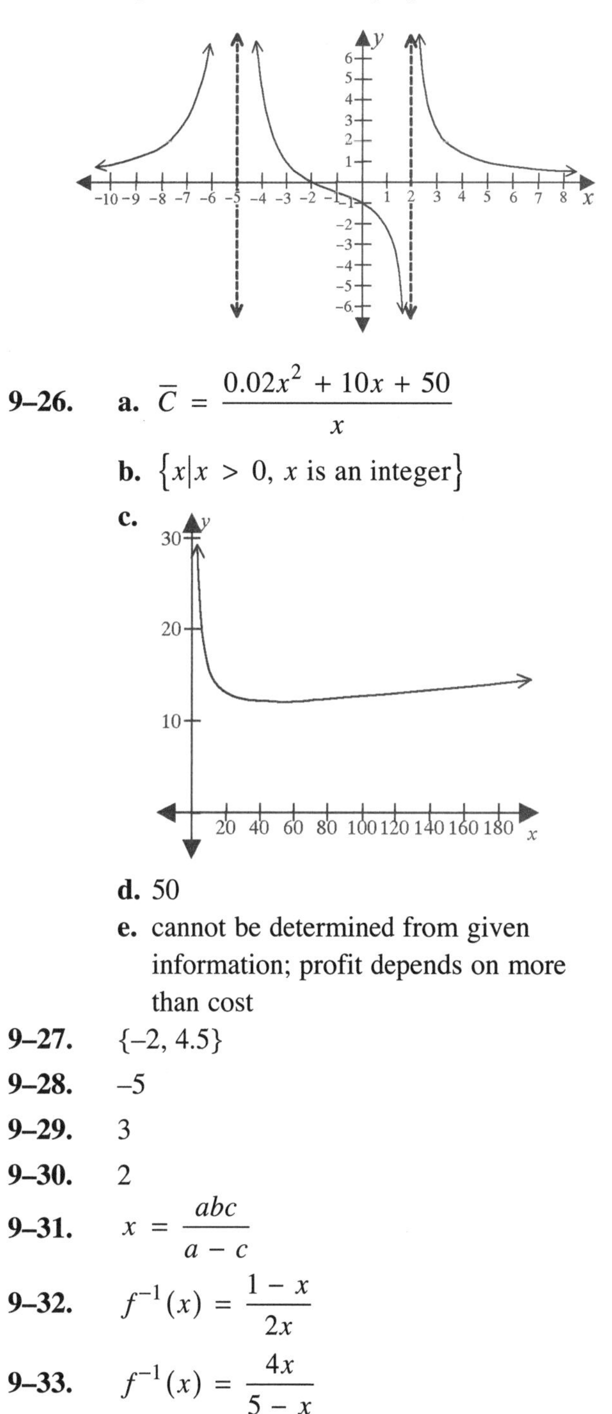

9–26. **a.** $\overline{C} = \dfrac{0.02x^2 + 10x + 50}{x}$

b. $\{x \mid x > 0,\ x \text{ is an integer}\}$

c.

d. 50

e. cannot be determined from given information; profit depends on more than cost

9–27. $\{-2, 4.5\}$

9–28. -5

9–29. 3

9–30. 2

9–31. $x = \dfrac{abc}{a-c}$

9–32. $f^{-1}(x) = \dfrac{1-x}{2x}$

9–33. $f^{-1}(x) = \dfrac{4x}{5-x}$

9–34. $(0, 3)$

9–35. $[-6,-5) \cup [1,\infty)$

9–36. $(-9,0) \cup (0,4)$

9–37. $(-4,2] \cup (5,8]$

C9–1. $\{-2, 0, 2\}$

C9–2. $\dfrac{-(1+a)}{a-3}$

C9–3. $\dfrac{2x}{(x-1)(2-x)}$

C9–4. $\dfrac{x-1}{x}$

C9–5. $\dfrac{1}{x-4}$

C9–6. $\dfrac{-x}{x+2}$

C9–7. $\dfrac{y^2-x^2}{xy}$

C9–8. $\dfrac{25x^4}{9}$

C9–9. $\dfrac{y^5}{4x^3}$

C9–10. 6.5

C9–11. y-intercept: none
roots: $x = -1$, $x = 2$
vertical asymptote: $x = 0$, odd; $x = 4$, even
end behavior: $y = 0$ is a horizontal asymptote; on the right, the graph is above; on the left, the graph is below

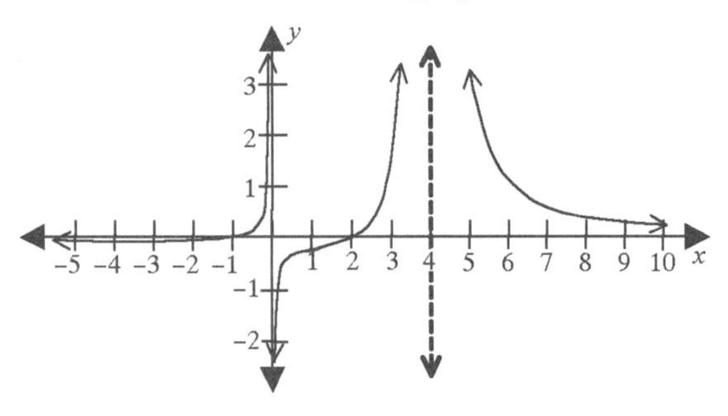

C9–12. y-intercept: $y = 0$
roots: $x = 0$, $x = 5$
vertical asymptote: $x = 2$, odd;
end behavior: $y = 0$ is a horizontal asymptote; on the right, the graph is below; on the left, the graph is below

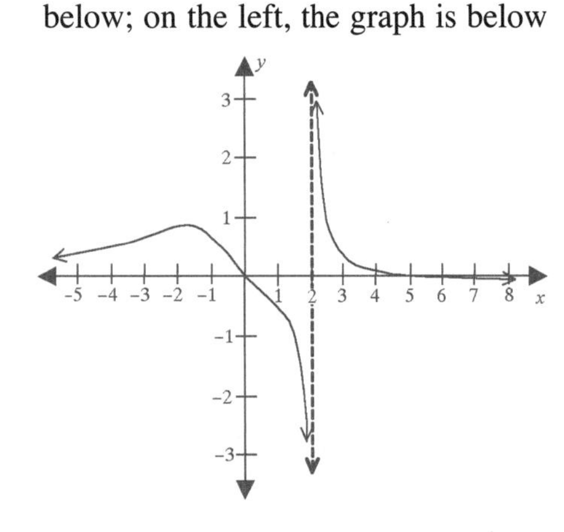

C9–13. $f(x) = \dfrac{-75(x-1)}{(x+3)(x-5)^2}$

C9–14. $f(x) = 2 + \dfrac{-11x - 10}{x^2 + 4x + 4}$

C9–15. –7

C9–16. –8

C9–17. $x = \dfrac{ab}{bc - 1}$

C9–18. $f^{-1}(x) = \dfrac{2}{x} - 3$

C9–19. $f^{-1}(x) = \dfrac{4x+1}{1-x}$

C9–20. **a.** 50 deer; this makes sense since it is the number introduced
b. $\approx$ 167 deer
c. 25 years
d. 750 deer

C9–21. $[-1,3) \cup (3,4]$

C9–22. $[-5,-3) \cup (0,2] \cup (3,\infty)$

10 Exponential Functions

10–1. **a.** linear
b. exponential
c. neither

10–2. No. When $x = 0$, $4(2)^0 = 4$ and $8^0 = 1$.

10–3. **a.** 7.9 billion
b. 0.084 billion people per year

10–4. 88.2°F

10–5. **a.** $\{x \mid x \in \mathbb{R}\}$
b. $\{y \mid y > 0\}$
c. 1
d. none
e. $a > 1$
f. $0 < a < 1$
g. reflection over the y-axis
h. translation 2 units down
i. translation 2 units right
j. vertical dilation by factor of 2

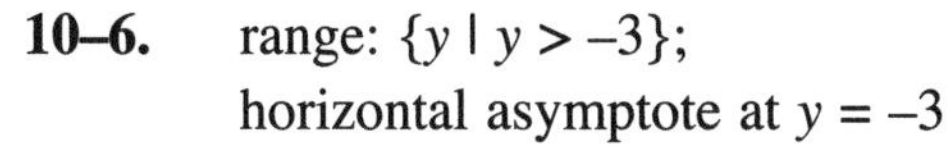

10–6. range: $\{y \mid y > -3\}$; horizontal asymptote at $y = -3$

10–7. **a. and c.**

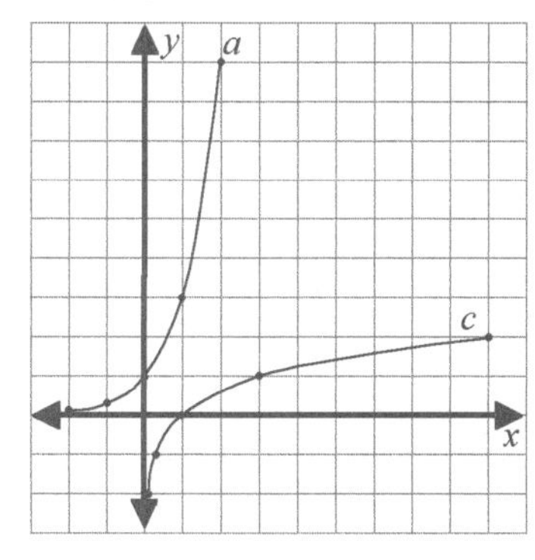

b. ≈ 5.2

10–8. **a.**

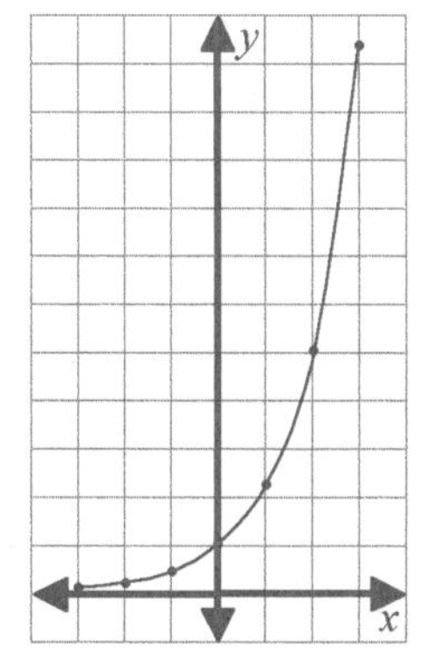

b. ≈ 1.7 **c.** ≈ 2.2

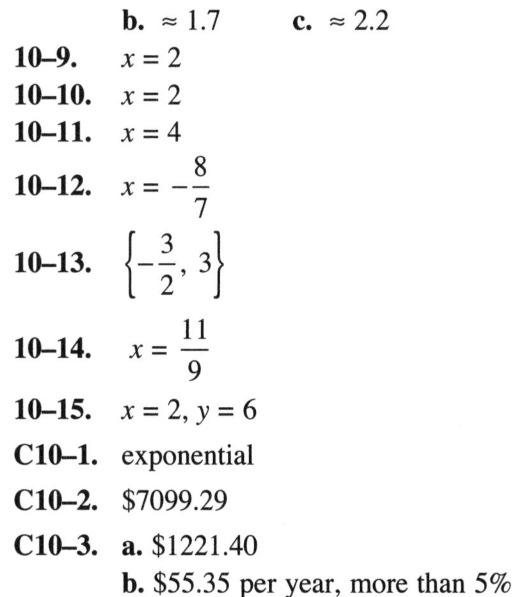

10–9. $x = 2$

10–10. $x = 2$

10–11. $x = 4$

10–12. $x = -\frac{8}{7}$

10–13. $\left\{-\frac{3}{2}, 3\right\}$

10–14. $x = \frac{11}{9}$

10–15. $x = 2, y = 6$

C10–1. exponential

C10–2. \$7099.29

C10–3. **a.** \$1221.40
b. \$55.35 per year, more than 5%

C10–4. **a., b., and d.**

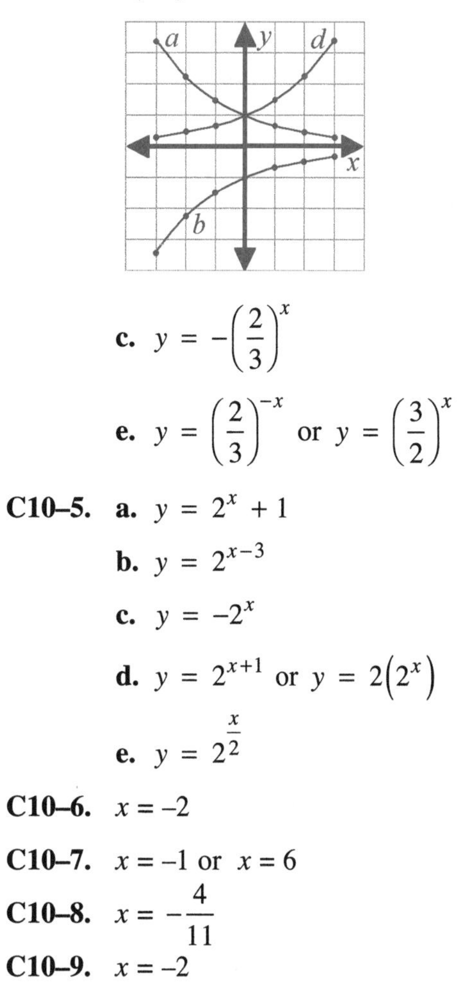

c. $y = -\left(\frac{2}{3}\right)^x$

e. $y = \left(\frac{2}{3}\right)^{-x}$ or $y = \left(\frac{3}{2}\right)^x$

C10–5. **a.** $y = 2^x + 1$

b. $y = 2^{x-3}$

c. $y = -2^x$

d. $y = 2^{x+1}$ or $y = 2\left(2^x\right)$

e. $y = 2^{\frac{x}{2}}$

C10–6. $x = -2$

C10–7. $x = -1$ or $x = 6$

C10–8. $x = -\frac{4}{11}$

C10–9. $x = -2$

11 Logarithmic Functions

11–1. $x = \log_4 y$

11–2. $k = \log_b N$

11–3. $2^y = x$

11–4. $a^k = x$

11–5. $f^{-1}(x) = \log_6 x$

11–6. $f^{-1}(x) = e^x$

11–7. 5

11–8. $-2x$

11–9. $3\log x - \log y - 2\log z$

11–10. $3(\log x + \log y)$

11–11. $3 + 3\log x$

11–12. $2\ln x - \frac{1}{2}\ln(x^2 - 1)$

11–13. $\ln\left(\frac{x^2}{3}\right)$

11–14. $\log\left(\sqrt{x}(x-1)^2\right)$

11–15. $\log\left(\frac{x^3\sqrt{y}}{10}\right)$

11–16. $\log\left(\frac{a^2b^2}{c}\right)$

11–17. (2)

11–18. **a.** 1.6 **b.** −1

11–19. **a.** $y = \log_3(-x)$

b. $y = \log_3(x+2)$

c. $y = -\log_3 x$

11–20. domain: $x > 2$
range: all real numbers
vertical asymptote: $x = 2$
y-intercept: none
x-intercept: $x = 3$

11–21. domain: $x < 8$
range: all real numbers
vertical asymptote: $x = 8$
y-intercept: $y = \ln 8 \approx 2.1$
x-intercept: $x = 7$

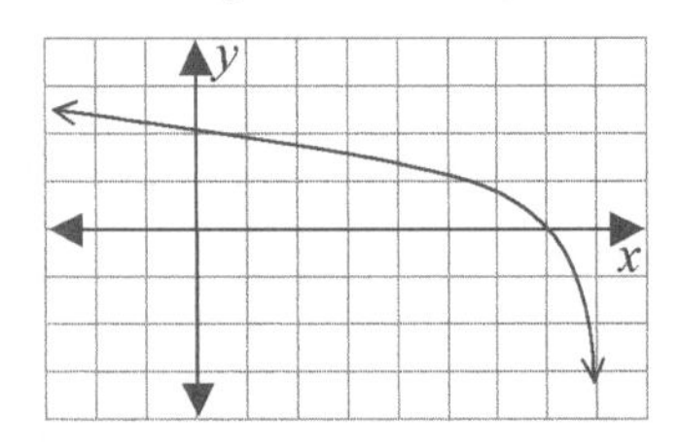

11–22. (1, 0)

11–23. $0 < x < 1$

11–24. (4)

11–25. 1.544

11–26. −2.524

11–27. 211.933

11–28. 11.610

11–29. 2.236 or $\sqrt{5}$

11–30. 0.045

11–31. 9.965

11–32. 2

11–33. $\left\{\frac{1}{4}, \frac{1}{2}\right\}$

11–34. 31

11–35. 13.9

11–36. $x > -3$

11–37. **a.** $\frac{\ln 64}{\ln 2}$ **b.** $\frac{\ln 0.6}{\ln 10}$

11–38. $\frac{\log 72}{\log 6}$

C11–1. $4 = \log_5 625$

C11–2. $x^n = y$

C11–3. $f^{-1}(x) = 8^x$

C11–4. $f^{-1}(x) = \ln x$

C11–5. $2x - 1$

C11–6. $f(x)$

C11–7. $\frac{1}{2}(\log x + \log y) - \log z$

C11–8. $\frac{1}{2}\log x - 2$

C11–9. $\log\left(\frac{a^3}{\sqrt{b}}\right)$

C11–10. $\log\left(\frac{x^3}{\sqrt{yz}}\right)$

C11–11. $\log\left(10x^2\right)$

C11–12. (4)

C11–13. 1.4

C11–14. **a.** $y = \log_3 x + 2$

b. $y = -\log_3 x + 1$

c. $y = \log_3(x+2) + 1$

C11–15. domain: $x > -5$
range: all real numbers
vertical asymptote: $x = -5$
y-intercept: $y = 1 + \log_2 5 \approx 3.3$
x-intercept: $x = -4.5$

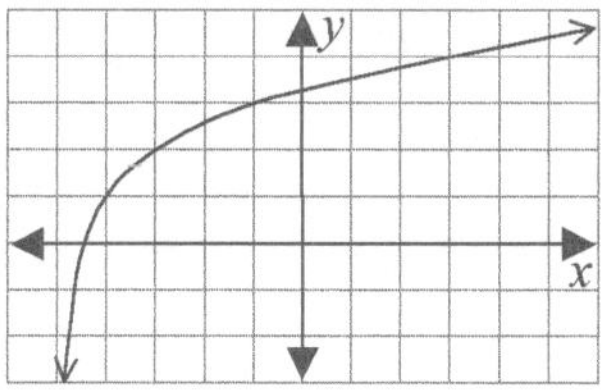

C11–16. $x < 2$

C11–17. **a.** $x > -6$ **b.** $y = 0.5$ **c.** $x = -4$ **d.** $x = -6$

C11–18. 1.262

C11–19. 0.255

C11–20. 15.588

C11–21. 8

C11–22. $\{-4, 5\}$

C11–23. 26.6

C11–24. 26.55

C11–25. **a.** $\frac{\ln 12}{\ln 3}$ **b.** $\frac{\ln\left(\frac{1}{32}\right)}{\ln 16} = -\frac{\ln 32}{\ln 16}$

12 Trigonometric Functions

12–1. $\frac{5}{18}\pi$

12–2. $\frac{10}{9}\pi$

12–3. 162°

12–4. 171.9°

12–5. 3 inches

12–6. 20 cm

12–7. **a.** 9.6 inches **b.** 0.8125 rad

12–8. 195°

12–9. $\frac{3}{4}\pi + 2\pi n$, n is an integer

12–10. **a.** 240° **b.** position 7

12–11. **a.** 0.940
b. $\cos\theta = -0.341$; $\sin\theta = 0.940$; $\tan\theta = -2.757$

12–12. Quadrant II or III

12–13. Quadrant III

12–14. **a.** 75° **b.** 40°

12–15. **a.** 145° **b.** 325°

12–16. **a.** $\frac{1}{5}\pi$ **b.** $\frac{2}{9}\pi$

12–17. $\frac{1}{2}$

12–18. -2

12–19. $-\frac{\sqrt{2}}{2}$

12–20. $\frac{1}{2}$

12–21. $\frac{\sqrt{3}}{2}$

12–22. 0

12–23. 3

12–24. -2

12–25. -1

12–26. $\frac{-\sqrt{3}}{3}$

12–27. $-2\sqrt{2}$

12–28. 1

12–29. -1

12–30. $\sqrt{3} - \frac{1}{2}$

12–31. $\cos\theta = \frac{\sqrt{21}}{5}$, $\tan\theta = \frac{2\sqrt{21}}{21}$

12–32. $\sin\theta = -\frac{24}{25}$

12–33. $\frac{2\sqrt{3}}{3}$

12–34. **a.** $\frac{8}{7}$ **b.** $\frac{\sqrt{15}}{7}$ **c.** $\frac{8\sqrt{15}}{15}$
d. $\frac{8}{7}$ **e.** $\frac{8\sqrt{15}}{15}$ **f.** $\frac{7\sqrt{15}}{15}$

12–35. $\frac{7\sqrt{2}}{10}$

12–36. $\frac{\sqrt{3}}{2}$

12–37. $\frac{\sqrt{2}}{2}$

12–38. $-\frac{\sqrt{2}}{2}$

12–39. $-\frac{12}{5}$

12–40. $-\frac{304}{425}$

C12–1. $\frac{11}{6}\pi$

C12–2. 157.5°

C12–3. 1.24 rad

C12–4. 1,833°

C12–5. $-\frac{3}{5}\pi, \frac{17}{5}\pi$

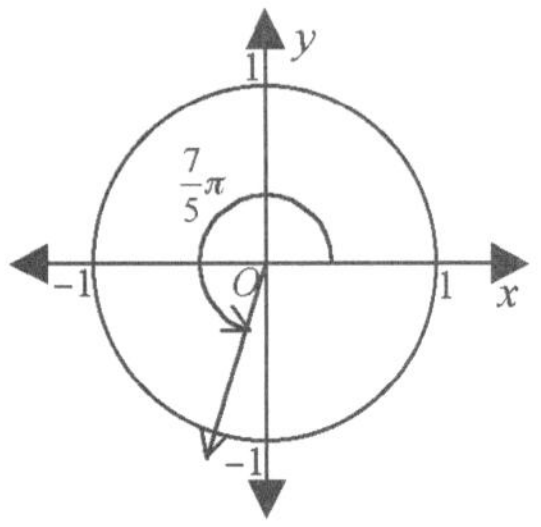

C12–6. Quadrant IV

C12–7. 65°

C12–8. 215°

C12–9. $\frac{3}{10}\pi$

C12–10. $-\frac{\sqrt{3}}{2}$

C12–11. $-\frac{\sqrt{3}}{3}$

C12–12. undefined

C12–13. –1

C12–14. $-\frac{\sqrt{3}}{2}$

C12–15. 0

C12–16. –2.5

C12–17. $-\frac{3}{5}$

C12–18. $\frac{\sqrt{6}-\sqrt{2}}{4}$

C12–19. (1)

C12–20. $\frac{\sqrt{2}}{2}$

C12–21. $-\frac{1}{3}$

C12–22. $\frac{33}{65}$

13. Trigonometric Graphs and Equations

13–1. (1)

13–2. 4

13–3. 4

13–4. 3

13–5. π

13–6. 4

13–7. $-5 \le y \le 5$

13–8. $50 \le y \le 100$

13–9. $a = 9$ or -9, $d = 4$

13–10. **a.**

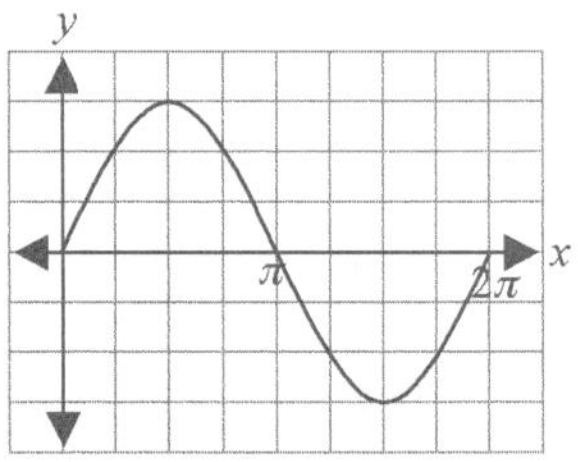

b. a vertical stretch by a factor of 3

c.

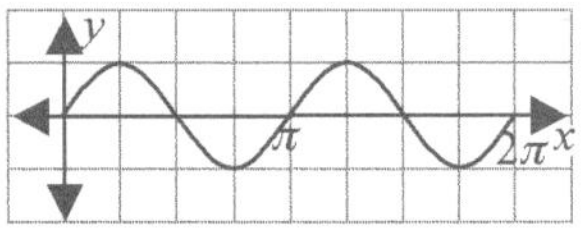

d. a horizontal shrink by a factor of $\frac{1}{2}$

13–11. **a.** 5 **b.** π

c.

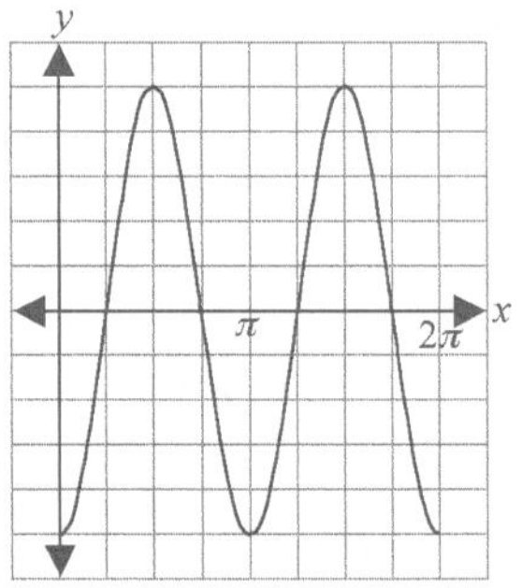

13–12. reflection over the x-axis; vertical stretch by a factor of 2; shift up 3

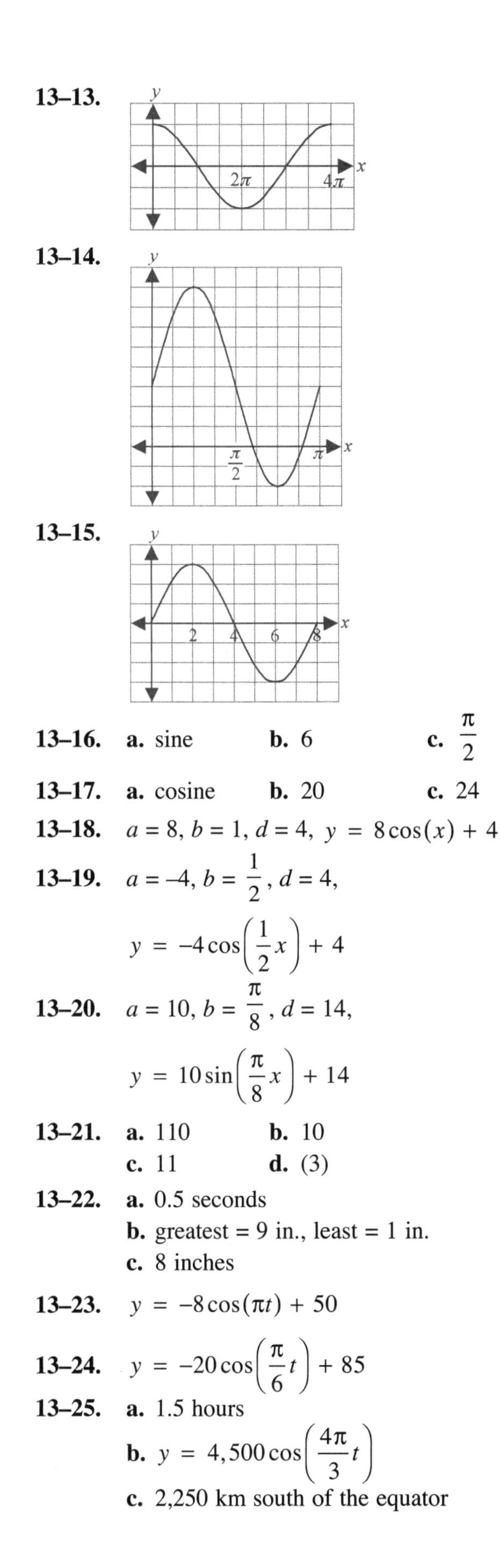

13–13.

13–14.

13–15.

13–16. **a.** sine **b.** 6 **c.** $\frac{\pi}{2}$

13–17. **a.** cosine **b.** 20 **c.** 24

13–18. $a = 8, b = 1, d = 4,\ y = 8\cos(x) + 4$

13–19. $a = -4, b = \frac{1}{2}, d = 4,$

$y = -4\cos\left(\frac{1}{2}x\right) + 4$

13–20. $a = 10, b = \frac{\pi}{8}, d = 14,$

$y = 10\sin\left(\frac{\pi}{8}x\right) + 14$

13–21. **a.** 110 **b.** 10
c. 11 **d.** (3)

13–22. **a.** 0.5 seconds
b. greatest = 9 in., least = 1 in.
c. 8 inches

13–23. $y = -8\cos(\pi t) + 50$

13–24. $y = -20\cos\left(\frac{\pi}{6}t\right) + 85$

13–25. **a.** 1.5 hours
b. $y = 4{,}500\cos\left(\frac{4\pi}{3}t\right)$
c. 2,250 km south of the equator

13–26. **a.** $-\frac{\pi}{4}$ or $-45°$ **b.** $\frac{2\pi}{3}$ or $120°$
c. $\frac{\pi}{6}$ or $30°$

13–27. **a.** π or $180°$ **b.** $\frac{\pi}{2}$ or $90°$

13–28. {65°, 115°}

13–29. {145°, 215°}

13–30. 210°

13–31. **a.** {135°, 225°} **b.** {120°, 300°}

13–32. **a.** {217°, 323°} **b.** {87°, 267°}

13–33. {120°, 240°}

13–34. {30°, 150°}

13–35. {199.5°, 340.5°}

13–36. {53.1°, 233.1°}

13–37. $\left\{\frac{3\pi}{4}, \frac{5\pi}{4}\right\}$

13–38. $\left\{\frac{5\pi}{6}, \frac{11\pi}{6}\right\}$

13–39. {90°, 450°}

13–40. {24°, 66°}

C13–1. (4)

C13–2. **a.** (3) **b.** (2)

C13–3. (4)

C13–4. 6

C13–5. 8π

C13–6. $-2 \le y \le 8$

C13–7. $a = \pm\frac{3}{2}, b = 3$

C13–8.

C13–9.

C13–10. $y = 10\cos(2x)$

C13–11. $y = -6\cos(x) + 10$

C13–12. $y = 5\sin\left(\frac{\pi}{6}x\right) + 3$

C13–13. greatest = 19.37, least = 5.45

C13–14. $A = \pm 115$, $B = 120\pi$

C13–15. **a.** 3 **b.** 12 hours **c.** $y = 3\cos\left(\frac{\pi}{6}t\right) + 5$

C13–16. {30°, 150°}

C13–17. {26°, 334°}

C13–18. {135°, 315°}

C13–19. {14.5°, 165.5°}

C13–20. $\{0, \pi\}$

C13–21. {20°, 40°}

14. Sequences and Series

14–1. **a.** $\frac{2}{4}, \frac{4}{5}, \frac{6}{6}, \frac{8}{7}$ **b.** $\frac{-1}{2}, \frac{1}{3}, \frac{-1}{4}, \frac{1}{5}$

14–2. 16,000; 12,800; 10,240; 8,192

14–3. 95, 90, 85, 80

14–4. 3, 2, 1.5, 1.25

14–5. 3, 5, 11, 21

14–6. $2^0 + 2^1 + 2^2 + 2^3 + 2^4 = 31$

14–7. $1^2 + 2^2 + 3^2 + 4^2 = 30$

14–8. $12\left(\frac{1}{1} + \frac{1}{2} + \frac{1}{3} + \frac{1}{4} + \frac{1}{5}\right) = 27.4$

14–9. $\frac{5}{4} + \frac{6}{6} + \frac{7}{8} + \frac{8}{10} = 3.925$

14–10. $\sum_{k=1}^{5} \frac{1}{k^2}$ or equivalent

14–11. $\sum_{k=1}^{20} \frac{12}{k}$ or equivalent

14–12. $\sum_{k=1}^{n} \frac{2k-1}{2^{k-1}}$ or equivalent

14–13. **a.** On February 5, he watched 2 hours.
b. The total number of hours he watched during February
c. He watched an average of 1.5 hours each day.

14–14. **a.** yes **b.** no **c.** yes **d.** no

14–15. **a.** $a_n = -7 + 4(n-1)$
b. $a_n = 6 + 14(n-1)$
c. $a_n = 8 - 2(n-1)$

14–16. **a.** 2, –1, –4, –7, –10; $a_n = 2 - 3(n-1)$
b. 200, 175, 150, 125, 100; $a_n = 200 - 25(n-1)$

14–17. 880

14–18. **a.** $a_n = 25 + 10(n-1)$ **b.** $1,980

14–19. **a.** no **b.** yes **c.** yes **d.** yes

14–20. **a.** 16, 8, 4, 2, 1; $a_n = 16\left(\frac{1}{2}\right)^{n-1}$
b. 4, –12, 36, –108, 324; $a_n = 4(-3)^{n-1}$

14–21. **a.** $\sum_{k=1}^{n} 5(3)^{k-1}$ **b.** $54.65
c. $15.4 trillion (dollars, not cents)

14–22. **a.** $15 extra **b.** $1,810
c. 266 **d.** $1,005,575

14–23. **a.** 16 feet **b.** 5.24 feet
c. $b_n = 16(0.8)^{n-1}$ or $20(0.8)^n$

14–24. 1,000 feet

14–25. **a.** geometric, does exist
b. geometric, does not exist
c. arithmetic, does not exist
d. geometric, does exist

14–26. sum = $\frac{16}{3}$

14–27. sum = 27

14–28. does not exist, $r = 2$

14–29. sum = 300

C14–1. $\frac{1}{2}, \frac{2}{4}, \frac{3}{8}, \frac{4}{16}, \frac{5}{32}, \frac{6}{64}$

C14–2. 4, 5, 7, 11, 19, 35

C14–3. $3\left(\frac{1}{2} + \frac{1}{3} + \frac{1}{4} + \frac{1}{5}\right) = 3.85$

C14–4. $\sum_{k=1}^{6} \frac{k}{(k+1)^2}$ or $\sum_{k=2}^{7} \frac{k-1}{k^2}$ or equivalent

C14–5. **a.** In the 10th game, Hoop scored 7 points.
b. Hoop scored a total of 126 points over all 15 games
c. $\frac{1}{15}\sum_{i=1}^{15} x_i$

C14–6. **a.** $a_1 = 2$, $a_n = a_{n-1} + 3$
b. $a_n = 2 + 3(n-1)$

C14–7. **a.** $a_n = 8 + 12(n-1)$, $S_{10} = 620$
b. $a_n = 50 - 12(n-1)$, $S_{25} = -2,350$

C14–8. **a.** $a_1 = 2$, $a_n = 3a_{n-1}$
b. $a_n = 2(3)^{n-1}$

C14–9. **a.** $a_n = 2(5)^{n-1}$; 195,312
b. $a_n = 8\left(-\frac{1}{2}\right)^{n-1}$; 5.25

C14–10. 234

C14–11. 20 rows, each row increases by 3 seats

C14–12. sum $= \frac{3}{5}$

C14–13. does not exist, $r = 1.02 > 1$

15. Statistics

15–1. **a.** qualitative **b.** quantitative
c. qualitative **d.** quantitative

15–2. **a.** univariate **b.** bivariate
c. bivariate **d.** univariate

15–3. **a.** sample **b.** population
c. population **d.** sample

15–4. **a.** random **b.** not random
c. not random **d.** not random

15–5. **a.** 42 **b.** 5.25 **c.** 2.5 **d.** 3.0 **e.** 3.0
f. 3 have to get a 2.0, the other student could not get a 3.0
g. 3 have to get a 4.0, the fourth could get a 3.0 or 4.0

15–6. **a.** 83.3 **b.** 82 **c.** 7.22
d. 43% **e.** 68%

15–7. **a.** 5 **b.** 9.5 **c.** 2.14

15–8. **a.** 2 **b.** 3 **c.** 3.9 **d.** 2.5
e. 1 **f.** 68%

15–9. **a.** Hoops **b.** Hoops

15–10. **a.** (3) **b.** (1) **c.** (1) **d.** (2) **e.** (1)

15–11. **a.** 0.067 **b.** 15.9% **c.** (2)
d. (4) **e.** none

15–12. **a.** from 63.7 to 75.7 inches **b.** 850
c. 97% **d.** 78th percentile

15–13. **a.** 88 **b.** 6.5 **c.** 80

15–14. 0.0668

15–15. **a.**

b. $P = -0.43h + 174.60$ **c.** 174.60 psi
d. 0.43 psi/ft **e.** 371.9

15–16. **a.** $P = 32.46(0.97)^t$ **b.** 15.2

15–17. **a.** $y = 2x^2 - 28.45x + 109.27$
b. 24.8 **c.** 8.3

15–18. **a.** $y = 2.36x - 2.03$ **b.** 2.7 **c.** 5.9

C15–1. **a.** 36.3 **b.** 3.2 **c.** 69.6%
d. 95.7% **e.** yes

C15–2. **a.** $36,800 **b.** $24,000
c. $1,104,000 **d.** yes **e.** no

C15–3. **a.** 8.1 **b.** 1.26 **c.** $\frac{8}{12}$

C15–4. (2) smallest standard deviation

C15–5. (1)

C15–6. **a.** 95.4%
b. mean = 64.5, standard deviation = 2.7
c. 27

C15–7. 91%

C15–8. **a.** mean = 2.5, standard deviation = 0.1
b. 383

C15–9. **a.**

b. $T = -1.24L + 111.29$ **c.** 64.5 **d.** 63.9
e. the average temperature decreases 1.24 degrees for each increase of one degree in latitude
f. decreases 6.2 degrees

C15–10. **a.** $y = 14063(0.868)^x$ **b.** $6,014
c. almost 9 years old

C15–11. **a.** $y = 0.052x^2 + 1.040x + 3.177$
b. 128 feet **c.** about 60 mph

C15–12. **a.** $y = 4.83x^{0.49}$ **b.** 73 mph

Formulas, Graphs, and Theorems

Lines

Standard Form
$Ax + By = C$
A, B not both 0

Slope-Intercept Form
$y = mx + b$
m is the slope,
b is the y-intercept

Point-Slope Form
$y = m(x - x_1) + y_1$
m is the slope,
(x_1, y_1) is a point on the line

$$\textbf{Slope} = m = \frac{\Delta y}{\Delta x} = \frac{y_2 - y_1}{x_2 - x_1}$$

Direct and Inverse Variation

Direct Variation

$\frac{y_1}{x_1} = \frac{y_2}{x_2}$ or $y = kx$

Inverse Variation

$x_1 y_1 = x_2 y_2$ or $y = \frac{k}{x}$

Basic Graphs

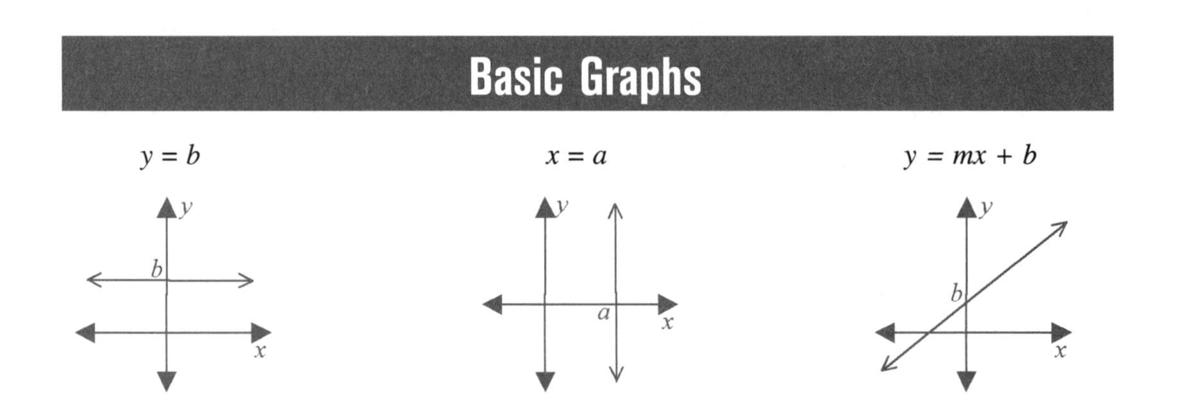

$y = |x|$

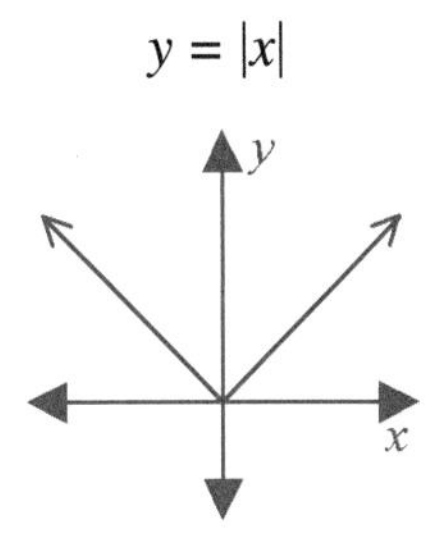

$y = x^2$

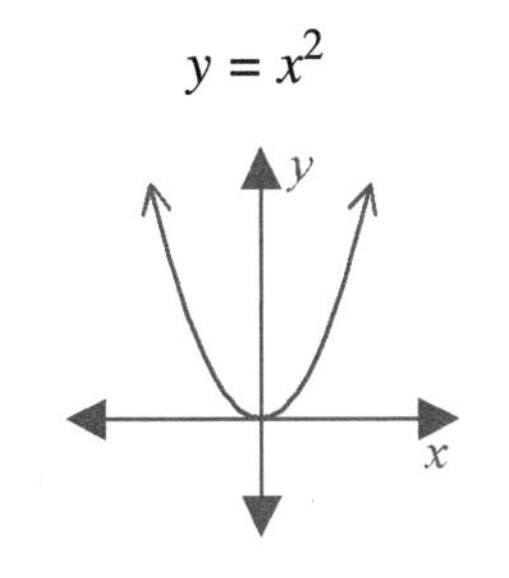

$y = x^3$

$y = \sqrt{x}$

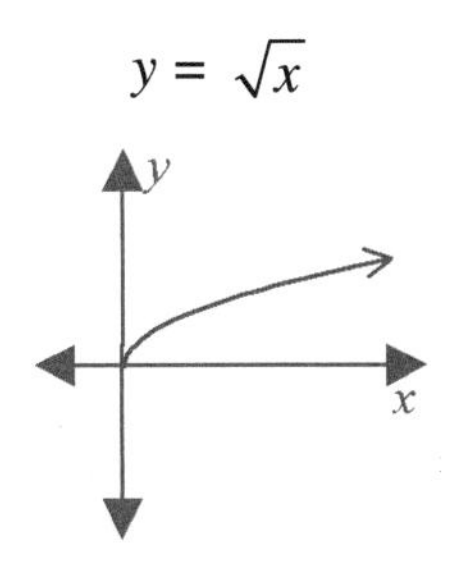

$y = \sqrt[3]{x}$

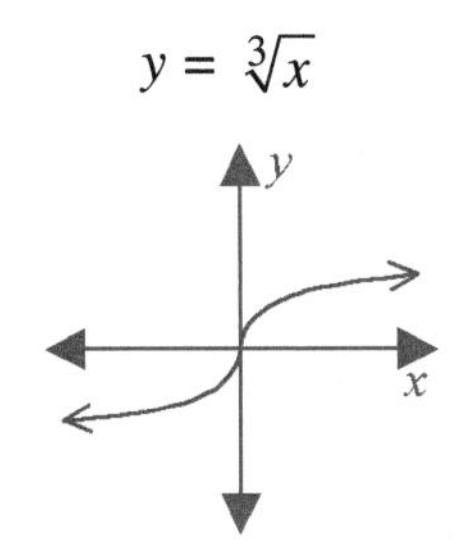

$y = e^x$

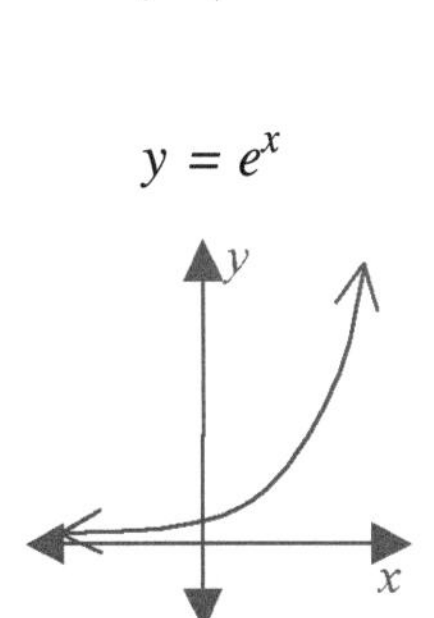

$y = \ln x$

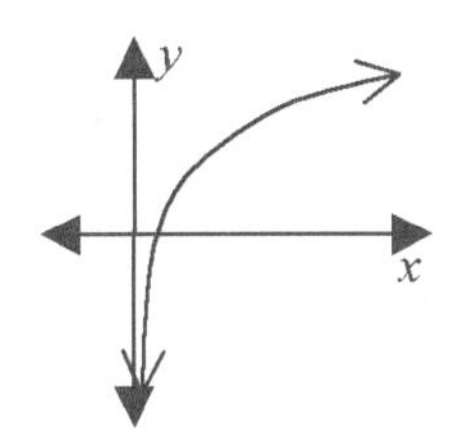

$y = \sin x$

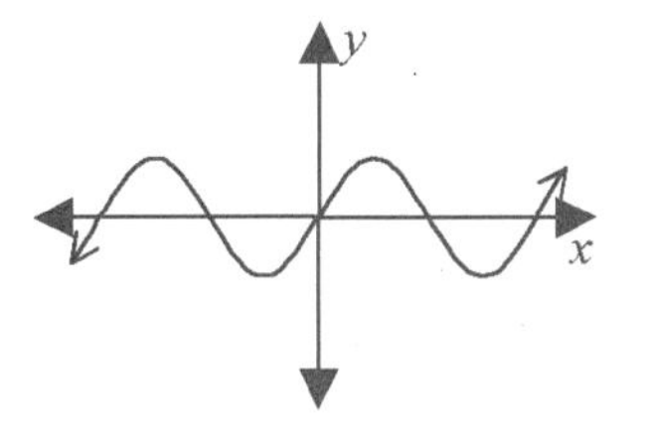

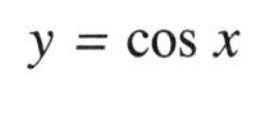
$y = \cos x$

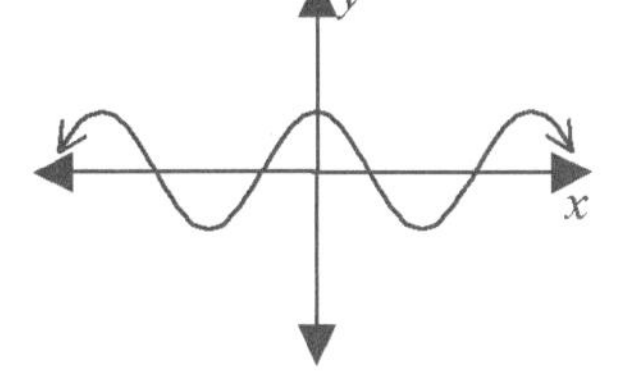

$y = \tan x$

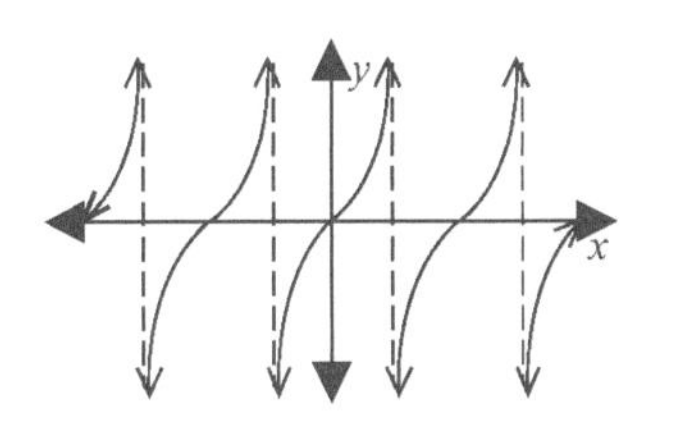

$y = \dfrac{1}{x}$

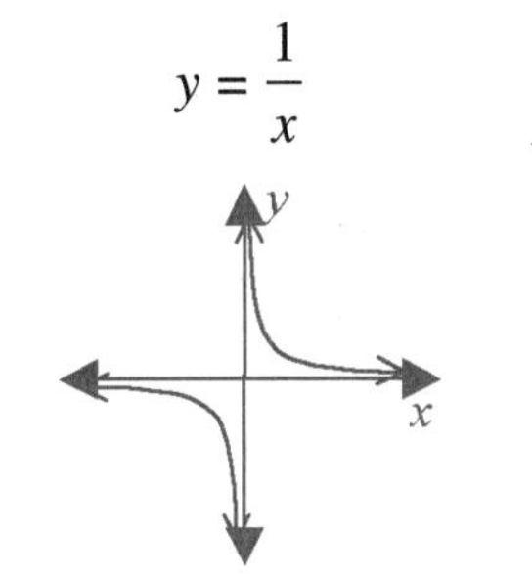

$y = \lfloor x \rfloor$

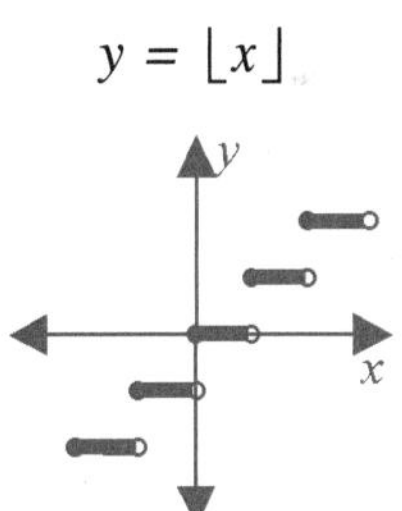

$y = \lceil x \rceil$

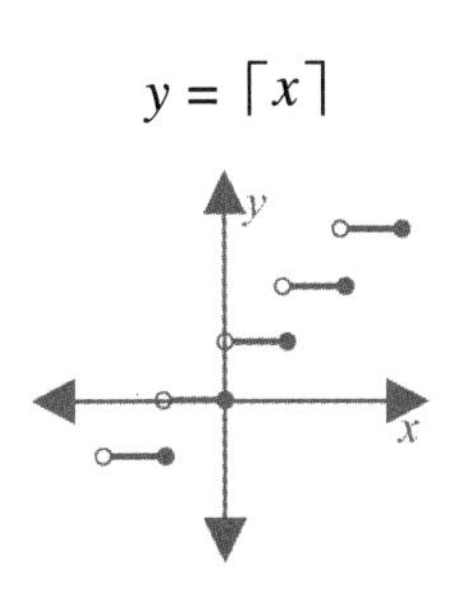

Symmetry of Functions

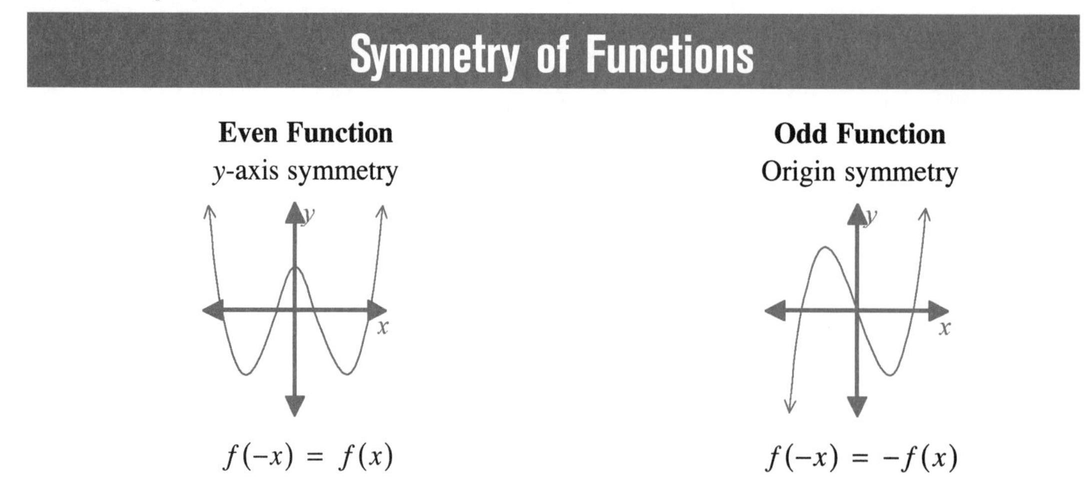

Transformations of Functions

Transformation ($a > 0$)	Description	Example Graph and Equation
$y = f(x + a)$ $y = f(x - a)$	Horizontal translation a units; ($x + a$) translates graph left; ($x - a$) translates graph right	$y = f(x + a)$, $y = f(x)$, $y = f(x - a)$
$y = f(x) + a$ $y = f(x) - a$	Vertical translation a units; $+ a$ translates graph up; $- a$ translates graph down	$y = f(x) + a$, $y = f(x)$, $y = f(x) - a$
$y = f(-x)$	Reflection over the y-axis	$y = f(-x)$, $y = f(x)$

Transformation ($a > 0$)	Description	Example Graph and Equation
$y = -f(x)$	Reflection over the x-axis	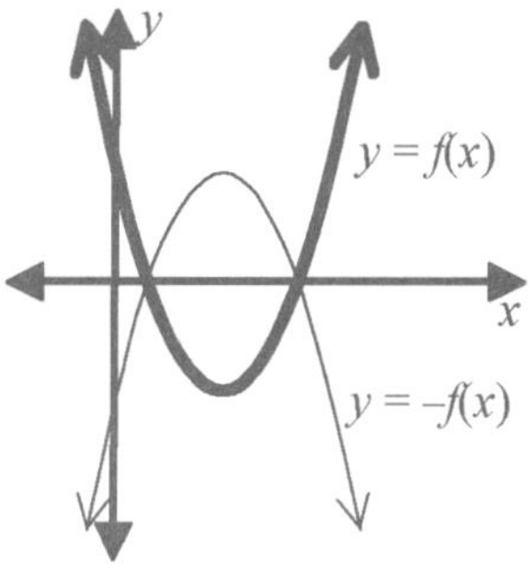 y; $y = f(x)$; x; $y = -f(x)$
$y = af(x)$	Vertical dilation by a factor of a	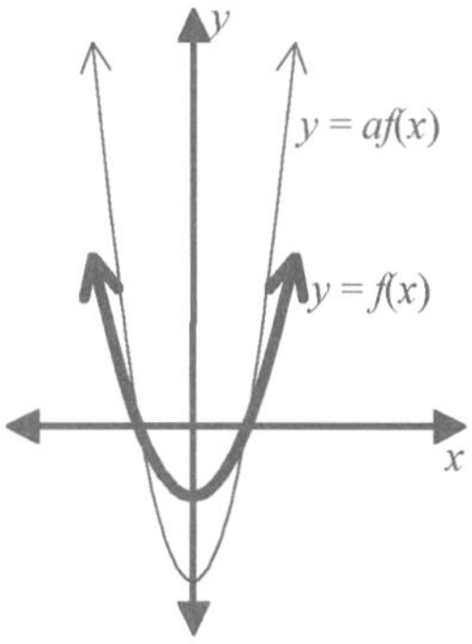 y; $y = af(x)$; $y = f(x)$; x
$y = f\left(\frac{x}{a}\right)$	Horizontal dilation by a factor of a	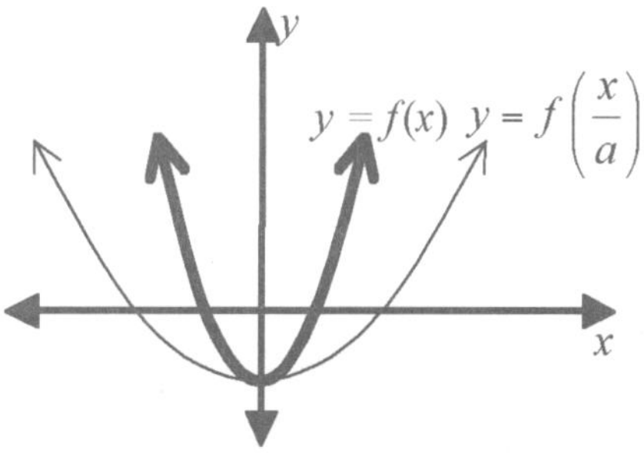 y; $y = f(x)$; $y = f\left(\frac{x}{a}\right)$; x

Quadratic Functions and Parabolas

General Form

$$y = ax^2 + bx + c$$

Axis of symmetry is $x = \frac{-b}{2a}$

Vertex is $\left(\frac{-b}{2a}, c - \frac{b^2}{4a}\right)$

Vertex Form

$$y = a(x - h)^2 + k$$

Axis of symmetry is $x = h$

Vertex is (h, k)

Quadratic Formula

The solutions to $ax^2 + bx + c = 0$ are $x = \frac{-b \pm \sqrt{b^2 - 4ac}}{2a}$.

Binomials

Powers of Binomials

$$(a+b)^2 = a^2 + 2ab + b^2$$

$$(a+b)^3 = a^3 + 3a^2b + 3ab^2 + b^3$$

$$(a+b)^4 = a^4 + 4a^3b + 6a^2b^2 + 4ab^3 + b^4$$

$$(a+b)^5 = a^5 + 5a^4b + 10a^3b^2 + 10a^2b^3 + 5ab^4 + b^5$$

$$(a+b)^n = {}_nC_0a^nb^0 + {}_nC_1a^{n-1}b^1 + {}_nC_2a^{n-2}b^2 + \ldots + {}_nC_na^0b^n$$

$$= \sum_{r=0}^{n} {}_nC_ra^{n-r}b^r$$

$${}_nC_r = \frac{n!}{r!(n-r)!}$$

Factoring Special Binomials

$$a^2 - b^2 = (a+b)(a-b)$$

$$a^3 - b^3 = (a-b)\left(a^2 + ab + b^2\right)$$

$$a^3 + b^3 = (a+b)\left(a^2 - ab + b^2\right)$$

Complex Numbers

Powers of i		
i^0	=	1
i^1	=	i
i^2	=	–1
i^3	=	$-i$
i^4	=	1
i^5	=	i
i^6	=	–1
i^7	=	$-i$
i^8	=	1

Properties of a Polynomial Function of Degree n

$$y = a_n x^n + a_{n-1}x^{n-1} + \ldots + a_2 x^2 + a_1 x + a_0$$

Number of Real Zeros

n odd: At least one zero, at most n

n even: Any number of zeros from 0 to n

Number of Turning Points

n odd: Even number of turning points, at most $n - 1$

n even: Odd number of turning points, at most $n - 1$

End Behavior

n odd: Ends go in opposite directions (one up, one down)

n even: Ends go in same direction (both up or both down)

a_n positive: Right end goes up

a_n negative: Right end goes down

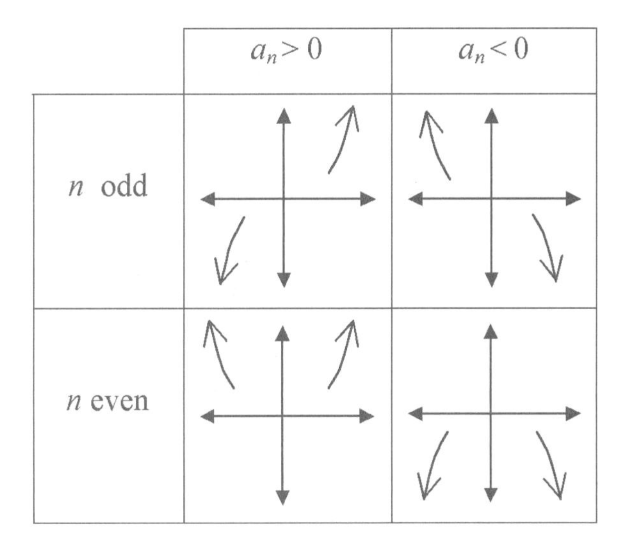

Theorems

- Factor theorem: A polynomial function, $f(x)$, has a factor $x - r$ if and only if $f(r) = 0$.
- Fundamental theorem of algebra: A polynomial of degree $n \geq 1$ will always have exactly n total zeros.

Radicals

Properties of Radicals

If $\sqrt[n]{x}$ and $\sqrt[n]{y}$ are both real:

1. $\sqrt[n]{xy} = \sqrt[n]{x}\sqrt[n]{y}$
2. $\sqrt[n]{\dfrac{x}{y}} = \dfrac{\sqrt[n]{x}}{\sqrt[n]{y}}$, $(y \neq 0)$
3. $\sqrt[m]{\sqrt[n]{x}} = \sqrt[mn]{x}$
4. $x^{\frac{m}{n}} = \sqrt[n]{x^m} = \left(\sqrt[n]{x}\right)^m$
5. $\left(\sqrt[n]{x}\right)^n = x$
6. $\sqrt[n]{x^n} = \begin{cases} |x| & \text{if } n \text{ is even} \\ x & \text{if } n \text{ is odd} \end{cases}$

Exponents

Properties of Exponents

For positive bases a and b, the following rules of exponents hold for all real numbers m and n:

1. $b^m b^n = b^{m+n}$
2. $\dfrac{b^m}{b^n} = b^{m-n}$
3. $\left(b^m\right)^n = b^{mn}$
4. $b^0 = 1$
5. $b^{-n} = \left(\dfrac{1}{b}\right)^n = \dfrac{1}{b^n}$
6. $b^{m/n} = \left(\sqrt[n]{b}\right)^m = \sqrt[n]{b^m}$
7. $(ab)^n = a^n b^n$
8. $\left(\dfrac{a}{b}\right)^n = \dfrac{a^n}{b^n}$

Properties of Logarithms

For a positive base $b \neq 1$; x and $y > 0$:

1. $\log_b 1 = 0$
2. $\log_b b = 1$
3. $\log_b b^x = x$
4. $b^{\log_b x} = x$
5. $\log_b (xy) = \log_b x + \log_b y$
6. $\log_b \left(\dfrac{x}{y}\right) = \log_b x - \log_b y$
7. $\log_b x^n = n\log_b x$

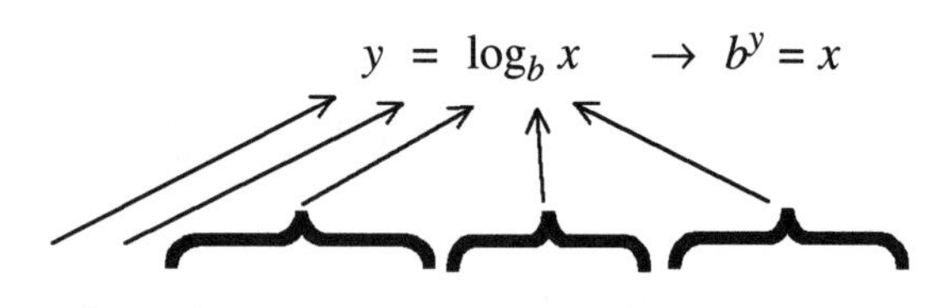

Change of base: $\log_b N = \dfrac{\log_a N}{\log_a b}$

Trigonometry

Radians

$$\theta = \frac{s}{r}$$

$360° = 2\pi$ radians or $180° = \pi$ radians

Right Triangle Trigonometry

$$\sin\theta = \frac{\text{Opposite}}{\text{Hypotenuse}}$$

$$\cos\theta = \frac{\text{Adjacent}}{\text{Hypotenuse}}$$

$$\tan\theta = \frac{\text{Opposite}}{\text{Adjacent}}$$

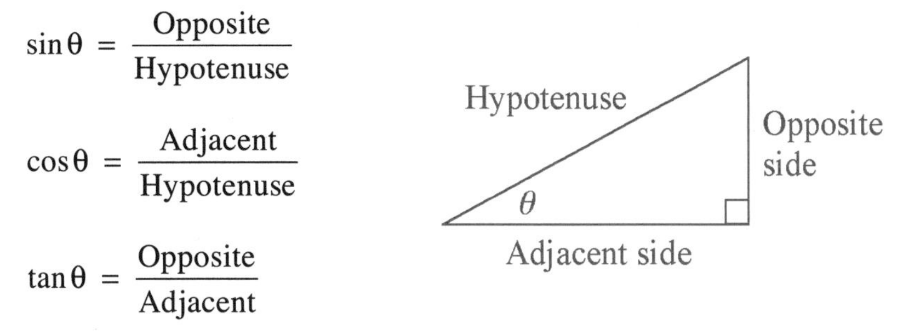

Unit Circle Definitions

$$\sin\theta = y$$

$$\cos\theta = x$$

$$\tan\theta = \frac{y}{x},\ x \neq 0$$

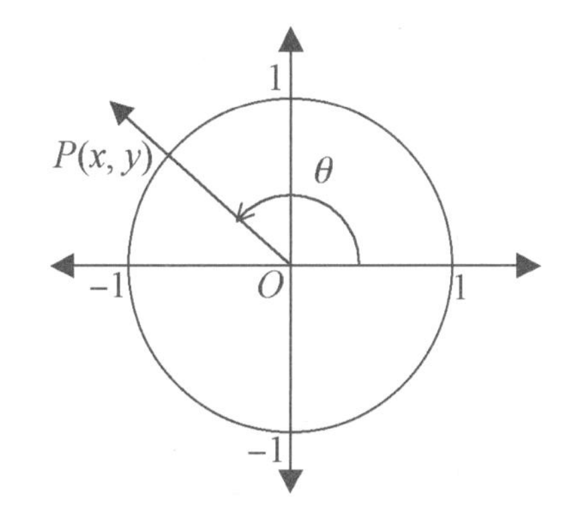

Trigonometric Function Values of Quadrantal Angles

	0° (0 rad)	**90° $\left(\frac{\pi}{2}\text{ rad}\right)$**	**180° (π rad)**	**270° $\left(\frac{3\pi}{2}\text{ rad}\right)$**	**360° (2π rad)**
$\sin\theta = y$	0	1	0	−1	0
$\cos\theta = x$	1	0	−1	0	1
$\tan\theta = \frac{y}{x}$	0	Undefined	0	Undefined	0

Trigonometric Function Values of Special Angles

	$30° \left(\frac{\pi}{6} \text{ rad}\right)$	$45° \left(\frac{\pi}{4} \text{ rad}\right)$	$60° \left(\frac{\pi}{3} \text{ rad}\right)$
$\sin\theta$	$\frac{1}{2}$	$\frac{\sqrt{2}}{2}$	$\frac{\sqrt{3}}{2}$
$\cos\theta$	$\frac{\sqrt{3}}{2}$	$\frac{\sqrt{2}}{2}$	$\frac{1}{2}$
$\tan\theta$	$\frac{\sqrt{3}}{3}$	1	$\sqrt{3}$

Signs of Trigonometric Functions by Quadrant

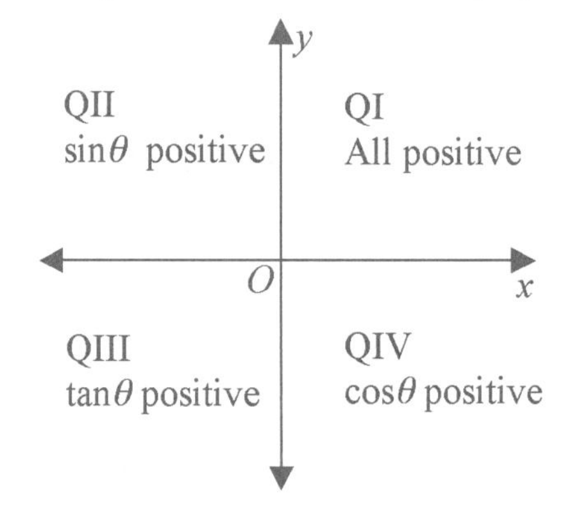

Trigonometric Identities

Reciprocal Identities

$$\csc\theta = \frac{1}{\sin\theta} \qquad \sec\theta = \frac{1}{\cos\theta} \qquad \cot\theta = \frac{1}{\tan\theta}$$

Quotient Identities

$$\tan\theta = \frac{\sin\theta}{\cos\theta} \qquad \cot\theta = \frac{\cos\theta}{\sin\theta}$$

Pythagorean Identities

$$\sin^2\theta + \cos^2\theta = 1$$

$$\text{or } \sin^2\theta = 1 - \cos^2\theta$$

$$\text{or } \cos^2\theta = 1 - \sin^2\theta$$

$$1 + \cot^2\theta = \csc^2\theta$$

$$\tan^2\theta + 1 = \sec^2\theta$$

Functions of the Sum of Two Angles

$$\sin(A + B) = \sin A \cos B + \cos A \sin B$$

$$\cos(A + B) = \cos A \cos B - \sin A \sin B$$

$$\tan(A + B) = \frac{\tan A + \tan B}{1 - \tan A \tan B}$$

Functions of the Difference of Two Angles

$$\sin(A - B) = \sin A \cos B - \cos A \sin B$$

$$\cos(A - B) = \cos A \cos B + \sin A \sin B$$

$$\tan(A - B) = \frac{\tan A - \tan B}{1 + \tan A \tan B}$$

Trigonometric Graphs

$$y = a\sin(bx) + d \text{ or } y = a\cos(bx) + d$$

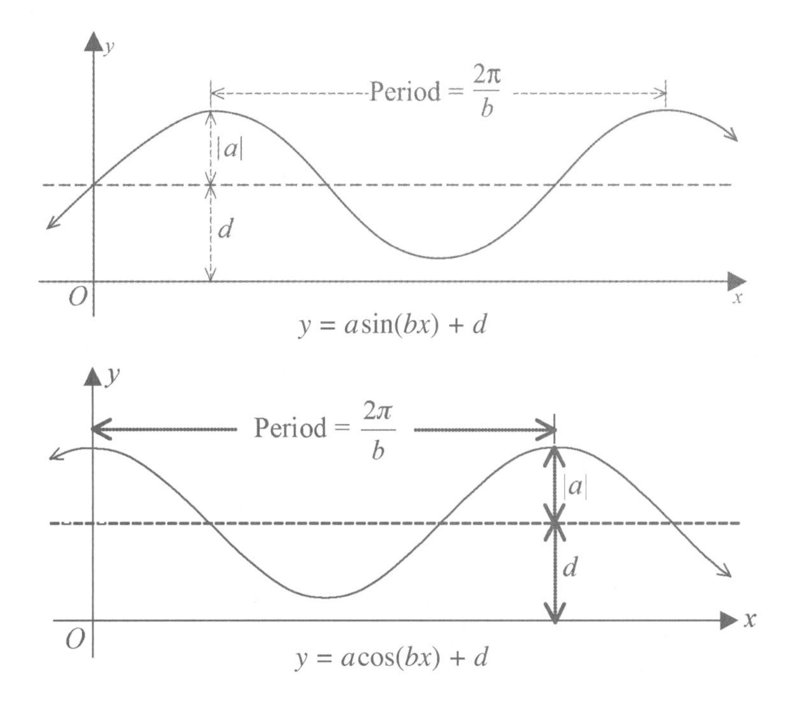

$|a|$ is the amplitude

$$|a| = \frac{\max - \min}{2}$$

or $|a| = \max - d$

or $|a| = d - \min$

$\frac{2\pi}{b}$ is the period

$$b = \frac{2\pi}{\text{period}}$$

d is the vertical shift

$$d = \frac{\max + \min}{2}$$

Sequences and Series

Arithmetic Sequence

Recursive Formula

a_1 is given,

$a_n = a_{n-1} + d$ for $n \geq 2$

Explicit Formula

$a_n = a_1 + d(n - 1)$ for $n \geq 1$

a_1 is the first term

a_n is the nth term

d is the common difference

$d = a_n - a_{n-1}$ for all $n \geq 2$

Geometric Sequence

Recursive Formula

a_1 is given,

$a_n = ra_{n-1}$ for $n \geq 2$

Explicit Formula

$a_n = a_1 r^{n-1}$ for $n \geq 2$

a_1 is the first term

a_n is the nth term

r is the common ratio

$$r = \frac{a_n}{a_{n-1}} \text{ for all } n \geq 2$$

Arithmetic and Geometric Series

Finite arithmetic series: $$S_n = \sum_{k=1}^{n} a_k = \frac{n(a_1 + a_n)}{2}$$

Finite geometric series: $$S_n = \sum_{k=1}^{n} a_k = a_1 \frac{(1 - r^n)}{(1 - r)}$$

Infinite geometric series: $$S = \sum_{k=1}^{\infty} a_k = \frac{a_1}{1 - r}, \text{ provided } |r| < 1$$

Basic Statistical Measures

Mean (Average)

$$\mu \text{ or } \overline{x} = \frac{\sum_{i=1}^{n} x_i}{n}$$

Standard Deviation

Population

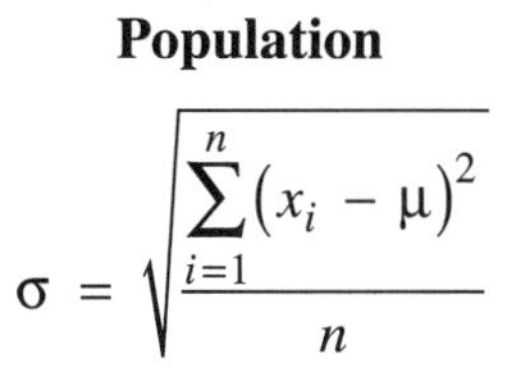

$$\sigma = \sqrt{\frac{\sum_{i=1}^{n}(x_i - \mu)^2}{n}}$$

Sample

$$s = \sqrt{\frac{\sum_{i=1}^{n}(x_i - \bar{x})^2}{n - 1}}$$

Variance

Population

$$\sigma^2 = \frac{\sum_{i=1}^{n}(x_i - \mu)^2}{n}$$

Sample

$$s^2 = \frac{\sum_{i=1}^{n}(x_i - \bar{x})^2}{n - 1}$$

Normal Distribution

Empirical Rule

About 68% of the scores are within one standard deviation of the mean.
About 95% of the scores are within two standard deviations of the mean.
Over 99% of the scores are within three standard deviations of the mean.

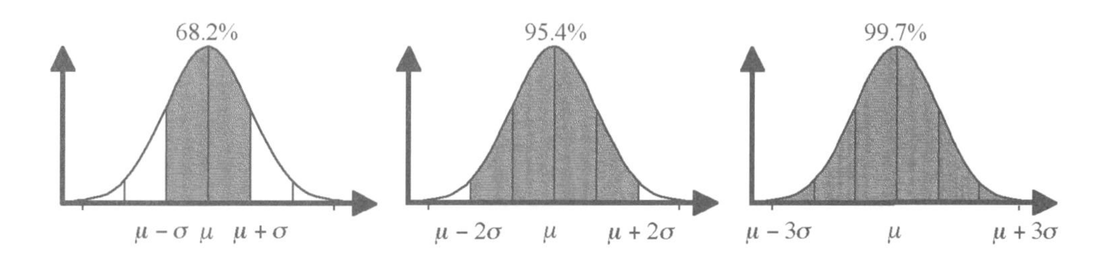

Percentiles

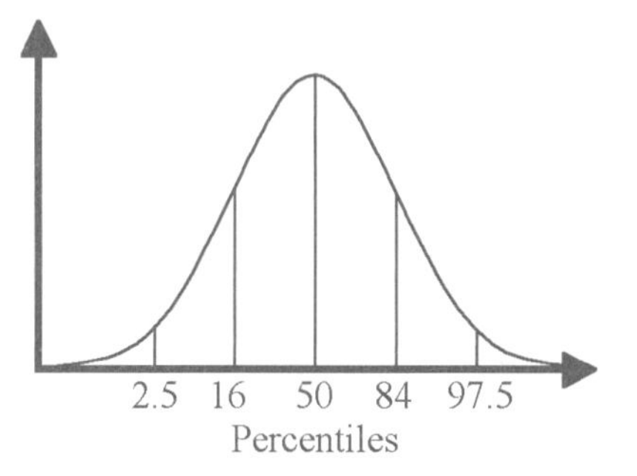

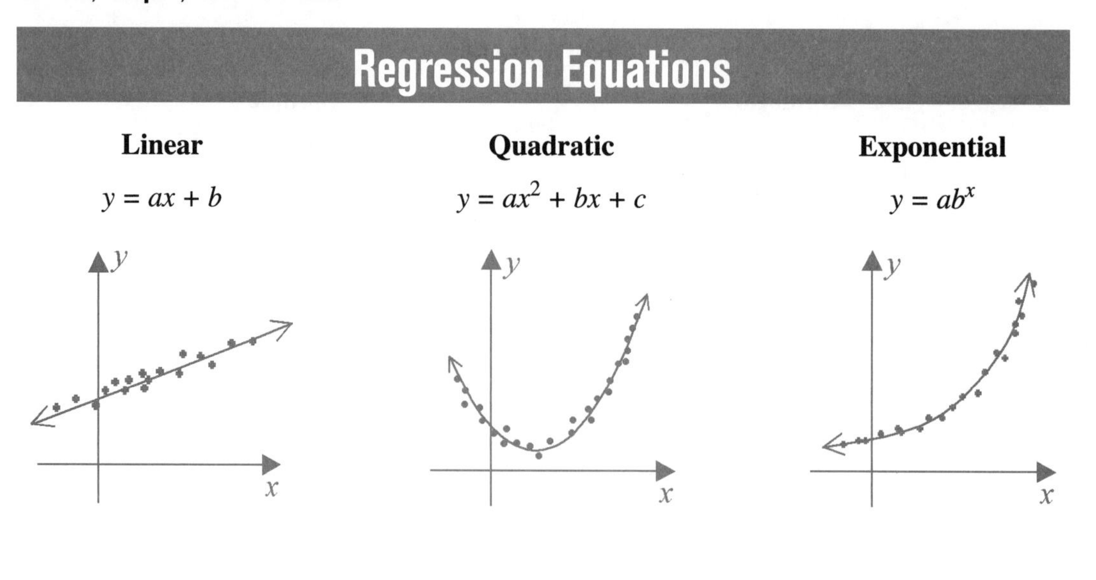
Regression Equations
Linear
$y = ax + b$
Quadratic
$y = ax^2 + bx + c$
Exponential
$y = ab^x$
y
x
y
x
y
x

Graphing Calculator Tips

Many brands and models of graphing calculators are available. To illustrate these tips, this chapter uses instructions and screens for the TI-84 Plus calculator with the 2.55 operating system. If you have a different calculator, you should consult your manual for instructions. This is not an instruction manual on how to use your graphing calculator. Rather, these are ideas on how to use a graphing calculator to check your work and/or to find answers. These tips are arranged in the same order as the chapters in the book.

1. Linear Functions

Checking a solution on a graphing calculator can be quick and accurate using the store feature. For a numerical check, store your answer in X or the letter used for the variable in your equation. Then type the left side of the original equation and compare it to the right side of the original equation. The store key looks like STO▸ but appears on the calculator screen as an arrow.

TIP 1A

Solve $4x - 2(x + 6) = 3(2x + 7) + 5$ for x. (Answer: $x = -9.5$)

Check the correct answer, $x = -9.5$. The value for the expression on the left equals the value for the expression on the right.

```
-9.5→X
                 -9.5
4X-2(X+6)
                  -31
3(2X+7)+5
                  -31
```

A common incorrect answer is $x = -3.5$. If this is checked, the value for the expression on the left does not equal the value for the expression on the right.

```
-3.5→X
                -3.5
4X-2(X+6)
                 -19
3(2X+7)+5
                   5
```

You should understand how to solve an equation graphically on your calculator. Type the left side of the original equation in Y1 and the right side of the original equation in Y2. After you find the best viewing window, you can use the intersect feature to find the solution to the equation.

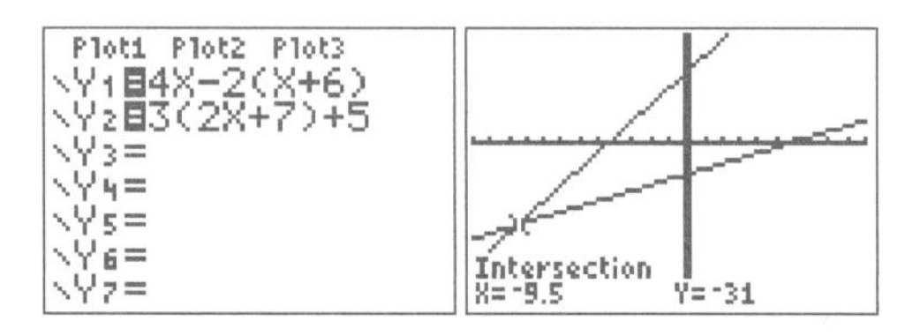

If a linear equation has no solution, the two sides of the equation will graph as parallel lines. Keep in mind that looking at a calculator graph is not conclusive. Two lines are parallel only if they have exactly the same slope. If the slopes are only close to equal, the lines may look parallel on the screen but intersect off the screen.

TIP 1B

Solve $4 - (3 - x) = x + 5$ for x. (Answer: no solution)

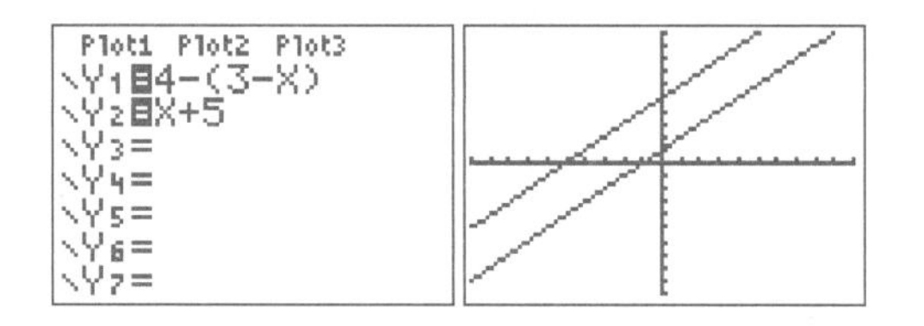

If the two sides of the equation graph as exactly the same line, the equation is an identity. Again, do not rely solely on the graph. Check the table to see if the two sides always have the same value.

TIP 1C

Solve $2(x + 1) - (x - 2) = x + 4$ for x. (Answer: identity, all real numbers)

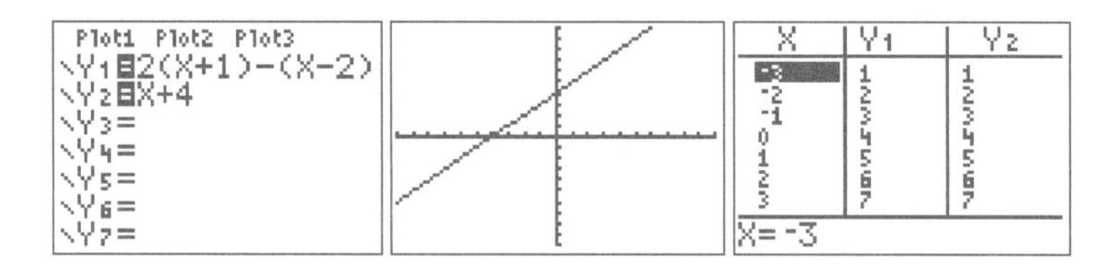

To check a problem that asks you to write an equation of a line, type your answer into Y1 and check the points in the table to see if the given points are listed. You can use this to check if alternate forms of an equation are equivalent.

TIP 1D

Write the equation of the line passing through the points (–3, 6) and (9, –10). (There are three correct solutions: $y = -\frac{4}{3}(x + 3) + 6$, $y = -\frac{4}{3}(x - 9) - 10$, or $y = -\frac{4}{3}x + 2$.)

```
Plot1 Plot2 Plot3
\Y1 = -4/3(X+3)+6
\Y2 = -4/3(X-9)-10
\Y3 = -4/3X+2
\Y4 =
```

X	Y1	Y2
3	-2	-2
4	-10/3	-10/3
5	-14/3	-14/3
6	-6	-6
7	-22/3	-22/3
8	-26/3	-26/3
9	-10	-10

Y2 = -(4/3)(X-9)-...

X	Y2	Y3
3	-2	-2
4	-10/3	-10/3
5	-14/3	-14/3
6	-6	-6
7	-22/3	-22/3
8	-26/3	-26/3
9	-10	-10

Y3 = -(4/3)X+2

2. Polynomial Operations

Unless you have a calculator that has Computer Algebra System (CAS) capabilities, your graphing calculator will not simplify polynomials. However, you can use your calculator's table to check the answer you obtained using your algebra skills. This works as long as the polynomial has only one variable, like x or t. Type the original problem in Y1, and type your answer in Y2. Look at the table values for Y1 and Y2, and see if they are equivalent. This is not a rigorous method but rather a check; it is particularly useful for multiple-choice questions.

TIP 2A

Simplify: $4x^3 - 5x^2 + 7 - \left(5x^3 + 3x^2 - 8x - 9\right)$

(Answer: $-x^3 - 8x^2 + 8x + 16$)

```
Plot1 Plot2 Plot3
\Y1 = 4X³-5X²+7-(5X³+3X²-8X-9)
\Y2 = -X³-8X²+8X+16
\Y3 =
\Y4 =
\Y5 =
```

X	Y1	Y2
-2	-24	-24
-1	1	1
0	16	16
1	15	15
2	-8	-8
3	-59	-59
4	-144	-144

Press + for ΔTbl

You can use this method to double-check your long division. Remember this does not give you the answer. Instead, it checks your answer to make sure it is equivalent to the original expression.

TIP 2B

Find the quotient: $\dfrac{6x^3 - 19x^2 + 25}{3x - 5}$ (Answer: $2x^2 - 3x - 5$)

```
Plot1 Plot2 Plot3
\Y1 = (6X³-19X²+25)/(3X-5)
\Y2 = 2X²-3X-5
\Y3 =
\Y4 =
\Y5 =
```

X	Y1	Y2
-2	9	9
-1	0	0
0	-5	-5
1	-6	-6
2	-3	-3
3	4	4
4	15	15

Press + for ΔTbl

Powers of binomials can be expanded by using coefficients found in Pascal's triangle. However, you must construct the entire triangle to get the row you want. These same coefficients can also be found using combinations. You can use the formula for $_nC_r = \frac{n!}{r!(n-r)!}$, but calculators do it easily. You will find $_nC_r$ under the MATH key, PRB menu 3: $_nC_r$.

TIP 2C

Find the third term of $(x + 5)^8$. (Answer: $700x^6$)

$$_nC_r a^{n-r} b^r = {}_8C_2 a^{8-2} b^2 = 28a^6b^2$$

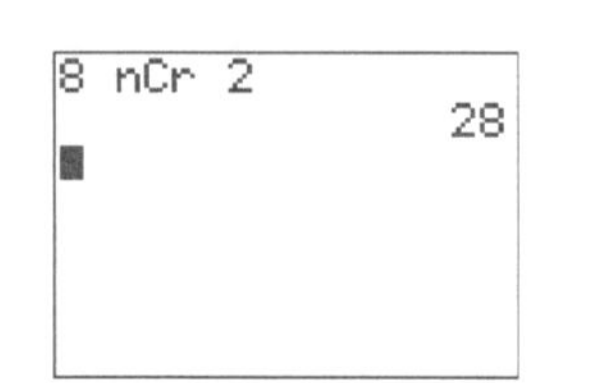

Then substitute $a = x$ and $b = 5$. $28a^6b^2 = 28x^6(5)^2 = 28x^6(25) = 700x^6$

Many students find factoring polynomials difficult. Factoring can be difficult if you cannot think of numbers that fit the criteria and get stuck. You can use your calculator to help you find numbers that multiply and add to certain values. In the following example, we need to find two numbers that multiply to 120 (product) and add to –29 (sum). In Y1, type the product number divided by x. This gives all the numbers, X and Y1, that multiply to your number. In Y2, type x + the product number divided by x (no parentheses). This gives the sum. Look in the table at Y2 for the sum you need; your numbers are in the first two columns.

TIP 2D

Factor: $x^2 - 29x + 120$. (Answer: $(x - 5)(x - 24)$)

```
Plot1 Plot2 Plot3
\Y1=120/X
\Y2=X+120/X
\Y3=
\Y4=
\Y5=
\Y6=
\Y7=
```

X	Y1	Y2
-5	-24	-29
-4	-30	-34
-3	-40	-43
-2	-60	-62
-1	-120	-121
0	ERROR	ERROR
1	120	121

Y2=-29

Under Y2, find –29. The first two numbers in that row are your numbers. –5 and –24 both multiply to 120 and add to –29.

This tip works for split the middle factoring too. Just multiply the values of a and c to find the product number.

TIP 2E

Factor: $15x^2 - 38x + 24$ (Answer: $(3x - 4)(5x - 6)$)

```
Plot1 Plot2 Plot3
\Y1=360/X
\Y2=X+360/X
\Y3=
\Y4=
\Y5=
\Y6=
\Y7=
```

X	Y1	Y2
-18	-20	-38
-17	-21.18	-38.18
-16	-22.5	-38.5
-15	-24	-39
-14	-25.71	-39.71
-13	-27.69	-40.69
-12	-30	-42

Y2=-38

In this example, $a = 15$, $c = 24$, and $ac = 360$. Under Y2, find –38. The first two numbers in that row are your numbers. –18 and –20 both multiply to 360 and add to –38. In split the middle factoring, these numbers are not in the actual factors but, instead, tell you to split $-38x$ into $-18x + -20x$ and then factor by grouping.

3. Functions and Relations

Calculators can use function notation for evaluating expressions. You must remember that on a calculator, the independent variable is always x (even if the problem says it is t or something else) and the function name is always Y1, Y2, etc. (even if the problem says it is f).

To use the evaluation feature, you will need to find out how to type Y1. This menu is displayed by pressing the F4 key. You may need to consult your manual if your calculator is different.

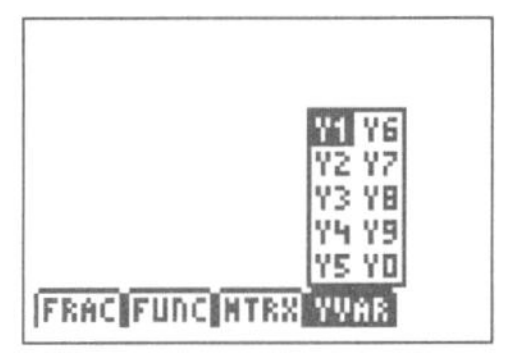

Type your function into Y1, and then evaluate your function on the home screen.

TIP 3A

The function f is defined by $f(x) = x^2 - 2x - 3$. Evaluate $f(-2)$. (Answer: 5)

```
Plot1 Plot2 Plot3
\Y1=X²-2X-3
\Y2=
\Y3=
\Y4=
\Y5=
\Y6=
```

```
Y1(-2)
                5
```

Unless you have a Computer Algebra System (CAS), your calculator will not simplify functions of variable expressions. However, you can check your answer with the table.

TIP 3B

$f(x) = x^2 - 2x + 3$, find $f(x + 5)$ and simplify it.

(Answer: $f(x + 5) = x^2 + 8x + 18$)

Shut off the column for Y1 by deselecting the "=" in the equation.

Finding domain on a graphing calculator can be tricky. Many students are more accurate using the algebraic rules explained in Chapter 3. Here are a few examples with an explanation of how to interpret the calculator. A good starting place is to look at the table. Then use the graph to interpret the values in the table.

TIP 3C

Find the natural domain of the function $y = \dfrac{3x}{2x - 4}$.

(Answer: $(-\infty, -2) \cup (2, \infty)$)

The table displays ERROR when a function value is undefined or imaginary. In this example, the function is undefined at $x = 2$ and the domain is all real numbers except $x = 2$.

TIP 3D

Find the natural domain of the function $y = \sqrt{2x - 6}$. (Answer: $x \geq 3$)

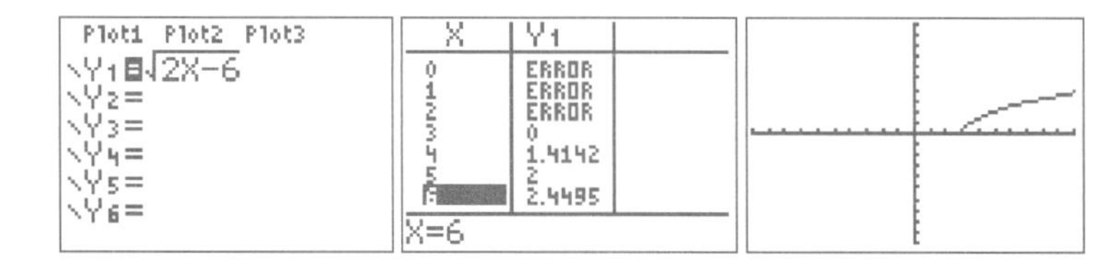

In this example, the domain is $x \geq 3$. It is easily read from the table and confirmed on the graph.

TIP 3E

Find the natural domain of the function $y = \dfrac{3}{\sqrt{2x - 6}}$. (Answer: $x > 3$)

This example shows why finding domain on a calculator can be tricky. The table might lead you to think the domain is $x \geq 4$. The graph is difficult to interpret. Algebraically, it is not hard to find that the domain is $x > 3$. Strict inequalities are difficult to see on the table or graph.

You can use Y1 and Y2 to evaluate the composition of functions.

TIP 3F

Let $f(x) = x^2 - 3x + 2$ and $g(x) = \sqrt{x - 1}$. Evaluate $(g \circ f)(3)$. (Answer: 1)

Pay careful attention to which Y function is $f(x)$ and which is $g(x)$ so that you type the composition order correctly.

You can use composition of functions to verify that you have found an inverse correctly. In Chapter 3, you learned that $f\left(f^{-1}(x)\right) = f^{-1}(f(x)) = x$ is true for inverses. Type the original function in Y1 and your inverse in Y2. Then graph Y1(Y2) and Y2(Y1). Both graphs should just give the diagonal line $y = x$, which can be confirmed in the table.

TIP 3G

Find the inverse of the function $f(x) = \dfrac{2x}{x - 5}$, $x \neq 5$. (Answer: $f^{-1}(x) = \dfrac{5x}{x - 2}$)

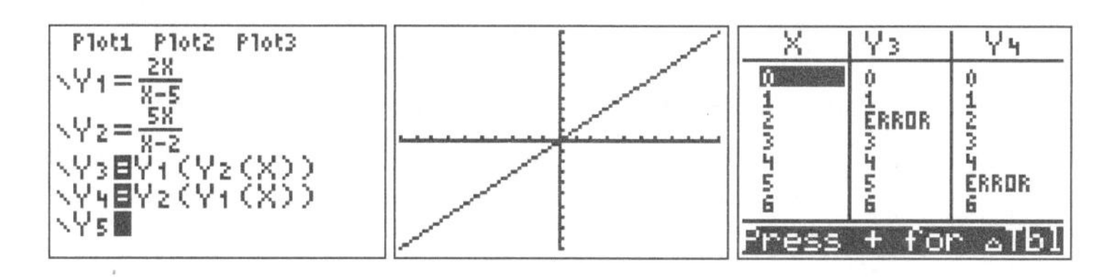

The ERRORs in the table occur at the x-values where each function is undefined. You are confirming that all the nonerror values are the same.

You can also use the table to confirm that you have found the correct inverse function, especially if you took the square root and needed to choose between the positive and the negative root. The table for the inverse should have the x and y columns reversed from the table for the original function.

To easily compare the tables, use the Table Ask feature. On the Table Setup screen, found by pressing the TBLSET key, change the independent variable from Auto to Ask. Remember to change it back when you are done or you will be surprised the next time you use the table. After you set the table to Ask, go to the table (which is empty) and type in the y-values from the table for the original function as the x-values for the inverse function.

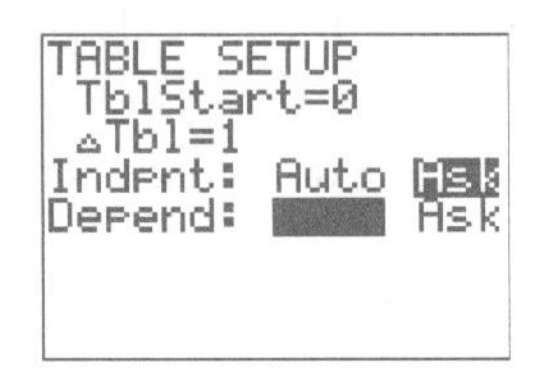

TIP 3H

Find the inverse of the function $f(x) = 2x^2 - 8$, $x \geq 0$. (Answer: $f^{-1}(x) = \sqrt{\frac{x+8}{2}}$)

Original Function Inverse Function

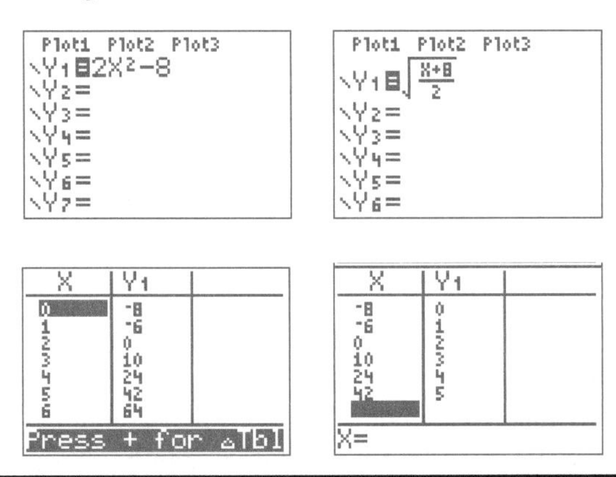

A graphing calculator has drawing, not graphing, features that can help you visualize the graph of an inverse. Drawn graphs are just pictures. You cannot use the calculator's features to evaluate them, or to find their zeros, intersections, or minimums. Nevertheless, a picture can be worth a thousand words.

Before drawing the inverse, you must have a good graph of the original function with enough window space to see the graph of the inverse, a reflection over the line $y = x$. The Draw Inverse command is found under the Draw key at 8: DrawInv. You must input Y1 from the F4 menu to specify the function. 1:ClrDraw will erase the picture when you are done.

DRAW POINTS STO
3↑Horizontal
4:Vertical
5:Tangent(
6:DrawF
7:Shade(
8:DrawInv
9↓Circle(

DrawInv Y1

DRAW POINTS STO
1:ClrDraw
2:Line(
3:Horizontal
4:Vertical
5:Tangent(
6:DrawF
7↓Shade(

Sketch the graph of the inverse of $f(x) = \sqrt{2x - 4}, x \geq 2$.

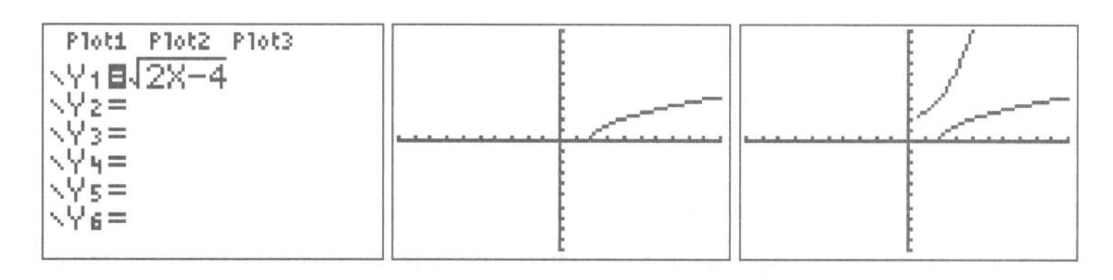

---

4. Absolute Value Functions

You should know how to graph absolute value functions on your calculator. Solving absolute value equations can introduce extraneous solutions, so checking either numerically or graphically is required. Equations with two absolute value expressions can be tedious to solve algebraically but straightforward to solve on a graphing calculator using its intersect feature.

TIP 4A

Solve for x: $|3x + 5| + |x - 3| = 12$. (Answer: $x = -3.5$ or $x = 2$)

Plot1 Plot2 Plot3
\Y1=|3X+5|+|X−3|
\Y2=12
\Y3=
\Y4=
\Y5=
\Y6=
\Y7=

Intersection X=-3.5 Y=12

Intersection X=2 Y=12

Your graphing calculator can graph piecewise defined functions, but they may take a lot more keystrokes than a simpler function. The following example is more easily done without a calculator. In order to graph this function on a graphing calculator, you must find the inequality test symbols in the TEST menu. These symbols test if x satisfies the inequality. Type the piecewise function in parentheses with its domain, also in parentheses, right after it. Add the different pieces together to put them in one function.

TEST LOGIC
1:=
2:≠
3:>
4:≥
5:<
6:≤

TIP 4B

Graph the function: $y = \begin{cases} -x & \text{if } x \leq -1 \\ 2x & \text{if } x > -1 \end{cases}$

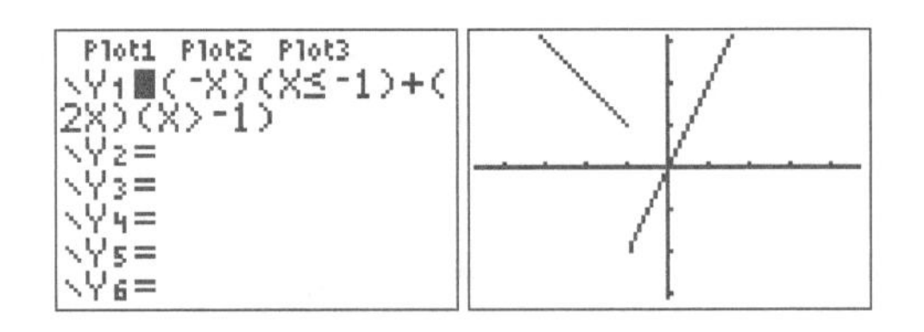

ZoomDecimal was used for the graph. Notice that the calculator does not convey whether the endpoints are open or closed. Using the table is also misleading because it is not clear if or where the function is not continuous.

Graphing a piecewise defined function with three definitions is even trickier. In addition to using the inequality symbols, you also need "and" found in the TEST key, LOGIC menu.

TIP 4C

Graph the function: $y = \begin{cases} -x + 2 & \text{if } x < 0 \\ 2 & \text{if } 0 \leq x < 3 \\ 2x - 4 & \text{if } x \geq 3 \end{cases}$

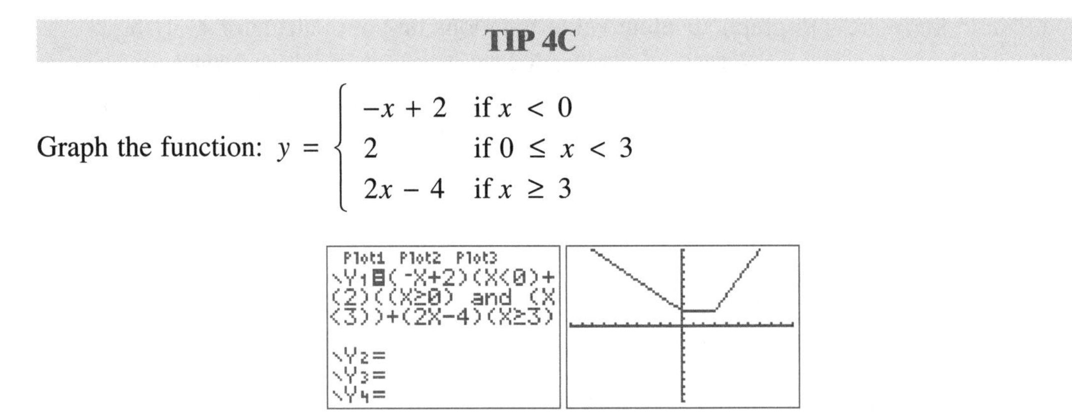

Note that you can type (0 ≤ X ≤ 3) in your calculator, but the TI-84 will not interpret it correctly. You must use "and." Many students find it easier to learn to graph piecewise defined functions by hand because the graphing calculator method can be so complicated.

Step functions can be graphed on calculators. The floor function, $y = \lfloor x \rfloor$ or $y = [x]$, is graphed using the MATH key, NUM menu 5: int.

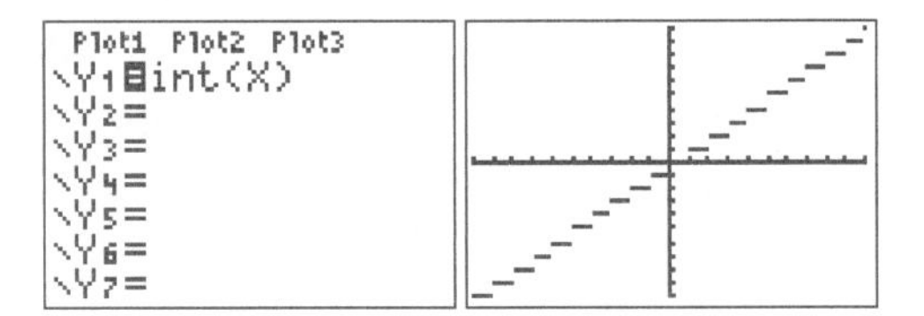

If your calculator does not have separate floor and ceiling functions, the ceiling function can be constructed from the floor function. The ceiling function, $y = \lceil x \rceil$ or $y =]x[$, is equivalent to $y = -\lfloor -x \rfloor$ or $-[-x]$.

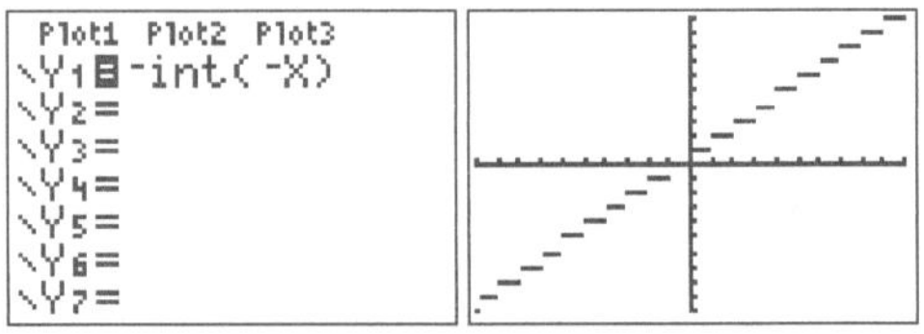

5. Quadratic Functions

Using a graphing calculator to find features of quadratic graphs is straightforward. You should know how to use a calculator to graph any parabola.

A graphing calculator can be used to solve quadratic equations. Set the equation equal to zero, and then use CALC 2: zero to find the zeros. You should spend time to get a good graph first. You want to see all the zeros in the window.

TIP 5A

Solve: $2x^2 = 12 - 5x$. (Answer: The solution set is {–4, 1.5}.)

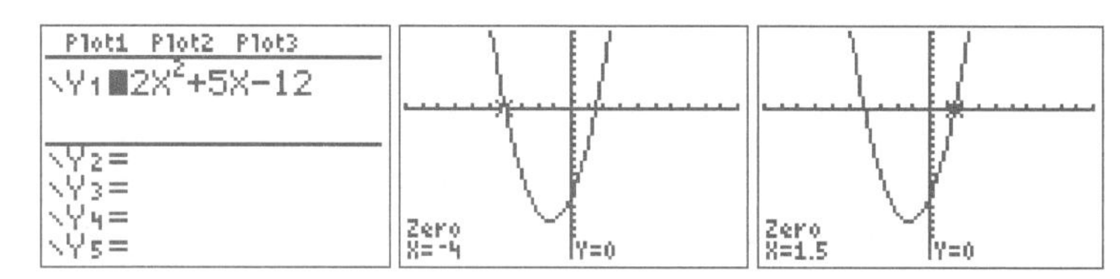

You can move the cursor with the left and right arrows. However, it is quicker just to type in a value to the left and then a value to the right. The value for the guess can be the same as the value for the right.

You can use the zero feature to help you factor quadratic polynomials. Chapter 5 includes a section on how to write a quadratic equation from its zeros. In this example, $2x^2 + 5x - 12$ factors into $(x + 4)(2x - 3)$.

TIP 5B

Sketch the graph of $y = -2x^2 - 4x + 6$. Identify the axis of symmetry, the zeros, the vertex, and the y-intercept. Describe where the graph is increasing and decreasing. Describe the end behavior.

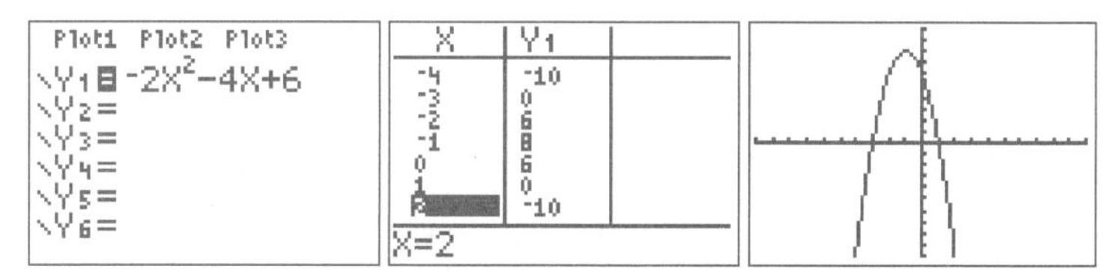

In many parabolas, you can use the table to find the axis of symmetry and vertex. Align the table so that the y-values are symmetric about the middle row, as shown in the table above. The vertex is the ordered pair in the middle row, (–1, 8). The axis of symmetry is the vertical line through this coordinate, $x = -1$. The y-intercept is found in the row where $x = 0$. Here, the y-intercept is 6. The zeros are found in the rows where the y-values equal 0. Here the zeros are $x = -3$ and $x = 1$. The graph changes from increasing to decreasing (because a is negative) at its axis of symmetry. The graph is easy to construct with the seven points shown in the table. End behavior is easy to see on the graph.

Not all parabolas will show their symmetry so obviously in a table like the example above, but many in Algebra 2 problems do. The graphing calculator has other features to help find the specific values for all quadratic equations.

TIP 5C

A parabola has equation $y = 2x^2 + 5x - 12$.

a. Does the parabola open up or down?
b. What is the y-intercept of the parabola?
c. What is the equation of the axis of symmetry of the parabola?
d. What are the coordinates of the vertex of the parabola? Is it a maximum or a minimum?
e. What are the zeros of the parabola?
f. Sketch the parabola.

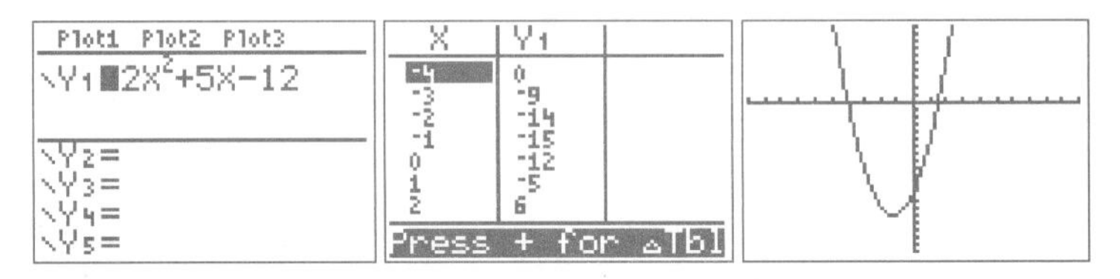

You can see from the graph that the parabola opens up. From the table, you can see that the y-intercept is –12. The table does not show the axis of symmetry. From the graph, the turning point is a minimum so use CALC 3: minimum to find the vertex.

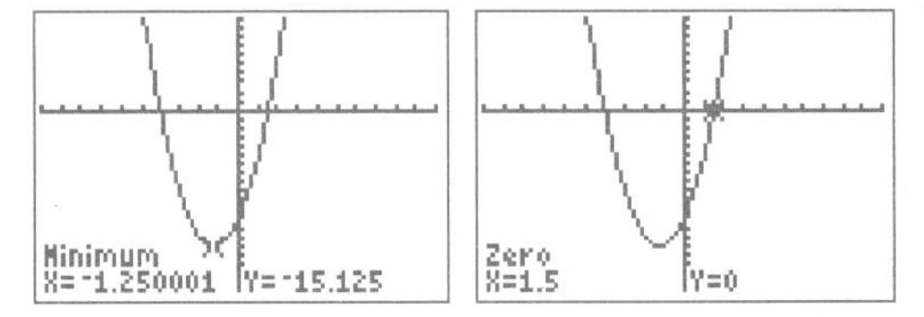

The vertex is (–1.25, –15.125). Knowing the value of the vertex's x-coordinate means the axis of symmetry is $x = -1.25$. From the table, one zero is $x = -4$. You cannot see the other root in the table. Use the CALC 2: zero to find the other zero. Now we know the two zeros are $x = -4$ and $x = 1.5$. Sketch the parabola using values from the table and the coordinates of the vertex and zeros.

You should be aware that the calculator does not use algebra to find roots and minimum values. It uses numerical techniques to find an answer that is "close enough" and displays that answer. Many times, the answer is exactly correct as in the case of the root above. However, sometimes an answer will appear as X = –1.250001 or X = –1.249997 when the true value is really –1.25. This is what happened on the calculation of the minimum above. In Algebra 2, it is usually safe to assume those are calculator errors and round appropriately.

These same techniques are used to solve word problems modeled with quadratic functions.

TIP 5D

After a baseball is hit by a bat, it follows a path given by $y = 3 + 0.4x + ax^2$, where y is the vertical height of the ball and x is the horizontal distance of the ball from the batter, both in feet.

a. How high off the ground was the ball when it was hit? (Answer: 3)
b. Suppose $a = -0.002$. What is the greatest height reached by the ball? (Answer: 23)

c. If the ball was not caught, how far from the batter, to the nearest tenth of a foot, did the ball first hit the ground if $a = -0.002$. (Answer: 207.2)

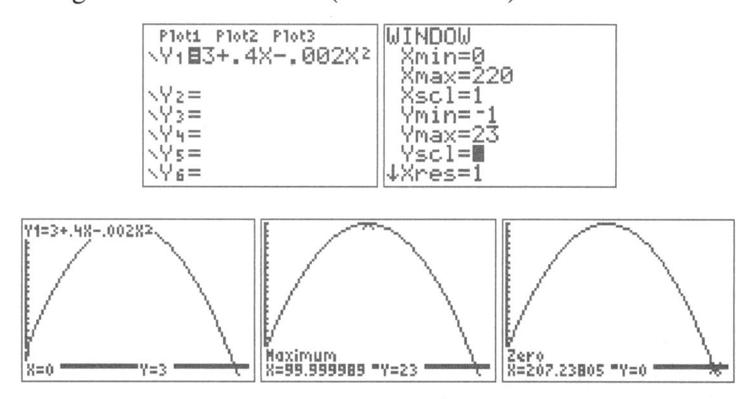

Spend time to get a good window for the graph before you start to use the CALC menu features. For this graph, $x = 100$ is the axis of symmetry, or the middle of the parabola. So set the x-values in the window from 0 to more than 200 to make sure you see where the ball hits the ground. Use ZoomFit to set the y-values for the window. Adjust your window until you see both the maximum and the right zero for the parabola. Use the TABLE or TRACE (type 0 for the x-value) to find the initial height of the ball. Use CALC 4: maximum to find the greatest height of the ball. Use CALC 2: zero to find where the ball hits the ground.

These same techniques are used to solve quadratic equations. Set the equation equal to zero, and then use a calculator to find the zeros. You should spend time to get a good graph first. You want to see all the zeros in the window.

Using the minimum or maximum feature of the calculator, we can rewrite a standard form equation in vertex form.

TIP 5E

Rewrite $y = 2x^2 - 8x + 9$ in vertex form. (Answer: $y = 2(x - 2)^2 + 1$)

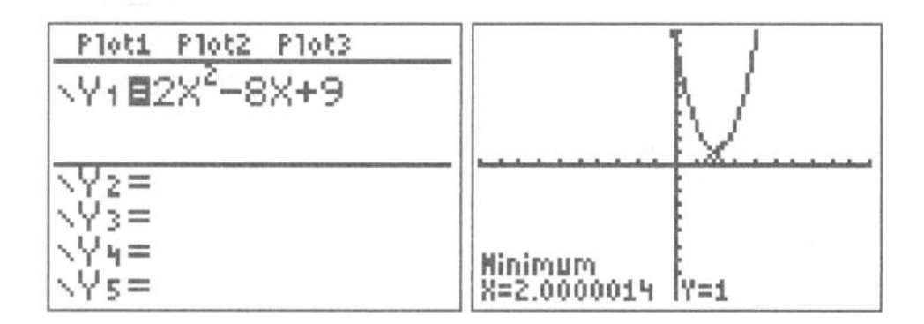

From the calculator, we find the vertex is (2, 1). The vertex form of a quadratic equation is $y = a(x - h)^2 + k$. The value of a stays the same in vertex form, so $y = 2x^2 - 8x + 9$ changes to $y = 2(x - 2)^2 + 1$.

You can solve quadratic-linear systems on your graphing calculator easily using CALC 5: intersection. If a problem requires that the system be solved algebraically, use your graphing calculator to double-check your solution. You may need to solve one or both equations for y first.

TIP 5F

Solve: $\begin{cases} 3x - 2y = 4 \\ y = \dfrac{1}{2}x^2 + x - 3 \end{cases}$. (Answer: {(–1, –3.5), (2, 1)})

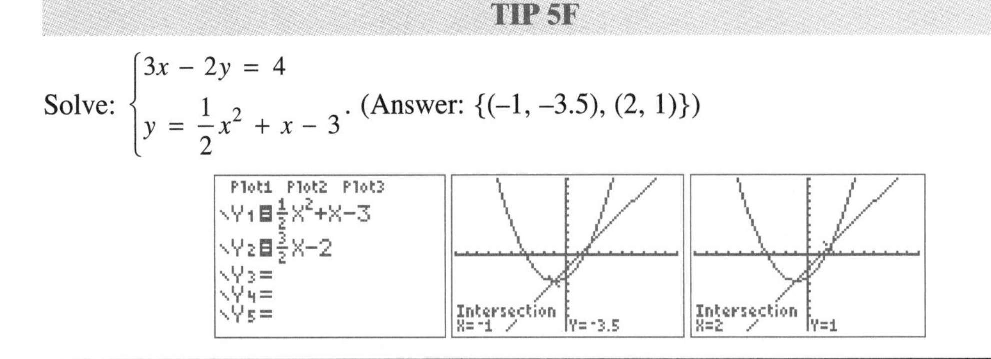

---

6. Complex Numbers

Under the MODE key, you can set your calculator to $a + bi$ mode. This will allow you to use your calculator to do calculations with complex numbers. Sometimes the results displayed are confusing, so you should practice on examples with known answers to see how to interpret results in this mode.

First evaluate the square root of a negative number in both REAL and $a + bi$ modes to see the difference in the calculator's response.

TIP 6A

Rewrite the following in terms of i: $\sqrt{-16}$. (Answer: $4i$)

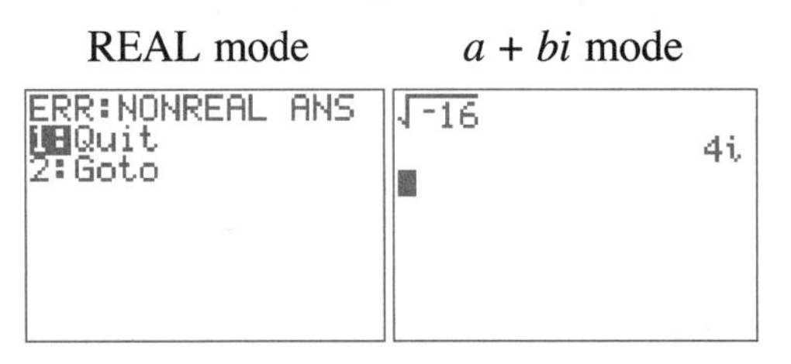

Powers of i can be computed on the calculator, but the answer sometimes needs interpreting. Look for the i symbol above the decimal key.

TIP 6B

Evaluate: i^{11} (Answer: $-i$)

Evaluate: i^{126} (Answer: -1)

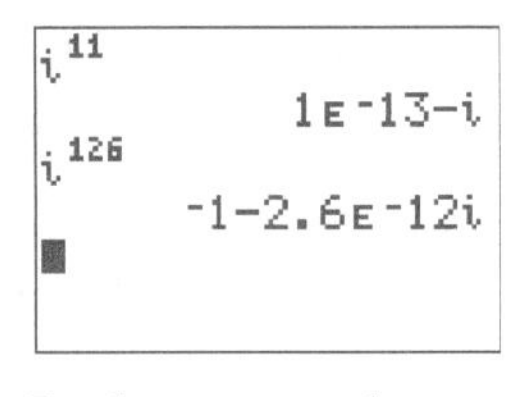

Computations with i can involve very small calculator errors. In the screen above, 1E-13 is calculator notation for 1×10^{-13}, which is the same as 0.0000000000001, very close to 0. When using $a + bi$ mode, you should interpret such numbers as calculator error and round appropriately. So 1E-13–i really means $0 - i$ or $-i$. Interpret the next line, –1–2.6E-12i as $-1 - 0i$ or -1.

Except for interpreting the numbers carefully, using a calculator to check your work for complex numbers is generally easy.

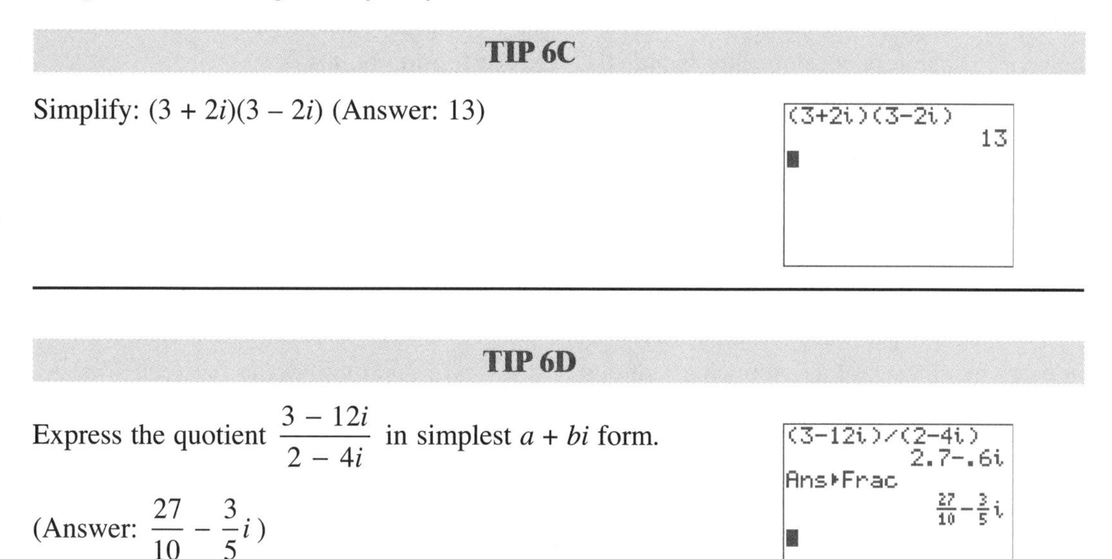

TIP 6C

Simplify: $(3 + 2i)(3 - 2i)$ (Answer: 13)

TIP 6D

Express the quotient $\frac{3 - 12i}{2 - 4i}$ in simplest $a + bi$ form.

(Answer: $\frac{27}{10} - \frac{3}{5}i$)

TIP 6E

Simplify: $(2 + 3i)^5$ (Answer: $122 - 597i$)

(2+3i)⁵
122-597i

Be sure you return to REAL mode when you are done. Sometimes $a + bi$ mode can give surprising results for functions of real numbers. For example, logarithms of negative numbers have complex values. However, in Algebra 2, we use only logarithms of positive numbers.

7. Polynomial Functions

Getting a good graph of a polynomial function on a graphing calculator can be trickier than you might expect. You need to find the best window to display the important features of the graph. If the y-scale is too small, the graph may end up looking like a series of very steep lines. If the y-scale is too large, you may not be able to see the individual zeros and turning points. You want to find an appropriate window between these two extremes. Using the table can help you set an appropriate scale.

TIP 7A

Factor and find the zeros of the polynomial function: $y = 6x^3 - 23x^2 - 13x + 70$

(Answer: Zeros: $x = 2$, $x = -\frac{5}{3}$, $x = \frac{7}{2}$)

You want to graph the polynomial to help answer this question. The graph on the left below does indicate the zeros; however, the graph does not look like a polynomial function. The graph on the right below looks like a continuous curve and shows the end behavior accurately but does not display the zeros or turning points.

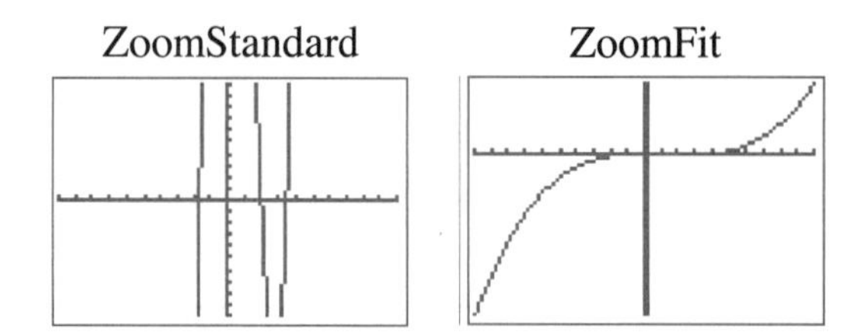

To find an appropriate window, first choose your x-values. From the graphs above, it appears that most of the interesting features of this polynomial will occur between $x = -3$ and $x = 5$. Enter those values in WINDOW and then use ZOOM 0: ZoomFit.

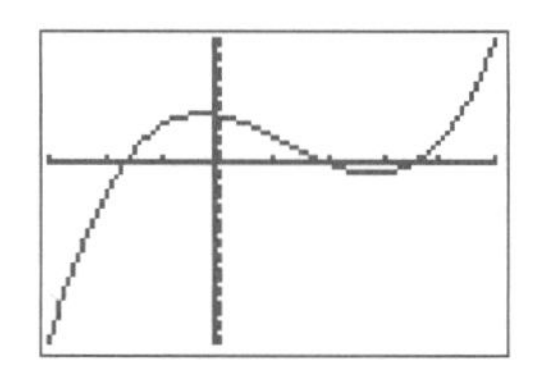

This graph may already be good enough for your purposes. If necessary, you can refine it using the table to help select values for Ymin and Ymax in the window. The table shows that for $-3 \le x \le 5$, the y-values range from -260 to 180. If you ignore those two values that occur at the endpoints of the x-interval, the next smallest y-value is -44 and the next largest y-value is 70. Setting the Ymin = -50 and the Ymax = 80 gives a very good graph of the polynomial function.

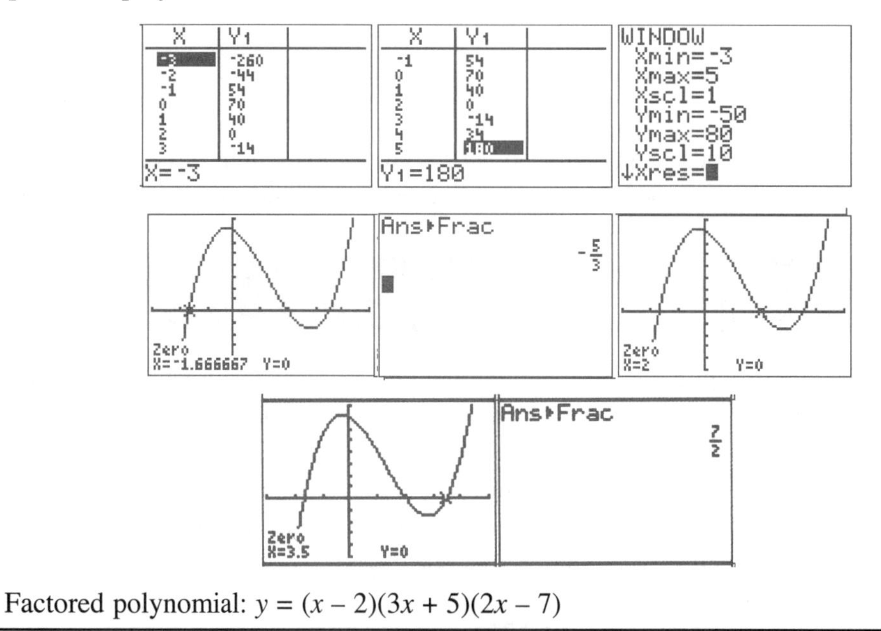

Factored polynomial: $y = (x - 2)(3x + 5)(2x - 7)$

Using the TableAsk feature can help you check values in different intervals to solve an inequality algebraically.

TIP 7B

Solve $2x^3 - 3x^2 - 18x + 27 > 0$ algebraically. (Answer: $-3 < x < 1.5$ or $x > 3$)

After you find the zeros by factoring the polynomial, you construct a number line to show the intervals between the zeros.

Then you check values in each interval to see if that interval satisfies the original inequality. Use TableAsk to check a value in each interval. For example, for the interval on the left, $x < -3$, check $x = -4$.

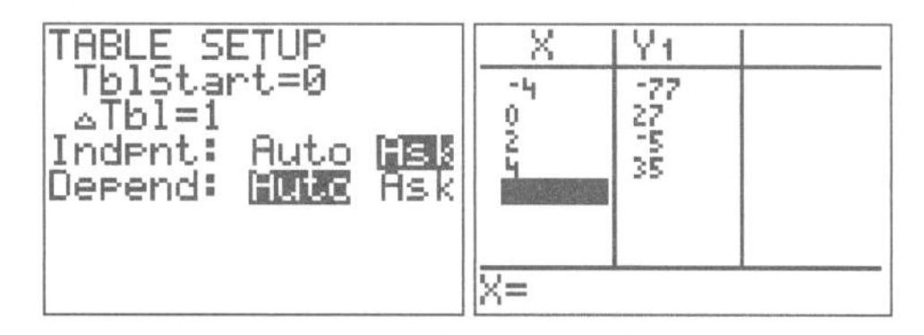

The inequality is satisfied for values of y that are greater than 0. At $x = 0$ and $x = 4$, the y-values are positive, so you can sketch the solution on the number line on those intervals.

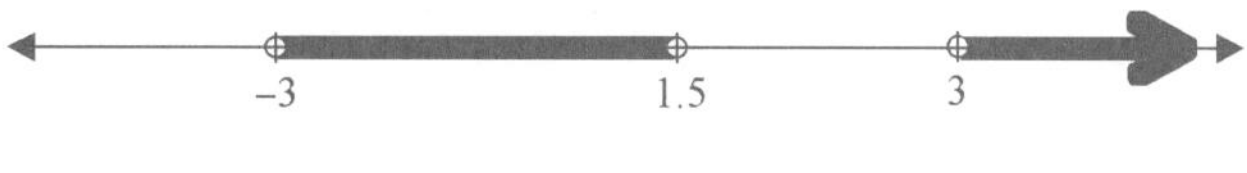

The solution is $-3 < x < 1.5$ or $x > 3$.

---

8. Radical Functions

Simplifying radicals requires that you be able to construct a prime factor tree of a number to recognize perfect squares and cubes as factors. When the numbers are small and familiar, many students can do this mentally. When the numbers are larger, though, it would be helpful to know how to use the calculator efficiently. Use this technique to factor numbers into prime factors or to check if a number is already prime.

Type the number divided by x in Y1. Scan the table looking for integer factors.

TIP 8A

Simplify: $\sqrt{60}$ (Answer: $2\sqrt{15}$)

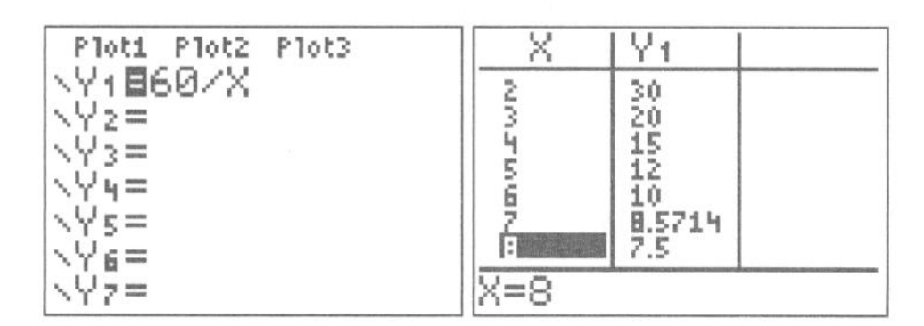

You can use any of the integer factors in the table as the start of the prime factor tree. For example, if you use 6 × 10, then you can continue and get the prime factor tree shown at right.

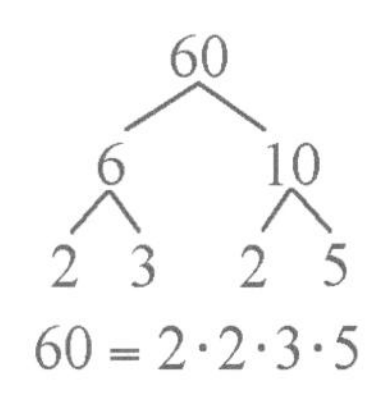

Here is a much harder example. A calculator is very helpful for such large numbers. Factor 539 into prime factors.

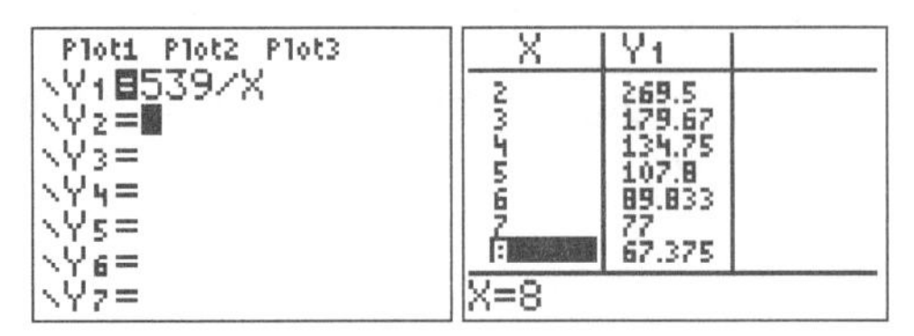

From the table, you can see that 7(77) = 539. You probably know that 77 factors into 7 and 11, so 539 = 7 · 7 · 11.

You need to look at only the x-values from 2 to the square root of your number. A number with no integer factors smaller than its square root is a prime number and cannot be factored further.

Factor 149 into prime factors.

Plot1 Plot2 Plot3
\Y1=149/X
\Y2=
\Y3=
\Y4=
\Y5=
\Y6=
\Y7=

X	Y1
2	74.5
3	49.667
4	37.25
5	29.8
6	24.833
7	21.286
8	18.625

Press + for ΔTbl

X	Y1
6	24.833
7	21.286
8	18.625
9	16.556
10	14.9
11	13.545
12	12.417

X=6

From the table, you can see that 149 has no integer factors. You need to check up to only $x = 12$ because $\sqrt{149} \approx 12.2$.

Pay attention to the wording of the problem before using your calculator to convert radicals to decimal form. The decimal given by your calculator is almost never the exact value. If a problem specifies "simplest radical form" or "exact value," do not write the decimal approximation as your answer. On the other hand, using decimal approximations can be very helpful for checking multiple-choice questions. Be sure you always check all the choices; there may be equivalent values where you need to choose the simplest form.

TIP 8B

Which of the following is $\frac{\sqrt{72}}{\sqrt{12}}$ written in simplest radical form? (Answer: Choice (4))

(1) $2\sqrt{3}$ (2) $3\sqrt{2}$ (3) $\frac{3\sqrt{2}}{\sqrt{3}}$ (4) $\sqrt{6}$

Evaluate the original problem and then each answer on your calculator, rounding to the nearest hundredth: $\frac{\sqrt{72}}{\sqrt{12}} \approx 2.45$

(1) 3.46 (2) 4.24 (3) 2.45 (4) 2.45

Clearly answers (1) and (2) are not correct. Answers (3) and (4) are equivalent numbers, but answer (3) is not in simplest radical form because there is a radical in the denominator. Answer (4) is correct.

You should know how to graph radical functions on your calculator. Solving radical equations algebraically can introduce extraneous solutions, so checking either numerically or graphically is required. Extraneous solutions do not occur graphically.

TIP 8C

Solve $x + 2\sqrt{7 - 2x} = 1$ graphically. (Answer: $x = -9$)

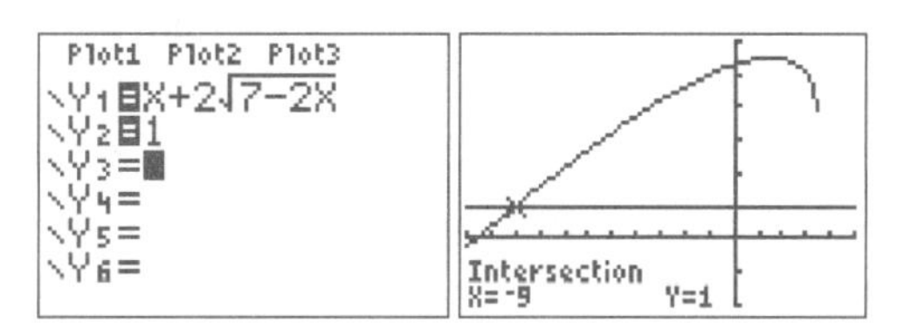

To find inverses of functions with square roots, you need to know the range of the original function to be able to state the restricted domain on the inverse. Use your graphing calculator and your knowledge of transformations of square root functions to find the range. Double-check your answer by comparing the graph displayed by DrawInverse with your inverse function.

TIP 8D

Find the inverse: $f(x) = \frac{1}{2}\sqrt{x + 1} + 6$ (Answer: $f^{-1}(x) = 4(x - 6)^2 - 1, x \geq 6$)

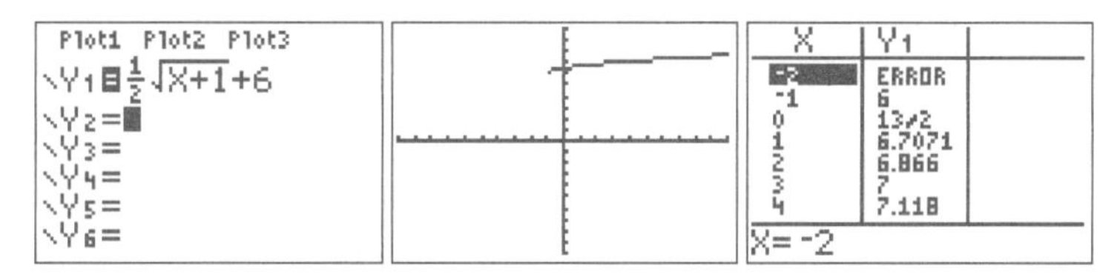

Your knowledge of transformations should tell you that the graph has been moved up 6 units and the range is $y \geq 6$. You can confirm this on the graph and table shown above. Algebraically, you find that the inverse function is $f^{-1}(x) = 4(x - 6)^2 - 1, x \geq 6$. The inverse graph below right was graphed using DrawInverse on the original function.

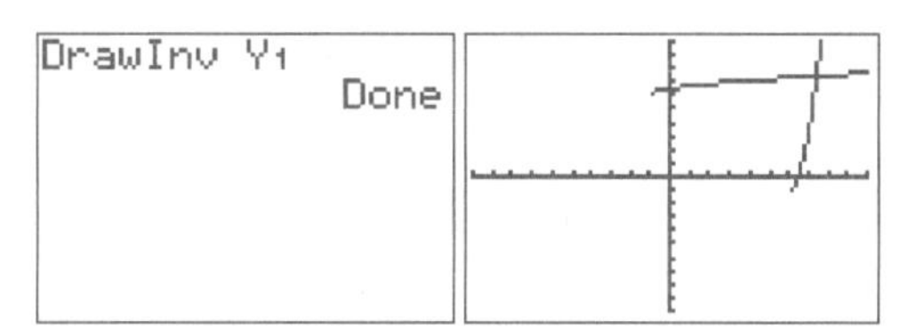

Compare the DrawInverse graph to the actual graph of the inverse function. To graph the inverse function, in Y2, enclose the entire inverse function in parentheses, and then type the domain restriction, also in parentheses. See the tips from Chapter 3 if you have not done inverses on your calculator before.

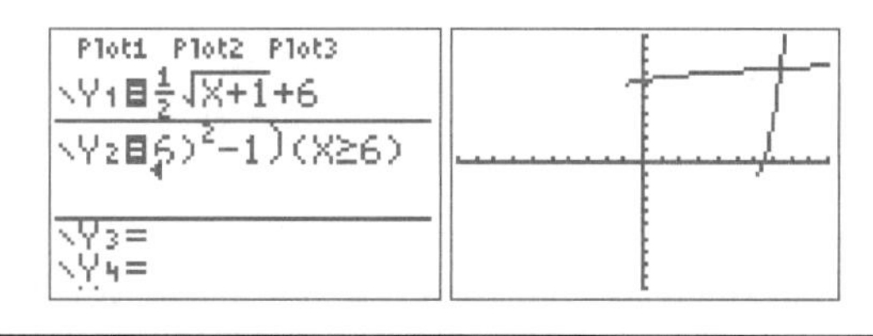

9. Rational Functions

Remember, your calculator will not actually simplify rational expressions unless it has Computer Algebra System (CAS) capabilities. However, you can use your calculator to check your answers to any simplifications as long as the expressions have a single variable (only x, not x and y). This technique is particularly helpful on multiple-choice questions. Be sure to check all choices; you need to choose the simplest form.

TIP 9A

Simplify: $\dfrac{8-4x}{x^2-4}$ (Answer: $\dfrac{-4}{x+2}$)

Plot1 Plot2 Plot3
$\backslash Y_1 = \dfrac{8-4X}{X^2-4}$
$\backslash Y_2 = \dfrac{-4}{X+2}$
$\backslash Y_3 =$
$\backslash Y_4 =$
$\backslash Y_5 =$

X	Y1	Y2
-3	4	4
-2	ERROR	ERROR
-1	-4	-4
0	-2	-2
1	-4/3	-4/3
2	ERROR	-1
3	-4/5	-4/5

Press + for △Tbl

When comparing the table for rational expressions, some x-values may be undefined in the original expression but not in the simplified version. These x-values were in the factors that were reduced out. Just check rows where Y_1 has a real value.

The screen shots below are what you would see if you forgot the –1 factor. Notice how the values in the table do not match; they have opposite signs.

Plot1 Plot2 Plot3
$\backslash Y_1 = \dfrac{8-4X}{X^2-4}$
$\backslash Y_2 = \dfrac{4}{X+2}$
$\backslash Y_3 =$
$\backslash Y_4 =$
$\backslash Y_5 =$

X	Y1	Y2
-3	4	-4
-2	ERROR	ERROR
-1	-4	4
0	-2	2
1	-1.333	1.3333
2	ERROR	1
3	-.8	.8

Press + for △Tbl

As with polynomial functions, you want to get an appropriate window when graphing rational functions on your calculator. Even with a good window, you may need to interpret the graph carefully. Use the graph, the table, and your knowledge of rational functions to help find all the important features of the graph.

TIP 9B

Identify the vertical asymptote(s) for $y = \dfrac{x^2+x-2}{x^2-4} = \dfrac{(x+2)(x-1)}{(x+2)(x-2)}$.

Plot1 Plot2 Plot3
$\backslash Y_1 = \dfrac{X^2+X-2}{X^2-4}$
$\backslash Y_2 =$
$\backslash Y_3 =$
$\backslash Y_4 =$
$\backslash Y_5 =$

X	Y1
-3	.8
-2	ERROR
-1	.66667
0	.5
1	0
2	ERROR
3	2

Press + for △Tbl

Factoring the denominator suggests possible vertical asymptotes at $x = \pm 2$, and the table confirms the function is undefined at both those points. However, a graph of the function in ZoomStandard, following, shows no vertical asymptote at $x = -2$. Instead, the function has a removable discontinuity. However, it does not show on the calculator graph unless you switch to the ZoomDecimal window.

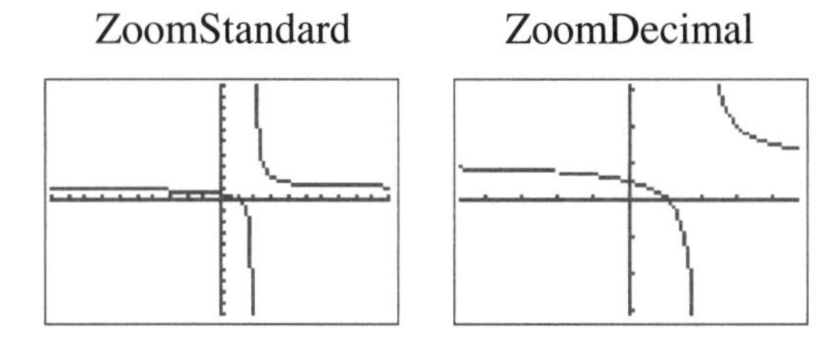

TIP 9C

Find the vertical asymptotes and the zeros of the function: $y = \dfrac{x^2 - x - 12}{x^3 - 4x}$

(Answer: The vertical asymptotes are $x = -2$, $x = 0$, and $x = 2$. The zeros are $x = -3$ and $x = 4$.)

On the calculator, examine the table to find zeros where the y-value equals 0 and to find potential vertical asymptotes where the y-value says ERROR. The function has roots at $x = -3$ and $x = 4$ and has possible vertical asymptotes at $x = \pm 2$ and $x = 0$.

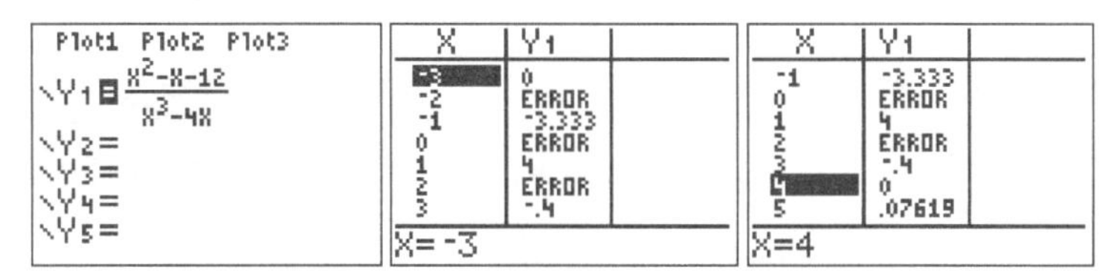

Check the asymptotes by graphing the function. If your rational equation is not fully reduced, the rational function may have a removable discontinuity instead of a vertical asymptote. In this case, the graph confirms vertical asymptotes at all three values.

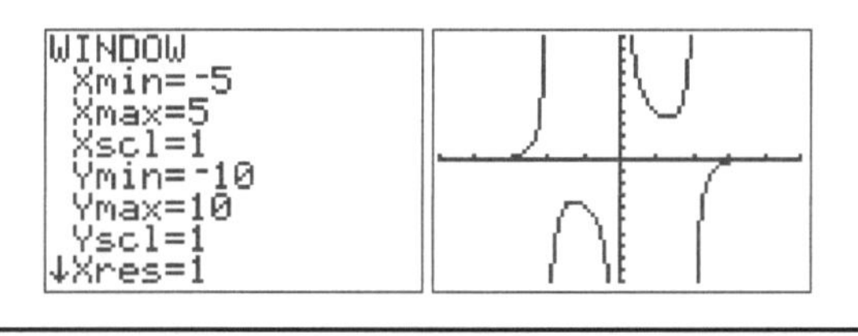

TIP 9D

Describe the end behavior of the rational function: $y = \dfrac{2x^2 - 8}{x^2 - 2x + 1}$

(Answer: Horizontal asymptote at $y = 2$)

In addition to identifying the vertical asymptotes and zeros, you need to know the end behavior of the function. The graph below left with the ZoomDecimal window suggests a horizontal asymptote around $y = 2$. Changing the scale in the x-direction to Xmin = –20 and Xmax = 20, below right, gives even stronger evidence of a horizontal asymptote.

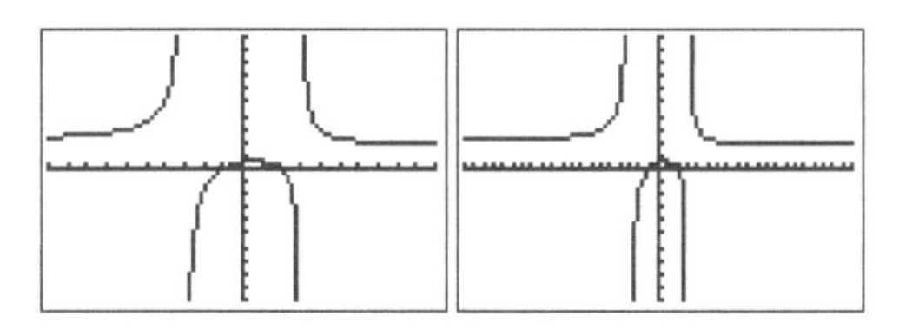

You can use the table to get further confirmation by evaluating very large positive x-values and very large negative x-values and see what happens to the y-values. One easy way to do this is to use the Table Ask feature, found on the Table Setup screen. After you set the table to Ask, go to the table (which is empty), and type in ever larger positive x-values. Look at the pattern in the y-values.

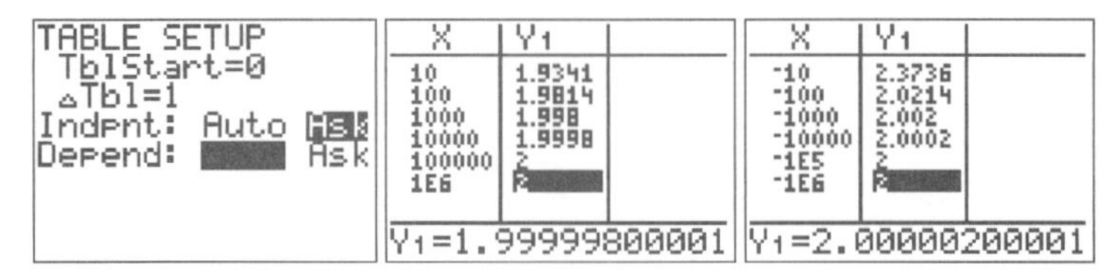

In this example, the y-values appear to be converging to 2. (In fact, in the table it appears that $Y_1(100{,}000) = 2$. Placing the cursor on a value in the Y_1 column will give you more decimals of accuracy.) More specifically, on the right side of the graph, the function approaches $y = 2$ from below. Repeating the process for large negative x-values shows that the left side of the graph approaches $y = 2$ from above.

---

10. Exponential Functions

Exponential functions are quite clear on a graphing calculator. It is easy to sketch the graph. Be careful when interpreting the table. Exponential functions grow quickly, so the table will show rounded values. Placing the cursor on a rounded value can show more decimals.

TIP 10A

Find the horizontal asymptote and range: $y = e^{-x} + 3$. (Answer: The horizontal asymptote is $y = 3$, and the range is $(3,\infty)$.)

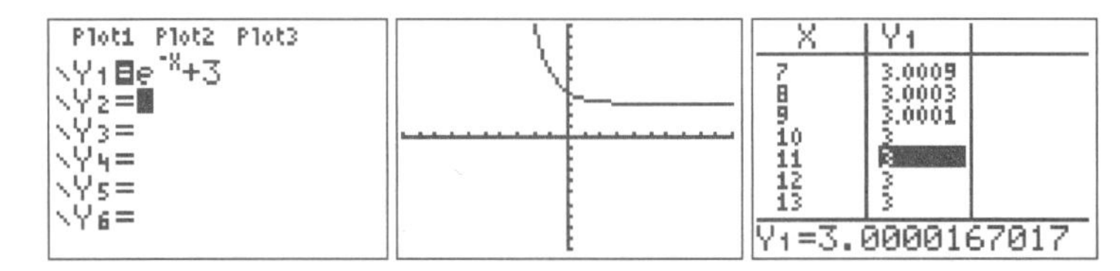

By using the table and graph, the horizontal asymptote is $y = 3$ and the range is $y > 3$. Be careful to use your knowledge about exponential functions. Despite the table values, y never equals 3 so the range cannot be $y \geq 3$.

You can solve exponential equations easily with a graphing calculator. If a problem requires that the solution be shown algebraically, use your calculator to double-check your answer.

TIP 10B

Solve: $\left(\frac{1}{4}\right)^{x+2} = 8^{\frac{2x}{3}}$ (Answer: $x = -1$)

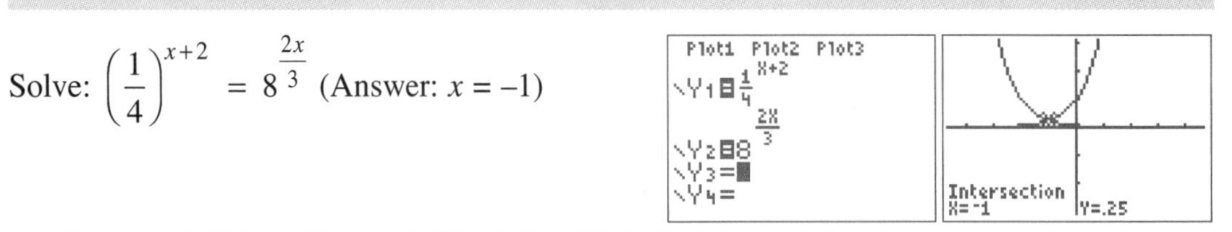

---

11. Logarithmic Functions

You need to use your knowledge of logarithmic functions to sketch a log graph accurately from a calculator screen.

TIP 11A

For $f(x) = \ln(x - 3) + 1$, identify its domain, range, and vertical asymptote. (Answer: Domain is $x > 3$. Range is all real numbers. The vertical asymptote is $x = 3$.)

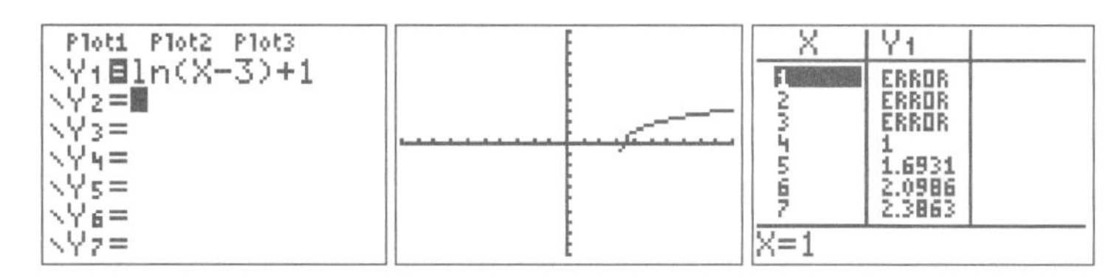

The graph appears to just end in Quadrant IV, like where a square root graph might end. By looking at the graph, you might think the range is limited, but logarithm functions have a range of all real numbers. You must be careful when interpreting graphs of logarithmic functions on a calculator.

Most log graphs have a vertical asymptote at one edge of their domain. The calculator screen does not sketch the graph well as it approaches its vertical asymptote. The graph of $y = \ln x$ has a vertical asymptote at the y-axis ($x = 0$). This graph is a translation of $y = \ln x$ 3 units to the right, so this vertical asymptote is at $x = 3$. Be careful interpreting the table to find domain. You might be tempted to say the domain is $x \geq 4$, but actually the domain is $x > 3$.

You can solve both exponential and logarithmic equations easily with a graphing calculator. If a problem requires that the solution be shown algebraically, use your calculator to double-check your answer.

TIP 11B

Solve for x to the nearest thousandth: $8e^{0.3x} = 300$ (Answer: 12.081)

The most difficult part of solving equations on the calculator is setting up the window for a good graph. A check of the table shows the solution will be near $x = 12$. Set the window accordingly. The window for the graph at right is $0 \leq x \leq 25$ and $0 \leq y \leq 600$, which puts the intersection near the center of the window.

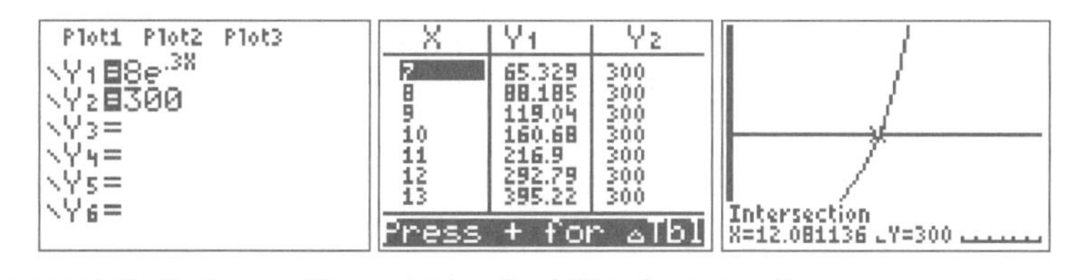

TIP 11C

Solve for x: $3\ln(2x + 1) = 5$ (Answer: $x = \dfrac{e^{\frac{5}{3}} - 1}{2} \approx 2.147$)

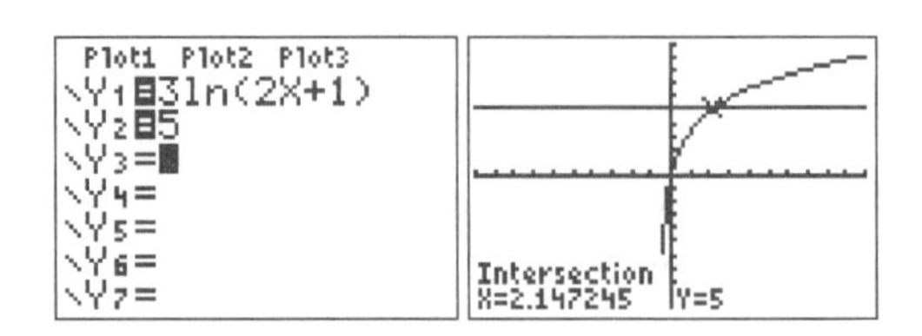

Using a calculator to solve equations will give you the decimal approximation of any irrational solution. If a problem requires an exact solution, you can use this decimal to double-check your algebraic solution.

TIP 11D

Solve for x: $2\log_4 x - 1 = \log_4 (24 - x)$ (Answer: $x = 8$)

When this equation was solved algebraically, $x = -12$ was an extraneous solution that had to be checked and rejected. It is clear by looking at the graph that the curves will not intersect at $x = -12$ and that $x = 8$ is the only solution.

Some calculators can evaluate logarithms in any base. Even if your calculator does not have this feature, you can evaluate logarithms in any base using the change of base formula: $\log_b N = \dfrac{\log_a N}{\log_a b}$

TIP 11E

Evaluate: $\log_2 50$ (Answer: $\log_2 50 = \dfrac{\log 50}{\log 2} \approx 5.644$)

Built-In Function

Change-of-Base Formula

12. Trigonometric Functions

When working with trigonometric functions, you need to be careful to set your calculator in the correct mode for your angle measure. Angles can be measured in either radians or degrees. To use a calculator to convert between degrees and radians, set the mode to the units you are converting to. Then type in the angle using the label for the units you are converting from, found in the ANGLE menu. 1: ° is for degrees, and 3: r is for radians.

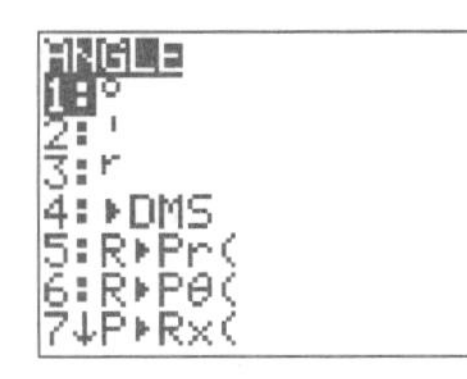

TIP 12A

Convert 75° to radians.
(Answer: $\frac{5\pi}{12} \approx 1.309$)

```
NORMAL SCI ENG
FLOAT 0123456789
RADIAN DEGREE
FUNC PAR POL SEQ
CONNECTED DOT
SEQUENTIAL SIMUL
REAL a+bi re^θi
FULL HORIZ G-T
↓NEXT↓
```

```
75°
             1.308996939
```

TIP 12B

Convert 2 radians to degrees. Round your answer to the nearest thousandth.
(Answer: 114.592°)

```
NORMAL SCI ENG
FLOAT 0123456789
RADIAN DEGREE
FUNC PAR POL SEQ
CONNECTED DOT
SEQUENTIAL SIMUL
REAL a+bi re^θi
FULL HORIZ G-T
↓NEXT↓
```

```
2r
              114.591559
```

Sometimes angles in degree measure are given in degrees, minutes, seconds (DMS) instead of decimal degrees. The calculator will convert between these types of measures using 4: ▶ DMS.

Express 50.16° to the nearest second.

```
50.16▸DMS
              50°9'36"
```

You can also input degrees given in DMS notation using the degree symbol in the ANGLE menu, 1: °; the minute symbol in the Angle menu 2: ′; and the seconds symbol found above the + key, ″.

Some angles have exact trigonometric function values that students are expected to memorize (or be able to derive). The most common ones are 30°, 45°, and 60°. Remember that if a problem asks for the exact value, a decimal approximation from the calculator of an irrational number will not receive full credit.

TIP 12C

Write the exact values for sin 150°, cos 150°, and tan 150°.

Many Algebra 2 teachers expect their students to be able to find exact values for the trigonometric function of angles where the reference angle is 30°, 45°, or 60°. Some students simply memorize them all or remember how to construct Table 12.2 quickly. Other students memorize the two special triangles, the 30°-60°-90° with sides 1, $\sqrt{3}$, and 2 (Figure 12.17), and the 45°-45°-90° with sides 1, 1, and $\sqrt{2}$ (Figure 12.16), and use them to find the exact values. Still other students prefer to memorize the decimal approximations of the special irrational trigonometric function values and then use their calculators. Here are the three common special angles and their trigonmetric function values.

$$\sin 30° = \frac{1}{2} \qquad \sin 45° = \frac{\sqrt{2}}{2} \qquad \sin 60° = \frac{\sqrt{3}}{2}$$

$$\cos 30° = \frac{\sqrt{3}}{2} \qquad \cos 45° = \frac{\sqrt{2}}{2} \qquad \cos 60° = \frac{1}{2}$$

$$\tan 30° = \frac{\sqrt{3}}{3} \qquad \tan 45° = 1 \qquad \tan 60° = \sqrt{3}$$

```
sin(30)                  sin(45)                  sin(60)
             .5             .7071067812              .8660254038
cos(30)                  cos(45)                  cos(60)
    .8660254038             .7071067812                       .5
tan(30)                  tan(45)                  tan(60)
    .5773502692                       1              1.732050808
```

You can see there are four irrational values that might be worth memorizing:

$$\frac{\sqrt{2}}{2} = 0.7071\ldots, \quad \frac{\sqrt{3}}{2} = 0.8660\ldots, \quad \frac{\sqrt{3}}{3} = 0.5773\ldots, \quad \sqrt{3} \text{ and } = 1.7320\ldots.$$

If you memorize these values, you can use your calculator to help do the example above.

```
sin(150)
             .5
cos(150
   -.8660254038
tan(150)
   -.5773502692
```

From this, you see that $\sin 150° = \frac{1}{2}$, $\cos 150° = -\frac{\sqrt{3}}{2}$, and $\tan 150° = -\frac{\sqrt{3}}{3}$

You can do the same thing in radian mode.

TIP 12D

Write the exact values for $\sin \frac{5\pi}{4}$, $\cos \frac{5\pi}{4}$, and $\tan \frac{5\pi}{4}$.

```
sin(5π/4)                tan(5π/4)
   -.7071067812                       1
cos(5π/4)
   -.7071067812
```

From this, you see that $\sin \frac{5\pi}{4} = \cos \frac{5\pi}{4} = -\frac{\sqrt{2}}{2}$ and $\tan \frac{5\pi}{4} = 1$.

For problems requiring an exact value, you can use the calculator to check your answer.

TIP 12E

Use the sum formula to find the exact value of sin 75°.

(Answer: $\dfrac{\sqrt{6}+\sqrt{2}}{4}$)

Having worked out the answer using the formula for sin(A + B), check it on the calculator.

```
sin(75)
           .9659258263
√6+√2
  4
           .9659258263
```

13. Trigonometric Graphs and Equations

Trigonometric functions are usually graphed in radians. Read the problem carefully, and be sure to set your calculator to the correct mode. Set the x-values in the WINDOW menu to match the requested interval. Use ZoomFit to set the y-values for sine or cosine graphs.

ZoomFit can give poor graphs for the tangent function. You need to sketch the vertical asymptotes accurately. The ZoomTrig window found under the ZOOM menu at 7: ZTrig is often a good choice for graphing the tangent function. Using ZoomFit is not advised for tangent because the y-values are unbounded over most intervals.

TIP 13A

Sketch the graph of $f(x) = \tan\left(\frac{1}{2}x\right)$ over the interval $-2\pi \le x \le 2\pi$.

Enter -2π and 2π into Xmin and Xmax. (The calculator will convert them to decimals.) Set Xscl to $\frac{\pi}{2}$. The graph on the left is the result of using ZoomFit. The calculator chose a range for y of about –29.9 to 29.9. Setting a smaller range for y, –5 to 5, gives the second graph. Both graphs are correct, but the second one might be considered more representative of what a tangent graph looks like.

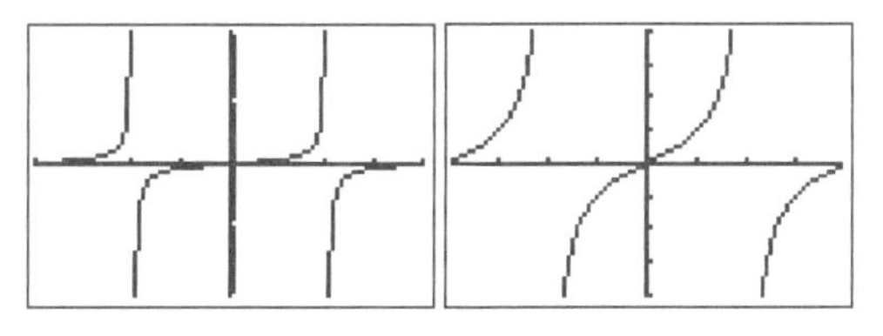

When making a table for a trig function, consider what x-values are of interest when setting ΔTbl. The table below uses ΔTbl = $\frac{\pi}{4}$. Other good choices might be $\frac{\pi}{6}$ or $\frac{\pi}{12}$.

X	Y1
0	0
.7854	.41421
1.5708	1
2.3562	2.4142
3.1416	ERROR
3.927	⁻2.414
[illegible]	⁻1

X=4.712388980385

X	Y1
18.064	⁻.4142
[illegible]	1E⁻13
19.635	.41421
20.42	1
21.206	2.4142
21.991	⁻1E12
22.777	⁻2.414

X=18.84955592154

Note that the calculator shows $y = 0$ at $x = 0$ where the graph has a zero and says ERROR at $x = \pi$ where the graph has a vertical asymptote. Be aware that as you scroll through the table, small amounts of calculator error build up. So the zeros and asymptotes will not always be as clear. At $6\pi \approx 18.85$, the table shows 1×10^{-13} instead of 0. At $7\pi \approx 21.991$, the table shows -1×10^{12} instead of undefined. Numbers close to 0, especially if displayed in scientific notation, suggest you are close to a root. Very large numbers, either positive or negative, suggest you are close to a vertical asymptote.

TIP 13B

Sketch a graph of $y = 5 + 3\sin(2x)$.

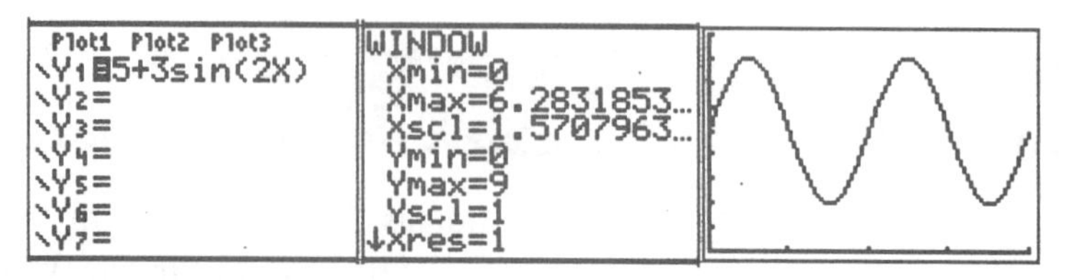

If you have already sketched the graph, use the calculator display to check your work. If you are using a calculator to plot points, you should change the table to increase incrementally by the scale on the x-axis. The scale is found by determining the interval between key points, $\frac{\pi}{2b}$. For this example, the x-scale is $\frac{\pi}{2(2)} = \frac{\pi}{4}$. Using this increment in the table will ensure that you plot the critical points to sketch an accurate graph. Recognizing the decimal approximations of π and 2π in the table is helpful.

TABLE SETUP
TblStart=0
ΔTbl=π/4
Indpnt: Auto Ask
Depend: Auto Ask

X	Y1
0	5
.7854	8
1.5708	5
2.3562	2
[illegible]	5
3.927	8
4.7124	5

X=3.14159265359

X	Y1
1.5708	5
2.3562	2
3.1416	5
3.927	8
4.7124	5
5.4978	2
[illegible]	5

X=6.28318530718

A calculator can also be useful for solving trigonometric equations or checking solutions to trigonometric equations found algebraically. Make sure your calculator is in degree mode because "solve for all θ in $0° \le \theta < 360°$" means the solutions should be expressed in degrees.

TIP 13C

Solve for all x in $0° \le x < 360°$ that satisfies $2\cos x + \sqrt{3} = 0$. (Answer: {150°, 210°})

Set the equation equal to 0 (if necessary), and type the equation in Y1. In the WINDOW menu, set the x-values from 0 to 360 to match the requested interval for the solutions. Use ZoomFit to set the y-values for the window. Then use the Zero feature of the calculator to find all the solutions.

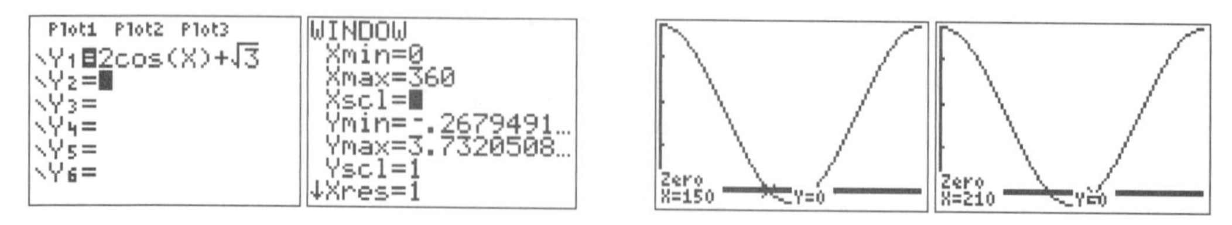

If all you want to do is check solutions you found algebraically, you have a couple of choices. The best one is to solve the equation graphically as previously and see if the calculator solutions match the ones you found. The advantage to this is that a graphical solution should not miss any correct answers.

Alternatively, you can check your answers in the original equation just as you would for any other type of equation. This will tell you whether the answers you found are correct. However, it will not tell you if you missed any answers. If you found only one of the answers previously, a simple check would not tell you there was another answer.

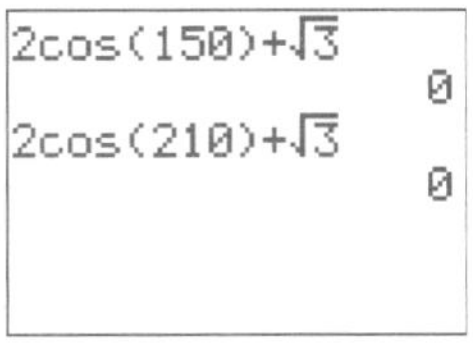

```
2cos(150)+√3
                0
2cos(210)+√3
                0
```

14. Sequences and Series

Sequences can be evaluated on calculators by setting the calculator in Sequence mode.

```
NORMAL SCI ENG
FLOAT 0123456789
RADIAN DEGREE
FUNC PAR POL SEQ
CONNECTED DOT
SEQUENTIAL SIMUL
REAL a+bi re^θi
FULL HORIZ G-T
↓NEXT↓
```

In sequence mode, the Y= screen is transformed into a screen for sequences. In this mode, $u(n)$ is the sequence function, which is the calculator's symbol for a_n, nMin is the smallest value of n for the sequence (normally 1), and $u(n\text{Min})$ is the value of the first term of the sequence. It is possible to have more than one initial value by separating the values, in order, with commas. The u symbol is found above the 7 key, and n is typed when you press the X,T,θ,n key in sequence mode.

This example has one initial value.

TIP 14A

For the recursive sequence, $a_1 = 3$, $a_n = 3a_{n-1} - 4$ for $n \geq 2$, write the first four terms. (Answer: {3, 5, 11, 29})

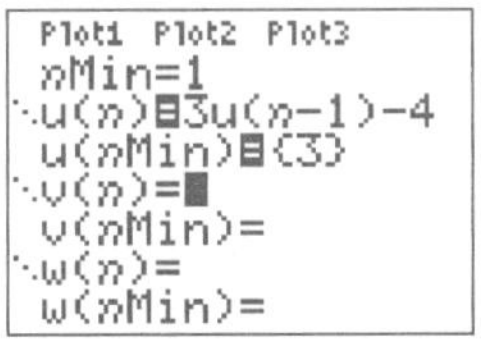

```
Plot1 Plot2 Plot3
nMin=1
u(n)=3u(n-1)-4
u(nMin)={3}
v(n)=
v(nMin)=
w(n)=
w(nMin)=
```

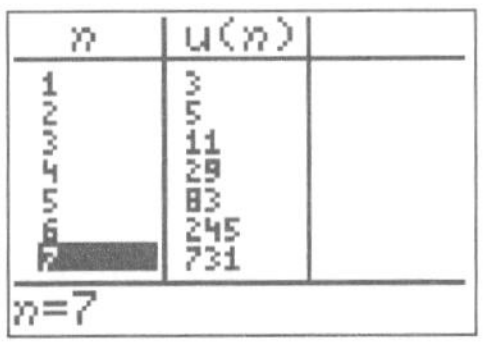

n	u(n)
1	3
2	5
3	11
4	29
5	83
6	245
7	731

n=7

Note that on the calculator, a_{n-1} is written $u(n-1)$. The curly braces around 3 were added automatically by the calculator. After entering the formula for $u(n)$ and the value for $u(n\text{Min})$, go to the table to see the terms of the sequence.

This example has two initial values.

TIP 14B

For the recursive sequence, $a_1 = 1$, $a_2 = 1$, $a_n = a_{n-1} + a_{n-2}$, write the first seven terms. (Answer: {1, 1, 2, 3, 5, 8, 13})

```
Plot1 Plot2 Plot3
nMin=1
·.u(n)=u(n-1)+u(n
-2)
 u(nMin)={1,1}
·.v(n)=
 v(nMin)=
·.w(n)=
```

```
n   u(n)
1   1
2   1
3   2
4   3
5   5
6   8
7   13
Press + for ΔTbl
```

You must type the curly braces around two or more initial values. The symbols are found above the parentheses keys. Multiple initial values are entered in *reverse order*. In this example with $a_1 = a_2 = 1$, that was not a problem. However, if you want to change it to $a_1 = 1$ and $a_2 = 2$, you need to set $u(n\text{Min}) = \{2, 1\}$, not $\{1, 2\}$.

TIP 14C

An explicit formula for the amount of money Aunt Beatrice gives you on each birthday is $a_n = 2.5(2)^n$. How much would Aunt Bea give you on your 18th birthday? (Answer: $655,360)

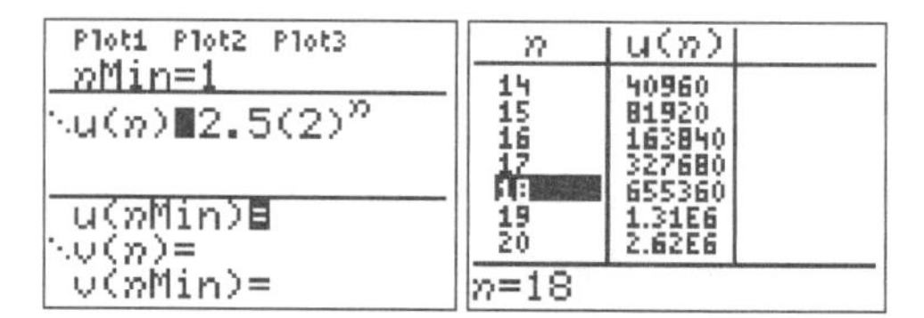

You can leave $u(n\text{Min})$ blank for an explicit sequence.

Some calculators have a built-in summation notation. This is a very accurate way to check your answer, particularly for a multiple-choice question. Be sure that you write down all the steps algebraically if you are required to show work.

TIP 14D

Evaluate: $\sum_{m=3}^{5} m(m-1)$ (Answer: 38)

Any letter can be used for the index.
Be careful that the entire expression is inside the parentheses.

```
 5
 Σ (M(M-1))
M=3
                38
```

15. Statistics

Finding the mean, variance, and standard deviation for a large data set can be tedious. Calculators make this chore quick and accurate. You should be able to use your calculator to find the mean, standard deviation, and variance for a data set given in a frequency table. You will need to enter the data in lists and then calculate the 1-variable statistics for the lists with the frequency list specified. Make sure you know the notation so you can find the answers on the calculator screen.

TIP 15A

Find the mean and median for the price of gas at a number of gas stations given in the table below. (Answer: mean = \$3.93, median = \$3.92)

Price of Gas (\$ per gallon)	**Frequency (number of gas stations)**
3.89	7
3.92	14
3.93	9
3.95	6
3.98	3
4.04	1

Lists for data are found under the STAT key, 1: Edit. If there is already data in the lists, you can clear it using STAT 4: ClrList. Type the lists to be cleared after the command, separated by commas. The lists are found above the 1–6 keys. Then enter the data in the lists. Enter the data values in L1 and the frequencies (if applicable) in L2.

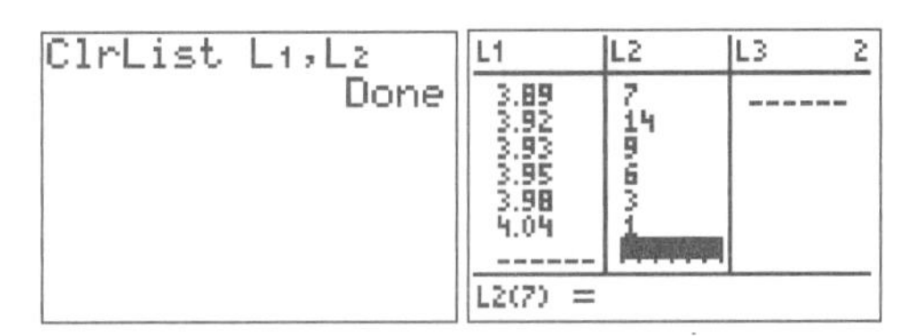

The mean, standard deviation, and other statistics can be found with the STAT key, under the CALC menu, 1: 1–Var Stats. For List, enter the name of the list with the actual data values. For FreqList, enter the name of the list with the frequencies. (If this list is omitted, the calculator assumes all frequencies are 1.) The first screen will give the mean, $\bar{x}$, the standard deviations Sx and σx, and the total number of data values n. Use the down arrow to get additional statistics including the median, abbreviated Med.

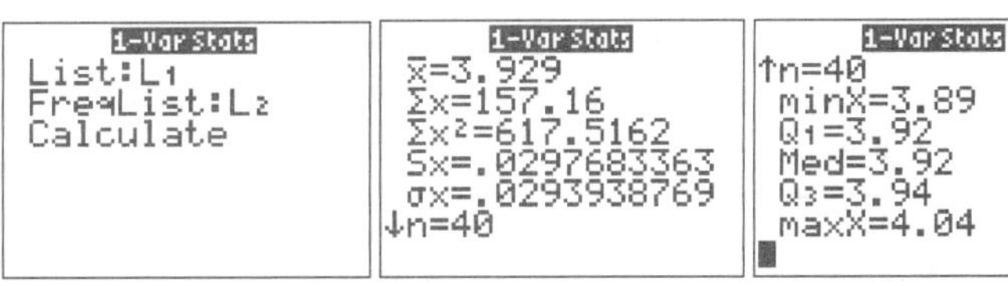

$\bar{x} = 3.929$ or ≈ \$3.93.

Med = \$3.92.

TIP 15B

Find the population variance and population standard deviation, each to the nearest thousandth, for the height of students given in the table below. (Answer: population variance = 8.933, population standard deviation = 2.989)

Height (inches)	**Frequency (number of students)**
59	1
62	4
65	12
67	7
70	4
72	2

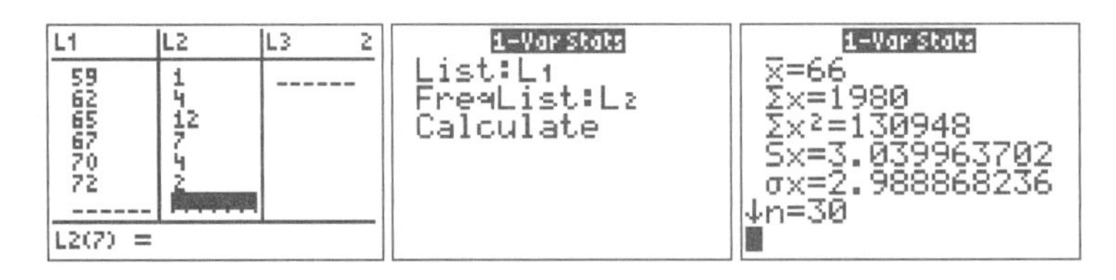

The population standard deviation is σx, and the sample standard deviation is Sx. For this example, the population standard deviation was requested, $\sigma x = 2.989$. To find the variance, square the standard deviation (symbol found in VARS menu 5: Statistics 4: σx). The population variance is 8.933.

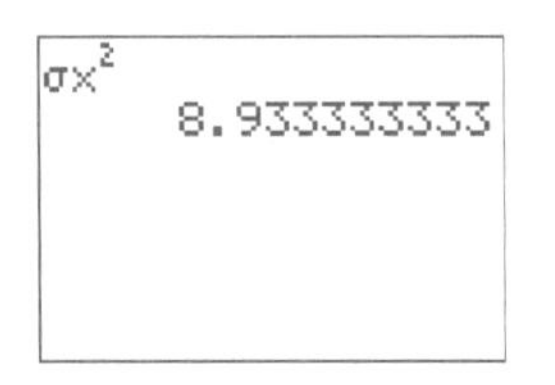

Normal distribution problems are sometimes done with reference tables. However, a calculator can do these problems more quickly and with greater precision. The normal distribution calculation is found in the DISTR menu, above the VARS key. Use 2: normalcdf(.

```
DISTR DRAW
1:normalpdf(
2:normalcdf(
3:invNorm(
4:invT(
5:tpdf(
6:tcdf(
7↓X²pdf(
```

You must specify the lower and upper values of the interval. Use –9,999,999 or 9,999,999 if the intervals are open on one end. You also need to specify the mean, μ, and the standard deviation, σ.

TIP 15C

The prices of existing homes in a certain suburban neighborhood are approximately normally distributed with a mean of $187,500 and a standard deviation of $12,500. What percent of the homes in the city are priced between $150,000 and $225,000? (Answer: 99.7%)

```
normalcdf
lower:150000
upper:225000
μ:187500
σ:12500
```

```
normalcdf(150000▸
      .9973000656
```

Convert 0.997 to percent form to find that 99.7% of the homes are priced between $150,000 and $225,000.

In the next example, the hours are greater than or equal to 6 and there is no upper bound.

TIP 15D

The battery life of a certain laptop computer is normally distributed with a mean of 5.5 hours and a standard deviation of 0.4 hours. If you take this laptop on a 6-hour airline flight, what is the probability you will be able to use it for the whole flight? (Answer: 0.106)

```
normalcdf
lower:6
upper:9999999
μ:5.5
σ:.4
Paste
```

```
normalcdf(6,999▸
         .105649839
```

In this example, the lower bound is 6 hours and there is no upper bound. So type several 9s to represent a very large number. Rounded to the nearest thousandth, the probability is 0.106 or 10.6% the battery will last the whole flight.

In the next example, the weights are less than or equal to 2 and there is no lower bound.

TIP 15E

The weights of the eggs laid by hens at a particular poultry farm are normally distributed with a mean of 2.1 ounces and a standard deviation of 0.25 ounces. Eggs weighing over 2.0 ounces are labeled large or bigger. What percent of the eggs from this farm are smaller than large? (Answer: 34.5%)

```
normalcdf
lower:-99999
upper:2
μ:2.1
σ:.25
Paste
```

```
normalcdf(-9999▸
        .3445783029
```

34.5% of the eggs are smaller than large.

TIP 15F

The size of a single scoop of ice cream at the Frozen Cow is approximately normally distributed with a mean of 0.9 of a cup and a standard deviation of 0.1 of a cup. Dorothy orders a single scoop of ice cream. She will eat up to 0.75 of a cup of ice cream; any leftover she feeds to her dog, Toto. Suppose Dorothy's scoop of ice cream is in the 80th percentile for size. How much ice cream does Toto get? (Answer: 0.234 cup)

```
DISTR DRAW
1:normalpdf(
2:normalcdf(
3:invNorm(
4:invT(
5:tpdf(
6:tcdf(
7↓χ²pdf(
```

```
invNorm
area:.80
μ:0.9
σ:0.1
Paste
```

```
invNorm(.80,0.9▸
        .9841621233
Ans-.75
        .2341621233
```

If you know the percentile, use the inverse normal distribution function to find the amount. Enter the percentile in decimal form; 80th percentile is 0.80.

We find that the 80th percentile is 0.984 of a cup. After Dorothy eats her 0.75 of a cup, Toto gets 0.984 – 0.75 = 0.234, or slightly less than a quarter of a cup.

Regression equations are found using technology. Many students use graphing calculators to find these equations, but computer software can also be used. You must type the data into the calculator's list and then calculate the type of regression equation you want.

TIP 15G

Find the linear regression equation for the following data. (Answer: $y = 0.435x + 14.662$)

Depth (ft.)	Pressure (psi)
5	16.1
10	20.2
15	20.9
20	22.6
25	26.3
30	27.0
40	33.4
50	35.6

Enter the x-values in List1 and the y-values in List2. (Note that both lists contain one more data value, 50 and 35.6, than shown in the screen below.) If you want to see the scatter plot, go to STAT PLOT (2nd Y=), select 1: Plot1 . . . On (if you see 1: Plot 1 . . . off, turn the plot on), specify the Xlist and Ylist, and select a Mark. Then use ZOOM 9: ZoomStat to see the scatter plot.

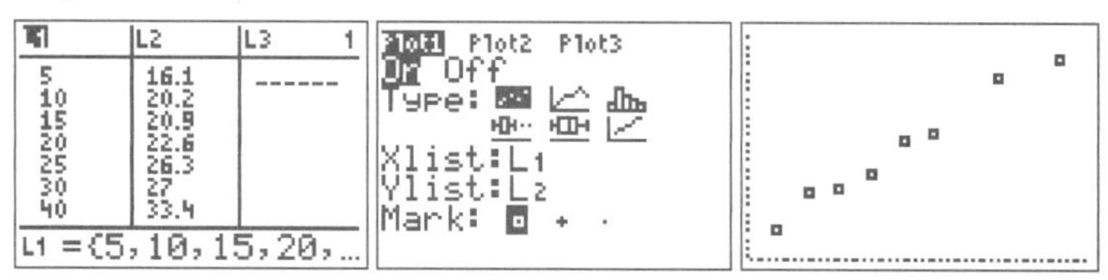

To get a linear regression equation, go to STAT CALC 4: LinReg. Specify the Xlist and Ylist. The FreqList is usually left blank. Enter Y1 in the RegEQ space if you would like to graph your equation.

```
LinReg(ax+b)
Xlist:L1
Ylist:L2
FreqList:
Store RegEQ:Y1
Calculate
```

```
LinReg
y=ax+b
a=.4348747592
b=14.66242775
```

With coefficients rounded to three decimal places, the regression equation for this problem is $y = 0.435x + 14.662$. If you already did a scatter plot and typed Y1 in the StoreRegEQ line, the GRAPH key will graph the equation on the scatter plot.

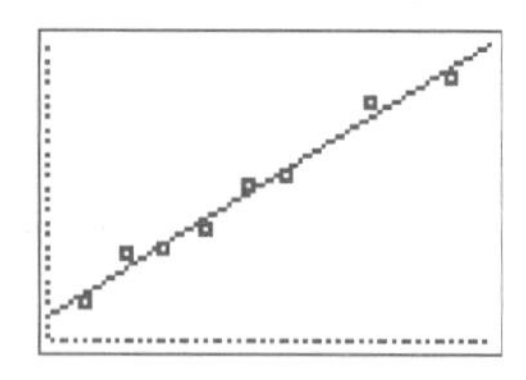

Note that when you are done, remember to turn your Stat Plot back off. Otherwise, your calculator will continue graphing the stat plot on top of future graphs you make.

INDEX